BASIC TECHNICAL MATHEMATICS—EXPLAINED

BASIC TECHNICAL MATHEMATICS–EXPLAINED

THOMAS P. ZURFLIEH

Associate Professor
Department of Engineering Technology
St. Petersburg Junior College

McGRAW-HILL BOOK COMPANY

New York *St. Louis* *San Francisco* *Düsseldorf* *Johannesburg* *Kuala Lumpur* *London* *Mexico* *Montreal* *New Delhi* *Panama* *Rio de Janeiro* *Singapore* *Sydney* *Toronto*

Library of Congress Cataloging in Publication Data
Zurflieh, Thomas P.
Basic technical mathematics—explained.
1. Mathematics—1961–   I. Title.
QA39.2.Z87   512'.1   73-6668
ISBN 0-07-073595-6

**This book is dedicated to
Jason A. Zurflieh
and Frances Zurflieh,
my father and mother.**

BASIC TECHNICAL MATHEMATICS—EXPLAINED

1234567890MURM7987654

*The editors for this book were Robert Buchanan and Suzanne Kistler, the designer was Marsha Cohen, and its production was supervised by James E. Lee and Laurence Charnow.
It was set in Garamond by Progressive Typographers.
It was printed by The Murray Printing Company and bound by Rand McNally & Company.*

CONTENTS

PREFACE

I. PURPOSE AND PHILOSOPHY

A. TYPE OF COURSE AND PROGRAM

This book is intended primarily for the technical-vocational or engineering technology student. It can be used for first- and second-semester courses in precalculus mathematics in either a two- or a four-year technology or vocational program. Although the examples and exercises stress applications for technology, the algebra and trigonometry are general enough to be used for other precalculus courses if "modern" mathematics is not of primary concern.

B. TWO MAJOR OBJECTIVES

One major objective of this book is to make the student proficient in basic algebraic manipulations. Students at this level seem to be universally deficient in manipulative skills. Thus they do poor work in all subjects that use basic mathematics. This book seeks to remedy such a situation by providing the student with a good grounding in precalculus mathematics, particularly fundamental algebra and trigonometry.

The other major objective is to develop the student's ability to use mathematical symbology and to think analytically. Problem analysis is stressed in the text, in the examples, and in the exercises. Pattern recognition is emphasized as part of the student's thinking process, since a large part of mathematics is nothing more than pattern recognition. Lack of ability to generalize or categorize algebraic expressions is one of the root causes of failure in mathematics. Therefore this book stresses the interpretation and significance of the mathematics.

C. STYLE AND APPROACH

The style is informal in order to aid the student's comprehension and to hold his interest. A major feature of the style is the way in which definitions, formulas, and examples are explained. In addition to presenting the usual symbolic mathematics, I try to guide the student's thinking and to interpret the symbology with everyday applications and with explanations in simple English. I have attempted to draw upon the student's intuition and experience to interpret the mathematics he is asked to learn.

However, after introducing topics in an intuitive and elementary way, I try to preserve the integrity of the mathematics by defining terms, symbols, operations, and so on, with precision and rigor. Geometric and physical arguments are freely employed to overcome abstractness and thereby make the mathematics more meaningful. In addition, wherever practicable, engineering and other applications are used to emphasize the relevance of the mathematics. A unique feature is the "Comment" sections which focus attention on important points or possible pitfalls.

The emphasis of the book is on understanding principles and applying them to solve practical problems. By immediately and repeatedly using mathematics for something that seems reasonable and useful, the student should develop an appreciation for this mathematics and should be motivated to learn it.

II. ORGANIZATION AND UTILIZATION OF THE BOOK

A. ORGANIZATION

The book is divided into four major units with at least four chapters in each unit. Unit 1, Chaps. 1–6, is an introduction to elementary concepts, definitions, notations, and operations used in algebra, geometry, and trigonometry. This unit includes many topics but does not delve deeply into them. The objective is to familiarize the student, particularly the rank beginner, with sufficient material in a short enough time to let him solve meaningful problems and to enable him to function in other courses that use mathematics.

Unit 2, Chaps. 7–12, takes up algebraic operations and equations in more detail. The student is now expected to master the operations of algebra. He should become proficient in manipulative skills and in solving problems, including word problems.

Unit 3, Chaps. 13–16, continues with trigonometry. Like Unit 2 it goes into more detail and covers more topics than do the introductory chapters in Unit 1. Much of this unit equips the student with the trigonometry required for calculus.

Unit 4, Chaps. 17–20, contains the more advanced topics of complex numbers, logarithms, inequalities, and matrices.

There are two major reasons for arranging the topics in this way. First, as mentioned above, the general survey of Unit 1 gives the student enough background to begin using mathematics in other courses. In some vocational and technology programs, the first mathematics course is taken concurrently with courses that use mathematics. The student thus needs to rapidly acquire some skill in algebraic manipulations and in reading mathematics.

The second objective in this arrangement of topics is to provide repetition of the subject matter. Many students fail to master mathematics quickly, if at all, because of the strangeness of the terminology and the symbology. After meeting various symbols, definitions, and so on in Unit 1, the student encounters these entities again, sometimes for the third or fourth time, in Units 2 and 3. By this time the strangeness of the subject should have worn off, and the student can concentrate on learning the finer points.

B. UTILIZATION

This book can be used in several different ways, depending upon the background and ability of the student. The first two units plus most of the third unit can be a basic course for students with little or no background in geometry and algebra. The entire first three units will require about 75 class meetings. Thus a 5-hour, 1-semester course, or two 4-hour quarters would be suitable.

A class of students with some background in geometry and algebra could omit or quickly review much of Units 1 and 2. Such a class might possibly begin with Chap. 5 on trigonometry and right triangles, progress rapidly to Chap. 13 on trigonometric functions of a general angle, and work into Unit 4. This approach would be suitable for one 3-hour semester course or one 5-hour quarter course.

A precalculus course for more advanced students could be built entirely around Units 3 and 4. The preliminary chapters would be useful for review assignments for students having difficulty in a particular subject area. In any case, the instructor will have a great deal of freedom in arranging a sequence of topics from this book. Within the limitations dictated by the subject matter, I have tried to make each chapter independent of the others. For example, Chap. 13 on trigonometric functions of a general angle could immediately follow Chap. 5 if the instructor so desires. Some instructors might wish to go to Chap. 20 on matrices immediately after Chap. 8, which contains the material on determinants. Many other regroupings of chapters are possible.

To aid the instructor in planning, each section (or subsection) has been designed to require one class meeting. Most sections have an excess of exercises. Thus assignments can be varied from one class to another.

The author is grateful for the help of several people who have assisted in preparing the manuscript. In particular, I wish to thank Bill Rice, chairman of the St. Petersburg Junior College mathematics department, for his encouragement and suggestions, and Kathryne Steffenson for typing the manuscript and helping with innumerable details. The assistance of the McGraw-Hill editors and their staff is greatly appreciated. A special note of thanks goes to my wife, Jane Zurflieh, for editing the entire manuscript.

Tom Zurflieh

ELEMENTARY CONCEPTS AND OPERATIONS OF ALGEBRA, GEOMETRY, AND TRIGONOMETRY

UNIT 1

UNDERSTANDING MATHEMATICS

CHAPTER 1

1.1 THE LANGUAGE OF MATHEMATICS

Mathematics is a language. It is similar in several respects to English, shorthand systems, music, and other codes. Like these other languages, mathematics conveys information by means of its own special symbols, vocabulary, and grammar.

The symbols of a language have meanings, and both the sender and the receiver of a message must agree on them ahead of time. Otherwise the message is meaningless. For example, we agree that the English word "cat" is a symbol for a certain kind of furry animal. Similarly, in mathematics, the symbol "sin x" is used to convey a certain meaning (which you are not expected to understand just now).

If you are to learn mathematics, then you must learn to read it much as you would any other language. The "words" and "sentences" of mathematics may appear strange at first, but they convey meanings much like those of English words and sentences. Too often, a beginner literally does not know how to read mathematics. Even worse, he may not realize that it is *supposed* to be read. He ignores the *significance* of the mathematics and concentrates on learning formulas by rote. As a result, he gains little understanding of mathematics and finds that he cannot progress in it or use it to solve problems. For instance, he may fail to make a distinction between two "words" such as 2 sin x and sin $2x$. Such a lack of discrimination is comparable to that of a beginning reader confusing "scat" with "cats."

One aspect of reading mathematics is learning the special, technical definitions of certain words. Words like "power," "rational," and "imaginary" have special meanings that are entirely different from their everyday meanings. The student must learn the special meanings of such words and use them correctly.

Another aspect of reading mathematics is interpreting its shorthandlike symbols. Mathematics uses special combinations of letters, numerals, and other symbols to express ideas in a precise, highly condensed form. These combinations of symbols that form the "words" and "sentences" of mathematics often stand for quite complex ideas. As the student reads mathematical "sentences," he must be aware of their meanings; he must not just memorize formulas.

Perhaps the most difficult aspect of reading mathematics is recognizing patterns. Mathematics is a language of patterns. Often we express a rule by a certain arrangement of a set of letters. The letters themselves are not so important; the rule is conveyed by their arrangement into a specific pattern. For example, there is a certain rule expressed by the formula $c^2 = a^2 + b^2$; it could just as well be written $x^2 = y^2 + z^2$. Both formulas contain symbols arranged in the same pattern; thus both state the same rule. Furthermore, the formula $b^2 = c^2 - a^2$ is a variation of this same rule. If the reader is not conscious of the idea behind a rule, then he may not see how to use it when different sets of letters or variations are employed.

If you can remember that mathematics is a language with certain agreed-upon meanings and ways of stating them, you will be able to read it with deeper understanding. You will also be taking your first step toward mastery of mathematics.

1.2 TECHNIQUES FOR STUDYING MATHEMATICS

We have just emphasized that you should read and interpret mathematics. Now we shall discuss *how* to do it. If you practice the following techniques regularly, you should find the mathematics clear and meaningful. Try to review this section often, perhaps before you begin each chapter.

A. GENERAL READING

Sometimes a definition, rule, or the like is difficult to explain in one breath. (Remember, mathematics lets us say a lot in a short space.) If you do not understand something immediately, keep reading, study the examples that follow, and then reread the entire passage. Keep in mind that the main purpose of the examples is to clarify the explanation in the text. And so during the second reading, try to interpret the text as you look at the examples. Remember that you cannot always read mathematics rapidly and progressively as you would a novel. Sometimes you must read the text, keeping in mind the examples. You must employ repetition and feedback.

If you simply cannot understand something, make a note of it and question your teacher. Do not let a point pass you by, for mathematics is a progressive subject. That is, one concept or definition is used to develop the next one; each new concept is built upon earlier ones. Thus you frequently cannot understand a new point unless you understand most of the preceding material. If you have not mastered the earlier material, then you seriously impair your chances of making progress.

B. READING ILLUSTRATIVE EXAMPLES

We use examples in mathematics to help the reader grasp the meaning of a general statement (such as a rule or definition) or to show him how to apply the mathematics. A good way to use example problems is to try them yourself *before* you read the solutions in the book. Copy the problem, but *not* the solution, in a notebook. Solve it as best you can, and then read the solution in the book. You should see immediately how well you understood the text. If you cannot work an example problem, reread the text and the solution in the book. Then try again to work the example by yourself.

C. DOING EXERCISES

Do the exercises immediately after reading the corresponding section of the text. Chances are that the examples will show you how to do the exercises. Even if there is no direct example, the principles set forth in the text will apply. Remember that the exercises are related to the immediately preceding text, so try to apply *that* material.

A very good procedure is to write exercises in a neat, orderly, step-by-step manner. Group them by chapter and section in a notebook and clearly label them. For particularly difficult problems write notes to explain how you found the solution.

One purpose in writing exercises this way is to provide material for review. A good way to review for a test or final exam is to rework the previously assigned exercises. If you have a complete collection of assigned problems, then (1) you know what material was covered, and (2) you can rework a representative sample from each set and check the solutions against your original ones. If you don't do too well, then you know where to review further.

A more important purpose of writing exercises is being sure that you have mastered the mathematics. By doing the exercises thoroughly and conscientiously, you are almost certain to learn the mathematics. You cannot learn much by skimming over mathematics; you must get deeply involved in it.

D. WORD PROBLEMS

A *word problem,* as its name implies, is a problem presented by means of words rather than numerals or other mathematical symbols. It represents practical, applied mathematics. Solving word problems should be a major objective of students in technical mathematics. To employ a once popular expression, word problems are "where it's at." They are extremely important because they reflect industrial situations, and they require the type of analytical thinking needed by people in industry and business. Furthermore, by practicing certain procedures for solving word problems, the student will learn to organize his work and present it in clear, readable form. The lack of well-organized, clearly written reports is a common complaint among leaders in business, education, and industry.

If you use sound methods of attack, solving word problems can be an interesting and rewarding experience. All the preceding suggestions on working exercises apply to word problems as well. However, certain additional techniques will help you translate more easily the stated facts into mathematical equations. These techniques are discussed in Appendix 1 and at appropriate places in the text. You should review them each time word problems are assigned.

FUNDAMENTAL CONCEPTS AND DEFINITIONS

CHAPTER 2

The objective of this chapter is to make you familiar with some of the basic vocabulary of mathematics. Although this preliminary material may seem abstract, you must become familiar with it before you can understand the more useful mathematics that follows. In Chap. 1 we stated that mathematics builds each new concept upon the previous ones. The preliminary material of Chaps. 2 and 3 forms the basic foundation upon which all else depends.

2.1 NUMBERS

In using mathematics we shall encounter several kinds of numbers, each with its own unique characteristics and applications. For identification purposes, we give each type of number a name such as "integer" or "rational."

A. THE INTEGERS

The numbers we use for counting discrete (separate) items are called the *counting numbers* or the *positive integers.*

POSITIVE INTEGERS The counting numbers 1, 2, 3, 4, 5, . . .

When someone first tried to subtract a larger integer from a smaller one, as in 5 subtracted from 2, he needed *negative integers.* A good way to define the negative integers is by means of a *number line*—a line on which the numbers are represented as points. The standard number line is shown in Fig. 2.1. To show the integers in relation to each other, we choose an arbitrary point as *zero;* this point is called the *origin*—the starting point for counting or measuring. The number 0 is an integer, but it is neither positive nor negative.

Next, *to the right* of zero we measure a distance that represents one unit. The point at the end of this unit represents the positive integer +1. Similarly, we locate each succeeding positive integer, +2, +3, etc., one unit to the right of the previous integer.

Finally, we locate the integer negative one, denoted by −1, one unit *to the left* of zero; similarly we locate each succeeding negative integer, −2, −3, etc., one unit to the left of the previous integer.

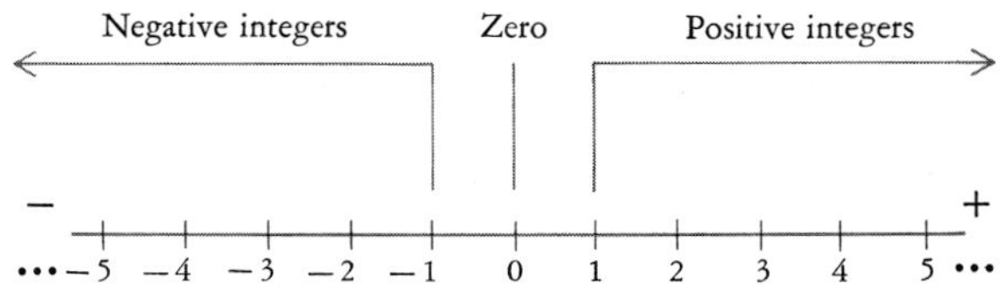

Fig. 2.1 The number line for integers.

Perhaps you have correctly observed that the only difference between positive integers and negative integers is the *direction* in which they are measured. Thus we can make the following definition:

> NEGATIVE INTEGERS Integers measured in the opposite direction from the positive integers. . . . −5, −4, −3, −2, −1

Therefore, a distance of negative 3 feet, −3 ft, means 3 ft measured opposite to whatever direction we choose as positive. For example, if we arbitrarily define eastward to be positive, then −3 ft means 3 ft opposite to eastward, or 3 ft westward.

At this point let us clearly understand that no physical quantity (such as length or weight) can have a negative *value.* No yardstick is −3 ft long (or +3 ft long, for that matter). But with a yardstick we can measure 3 ft in a direction which we choose to call "negative" or "positive." The negative sign or the positive sign that we attach to a quantity merely indicates the particular *direction* in which that quantity is being measured. A negative temperature, for example, is a temperature on one side of the mark which is called "zero degrees" on the thermometer. The zero mark is arbitrary, and the meaning of negative temperature is "colder than," or "less than," zero.

Prime Numbers

A *prime number* is any positive integer other than unity (+1) that can be evenly divided by only itself or unity. The first 10 prime numbers are 2, 3, 5, 7, 11, 13, 17, 19, 23, and 29. Test these according to the definition above. Can you list the next 10? The twentieth is 71.

B. RATIONAL AND IRRATIONAL NUMBERS

A *rational number* is any number formed by dividing any integer by any non-zero integer. Here are some examples of rational numbers:

$$\frac{1}{2} \qquad \frac{-5}{7} \qquad \frac{113}{-29} \qquad 7 \left(=\frac{7}{1}\right)$$

Notice that, just as 7 can be written in the form $^7/_1$, every integer is also a rational number. Also, just as $0.5 = {}^1/_2$ and $0.333\ldots = {}^1/_3$, every terminating decimal and every repeating decimal are rational numbers. We formally define a rational number as follows:

> RATIONAL NUMBER Any number that can be expressed in the form a/b, where a and b stand for integers and b is not zero.

We could also define a *rational* number as a *ratio* of two integers (or as a quotient of two integers) in which the divisor is not zero.

Operations with Zero

We keep emphasizing that the divisor is not zero because *division by zero is an undefined operation.* We can add or subtract zero, and we can multiply by zero, but division by zero (even zero divided by zero) is meaningless. In all following discussions, we shall assume that division by zero is automatically excluded; this fact will be reemphasized in important places.

Irrational Numbers

An *irrational number* is any number that is *not rational,* and thus *cannot* be expressed as the ratio of two integers. All nonterminating, nonrepeating decimal numbers are irrational. Some examples of irrational numbers are $\pi = 3.14159\ldots$, $\sqrt{3} = 1.732\ldots$, and, in general, the square root of any prime number.

The entire set of numbers composed of both the rational numbers (including the integers) and the irrational numbers is called the set of *real numbers.* If we say that a number is "real," we mean that it is one of these types of numbers.

Imaginary Numbers

Any number that contains a factor involving the square root of a negative number is called an *imaginary number.* Some examples are $\sqrt{-1}$, $\sqrt{-2/5}$, and $5\sqrt{-\pi}$. Imaginary numbers are not considered to be real numbers. We shall study imaginary numbers in Chap. 17. In the meantime, we shall confine all discussion to real numbers.

C. ABSOLUTE VALUE

The *absolute value* of any number is the size or magnitude of that number, regardless of its sign. Thus $+3$ and -3 have the same absolute value, 3, since each contains 3 units. We can think of absolute value in terms of distance from the origin of the number line. The absolute value of a number tells us how many units are between the number and the origin; the direction does not matter. For example, the distance between the ends of a yardstick is 3 ft, no matter which way it points or from which end we begin measuring.

In mathematics the *absolute value symbol* is a pair of vertical lines that enclose the number as follows:

$|n|$ absolute value of n

Example 2.1. Absolute values of numbers

a. $|+3| = |-3| = 3$

b. $|10 - 12| = |-2| = 2$

Notice that we *must* perform the arithmetic *inside* the absolute value sign first, *before* taking the absolute value.

c. $|20| - |5 - 35| = 20 - |-30| = 20 - 30 = -10$

D. FRACTIONS AND RECIPROCALS

When a division is implied by writing one number above the other, the entire expression is called a *fraction.* For example,

$$\frac{2}{3} \qquad \frac{16}{13} \qquad \frac{A}{B} \qquad \frac{2x-3}{y+1}$$

are all fractions.

In a division problem, the number to be divided is the *dividend.* In a fraction, the dividend—the expression above the dividing line—is called the *numerator.* Thus, in the fractions above, the expressions 2, 16, A, and $2x - 3$ are numerators. The *denominator* of a fraction is the *divisor*—the expression below the dividing line. Thus, in the fractions above, the expressions 3, 13, B, and $y + 1$ are denominators.

The *reciprocal* of a number is the fraction formed by dividing unity (+1) by that number. For example, the reciprocal of 2 is $1 \div 2 = 1/2$; and the reciprocal of $1/2$ is $1 \div 1/2 = 2$. (Compare this to Example 2.2*a* below.) In general, if N is any non-zero number, the reciprocal of N is $1/N$.

Example 2.2. Reciprocals of numbers

a. The reciprocal of 0.5 is

$$\frac{1}{0.5} = 2$$

b. The reciprocal of -4 is

$$\frac{1}{-4} = -0.25$$

c. The reciprocal of $2 + 3$ is

$$\frac{1}{2+3} = \frac{1}{5} = 0.2$$

This example makes an important point: *The reciprocal of a sum of two numbers is not equal to the sum of their reciprocals.* Note that $1/(2 + 3)$ is *not* the same as $1/2 + 1/3$.

d. The reciprocal of $2/5$ is:

$$\frac{1}{2/5} = \frac{1}{0.4} = 2.5 = \frac{5}{2}$$

This example makes an important point: *To find the reciprocal of a fraction, interchange the numerator and denominator.*

EXERCISE 2.1

1. List all classifications—*integer, rational, irrational, real, imaginary,* and *prime*—that apply to each of the following numbers.
   a. 16 b. $9/75$ c. $-33/3$ d. $\sqrt{16}$
   e. 2π f. $\sqrt{-8}$ g. $3 + 14$ h. $|-10/3|$
   i. -29 j. $0/2$

2. Evaluate each of the following expressions.
   a. $|75|$ b. $|-100|$ c. $|-4/5|$ d. $|16-23|$
   e. $-|25+3|+|22-50|$
3. In the Celsius temperature system (formerly Centigrade), 0°C is the freezing point of water. Describe the meaning of each of the temperatures: −17°C and +50°C.
4. In the Fahrenheit temperature system, +32°F is the freezing point of water. Describe the meaning of each of the following temperatures in relation to the freezing point of water: 0°F, −17°F, and +50°F.
5. Determine the reciprocal of each of the following expressions.
   a. 7 b. $1/3$ c. $5/2$ d. $1/5$
   e. 2.5 f. $1/\sqrt{3}$ g. 0.25 h. 1/0.75
   i. $(3+x)/(2y)$ j. 0.01
6. If x and y represent integers, what kinds of numbers—*integer, rational, odd,* and *even*—does each of the following expressions represent?
   a. x/y b. $x \cdot y$ c. $x+y$ d. $2x+1$
   e. $2y-1$

2.2 LITERAL NUMBERS AND EXPRESSIONS

Frequently in mathematics it is advantageous to write a formula or state a concept without using *explicit numbers* (specific numerical values). In such cases, we let letters represent the numbers. We call such letters *literal numbers*—letters that stand for numbers. We used literal numbers to define a rational number in Sec. 2.1B. That definition is a good example of the usefulness of literal numbers to convey a general idea.

By using literal numbers, we can make very general statements. For example, in the formula $P = 4S$ for the perimeter of a square, we use the letter S to represent the length of each side. S can be 3 in. or 89 yd, or any other value in any units of length. With this *literal formula,* we can talk about all squares in general; we do not have to confine ourselves to one specific square.

In mathematics, the word "expression" means any combination of literal and/or explicit numbers that is formed by any of the operations of addition, subtraction, multiplication, division, or extraction of roots. Some examples of expressions are $2(3)$, $3a+5$, a/b, and $5x/2 - 3T + 10/(3Z)$.

A. TERMS AND FACTORS

The expression $6 + 2ab - 5xy/z$ contains three distinct parts, 6, $2ab$, and $5xy/z$, called *terms.* A *term* is any part of an algebraic expression that is separated from the other parts by an addition sign or a subtraction sign.

Algebraic expressions are classified according to the number of terms they contain. An expression with *one term* is a *monomial;* an expression with *two* terms is a *binomial;* and an expression with *three* terms is a *trinomial.* The expression $6 + 2ab - 5xy/z$ is an example of a trinomial. Generally, if an expression has *many* terms (at least two), we call it a *multinomial.*

If two or more numbers are multiplied together, each number is called a *factor.* For example, the second term of the trinomial $6 + 2ab - 5xy/z$ contains three factors: 2, a, and b. Each factor or group of factors in a term is called the *cofactor* or *coefficient* of the others. Thus in the term $2ab$ above, $2a$ is the cofactor of b, and b is the cofactor of $2a$. An explicit numerical cofactor is called the *numerical coefficient.* Thus in the term $2ab$, 2 is the numerical coefficient of b, and a is the literal coefficient of b.

B. GROUPING SYMBOLS

Grouping symbols are symbols such as parentheses ( ), brackets [ ], and braces { } that enclose an expression. The objective of so enclosing an expression is to treat it as a *single number.* For example, the parentheses in the expression $2x - 5(x + 3y)$ indicate that the binomial factor $(x + 3y)$ acts as a single number. Also, the parentheses tell us the order in which we would combine the numbers if we knew the values of x and y. We would (1) multiply y by 3, (2) add x and $3y$, (3) multiply $x + 3y$ by 5, and (4) multiply x by 2 and subtract $5(x + 3y)$ from $2x$.

A common practice is to use different grouping symbols to indicate a group within a group, as in the expression

$$2\{3x - [5a + b][x - (a + b)]\}$$

Radical signs $\sqrt{\ }$, absolute-value signs | |, and bars — also act as grouping symbols.

2.3 EQUALITY, EQUIVALENCY, AND INEQUALITY

A. EQUALITY

An *equation* is a statement that says two expressions are equal, and by "equal" we mean that each expression represents the *same number.* For example, in the equation $2y = 3x$ the left side, $2y$, stands for a certain number, and the right side, $3x$, stands for the *same number.* Even though the number is unknown, the equation tells us that values of x and y must always be assigned so that $2y$ and $3x$ are equal. If x is 4, for example, then the "number" $3x$ is 12; and since $2y$ stands for the same number, y has to be 6.

If the two sides of an equation represent values of physical quantities such as distance, speed, weight, voltage, etc., then the equality has an important additional meaning. It means that both sides represent the *same units of measure.* For example, suppose that you are driving on a trip and you want to determine your average speed. As you know, average speed = distance ÷ time. With algebraic symbols, we can express this formula as

$$S = \frac{D}{T}$$

If in this equation S represents average speed in *miles per minute* (miles/min), then D *must have* units of miles, and T *must have* units of minutes. If you traveled 30 miles in 40 min, then

$$S\,\frac{\text{miles}}{\text{min}} = \frac{D\text{ miles}}{T\text{ min}} = \frac{30\text{ miles}}{40\text{ min}} = \frac{3}{4}\text{ miles/min}$$

If you want to know the speed in different units from the ones measured, then you can change units by multiplying S by a conversion factor. For instance, if you want speed in miles per hour (mph), then multiply S by the conversion factor 60 min/hr. If we let V be average speed in miles per *hour,* then

$$V = 60\,\frac{\text{min}}{\text{hr}} \times S\,\frac{\text{miles}}{\text{min}} = 60\,\frac{\cancel{\text{min}}}{\text{hr}} \times \frac{3}{4}\,\frac{\text{miles}}{\cancel{\text{min}}} = 45\text{ mph}$$

Note that the units cancel just like numbers do.

In this example, would we be correct in writing 45 mph = ¾ miles/min, or V mph = S miles/min? In everyday language we tend to be lax. We might say "45 mph *equals* ¾ miles/min," but this use is technically incorrect. What we mean is "45 mph is *equivalent* to ¾ miles/min." When we work with unfamiliar problems and units, carelessness with numbers and units can lead to misunderstanding, incorrect equations, and other errors. Therefore you should practice thinking and writing rigorously precise equations. The correct equations are $V = 60S$, or 45 mph = (60 min/hr)(¾ miles/min). For a detailed discussion of physical quantities and units, see Appendix 2.

B. EQUIVALENCY

If two different numbers (such as 45 mph and ¾ miles/min) represent values of the same quantity but in different units, we say that the numbers are *equivalent.* We express this idea symbolically with an equivalency sign—an arrow. Thus the equivalency statement

$$45 \text{ mph} \longleftrightarrow \frac{3}{4}\frac{\text{miles}}{\text{min}}$$

means "45 mph is equivalent to ¾ miles/min" or "45 mph implies ¾ miles/min" or "from 45 mph it follows that the speed in miles per minute is ¾ miles/min."

C. INEQUALITIES

We sometimes need to state the fact that one expression does *not equal* another expression. Thus one expression is either "greater than" or "less than" the other. The symbols and the precise mathematical definitions for these statements are as follows:

$x > y$	x is greater than y; x is to the right of y on the number line	(2.1)
$x < y$	x is less than y; x is to the left of y on the number line	(2.2)

These and other such statements are called *inequalities.* For example, $5 > 3$ reads "5 is greater than 3," and means that 5 is to the right of 3 on the number line. In Chap. 19 we shall take up inequalities in much more detail.

Example 2.3. Inequality statements

The following examples illustrate correct inequality statements.

a.	$3 < 5$	3 is less than 5
b.	$-5 < -4$	−5 is less than −4
c.	$0 > -2$	zero is greater than −2
d.	$-50 < 10$	−50 is less than 10

Inequalities such as those in Example 2.3*b*, *c*, and *d* involving negative numbers may *seem* incorrect. The *seemingly* larger, but negative, number is always less than the other number. The answers to the following questions show why this is true.

1. Which is the lesser temperature, −5°F or −4°F?
2. Does your checking account contain less money when the balance is \$10 or when the balance is \$−50 (meaning that the account is overdrawn by \$50)?

Comment: *These questions show how the abstract definitions of mathematics apply to practical matters and how they can help us sharpen our thinking.*

We combine the inequality signs with the equal sign to form other inequality statements such as

$$x \geqslant y \qquad x \text{ is greater than or equal to } y \tag{2.3}$$

Also, we use a slash mark to negate (give the opposite meaning to) symbols. For example,

$$x \nless y \qquad x \text{ is not less than } y \tag{2.4}$$

Notice that this statement conveys the same meaning as does $x \geqslant y$ above. That is, if x is not less than y, then x has to be either equal to y or greater than y.

The following examples show how we use inequality symbols to specify a range of values for a literal number or expression.

Example 2.4. We can specify that the number a is never negative by writing $a \nless 0$, or $a \geqslant 0$. Another translation of this statement is "a can be any positive number or zero."

Example 2.5. We wish to specify that current can run in either direction through a motor, but that the magnitude of the current must never exceed 10 amp. We can use the expression $|i| \leqslant 10$ amp. Another way to express this statement is to combine two inequality statements as follows:

$$-10 \text{ amp} \leqslant i \leqslant 10 \text{ amp}$$

EXERCISE 2.2

1. Make a correct statement for each pair of numbers by inserting one of the signs, =, >, or < between the numbers.
   a. 5 12 b. $7/8$ $14/16$ c. −2 2
   d. 10 −12 e. 0.75 $\frac{3}{8}(2)$ f. −1,000 0.001
   g. −32°C −50°C
2. In each part below, let S stand for average speed, D for distance, and T for time. Write an equation for S in units of
   a. miles per day (miles/day), if D is in miles and T is in hours
   b. miles per minute (miles/min), if D is in miles and T is in hours
   c. feet per minute (ft/min), if D is in miles and T is in minutes

d. meters per second (m/sec), if D is in kilometers (km) and T is in seconds

e. miles per hour (mph), if D is in feet and T is in seconds

3. Use the equivalency sign to write a correct statement to the effect that 100°C "equals" 212°F.

4. Describe in words, as in Examples 2.4 and 2.5, the set of numbers that z can represent if

a. $z < 0$	b. $z \geqslant 2$	c. $\lvert z \rvert \leqslant 1$
d. $-10 \leqslant z \leqslant 10$	e. $z \ngtr 0$	f. $z \neq 0$
g. $4 \leqslant z < 10$	h. $\lvert z \rvert \leqslant 0$	i. $\lvert z \rvert < 0$
j. $\lvert z \rvert \nleqslant 5$	k. $2z < -8$	l. $\lvert z \rvert \geqslant 1$
m. $z \ngtr \lvert -3 \rvert$	n. $\lvert z \rvert \leqslant -1$	

5. Using equal signs, inequality signs, and absolute-value signs, rewrite each statement of Prob. 4 in another form. For example, $z < 0$ could also be $z \geqslant 0$.

6. Rearrange the following numbers in numerical order, beginning with the smallest: $25, -2, \sqrt{36}, \pi, -26, 0, \lvert -3 \rvert, -\lvert -5 \rvert$.

CHAPTER 3

FUNDAMENTAL RULES AND OPERATIONS

In this chapter we shall examine certain fundamental rules and operations of algebra. They are, like the definitions in Chap. 2, basic foundations of algebra.

3.1 THE FUNDAMENTAL LAWS

The following fundamental laws are basic axioms of algebra. That is, they are universally accepted as truths. They cannot be proved or derived from any other more basic laws. The fundamental laws can be stated symbolically as follows:

Commutative Laws

$$A + B = B + A \tag{3.1a}$$

$$A \cdot B = B \cdot A \tag{3.1b}$$

Associative Laws

$$A + B + C = A + (B + C) = (A + B) + C \tag{3.2a}$$

$$A \cdot B \cdot C = A \cdot (B \cdot C) = (A \cdot B) \cdot C \tag{3.2b}$$

Distributive Law

$$A \cdot (B + C) = A \cdot B + A \cdot C \tag{3.3}$$

The *commutative laws* mean that the terms in a sum or the factors in a product can be *commuted* (their positions can be interchanged) without affecting the sum or the product. In other words, the order in which we add numbers or multiply numbers does not affect the outcome. For example,

$$4 + 9 = 9 + 4 = 13$$

and

$$4 \cdot 9 = 9 \cdot 4 = 36$$

Our minds are capable of adding only two numbers at a time. The *associative law* for addition tells us that the terms can be *associated* (paired or grouped together) in any manner without affecting the sum. For example,

$$2 + 9 + 7 = 2 + (9 + 7) = 2 + 16 = 18$$

or

$$2 + 9 + 7 = (2 + 9) + 7 = 11 + 7 = 18$$

Likewise, factors can be paired in any manner without affecting the product.

The *distributive law* means that either we can add terms within a multinomial factor first and then multiply by the cofactor, $A(B + C)$, or we can multiply each and every term by the cofactor and then add the resulting terms, $AB + AC$. In the latter procedure, we "distribute" the multiplication. For example,

$$3(2 + 5) = 3(7) = 21$$

or

$$3(2 + 5) = 3(2) + 3(5) = 6 + 15 = 21$$

3.2 RULES OF SIGNS

The plus sign and the minus sign have two meanings each: *direction* and *operation*. The *directional sign* or simply *the sign* of a number is the + or − that indicates whether the number is measured in the positive direction or in the negative direction. We usually omit the directional + sign, and every number without a sign is understood to be positive. The *operational sign* is the + or − we place between terms to indicate the operation of addition or subtraction. Usually the type of sign is self-evident, but in this section we shall sometimes emphasize a directional sign with parentheses, as in (+5).

Directional and operational signs combine with each other in the operations of addition, subtraction, multiplication, and division. The following rules tell us how they combine.

A. RULES OF SIGNS FOR ADDITION AND SUBTRACTION

The sign of the sum of two numbers which have *like signs* (either both positive or both negative) is their common sign. For example,

$$(+2) + (+5) = +7 \qquad \text{and} \qquad (-2) + (-5) = -7$$

Thus we have the following rule:

Rule 1: Addition with Like Signs	To add two numbers with *like* signs, add their absolute values and affix their common sign to the sum.

Example 3.1. **Let's see how Rule 1 applies to an everyday situation. Suppose your checking account is overdrawn by \$2. The balance is thus \$−2. If you now write a check for \$5, you in effect add \$−5 to the balance. The new balance is thus (\$−2) + (\$−5) = \$−7.**

The sum of two numbers with *unlike signs* (one positive and the other negative) takes the sign of the number with the larger absolute value. For example,

$$(-2) + (+7) = +5 \qquad \text{and} \qquad (+2) + (-7) = -5$$

Thus we have the following rule:

Rule 2: Addition with Unlike Signs	To add two numbers with *unlike* signs, *subtract* their absolute values and affix to the result the sign of the number with the larger absolute value.

Subtracting a number is equivalent to adding the negative of the number. The following rule gives the procedure for subtraction:

Rule 3: Signs in Subtraction	To subtract two numbers, change the directional sign of the number to be subtracted, change the operational sign from minus to plus, and proceed as in addition.

Example 3.2. **Subtraction**

change directional sign

a. $(+2) - (-7) = (+2) + (+7) = 9$

change operational sign

b. $(+2) - (+7) = (+2) + (-7) = -5$

The following example illustrates the reasoning for Rule 3 and particularly the fact that subtracting a negative number is equivalent to adding the corresponding positive number.

Example 3.3. Let's examine your checking account again. Suppose you owe the bank \$5. To take away or subtract the \$5 debt, you must deposit \$5. That is, you must add a positive \$5. Therefore, the statement $-(\$-5) = +(\$+5)$ represents the subtracting of a debt of \$5 by adding \$5.

Example 3.4. Rules of signs on the number line

The number line is an excellent means for visualizing the rules of signs. We can represent a positive number by an arrow that points to the right, and a negative number by an arrow pointing to the left. The length of an arrow is the absolute value of the number. To add two numbers, we place the tail of the first arrow at the origin, and then place the tail of the second arrow at the head of the first one. See Fig. 3.1. The sum is the arrow that reaches from the origin to the head of the second arrow. Notice how this procedure corresponds to Rules 1 and 2 above.

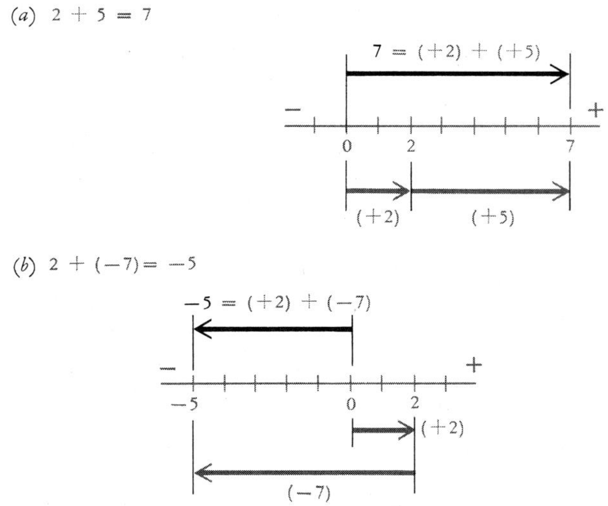

Fig. 3.1 Addition on the number line.

To subtract, place the first arrow at the origin as before, reverse the arrow to be subtracted, and proceed as in addition. See Fig. 3.2. Notice how this procedure corresponds to Rule 3 and Example 3.2.

B. RULES OF SIGNS FOR MULTIPLICATION AND DIVISION

Multiplication is a short form for addition. The expression $3(x)$, for example, means that x is to be a *term 3 times*. Thus $3(x) = x + x + x$. Similarly, $(-3)(x)$ means to use x as a term 3 times in *subtraction*. Thus $(-3)(x) = -(x) - (x) - (x)$, which by Rule 3 of signs is also $(-x) + (-x) + (-x)$.

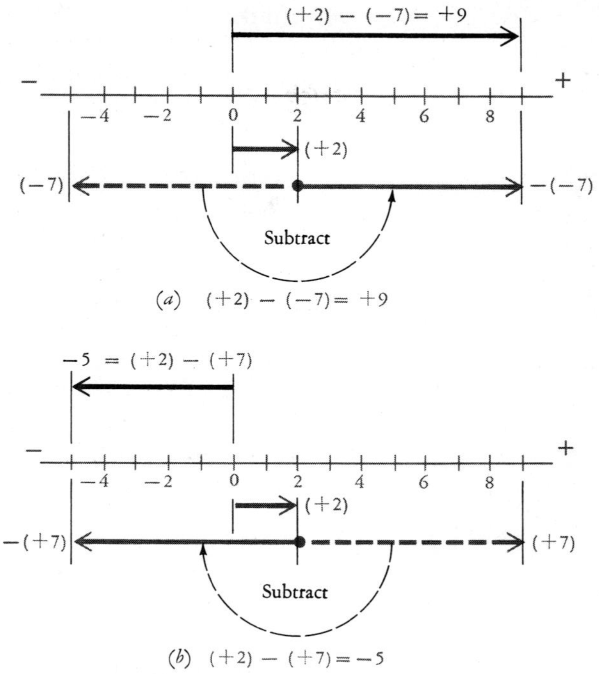

Fig. 3.2 Subtraction on the number line.

The signs of the terms implied by a multiplication are always alike, and Rule 1 therefore applies to multiplication. It also applies to division, since division is related to multiplication. For example, the division statement $^{15}/_{5} = 3$ is equivalent to $3(5) = 15$. For multiplication and division, Rule 1 can be simplified as follows:

Rule 4: Signs in Multiplication and Division	The product or quotient of two numbers with *like signs* is always *positive.* The product or quotient of two numbers with *unlike signs* is always *negative.*

The following example illustrates Rule 4 and all possible cases for signs in multiplication.

Example 3.5. Signs in multiplication

a. **Like signs:**

$$3(5) = 5 + 5 + 5 = +15$$
$$(-3)(-5) = -(-5) - (-5) - (-5)$$
$$= 5 + 5 + 5 = +15$$

b. Unlike signs:

$$3(-5) = (-5) + (-5) + (-5) = -15$$
$$-3(5) = -(+5) - (+5) - (+5) = -15$$

For more than two factors, you can predetermine the sign of any product or quotient if you can answer these questions:

What is the sign of the product of an even number of negative factors?
What is the sign of the product of an odd number of negative factors?

EXERCISE 3.1

In Probs. 1 to 26, use the fundamental laws and the rules of signs to evaluate each expression.

1. $15 - 12$
2. $16 + (-4)$
3. $2 - (-7)$
4. $-10 - 2$
5. $0.256 + (-0.312)$
6. $11.3 - (2.1 - 3.4)$
7. $19 - (7 - 12)$
8. $19 - (12 - 7)$
9. $-19 - (12 - 7)$
10. $-19 - (7 - 12)$
11. $3(5 - 2)$
12. $3(2 - 5)$
13. $(-3)(2 - 5)$
14. $(-3)(5 - 2)$
15. $6(x - 4 + a)$
16. $(-6)(x - 4 + a)$
17. $-2[x - 3(a - 2)]$
18. $5[4Z + 3 - 2(B - 10)]$
19. $6(-4) \div (-8)$
20. $(-6)(-4) \div 8$
21. $(-5)(-10) \div (-25)$
22. $5(10) \div (-25)$
23. $7(-12)(3)$
24. $-7(3)(12)$
25. $16(-1)(-5) \div [(-4)(-2)]$
26. $-32(2)(-3)(-5) \div [6(-4)]$

27. The initial temperature of a certain liquid is 100°F. What is the resulting temperature if
    a. the temperature is raised by 42°F?
    b. the temperature is lowered by 37°F?
    c. the temperature is multiplied by a factor of 2.5?
    d. the temperature is multiplied by a factor of −2?
28. Repeat Prob. 27 if the initial temperature is −100°F.

For Probs. 29 to 32, use Probs. 1 to 4. Work each problem with arrows on the number line, as in Figs. 3.1 and 3.2.

33. Demonstrate that Rule 4 works for division, by writing the division statement equivalent to each multiplication of Example 3.5. For example, if $3(5) = 15$, then $15/5 = 3$.

3.3 EXPONENTS AND RADICALS

We often need to use one number as a multiple factor, as in $2 \cdot 2 \cdot 2 \cdot 2 \cdot 2$ or $(x + 7)(x + 7)$. Mathematicians have invented a notation called the *exponent,* which eliminates the inconvenience of writing the same factor over and over.

A. DEFINITIONS AND RULES OF EXPONENTS

An exponent is a superscript affixed to a number to tell how many times the number is to be used as a *factor.* For example, the expression 2^5 means $2 \cdot 2 \cdot 2 \cdot 2 \cdot 2$. The exponent, 5, indicates that the number 2, called the *base,* is to be used as a factor 5 times. The general notation and terminology for exponents are as follows:

$$A^n = \underbrace{A \cdot A \cdot A \cdot A \cdot \cdots \cdot A \cdot A}_{n \text{ factors}} \tag{3.4}$$

where A is the base

n, a positive integer, is the *exponent* or *power* of A

A^n is called the "nth power of A," or "A to the nth"

Notice that the exponent must be a positive integer. In Chap. 12 we shall discuss negative exponents and rational exponents.

Example 3.6. **Exponents**

a. $5^3 = 5 \cdot 5 \cdot 5 = 125$ The third power of 5, or 5 ***cubed***

b. $(-4)^2 = (-4)(-4) = +16$ The second power of -4, or -4 ***squared***

c. $-4^2 = -(4)(4) = -16$

d. $(2 + 5)^3 = 7^3 = 343$

e. $2^3 + 5^3 = 8 + 125 = 133$

(d and e:) Note that $(2 + 5)^3$ is ***not*** equal to $2^3 + 5^3$

Comment: *Examples 3.6d and e illustrate a common error concerning exponents. Beginning students frequently write erroneous statements such as $(x + 7)^2$ equals $x^2 + 7^2$. This may seem reasonable, but* ***it does not adhere to the definition of an exponent.*** *There is simply no basis for it. You should read the expression $(x + 7)^2$, according to the definition, as "$(x + 7)$ times $(x + 7)$," and use the distributive law to perform the multiplication. The result is $x^2 + 14x + 49$,* ***not*** *$x^2 + 49$.*

The definition of an exponent produces the following rules for operations with exponents.

$$A^n \cdot A^k = A^{n+k} \tag{3.5}$$

$$\frac{A^n}{A^k} = A^{n-k} \qquad n > k \quad A \neq 0 \tag{3.6a}$$

$$\frac{A^n}{A^k} = \frac{1}{A^{k-n}} \qquad k > n, \quad A \neq 0 \tag{3.6b}$$

$$(A^n)^k = A^{nk} = (A^k)^n \tag{3.7}$$

$$(AB)^n = A^n \cdot B^n \tag{3.8a}$$

$$\left(\frac{A}{B}\right)^n = \frac{A^n}{B^n} \qquad B \neq 0 \tag{3.8b}$$

Rules of Exponents

Comment: *Remember to* ***learn the ideas*** *expressed by these equations; don't just memorize the formulas. For example, Eq. (3.5) means "when two (or more) numbers with the* ***same base*** *are multiplied, the resulting exponent is the* ***sum*** *of the individual exponents." To get the ideas behind the rules of exponents, you should translate Eqs. (3.6) to (3.8) in a similar fashion.*

Each of the exponent rules is based upon and can be derived from the definition, Eq. (3.4). For example, we can show why Eq. (3.5) is true, as follows:

$$A^n \cdot A^k = \underbrace{\underbrace{(A \cdot A \cdot A \cdot A \cdot \cdots \cdot A \cdot A)}_{n \text{ factors}}\underbrace{(A \cdot A \cdot A \cdot \cdots \cdot A \cdot A)}_{k \text{ factors}}}_{n+k \text{ factors}}$$

$$= A^{n+k}$$

You should derive the other rules in a similar fashion, always using the definition to obtain the results. See Examples 3.7 and 3.8.

Example 3.7. **Show that $A^2/A^5 = 1/A^3$.**

Solution:

$$\frac{A^2}{A^5} = \frac{\cancel{A} \cdot \cancel{A}}{\cancel{A} \cdot \cancel{A} \cdot A \cdot A \cdot A} = \frac{1}{A \cdot A \cdot A} = \frac{1}{A^3}$$

Example 3.8. Show that $(A^2)^3 = A^6$.

Solution: $(A^2)^3 = A^2 \cdot A^2 \cdot A^2 = (A \cdot A)(A \cdot A)(A \cdot A) = A^6$

Example 3.9. Applications of the rules of exponents:

a. $(-2)^2(-2)^4 = (-2)^{2+4} = (-2)^6 = +64$

b. $X^3 \cdot X^5 \cdot X^2 = X^{3+5+2} = X^{10}$

c. $(N^3 \cdot N^{15})/(N^7 \cdot N^{10}) = N^{3+15-7-10} = N$

d. $4^3 \cdot 3^3/6^3 = (4 \cdot 3/6)^3 = 2^3 = 8$

e. $(-Z^3)^3 = (-1)^3(Z^3)^3 = -Z^{3 \cdot 3} = -Z^9$

f. $(2Y)^4 = 2^4Y^4 = 16Y^4$

g. $9x^4/(y^6x^2) = (3x/y^3)^2$

h. $(27x)(x^2/y^6) = 3^3x^3/(y^2)^3 = (3x/y^2)^3$

EXERCISE 3.2

In Probs. 1 to 12, determine the value of each expression.

1. 7^4
2. 6^3
3. $(-3)^3$
4. $(-2)^5$
5. $2^3 \cdot 2^2/2^8$
6. $5^3 - 5^2$
7. $(4^2)(3^2)$
8. $9^2/3^2$
9. $2^6 - (-2)^5$
10. $(-3)^3 \cdot 3^2/3^4$
11. $(-5)^3 - (-5)^2$
12. $(-2)^2 + (-2)^3$

In Probs. 13 to 21, simplify each expression by using the rules of exponents.

13. $x^5(x^3 - x^2)$
14. $(2n/3)^3$
15. $(-3a^2x^2)^5$
16. $(-5ST^3)^2$
17. $-5(ST^3)^2$
18. $-2M^2(-2M)^2$
19. $-2M^2[(-2)^2M]^2$
20. $4X^2(4X)^2 \div (4X)^3$
21. $(2X + 3)^5(2X + 3)^7 \div (2X + 3)^9$

In each of the Probs. 22 to 29, combine the several bases into a single base, as in Examples 3.9*g* and *h*.

22. $4x^2$
23. $-8B^3$
24. $16G^4/81$
25. $16G^2/81$
26. $X^2 \cdot X^3/32$
27. $243X^7/X^2$
28. $(125U)(U^2/W^6)$
29. $49K^4/(K^2 \cdot N^6)$

In each of Probs. 30 to 32, write the given statement as an algebraic equation.

30. The power loss in an electric circuit equals the resistance multiplied by the square of the current. Let W be power loss, R be resistance, and I be current.
31. The area of a circle equals π (pi) multiplied by the square of one-half the diameter. Let A be the area and D be the diameter.
32. If a body starts at rest and falls freely due to gravity, the distance S that it falls is one-half the acceleration of gravity g multiplied by the square of the time t during which it falls.

B. RADICALS

If B equals the nth power of A, we can use the exponent notation to describe B: $B = A^n$. We shall also need a symbolic notation for writing A in terms of B and n. We can describe A as "a number which raised to the nth power equals B," and we call such a number an nth *root* of B. For example, 3 is a fourth root of 81, because $81 = 3^4$.

The general notation and meaning for an nth root are as follows:

$$A = \sqrt[n]{B} \qquad \text{means} \qquad B = A^n$$

For example, $3 = \sqrt[4]{81}$ means that $81 = 3^4$. The following terminology is used with this notation.

> In $A = \sqrt[n]{B}$
>
> A is the **nth root** of B. $\sqrt{B}$ is the **radical**,
> B is the **radicand**. $\sqrt{\ }$ is the **radical sign**.
> n is the **index** or **order** of the radical,
> and n must be a positive integer.

It is very important for you to understand that $A = \sqrt[n]{B}$ and $B = A^n$ mean the same thing; each equation expresses exactly the same relationship between the numbers A and B. As the following examples show, we can often evaluate radicals by using our knowledge of exponents.

Example 3.10. Radicals

a. Since $2^3 = 8$, then $2 = \sqrt[3]{8}$. The *cube root* of 8

b. Since $(-2)^3 = -8$, then $-2 = \sqrt[3]{-8}$. The *cube root* of -8

c. If $x^2 = 25$, then $x = \pm\sqrt{25} = \pm 5$, since $5^2 = 25$ and $(-5)^2 = 25$. Notice that we omit the index of 2 in the *square root* symbol.

d. If $x^4 = 81$, then $x = \pm\sqrt[4]{81} = \pm 3$, since $3^4 = 81$ and $(-3)^4 = 81$.

Principal Roots

Examples 3.10*c* and *d* show that a radical with an even-numbered index has two roots, one the negative of the other. To avoid this ambiguous situation, we define the principal root as follows:

> The *principal* nth *root* of a number is the real root whose sign matches the sign of the radicand.

Unless a different sign is specified, every radical, by definition, means the principal root.

Example 3.11. Principal roots

a. $\sqrt{25} = 5$ (but *not* -5)

b. $-\sqrt{25} = -(\sqrt{25}) = -5$

c. $\pm\sqrt{25} = \pm 5$

d. $\sqrt{-25}$, imaginary

e. $\sqrt[3]{8} = 2$

f. $\sqrt[3]{-8} = -2$

Comment: *A popular maneuver in the sport of jumping to conclusions is to take the nth root of a sum* $\sqrt[n]{A+B}$ *to be the sum of the nth roots* $\sqrt[n]{A} + \sqrt[n]{B}$. *But notice that* $\sqrt{9+16}$ *is* ***not*** *equal to* $\sqrt{9} + \sqrt{16} = 3 + 4 = 7$; *instead,* $\sqrt{9+16} = \sqrt{25} = 5$. *In general,* $\sqrt[n]{A+B}$ *is* ***not equal to*** $\sqrt[n]{A} + \sqrt[n]{B}$.

EXERCISE 3.3

Evaluate each expression given in Probs. 1 to 16.

1. $\sqrt{36}$
2. $\sqrt{-36}$
3. $\sqrt[3]{-125}$
4. $\sqrt[3]{125}$
5. $\sqrt[6]{-64}$
6. $\sqrt[4]{16}$
7. $\sqrt{4(25)}$
8. $4\sqrt{25}$
9. $\sqrt{16-25}$
10. $\sqrt{25-16}/\sqrt{16}$
11. $\sqrt[3]{8-27}$
12. $\sqrt[3]{8}-\sqrt[3]{27}$
13. $\dfrac{\sqrt{64}(3\cdot 4)^3}{(8\cdot 3)^2\sqrt{16}}$
14. $\dfrac{(7\cdot 3)^2\sqrt[5]{32}}{49\sqrt[3]{216}}$
15. $\sqrt{A^2}$ if $A = 4$
16. $\sqrt{A^2}$ if $A = -4$

17. Simplify the expression $\sqrt{A^2}$ if A is any positive number.
18. Simplify the expression $\sqrt{A^2}$ if A is any negative number. (*Hint:* See Prob. 16.)
19. Does $\sqrt{A^2}$ always equal A? Explain.

3.4 ADDITION AND SUBTRACTION OF ALGEBRAIC EXPRESSIONS

Algebraic terms with exactly the same literal factors are called *similar terms* or *like terms.* For example,

$$2x^2y \qquad 5x^2y \qquad \text{and} \qquad -16x^2y$$

are similar terms, but

$$2x^2y \qquad 5xy^2 \qquad \text{and} \qquad -16x^2y^2$$

are dissimilar terms. Notice that similar terms may differ only in their numerical coefficients.

The basis for adding or subtracting algebraic expressions is the distributive law. For example, since $(2 + 5)x = 2x + 5x$, the reverse is also true:

$$2x + 5x = (2 + 5)x = 7x$$

But $2x + 5y$ cannot be combined. In other words, to add similar terms, we take out the common literal factors and add the remaining numbers.

Example 3.12. Adding algebraic expressions

a. $16x - 3x = (16 - 3)x = 13x$

b. $2ax - 3ax = (2 - 3)ax = -ax$

c. $5x/y - 3x/y = (5 - 3)x/y = 2x/y$

d. $3ax + 5bx = (3a + 5b)x$

e. $5x^2 - 3x^2 + 12x^2 = (5 - 3 + 12)x^2 = 14x^2$

f. $7(x + y) + 8(x + y) = (7 + 8)(x + y) = 15(x + y)$

EXERCISE 3.4

Perform the indicated operations wherever possible.

1. $9a - 6a + 2a$
2. $4b - 7b$
3. $7ab + a - 2ab$
4. $1.9x + 2.7x$
5. $5y^3 + 2y^2 - y$
6. $(1/4)x + (3/4)x$
7. $3xy^2 - x^2y + 2xy^2$
8. $29A^2B - 13AB^2 - 11A^2B + 4AB^2$
9. $(3/8)K - (9/16)K + (3/32)K$
10. $4a(b - x) + 5a(b - x)$
11. $0.032M^3 - 0.014M^3$
12. $(2x + y)^2 - (2x + y)^2$
13. Mr. Jones, the grocer, has received three boxes of mixed fruit. One box contains 36 oranges, 29 grapefruit, and 43 tangerines; the second box contains 22 grapefruit, 56 oranges, and 30 tangerines; and the third box contains 13 tangerines, 31 grapefruit, and 64 oranges. Mr. Jones throws out a fourth box of outdated fruit containing 33 oranges, 39 grapefruit, and 36 tangerines. What is his net gain of each type of fruit?
14. In terms of the ultimate bending moment M_u, the ultimate load that a certain beam can carry is given by the expression $2M_u/L + 3M_u/L + 3M_u/(2L)$. Simplify this expression by adding the similar terms.

3.5 MULTIPLICATION OF ALGEBRAIC EXPRESSIONS

To multiply two or more monomials, we multiply the numerical coefficients and combine the literal coefficients according to the rules of exponents.

Example 3.13. Multiplication with monomial factors

a. $(2x)(3x^2) = (2 \cdot 3)x^{1+2} = 6x^3$

b. $(7a^3x^4)(8a^2x^5) = (7 \cdot 8)a^{3+2}x^{4+5} = 56a^5x^9$

c. $\dfrac{5A^4BZ^3}{3B^2Z^2}\,\dfrac{2AZ^7}{7B^3} = \dfrac{(5 \cdot 2)}{(3 \cdot 7)}\,\dfrac{A^{4+1}Z^{3+7-2}}{B^{2+3-1}} = \dfrac{10A^5Z^8}{21B^4}$

If a factor has more than one term, we apply the distributive law repeatedly. We must multiply each and every term of one factor by each and every term of the other factors. For example, to multiply $(c + d)$ by $(a + b)$, we treat $(a + b)$ as a single number and multiply it first by c and then by d:

$$(a + b)(c + d) = (a + b)c + (a + b)d$$

Next, we multiply $(a + b)$ by c and $(a + b)$ by d to obtain

$$(a + b)(c + d) = ac + bc + ad + bd$$

If the resulting product contains similar terms, we combine them as in Example 3.14*c* below.

Example 3.14. Multiplication with multinomial factors.

Try to work each of the following examples mentally or on scratch paper before you read the results given here; then compare your answers to these results.

a. $3(2 + x) = 3(2) + 3(x) = 6 + 3x$

b. $(2a)^3(2ay + a^2) = 8a^3(2ay) + 8a^3(a^2) = 16a^4y + 8a^5$

c. $(x - 7)^2 = (x - 7)(x - 7) = (x - 7)x - (x - 7)7$
$= x^2 - 7x - 7x + 49 = x^2 - 14x + 49$

Note that $(x - 7)^2$ is ***not*** $x^2 - 7^2 = x^2 - 49$.

If terms enclosed in grouping symbols are to be combined with similar terms outside the group, we must first remove the grouping symbols by performing all indicated operations. For example, to add $5a$ and $2(3a - x)$, we remove the parentheses by multiplying every term within parentheses by 2: $5a + 2(3a - x) = 5a + 6a - 2x = 11a - 2x$. In particular, if a set of grouped terms is to be subtracted, we remove parentheses by changing the sign of *every term* (not just the first term) within the parentheses. Thus: $3a - (5a - x) = 3a - 5a + x = -2a + x$.

Examples 3.15*b* and *c* below show one way to simplify groups within groups. We first remove the innermost grouping symbols and combine similar terms, if possible; then we continue removing grouping symbols until all terms are combined.

Example 3.15. **Removing grouping symbols**

a. $2(3x^2 - 5x) + 3(x - 5x^2) = 6x^2 - 10x + 3x - 15x^2 = -9x^2 - 7x$

b. $$\begin{aligned} 5[3ax - a(a - 2x)] &= 5[3ax - a^2 + 2ax] \\ &= 5[5ax - a^2] = 25ax - 5a^2 \end{aligned}$$

c. $$\begin{aligned} 3a\{2x[a - 5(x - 3a)] + x^2\} &= 3a\{2x[a - 5x + 15a] + x^2\} \\ &= 3a\{32ax - 10x^2 + x^2\} \\ &= 96a^2x - 27ax^2 \end{aligned}$$

EXERCISE 3.5

In Probs. 1 to 12, perform the indicated multiplication and combine similar terms wherever possible.

1. $2x^2 - 4 + (x - 2)^2 + 3x$
2. $7b - 3(a - 4b + 3c) + 7a + 5c$
3. $(x + 5)^2$
4. $(x - 5)^2$
5. $6(3a - 2b)^2$
6. $5(x + 2y)^3$
7. $(2a + 3)^3 - (2a + 3)^2$
8. $a[(a - b) - (a + b)]$
9. $2K\{3[K - (2 - K)]\}$
10. $3[2x(x - y + z) + x^2 - x(y + z)]$
11. $-2\{3y[x - y(y - x)] - 2y^3\}$
12. $\{2[a^2 + (a - b)^2 + b^2] - 3[a - b]^2\}^2$

13. An expression for the load that a certain riveted joint can carry is $t(w - d)\sigma_T$. Perform the indicated multiplication.
14. A certain fence section is 3 yd 2 ft 3 in. long. How long a fence would three such sections make, if a 4- by 4-in. post is used at each end and between sections?
15. The width of a certain rectangle is $a + b$ ft, and the length is $c + d$ ft.

a. What is the area of the rectangle in terms of the lengths $a, b, c,$ and d?
b. What is the perimeter of the rectangle?

16. The radius of a certain circle is $x + 8$ in. What is the area of the circle in terms of x?

3.6 DIVISION WITH ALGEBRAIC EXPRESSIONS

There are three distinct types of algebraic division problems. In order to do division problems efficiently, we must learn to recognize each type and use the appropriate method.

Monomials Divided by Monomials

To divide a monomial expression by another monomial, as in $6A^5 \div 3A^2$, divide the numerical coefficients and combine the literal coefficients by using the rules of exponents.

Example 3.16. *a.* $\dfrac{6A^5}{3A^2} = \dfrac{6}{3}A^{5-2} = 2A^3$

b. $\dfrac{27X^3Y}{3XY^2} = \dfrac{27}{3}\dfrac{X^{3-1}}{Y^{2-1}} = \dfrac{9X^2}{Y}$

c. Identical multinomial factors can be treated like monomial factors:

$$\frac{(x+a)^4}{(x+a)} = (x+a)^{4-1} = (x+a)^3$$

Multinomials Divided by Monomials

In a quotient such as

$$\frac{x^2 - 2ax + a^2}{2a}$$

the dividing line is a grouping symbol. It indicates that the *entire numerator* must be divided by the denominator. Therefore to divide a multinomial by a monomial: *Divide each and every term of the numerator by the denominator.*

Example 3.17. *a.* $\dfrac{10 + 5 + 25}{5} = \dfrac{10}{5} + \dfrac{5}{5} + \dfrac{25}{5} = 2 + 1 + 5 = 8$

Another method is

$$\frac{10 + 5 + 25}{5} = \frac{40}{5} = 8$$

But see Comment 2 below!

b. $$\frac{2x^5 - 6x^3 - 3}{3x^2} = \frac{2x^5}{3x^2} - \frac{6x^3}{3x^2} - \frac{3}{3x^2} = \frac{2x^3}{3} - 2x - \frac{1}{x^2}$$

Comment 1: *These examples illustrate an important point about reading mathematics. To read an example, analyze each step by comparing each term to the corresponding term of the preceding step. For instance, in the far right side of Example 3.17b above, the $2x^3/3$ comes from $2x^5/(3x^2)$ of the preceding step, the $2x$ comes from $6x^3/(3x^2)$, and the $1/x^2$ comes from $3/(3x^2)$. Mathematics authors purposely write this way to aid the student. But the student must make an effort to really read and understand. He must look, compare, and think about what has happened in the examples and other explanations.*

Comment 2: *A very frequent error in problems such as Example 3.17a is in canceling terms. You must remember that* ***in division you can cancel only factors, never terms!*** *For instance, many beginners would do Example 3.17a incorrectly by canceling the 5s, as follows:*

$$\frac{10 + \not{5} + 25}{\not{5}} = 10 + 1 + 25 = 36$$ *Example of* ***incorrect*** *procedure*

Compare this procedure and answer to those of Example 3.17a.

Polynomials Divided by Polynomials

A *polynomial* is a special kind of multinomial or monomial. Each term must be either an explicit number or of the form ax^n, where n is a positive integer. The degree of a polynomial is the greatest value of n that occurs.

Example 3.18. Polynomials

a. $x^2 + x + 1$ — a second-degree polynomial in x

b. $2 + Z^2 - 6Z^3 + 13Z^5$ — a fifth-degree polynomial in Z

c. y^3 — a third-degree polynomial in y

d. $x + 2\sqrt{x}$ — not a polynomial; $\sqrt{x}$ not in the form x^n

e. $x^2 - 2x + 1/x^2$ — not a polynomial; $1/x^2$ not in the form x^n

To divide one polynomial by another, we perform three specific steps over and over until no further division is possible. Example 3.19 below shows how. Notice that the procedure is exactly the same as that for long division with numbers.

Example 3.19. Divide $1 + 18x^3 - 17x$ by $3x - 2$.

Before dividing, arrange the dividend and the divisor in order of descending powers of x; if any power of x is missing, write it with a zero coefficient, as shown below.

Then perform the following three steps in order: (1) divide the *first term,* $18x^3$, of the dividend by the *first term,* $3x$, of the divisor, and place the quotient, $6x^2$, above the dividend; (2) multiply the *entire divisor* by this term of the quotient and align the product, $18x^3 - 12x^2$, under the dividend; and (3) subtract this product from the dividend to form a new dividend, $12x^2 - 17x$. Repeat this procedure, using the *new dividend,* until the degree of the final dividend is at least one less than the degree of the divisor.

	$6x^2 + 4x - 3 - \dfrac{5}{3x-2}$
Divide: $18x^3 \div 3x = 6x^2$	$3x - 2\,)\,18x^3 + 0x^2 - 17x + 1$
Multiply: $6x^2(3x - 2)$	$18x^3 - 12x^2$
Subtract:	$12x^2 - 17x$
Divide: $12x^2 \div 3x = 4x$	
Multiply: $4x(3x - 2)$	$12x^2 - 8x$
Subtract:	$-9x + 1$
Divide: $-9x \div 3x = -3$	
Multiply: $-3(3x - 2)$	$-9x + 6$
Subtract:	-5
Remainder: $-5 \div (3x - 2)$	

EXERCISE 3.6

Perform the indicated operations in Probs. 1 to 15.

1. $\dfrac{6a^3}{2a}$

2. $\dfrac{12f^{29}}{3f^7}$

3. $\dfrac{2a(a+b)^2}{a+b}$

4. $\dfrac{a^3b^2c}{ab^2c^3}$

5. $\dfrac{r + 2r + r^2}{3r}$

6. $\dfrac{x^2 + 2x + 2}{2x}$

7. $\dfrac{a(x + k) - b(x + k)}{c(x + k)}$

8. $\dfrac{a(x + k) - b(x + k)}{(a - b)(x + k)}$

9. $\dfrac{vt - \frac{1}{2}gt^2}{t}$

10. $\dfrac{x^2 + 2x + 1}{x + 1}$

11. $\dfrac{3x - y - 2x + 2y}{(x + y)^2}$

12. $\dfrac{2m^3 + m^2}{2m}$

13. $\dfrac{2(m - 3)^3 + (m - 3)^2}{2(m - 3)}$

14. $\dfrac{3n^4 - 5n^2 + 7n - 2}{n^2 - 1}$

15. $\dfrac{a^5 + 3a^4 - 5a^2 + a}{a + 2}$

16. The combined resistance of two resistors in parallel is given by the expression $R_1R_2/(R_2 + R_1)$. Take the reciprocal of this expression and perform any division that is possible.
17. For three resistors in parallel the combined resistance is $R_1R_2R_3/(R_2R_3 + R_1R_3 + R_1R_2)$. Take the reciprocal and divide, as in Prob. 16.
18. The image distance of a simple lens is given by the expression $d_f d_o/(d_o - d_f)$, where d_f is the focal length and d_o is the object distance. Take the reciprocal and divide, as in Prob. 16.
19. The strain along one axis of a certain strain gage is given by the expression $(e_x + e_y)/2 + (-\frac{1}{2})(e_x - e_y)/2$. Simplify this expression.
20. The efficiency of an ideal gas-cycle engine is given by the expression $(Q_2 - Q_1)/Q_2$. Perform the indicated division.
21. The isothermal compressibility of a gas is given by the expression $-(P/C)(-C/P^2)$. Simplify this expression.

3.7 SOLVING EQUATIONS

A major objective of mathematics is solving equations. And by "solving equations" we mean finding the value(s) of an unknown quantity that makes the equation true. For example, we would solve the equation $2x = 4$ by finding the value of x that makes the equation a true statement. In this case, $x = 2$.

Basic Ideas in Solving Equations

The general idea in solving an equation is to perform the same series of algebraic operations—addition, multiplication, and so on—on *both* sides of the equation, until the unknown quantity stands by itself. For example, to solve $2x = 4$, we divide both sides by 2 so that x will stand alone—that is, will have a coefficient of $+1$. The algebraic solution is as follows:

Given:	$2x = 4$	*Check:*	$2(2) \stackrel{?}{=} 4$
$\div$ 2:	$\frac{\not{2}x}{\not{2}} = \frac{4}{2}$		$4 = 4$
Answer:	$x = 2$		

As indicated above, we should always check the answer(s) to be sure that it does indeed make the equation a true statement. The answer is called a *solution* or *root* of the equation.

We can interpret the solving of an equation as "undoing what has been done to the unknown quantity." In the example above, x has been multiplied by 2; so we "undo" that multiplication by dividing $2x$ by 2. We must always perform exactly the same operation on both sides of the equation. When we divide the $2x$ by 2, we must also divide the 4 by 2; otherwise, we upset the balance of the equation—that is, we form a different statement, one that is not equivalent to the original equation. Failure to perform exactly the same operations on both sides is one of the most frequent errors made by beginners.

Procedures for Solving Equations

We use just three algebraic operations to solve equations: (1) add (or subtract) the same quantity to both sides, (2) multiply or divide both sides by the same non-zero quantity, and (3) take the same power or root of both sides. Observe in the following examples how we apply these procedures in order to isolate the unknown quantity.

Example 3.20. Solving equations

a. Solve for x: $3 + 2x = 3x - 5$ *Check:*

Subtract 3:	$\not{3} + 2x - \not{3} = 3x - 5 - 3$	$3 + 2(8) \stackrel{?}{=} 3(8) - 5$
	$2x = 3x - 8$	$3 + 16 \stackrel{?}{=} 24 - 5$
Subtract $3x$:	$2x - 3x = \not{3x} - 8 - \not{3x}$	$19 = 19$
	$-x = -8$	
Multiply by -1:	$-1(-x) = -1(-8)$	
Answer:	$\boxed{x = 8}$	

b. Solve for N: $\frac{N^2}{2} = 28$

Multiply by 2: $(\not{2})\left(\frac{N^2}{\not{2}}\right) = (2)(28)$

$N^2 = 56$

Take $\sqrt{\ }$: $\sqrt{N^2} = \pm\sqrt{56} = \pm\sqrt{4(14)}$

Answer: $\boxed{N = \pm 2\sqrt{14}}$

Check:

$$\frac{(\pm 2\sqrt{14})^2}{2} \stackrel{?}{=} 28$$

$$\frac{4(14)}{2} \stackrel{?}{=} 28$$

$$28 = 28$$

Note that we must specify both square roots, $+\sqrt{56}$ and $-\sqrt{56}$, not just the principal root $\sqrt{56}$; otherwise we would overlook one solution.

c. Solve for x in terms of a: $\sqrt{x^2+1} - 2 = 3 - a$

Add 2: $\sqrt{x^2+1} - \not{2} + \not{2} = 3 - a + 2$

$\sqrt{x^2+1} = 5 - a$

Square both sides: $(\sqrt{x^2+1})^2 = (5-a)^2$

$x^2 + 1 = 25 - 10a + a^2$

Subtract 1: $x^2 + 1 - 1 = 25 - 10a + a^2 - 1$

$x^2 = 24 - 10a + a^2$

Take $\sqrt{\ }$: $\sqrt{x^2} = \pm\sqrt{24 - 10a + a^2}$

Answer: $\boxed{x = \pm\sqrt{24 - 10a + a^2}}$

The check is left as an exercise for the student.

Comment: *Note in Example 3.20c that we squared the side $5 - a$ by squaring the quantity $(5 - a)$; we did* ***not*** *square each individual term. To take the square root of the side $24 - 10a + a^2$, we took the root of the quantity $(24 - 10a + a^2)$; we did* ***not*** *take the root of each individual term. Failure to observe such simple principles results in the error of not performing the same operation on both sides.*

Sometimes the scheduling of operations is critical. For example, to solve $x^2 + 1 = 26$, we must subtract 1 *before* taking the root. Otherwise we end up with $\sqrt{x^2+1} = \sqrt{26}$, which is correct, but which does not help to isolate x. Judicious applications of common sense and foresight will help in deciding the sequence of operations.

Formulas

A *formula* is an equation that states a general physical or geometric fact. For example, the equation $A = \pi r^2$ is the *formula* for the area of a circle in terms of the radius. The procedure for solving formulas is the same as that for solving any other equation.

Example 3.21. Formulas

a. To find the area of a circle with a 2.6-in. radius, we use the formula $A = \pi r^2$ and solve for A:

$$A = 3.14(2.6)^2 = 3.14(6.76 \text{ in.}^2) = 21.2 \text{ in.}^2$$

(Note the convention *in.*^{2} meaning "square inches.")

b. To find the radius of a sphere whose volume is 28π in.^{3}, we use the formula $V = (4/3)\pi r^3$ and solve for r:

$$r = \sqrt[3]{\frac{3V}{4\pi}} = \sqrt[3]{\frac{3(28\pi \text{ in.}^3)}{4\pi}} = 2.76 \text{ in.}$$

EXERCISE 3.7

In Probs. 1 to 30, solve for the letter in each equation.

1. $2x + 3 = 23$
2. $5a - 7 = 18$
3. $n - 3 = 5 - 3n$
4. $b/2 - 1 = 5$
5. $4 - 2z/3 + z = 0$ (*Hint:* Multiply by 3.)
6. $x^2 + 2x - 3 = 2x + 1$
7. $(1/2)(2c - 5) = 4 + c$
8. $(b - 5)^2 = 25$
9. $x - (1/5)(10x - 25) = 5$
10. $\sqrt{G + 9} = 13$
11. $\sqrt[3]{I} + 10 = 7$
12. $a^5 + 1 = -3(2 - a^5/3)$
13. $(x + 7)^2 = 52 + 14x$
14. $\sqrt[3]{x^3 + 27} = -3$
15. $5 - \sqrt{x} = -2$
16. $R^3 + 2^3 = 72$
17. $\sqrt{S} - \sqrt{4} = 5$
18. $m^2 - 7m + 12 = 2(m - 4)$ (*Hint:* If $m - 4 \neq 0$, divide by $m - 4$.)
19. $\sqrt{R + 5} = 2$
20. $\sqrt{4x + 3} = 2 + x$
21. $\sqrt{4x + 5} = 2 + x$
22. $\sqrt{x} = x$
23. $\sqrt{-x} = x$
24. $\sqrt{x^2} = -x$
25. $\sqrt{x^2} = x$
26. $z^3 + 15 = 7$
27. $2P^5 - 2 = P^5 - 34$
28. $\sqrt{32 - Q^5} = -2$
29. $1 - y^4 = 3y^4 - 3$
30. $x^2 - 3(8 + x^2) = 0$

In Probs. 31 to 48, solve for x in terms of the other letter or letters. If any restrictions are needed to avoid imaginary numbers or division by zero, state these restrictions.

31. $2x + 4a = 6b$
32. $15L - 3x = 9m$
33. $x^2 - y^2 = 5^2$
34. $3x^2 + 2a^2 = 5x^2 + 3a$
35. $\sqrt{a - x} - 3 = b$
36. $\sqrt{x + n} = n$
37. $(5x - 10k)/(5k) = 3$
38. $(3R - 6x)/(3R) = 2$
39. $(2a - 2x)/x = 1$
40. $(2 + a^2x)/(ax) = a/2$
41. $\sqrt[3]{x^3 + b^3} = -2$
42. $\sqrt[5]{a^5 - x^5} = -a$
43. $ax + bx = a + b$
44. $(n_1x - n_2x)/x = k$
45. $(rx + sx + 3)/(3rs) = 1$
46. $(ax - bx)/(a + b) = (a - b)^2$
47. $(2x - ax)/x^2 = 3x + bx$
48. $\sqrt{mx^2 + nx^2} = a$

In Probs. 49 to 63, use the formula to solve for the indicated quantity. If no formula is given, it is a common one that you should know or can look up, or it has been used in one of the examples of this chapter. Many formulas from geometry are given in Appendix 4.

49. Find the surface area of a sphere with a 3.6-in. radius.
50. Find the radius of a sphere with a surface area of 144 in.^{2}.
51. Find the radius of a sphere with a volume of 36π ft^3.
52. Find the distance in yards that a falling object starting from rest drops in 12 sec. Acceleration due to gravity is 32.2 ft/sec^2. For feet, the formula is $s = (1/2)gt^2$.
53. Find the time in seconds for a falling object starting from rest to drop 6,440 ft. Acceleration due to gravity is 32.2 ft/sec^2. See Prob. 52 for a formula.
54. The formula for the exit speed in ft/sec of water flowing from the bottom of a tank is $V = \sqrt{2gd}$, where d is the depth of water in feet, and $g = 32.2$ ft/sec^2. Find the depth d, in inches, if the exit speed is 44 ft/sec.
55. The centrifugal force in pounds on a rotating object is $F = 4\pi^2RN^2W/g$, where R is the radius of rotation in feet, N is the speed of revolution in rps (revolutions per second), W is the weight of the object in pounds, and g is 32.2 ft/sec^2.
    a. Find the centrifugal force on a 2-lb object rotating at 600 rpm (revolutions per minute) on a 3-ft radius.
    b. Find the speed of revolution in rps.
56. The period (time for one oscillation), in seconds, of a simple pendulum is $T = 2\pi\sqrt{L/g}$, where L is the length in feet, and g is 32.2 ft/sec^2.
    a. What is the period of a 57-in. pendulum?
    b. What is the length in inches of a pendulum with a period of 1 sec?
57. The frequency (number of oscillations per unit of time) of the pendulum of Prob. 56 is the reciprocal of the period.
    a. What is the frequency of the pendulum in Prob. 56*a*?

b. What is the length of a pendulum with a frequency of 60 Hz (hertz, formerly cycles per second)?

58. The discharge rate of water flowing through a tube is stated by Poiseuille's law as $Q = R^4(P_1 - P_2)/(8NL)$. Solve for R, the radius of the tube, in terms of Q, P_1, P_2, N, and L.

59. $T = 4P/\pi d^2$ is the formula for *shear stress* in a coil spring of diameter d under a load P. What must be the minimum diameter to support a stress of 10,000 psi (pounds per square inch) under a load of 1,256 lb?

60. $\delta = PL^3/(3EI)$ is the *deflection* in inches at the free end of a cantilever beam L in. long, due to a load of P lb acting at the free end. What load would cause a deflection of 0.0025 in. at the end of a 3-in. beam for which $E = 29$ million psi and $I = 0.80$ in.^{4}?

61. The equivalent resistance for two resistors in parallel is $R = R_1R_2/(R_1 + R_2)$. If R is 5 ohms and R_2 is 10 ohms, what is the value of R_1?

62. Child's law for the *perveance* K of a diode in terms of current I and voltage E is $K^2 = I^2/E^3$. Find the voltage if $K = 0.12$ and $I = 5$ amp.

63. $A = KR/(r + R)$ is the formula for *gain* in a triode amplifier circuit, where K is an amplifier factor, R is the load resistance, and r is the plate resistance. Find R if $A = 1.2$, $K = 3.6$, and $r = 0.15$ ohm.

3.8 APPLICATIONS

Mathematics is a powerful means for solving practical problems, the so-called word problems. The general approach to solving such problems is (1) to translate the stated facts into algebraic equations, and (2) to solve the equations. In this way, difficult mental figuring is reduced to simple mechanical operations—the procedures of Sec. 3.7. A methodical, step-by-step procedure is a virtual necessity for solving problems. The following example illustrates such a procedure.

Example 3.22. A certain window has the shape of a square surmounted by a flush semicircle. The total area of the window is 10.6 in.^{2}. What are the dimensions of the window?

Notation:

Let s be the length of each side of the square, ft.

Sketch:

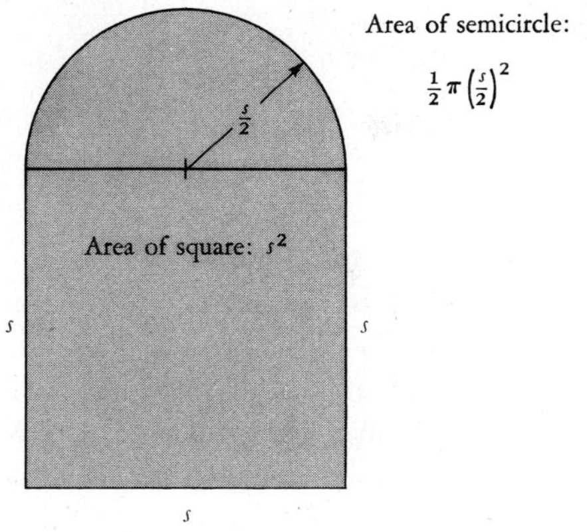

Fig. 3.3 Example 3.22. A square surmounted by a flush semicircle.

Notation (continued):

Then $s/2$ is the radius of the semicircle, ft.

s^2 is the area of the square, ft^2.

$\frac{1}{2}\pi(s/2)^2$ is the area of the semicircle, ft^2.

Equation: Area of square + area of semicircle = total area

Solution.

$$s^2 + \frac{\pi}{2}\left(\frac{s}{2}\right)^2 = 10.6$$

$$s^2 + \frac{\pi}{2}\left(\frac{s}{2}\right)^2 = 10.6$$

$$s^2 + \frac{\pi s^2}{8} = 10.6$$

$$8s^2 + \pi s^2 = 84.8$$

$$(8 + \pi)s^2 = 84.8$$

$$s^2 = \frac{84.8}{11.14}$$

$$s = 2.76 \text{ ft}$$

Answers:

$s = 2.76$ ft	each side of the square
$\frac{1}{2}s = 1.38$ ft	radius of the semicircle

Notice the following key steps in the example above: (1) a *sketch* to illustrate the geometric relationships, (2) a *notation* to specifically describe the meaning of the letter used and to explain the other algebraic expressions, and (3) a *translation* of the stated facts into an algebraic equation. With these steps, difficult problems can be made easy; without them, even the most elementary problems can be difficult. You should study Appendix 1 at this time. It contains a more detailed explanation of the procedure for solving problems. Then observe how the following example uses this procedure to solve a different kind of problem.

Example 3.23. By volume, a certain solution is 13.2 percent sulfuric acid and 86.8 percent water. A second solution is 20.9 percent acid and 79.1 percent water. How many cubic centimeters (cc) of each solution are needed to form 100 cc of a new solution containing 18 percent acid and 82 percent water?

Notation:

Let V = volume of first solution, cc

Then $100 - V$ = volume of second solution, cc

$0.132V$ = volume of acid in first solution, cc

$0.209(100 - V)$ = volume of acid in second solution, cc

$0.18(100)$ = volume of acid in final solution, cc

Equation: Volume of acid in first solution + volume of acid in second solution = total volume of acid

Solution.

$$0.132V + 0.209(100 - V) = 0.18(100)$$

$$0.132V + 0.209(100 - V) = 18$$

$$0.132V + 20.9 - 0.209V = 18$$

$$0.132V - 0.209V = 18 - 20.9$$

$$-0.077V = -2.9$$

$$V = 37.7 \text{ cc}$$

Answers:

$V = 37.7$ cc	needed from first solution
$100 - V = 62.3$ cc	needed from second solution

EXERCISE 3.8

Use the step-by-step procedure discussed in Appendix 1 and illustrated by the examples of this section to solve the following problems. Table A2.3 and Table A5.1 in the appendices will provide useful information for these problems.

1. In 6 years John will be one-third as old as his father. If John was born on the twenty-first anniversary of his father's birth, what are their present ages?
2. A certain earth satellite completes one circular orbit in 1 hr 30 min. If it is 200 miles above the earth (the earth's radius is 4,000 miles), how fast does it travel?
3. A bus made the 280-mile trip between two cities in 7 hr 15 min. The return trip took 30 min less. What speed did the bus average on the return trip?
4. The first-stage rocket engines of a Saturn V launch vehicle burn fuel at the rate of 15 tons/sec. How many pounds of fuel are consumed during the $2\frac{1}{2}$-min burn of these engines?
5. An astronaut's life-support pack allows him to walk 7.2 miles on the moon before supplies are exhausted. If he walks in a square path with the lunar base at one corner, how far can he walk in any one direction (before he must turn a corner)?
6. If the astronaut of Prob. 5 walks in a circle, what is the maximum distance he can be from the lunar base?
7. A rectangular swimming pool is 20 ft long. Every 30 gal of water increases the depth of water by $\frac{3}{16}$ in. What is the width of the pool? How many gallons are needed to fill the pool to a depth of 6 ft?
8. In 10 gal of a certain outboard motor fuel there is 2 percent oil. How much oil must be added to obtain a 9 percent oil mixture?
9. The total volume of fuel in the three stages of a Saturn V launch vehicle is 1,003,000 gal. The first stage carries 205,500 gal

more than the second stage, and the second stage carries 267,800 gal more than the third stage. How many gallons of fuel does each stage carry?

10. In a 250-mile road race, car 39 averaged 0.073 mph more than car 7, which finished in 2 hr 15 min 30.7 sec. How many seconds sooner did car 39 cross the finish line?
11. The outdoor temperature on a certain winter day dropped 56°F. The resulting temperature was $2\frac{1}{2}$ times as far below freezing as was the original temperature above freezing. What were the two temperatures?
12. A river cruise boat travels $2\frac{1}{2}$ miles upstream in 1 hr 40 min and makes the return trip in 1 hr 20 min. Assuming that the boat always runs at the same speed relative to the water, what is the speed of the current?
13. A car travels 1 mile at an average speed of 30 mph. What speed must it average for the next mile so that the average speed for the 2 miles will be 60 mph?
14. The amount of force needed to extend or compress a spring each inch is called the *spring rate.* What is the free, or unloaded, length of a spring with a rate of 6 lb/in., if with a 25-lb weight attached it is 19 in. long?
15. How far does the earth travel in its orbit around the sun in 1 day? What is the speed of the earth in mph? (Assume a circular orbit with a radius of 93 million miles.)
16. One bronze bar weighs 30 lb and is 35 percent copper, 65 percent zinc. How many pounds of a second bar of 43 percent copper, 57 percent zinc would be needed to melt with the first one to form a 38 percent copper, 62 percent zinc bronze bar?
17. The four batteries in a flashlight produce a total of $3\frac{3}{4}$ volts. The first produces $\frac{1}{2}$ volt more than the second, which produces $\frac{3}{4}$ volt less than the third. The fourth battery produces $1\frac{1}{4}$ volts more than the first one. What voltage does each battery produce?
18. On his way to work, Mario averages 30 mph for 18 min; then he covers the remaining distance in $2\frac{1}{4}$ min at 44 mph. What is his average speed for the entire trip?
19. If the radius of a certain circular carpet is increased by 10 percent, the area is increased by 9 yd^2. What was the original radius?
20. Train A is traveling 50 mph. Train B is headed toward train A on the same straight tracks and also travels 50 mph. Perched on the headlight of train A is Superbee, a bee that can fly 100 mph and can change his speed and direction of flight instantly. At the precise moment when the two trains are 100 miles apart, Superbee takes off at 100 mph from train A to train B, then turns and flies back to train A.

a. How long does Superbee take to fly from train A to train B and how far does he fly?
b. If Superbee continues flying back and forth between the trains, how far will he have flown when they meet?

3.9 GRAPHS OF EQUATIONS

Basic Ideas about Graphs

A graph is a special type of drawing that serves as a picture of some relationship. Figure 3.4 is such a "picture" of the relationship between world population and time.

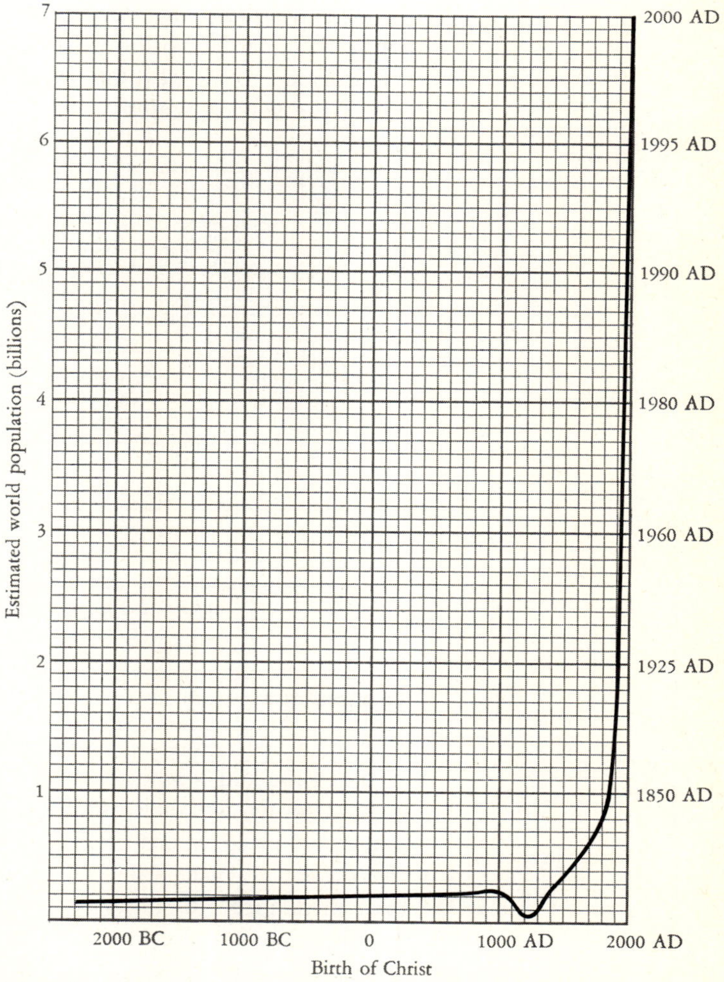

Fig. 3.4 A graph of estimated world population vs. time, from 2000 B.C. to 2000 A.D.

Graphs have many uses in business and industry. One is to show how one quantity changes as some related quantity changes. For example, with Fig. 3.4 we can literally see how world population has changed and is changing in relation to time. Even an approximate graph such as this can be a fascinating study when the reader is aware of the implications. For example, the rapidly decreasing amounts of time needed per increase of 1 billion people has dramatic social, agricultural, and ecological implications.

Many graphs are plotted from experimental or even estimated data, no equation being known. Figure 3.4 is an example. Figure 3.5 is more typical of an industrial graph. It shows how development time for photographic films is related to temperature, for two different developing solutions. To use the graph, the technician could decide on the allotted time for development and locate that time on the horizontal scale at the bottom of the graph. Then he would trace a line vertically up to his developer curve and read the temperature from one of the vertical scales.

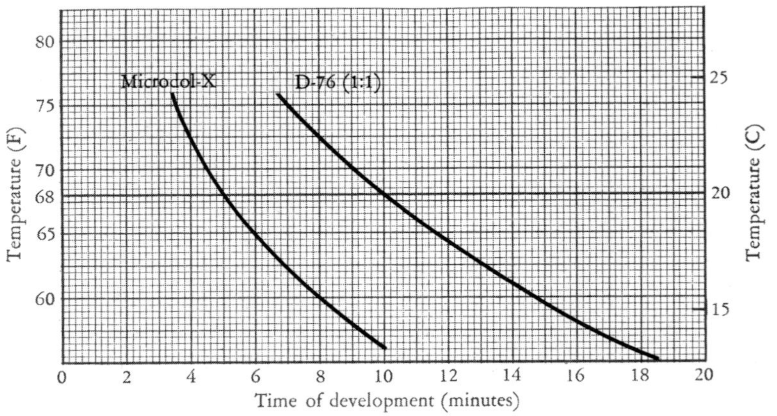

Fig. 3.5 Temperature vs. development time for two developers, Microdol-X and D-76. (Eastman Kodak Company, Rochester, N.Y.)

Terminology

A graph usually illustrates a relationship between physical quantities such as time and temperature. But in order to discuss graphs in more general terms, we shall use the abstract symbols x and y. It is customary to place values of x on a horizontal number line called the *abscissa* and to place values of y on a vertical number line called the *ordinate.* Figure 3.6 shows two such lines. In Fig. 3.5 the abscissa is the time line, and the ordinates are the temperature lines. The plane in which we draw these lines is called the *coordinate plane,* the abscissa

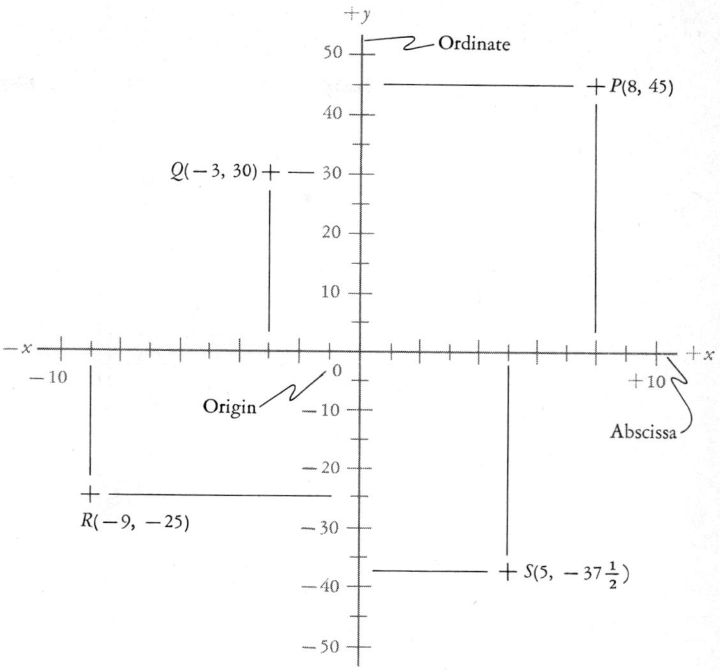

Fig. 3.6 Points in the coordinate plane.

and ordinate are called the *coordinate axes,* and the point where they cross is the *origin* for both number lines.

We can locate any point in the coordinate plane by specifying a pair of numbers (x,y) called the *coordinates* of the point. Figure 3.6 shows four points P, Q, R, and S and their coordinates. The notation (x,y) is called an *ordered pair* because the value of the abscissa x is always given first, and the value of the ordinate y is always given second.

The coordinates of every point on the graph of an equation *must* satisfy the equation. For the equation $y = x^2$, if $x = 2$, then y *must* be 4, and the point (2, 4) is on the graph of $y = x^2$. To graph the equation, we plot several such points and draw a *smooth curve* through them (usually with the aid of a French curve). Figure 3.7 is a graph of $y = x^2, -5 \leqslant x \leqslant 5$. The points labeled were used to plot the graph. This scheme where two numbers on perpendicular axes locate a single point in a plane is called a *rectangular coordinate system* (notice the rectangles in Fig. 3.6); it is also called a *cartesian coordinate system* in honor of the outstanding French mathematician René Descartes who invented it.

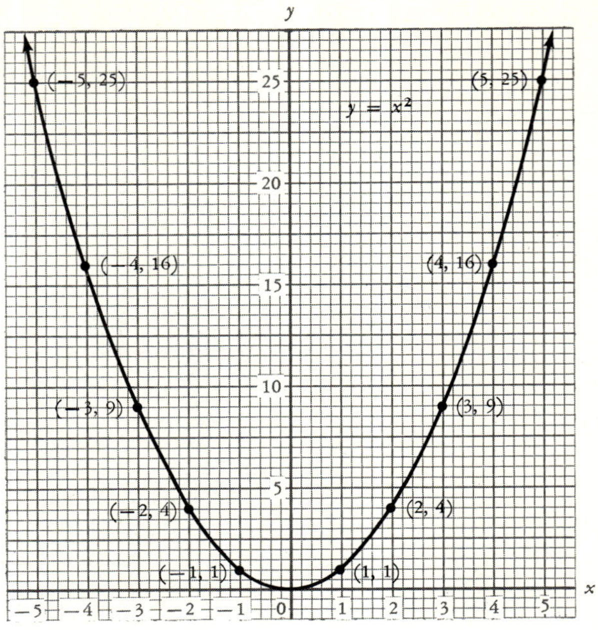

Fig. 3.7 A graph of y vs. x, for $y = x^2$.

Construction of Graphs

Notice in Figs. 3.4 to 3.7 that *values are labeled on the axes.* The graphs would be meaningless without values. If a graph represents physical quantities, as in Figs. 3.4 and 3.5, then the *units* (min, in., °F, etc.) *should be indicated.* The scales of the two axes need not be the same; we choose the scale of each axis to fit the paper and to suit the range of values needed.

Some types of equations, such as $y = 2x - 5$, always graph as a straight line. For this reason they are called *linear equations.* In a linear equation, the exponents (degrees) of both x and y must be unity (+1); hence a linear equation is also called a *first-degree equation.* We can graph any linear equation by plotting two well-separated points and drawing a straight line through them. A third point should be plotted as a check. See Fig. 3.8.

Figure 3.9 is a graph of the *third-degree* equation $y = x^3 - 2x^2 + 20$. A good method for computing points for graphing a complicated equation like this is to make a table of values. The table in Fig. 3.9 was constructed by working down the columns rather than across the rows. First, all values of x were listed. Then all values of x^2 were computed, and so on. Finally, each value of y was obtained by adding the values of $-2x^2$, x^3, and 20 in each row.

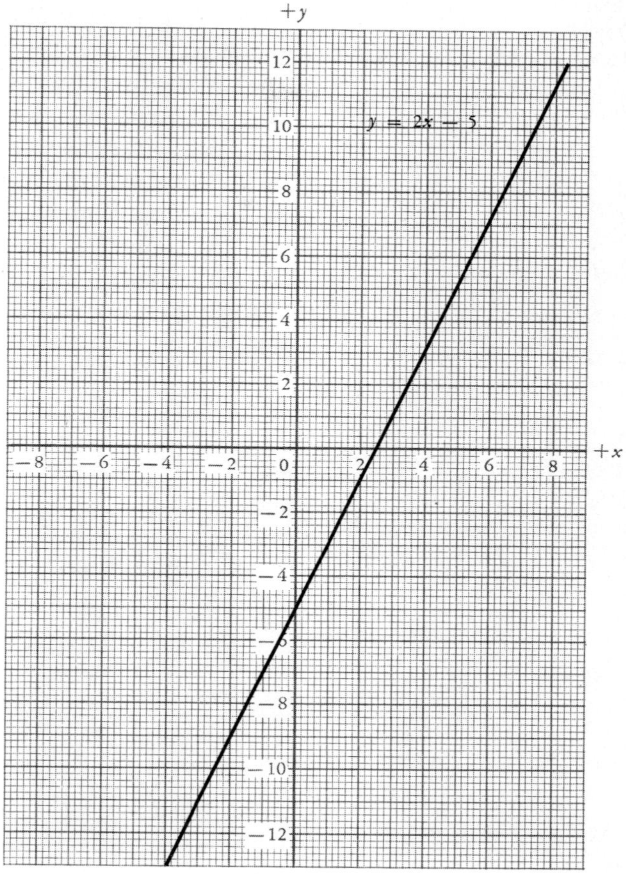

Fig. 3.8 A graph of y vs. x, for $y = 2x - 5$.

EXERCISE 3.9

Use rectangular graph paper for all problems involving plotting of points or construction of graphs.

1. Plot the following points.

a. $M(5,9)$	b. $N(9,5)$	c. $P(-5,9)$
d. $Q(-9,5)$	e. $R(-5,-9)$	f. $S(-9,-5)$
g. $T(5,-9)$	h. $U(9,-5)$	i. $V(2\frac{1}{2},-3\frac{1}{2})$
j. $W(-1.2,13.7)$	k. $X(0,6.3)$	l. $Y(-6.3,0)$
m. $Z(3.05,4.35)$		

2. Describe the graph of each of the following sets of points.
Example: All points whose ordinates equal their abscissas.

Answer: a straight line passing through the origin at a 45° counterclockwise (ccw) angle from the positive abscissa.

a. All points whose ordinates are the negative of their abscissas.
b. All points $(0,y)$ whose abscissas are zero.
c. All points $(x,0)$ whose ordinates are zero.
d. All points (x,y) such that $|y| = |x|$.
e. All points $(x,-6)$.
f. All points $(3,y)$.
g. All points (x,y) such that $x < 0$ and $y > 0$.

3. Construct a graph of each of the following *linear equations.*

a. $y = x$ b. $y = 2x$ c. $y = (1/2)x$

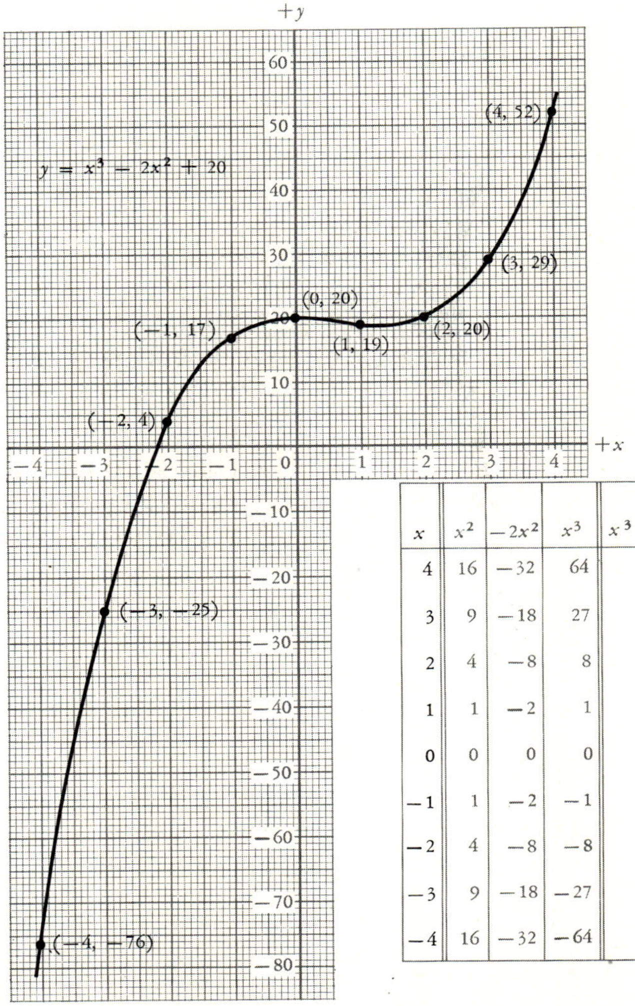

x	x^2	$-2x^2$	x^3	y $x^3 + (-2x^2) + 20$
4	16	−32	64	52
3	9	−18	27	29
2	4	−8	8	20
1	1	−2	1	19
0	0	0	0	20
−1	1	−2	−1	17
−2	4	−8	−8	4
−3	9	−18	−27	−25
−4	16	−32	−64	−76

Fig. 3.9 A graph of y vs. x, for $y = x^3 - 2x^2 + 20$.

d. $y = -2x$ e. $y = x + 2$ f. $y = x - 2$
g. $2y = 2x + 4$ h. $3y = 5x - 6$ i. $-5y = 10x - 8$
j. $y = 6$ k. $x = -6$

4. Observe the graphs of Prob. 3*a* to *d*. What effect does *a*, the coefficient of *x*, have on determining the position of the line $y = ax$?
5. Observe the graphs of Prob. 3*e* and *f*. What effect does the term *b* have in determining the position of the line $y = x + b$?
6. Determine the equation for each of the lines in Fig. 3.10. The answers to Probs. 4 and 5 will help.

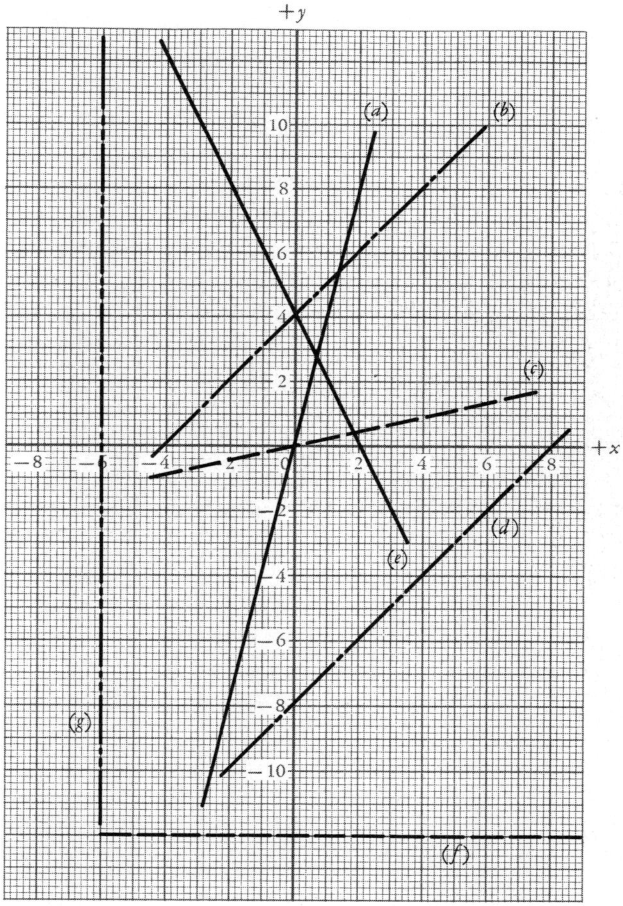

Fig. 3.10 Prob. 6.

7. Construct the graph of each of the following equations by first making a table of points, as in Fig. 3.9.
   a. $y = x^2$, $|x| \leqslant 10$ b. $y = -x^2$, $|x| \leqslant 10$
   c. $y = \sqrt{x}$, $0 \leqslant x \leqslant 100$ d. $y = -\sqrt{x}$, $0 \leqslant x \leqslant 100$

e. $y = x^2 + 2$, $|x| \leq 8$ f. $y = 3x^2 - 10$, $|x| \leq 5$
g. $y = x^3 - x^2$, $|x| \leq 5$
h. $2y = x^2 - (1/2)x^3 + 10$, $|x| \leq 10$
i. $y = x^4 - 3x^2 + 5$ j. $y = x - \sqrt[3]{x}$

8. Read from Fig. 3.7 the values of y corresponding to the following values of x.
a. 3.6 b. 4.8 c. -1.4
d. 9.6 (*Hint:* Use $x = 4.8$ and note that doubling x quadruples y.)
e. 460 (*Hint:* Use $x = 4.6$.)
f. -0.012 in. (*Hint:* Use $x = -1.2$.)

9. Read from Fig. 3.7 the values of x corresponding to the following values of y.
a. 17.0 b. 9.5
c. 72 (*Hint:* Use $y = 18$ and take twice the corresponding values of x.)

10. Estimate from Fig. 3.4 (a) the population at the time of Christ, (b) the present population, and (c) the date when the population will have increased by 1 billion over the 1980 population.

11. Read from Fig. 3.5 (a) the time for which film should be developed in D-76 at 74°F and (b) the temperature at which to develop film in Microdol-X for 6½ min.

ELEMENTS OF GEOMETRY: ANGLES, TRIANGLES, AND CIRCLES

CHAPTER 4

This chapter is a brief review of selected topics from plane geometry. Students who need a more thorough grounding should consult a specialized book such as *Principles and Problems of Plane Geometry* in the Schaum Outline series. Appendix 4 gives frequently used geometric formulas.

4.1 ANGLES ∠

Table 4.1 is a summary of various words, symbols, and definitions used with angles. We shall use most of these terms without further explanation, using the table as a reference.

In angle designations such as $\angle ACB$, the middle letter always refers to the vertex. Other designations such as $\angle C$ use only the vertex; and still others such as θ (the Greek letter theta) do not refer to any specific point. See line 1 of Table 4.1.

Units of Measure

Angles are frequently measured in *degrees, minutes,* and *seconds*. Decimal fractions of a degree are also commonplace. The following example shows how to convert from minutes and seconds to decimal degrees.

Example 4.1. **Change 17°34′12″ to degrees.**

Since minutes means "sixtieths of a degree," we divide 34′ by 60 to convert to decimal degrees. Likewise, we divide 12″ by 3,600.

$$17°34'12'' = 17° + \frac{34°}{60} + \frac{12°}{3{,}600}$$
$$= 17° + 0.5666\ldots° + 0.00333\ldots° = 17.570°$$

TABLE 4.1. TERMINOLOGY AND SYMBOLS FOR ANGLES

Angle: a figure formed by two line segments with a common endpoint (vertex).

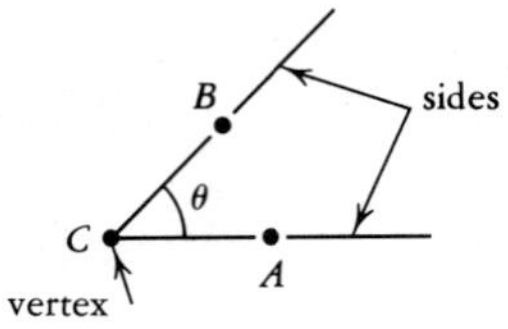

∠ACB or ∠C or θ

Measure: the amount of rotation needed to revolve one side about the vertex until it touches the other side.

Units of measure:

Degree, °: $\frac{1}{360}$ of 1 revolution: $360° \to 1$ rev
Minute, ′: $\frac{1}{60}$ of 1 degree: $60' \to 1°$
Second, ″: $\frac{1}{60}$ of 1 minute: $60'' \to 1'$
Radians, rad: 2π radians per revolution: 2π rad $\to$ 1 rev

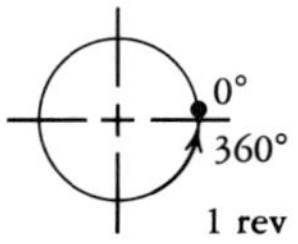

Initial side: the side from which a measurement begins.
Terminal side: the side where a measurement ends.

Right angle: one-quarter revolution: 90°

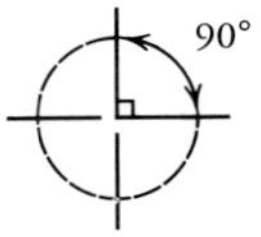

Straight angle: one-half revolution: 180°

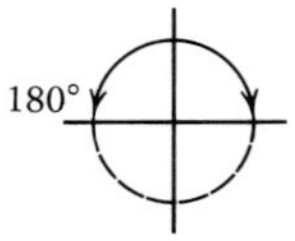

Complementary angles: $\alpha + \beta = 90°$
Supplementary angles: $\theta + \phi = 180°$

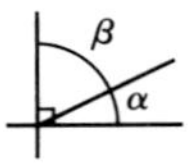

Acute angle: less than 90°: $|\phi| < 90°$
Obtuse angle: between 90° and 180°: $90° < \theta < 180°$

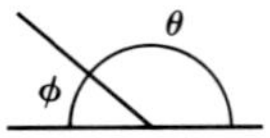

Standard Position 

Figure 4.1 illustrates angles in *standard position* on a rectangular coordinate plane. The features of the standard position are as follows: (1) the vertex of the angle is at the origin of the coordinate axes, (2) the initial side of the angle is on the positive abscissa ($+x$ axis), (3) positive angles correspond to ccw (counterclockwise) rotation from the $+x$ axis, and (4) negative angles correspond to cw (clockwise) rotation. In Fig. 4.1, α (alpha) is a positive angle in standard position, and β (beta) is a negative angle.

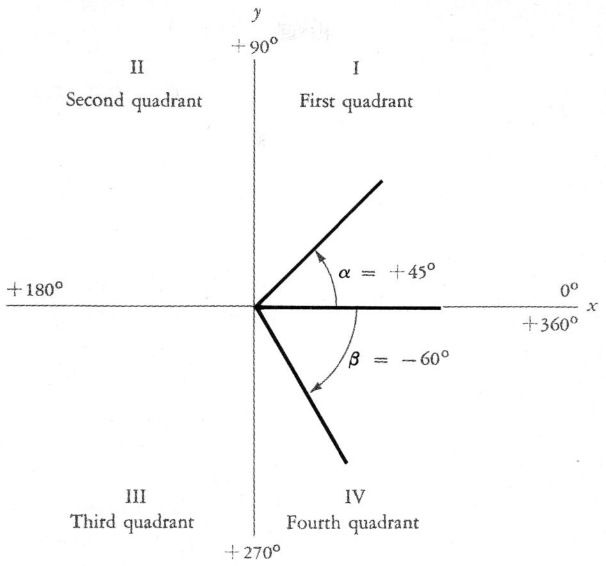

Fig. 4.1 Angles in standard position.

As shown in Fig. 4.1, the four quadrants of the coordinate plane are designated I, II, III, and IV, in ccw order. An angle in standard position is classified according to the quadrant in which the terminal side is found. Thus α of Fig. 4.1 is a *first-quadrant* angle, and β is a *fourth-quadrant* angle.

If two or more angles in standard position terminate on the same line, they are called *coterminal angles.* For example, $+240°$ and $-120°$ are coterminal, third-quadrant angles. In general, coterminal angles differ by an integral number of revolutions or a multiple of 360°. Thus if θ and ϕ (phi) are coterminal angles, then $\theta = \phi \pm k \cdot 360°$, where $k = 0, 1, 2, 3, \ldots$.

Example 4.2. **Coterminal angles**

a. **Determine the smallest positive angle coterminal with 30°.**

$$\theta = 30° + (1)(360°) = 390°$$

b. **Find the largest negative angle coterminal with 600°27′.**

$$\theta = 600°27' - 2(360°) = 600°27' - 720°$$

$$\theta = -(719°60' - 600°27') = -119°33'$$

Notice how we "borrow" 60′ in order to subtract.

EXERCISE 4.1

In Probs. 1 to 6, use a protractor to construct each angle in standard position.

1. $37°$ 2. $81°$ 3. $100°$
4. $-130°$ 5. $315°$ 6. $-575°$

In Probs. 7 to 12, determine the number of degrees corresponding to each revolution (rev).

7. $\frac{3}{8}$ rev 8. $\frac{1}{16}$ rev 9. $\frac{5}{6}$ rev
10. $1\frac{2}{3}$ rev 11. $\frac{11}{12}$ rev 12. 0.56 rev

13. For each revolution of the handle, the anvils of a micrometer move 0.025 in. relative to each other.
    a. How many degrees' rotation of the handle will produce a change of 0.001 in. between the anvils?
    b. How many degrees would you have to turn the handle to obtain a setting of 1.125 in. between the anvils if the initial setting is zero?
14. How many degrees does each of the three hands of a clock move between 1:07 p.m. and 1:13 p.m.?
15. If the hour hand of a clock has moved 9° since 12 noon, what time is it? What is the clockwise angle between the hour hand and the minute hand?

In Probs. 16 to 21, convert each given angle to an equivalent angle in decimal degrees.

16. $10°45'$ 17. $53°30'45''$ 18. $160°12'35''$
19. $-30°15'12''$ 20. $-212°20'20''$ 21. $0°1'58''$

In Probs. 22 to 27, use the angles given in Probs. 16 to 21, respectively. Assuming each angle is in standard position, determine the smallest positive and the largest negative coterminal angles for each given angle, and classify each angle according to quadrant (first quadrant, second quadrant, etc.).

In Probs. 28 to 33, convert each angle in degrees to an equivalent angle in degrees, minutes, and seconds.

28. $14.50°$ 29. $-34.92°$ 30. $-273.75°$
31. $1.462°$ 32. $0.02777\ldots°$ 33. $0.002777\ldots°$

4.2 ANGLES AND CROSSING LINES

If two lines cross as in Fig. 4.2, four angles are formed. Angles such as α (alpha) and β (beta) that are next to each other and share one side are called *adjacent angles; adjacent angles are supplementary:* $\alpha + \beta = 180°$.

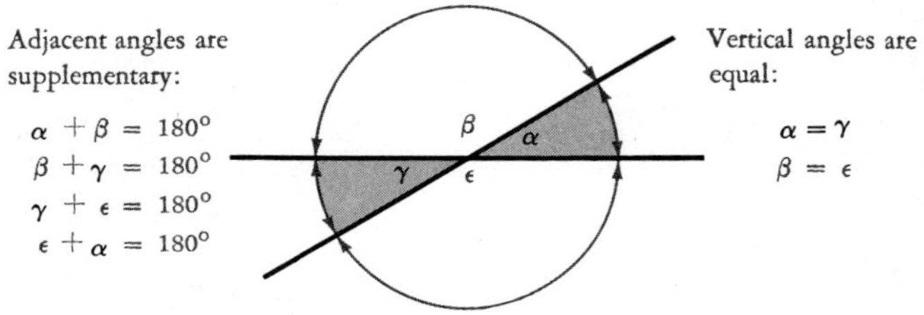

Fig. 4.2 Angles formed by crossing lines.

Angles such as β and ϵ (epsilon) on opposite sides of the vertex are called *vertical angles.* Because they are supplements of the same angle, *vertical angles are equal.* For example, $\alpha + \beta = 180°$ and $\alpha + \epsilon = 180°$; therefore $\beta = \epsilon$. For similar reasons, $\alpha = \gamma$ (gamma).

Figure 4.3 shows two parallel lines crossed by a *transverse* line, or *transversal.* The four angles *c, d, w,* and *x* between or inside the parallel lines are called *interior angles,* and the angles *a, b, y,* and *z* outside the parallel lines are called *exterior angles.* Angles such as *c* and *y* on the same side of a transversal and on corresponding sides of the parallels are *corresponding angles.* Interior angles, such as *d* and *x,* on opposite sides of the transversal are called *alternate interior angles;* likewise, angles such as *a* and *y* are *alternate exterior angles.*

Properties of Angles and Parallel Lines

In Fig. 4.3 several properties of angles and parallel lines are listed. These properties can be justified by the similarities of various parts of the figure, and by supplementary adjacent angles and equal vertical angles. The converse of each of these properties is also true. (The converse of a statement is a statement with the original premise and conclusion reversed. Most properties in mathematics have valid converses.) For example, if two exterior angles on the same side of a transversal of two lines are supplementary, then the two lines are parallel: If $a + z = 180°$, then $L_1 \parallel L_2$.

1. Exterior angles on the same side of the transversal are supplementary:

$$a + z = 180^\circ$$
$$b + y = 180^\circ$$

2. Interior angles on the same side of the transversal are supplementary:

$$c + x = 180^\circ$$
$$d + w = 180^\circ$$

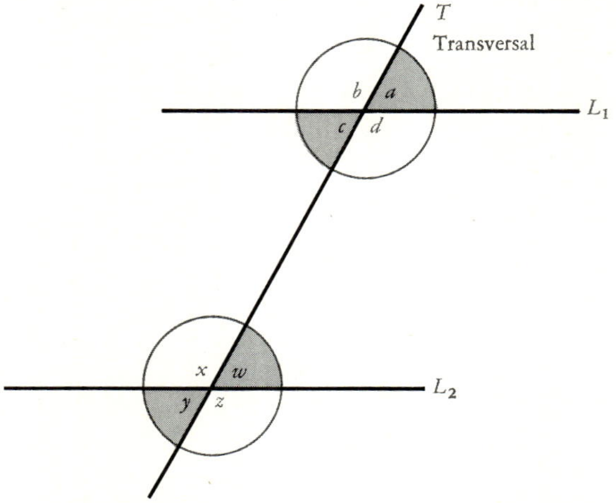

3. Corresponding angles are equal:

$$a = w$$
$$b = x$$
$$c = y$$
$$d = z$$

4. Alternate exterior angles are equal:

$$a = y$$
$$b = z$$

5. Alternate interior angles are equal:

$$c = w$$
$$d = x$$

Fig. 4.3 Properties of angles and parallel lines crossed by a transversal.

The following examples show how these properties can be used to solve problems.

Example 4.3. **Determine angle θ in the dovetail slot of Fig. 4.4.**

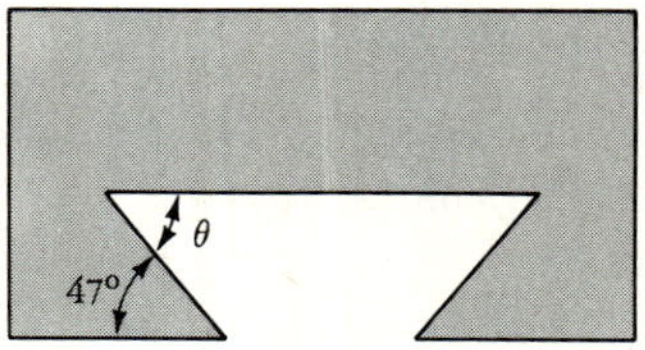

Fig. 4.4 Example 4.3. A dovetail slot.

Angle θ and the 47° angle are alternate interior angles of parallel lines crossed by a transversal. Therefore by Property 5 of Fig. 4.3, $\theta = 47^\circ$.

Example 4.4. **Determine α in the coordinate plane of Fig. 4.5. α and the 32° angle are vertical angles: $\alpha = 32^\circ$.**

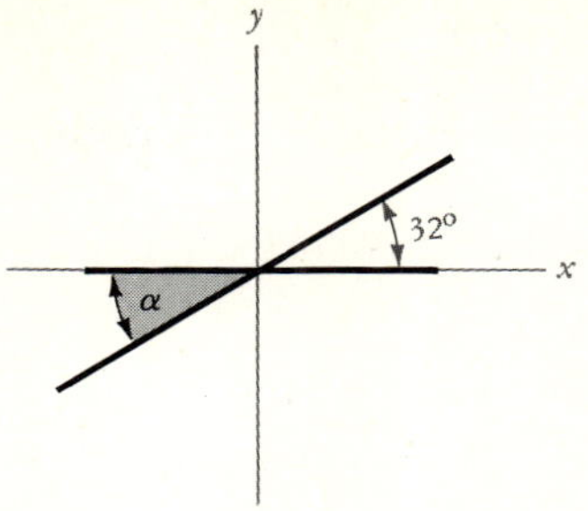

Fig. 4.5 Example 4.4.

Example 4.5. **Determine angle λ (lambda) of the parking lot stripe in Fig. 4.6.**

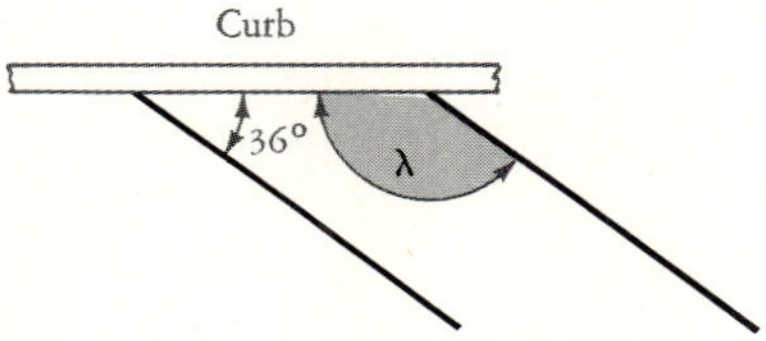

Fig. 4.6 Example 4.5. Parking lot stripes.

Angle λ and the 36° angle are interior angles on the same side of a transversal of two parallel lines. Therefore by Property 2 (see Fig. 4.3): $\lambda = 180 - 36 = 144°$.

EXERCISE 4.2

In Probs. 1 to 6, determine the sizes of angles α, β, and γ, and name the relationship of each to the given angle (supplement, corresponding angle, etc.).

1.

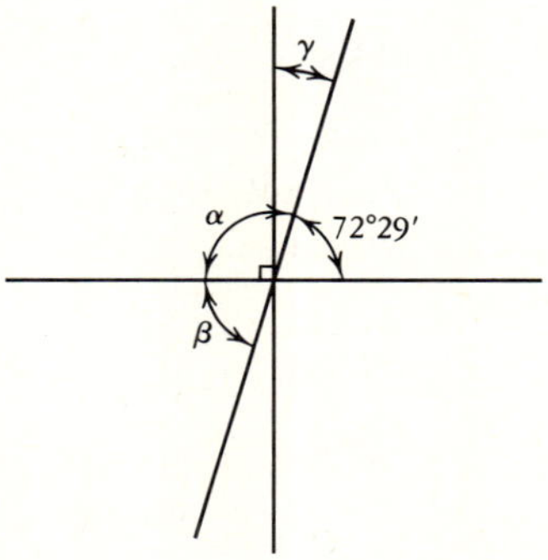

2.

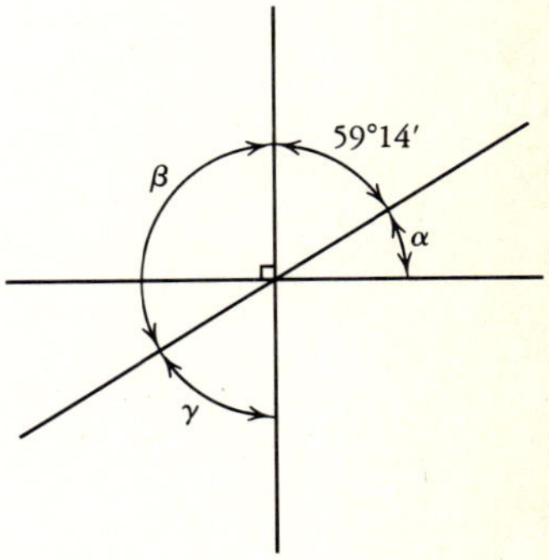

3.

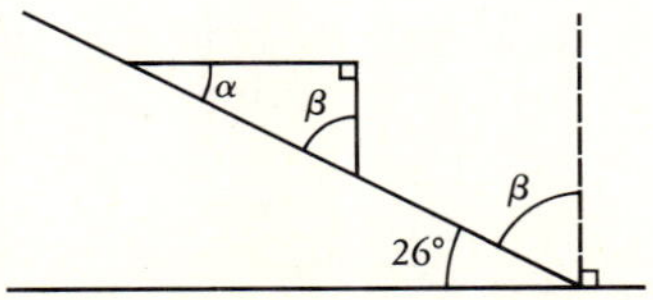

4.

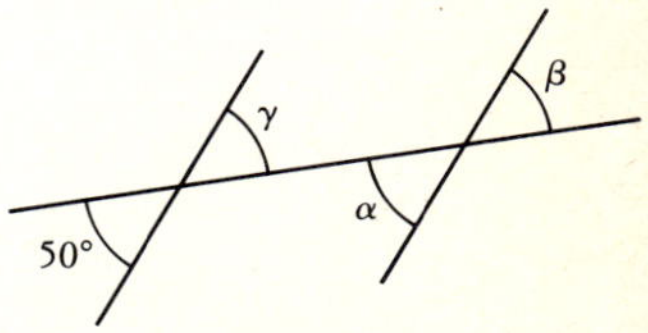

5.

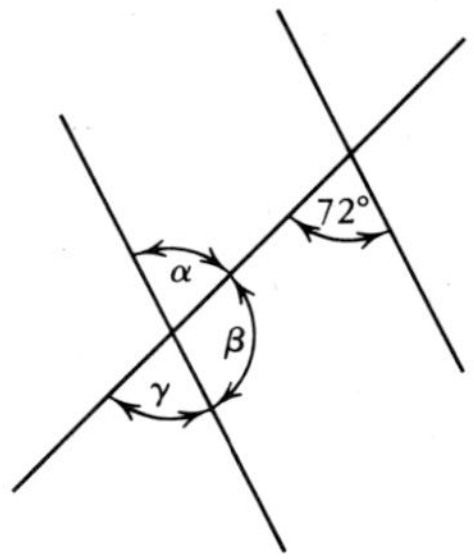

6. If the bearing of the line from A to B is N57°E (north 57° east), what is the bearing from B to A?

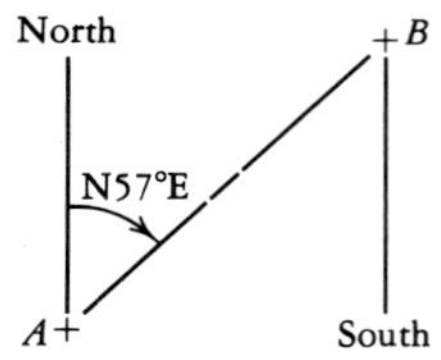

In Probs. 7 and 8, solve for angles θ and ϕ.

7.

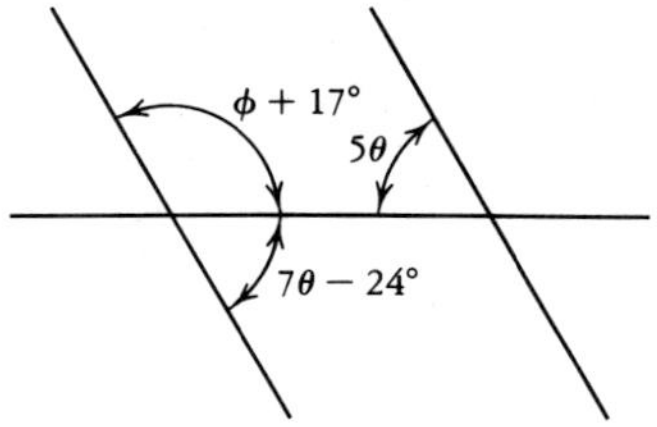

8.

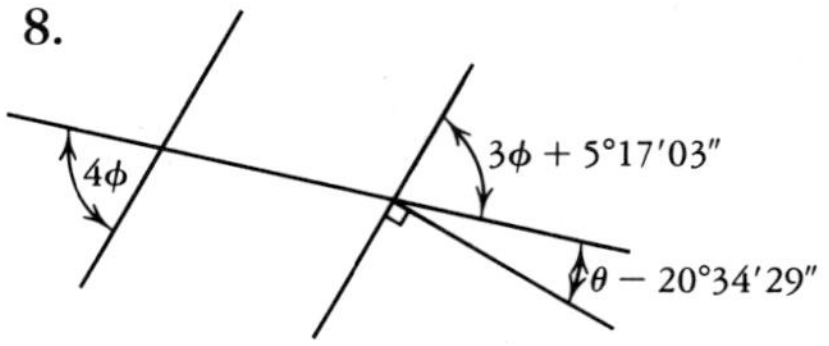

4.3 TRIANGLES △

Triangles are frequently used for solving problems in widely diverse areas of science and technology, such as surveying, navigation, astronomy, structures, and electronics. A knowledge of the properties of triangles is necessary for anyone involved in these fields.

A. BASIC DEFINITIONS AND PROPERTIES OF TRIANGLES

Table 4.2 illustrates several frequently used parts of triangles and their names. We shall not discuss these features, but we shall use the table as a reference.

The Sum of the Angles Is 180°

One of the most important properties of triangles is this: *The sum of the three angles is 180°*. We can demonstrate this fact as follows.

Draw any arbitrary triangle such as $\triangle ABC$ of Fig. 4.7. Extend line AC through C to a random point F; then $\angle ACF$ is a straight angle. Draw line CD parallel to AB. AB and CD are parallel lines crossed by the transversal AF. To show that $\alpha + \beta + \gamma = 180°$, observe the following facts:

TABLE 4.2. PARTS OF TRIANGLES

Vertex: the point where any two sides intersect.

Altitude: a line starting at any vertex and drawn perpendicular to the opposite side.

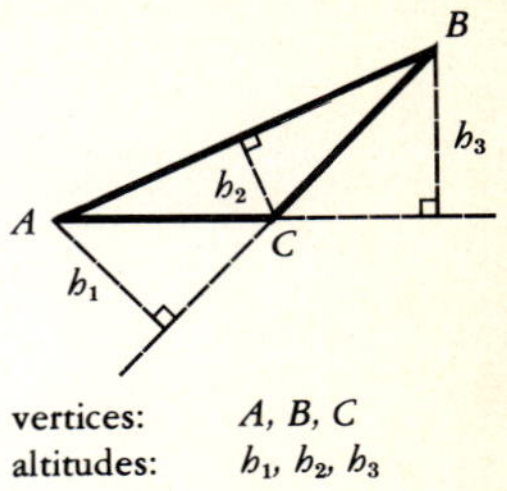

Median: a line drawn from any vertex to the midpoint of the opposite side.

Centroid: the point equidistant from all three vertices; it is two-thirds the distance from any vertex to the midpoint of the opposite side; it is the center of gravity.

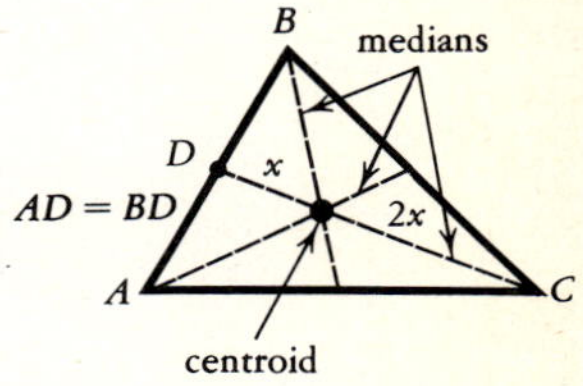

Perimeter: the sum of the lengths of the three sides: $P = a + b + c$

Area: one-half of any altitude × base: $A = \frac{1}{2}hb$

(The base is the side perpendicular to the altitude.)

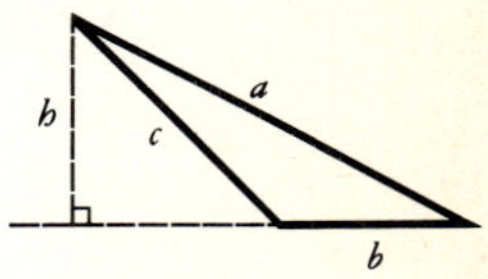

a. $\angle DCF = \angle BAC = \alpha$ *corresponding angles*
b. $\angle BCD = \angle ABC = \beta$ *alternate interior angles*
c. $\angle ACF = \alpha + \beta + \gamma = 180°$ *straight angle*

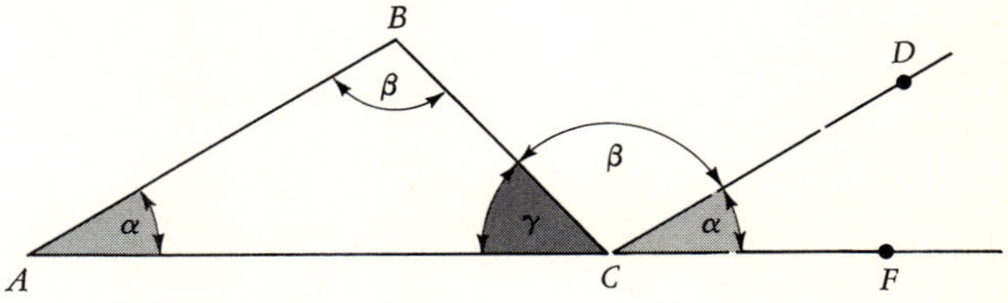

Fig. 4.7 The sum of the three angles of a triangle is 180°.

This fact is frequently employed to determine the third angle of a triangle when two angles are known. For example, if in Fig. 4.7 $\alpha = 30°$ and $\gamma = 45°$, then $\beta = 180 - (30 + 45) = 105°$.

Types and Properties of Triangles

Triangles are classified according to the relative lengths of the sides and also according to the types of angles. For example, (1) a triangle with two equal sides is called an *isosceles triangle;* (2) a triangle with all three equal sides is an *equilateral triangle;* (3) a triangle with three acute angles is an *acute triangle;* (4) a triangle with a right angle is a *right triangle;* and (5) a triangle with an obtuse angle is an *obtuse triangle.*

Table 4.3 shows several important types of triangles and lists certain special properties. The following examples demonstrate applications of these properties.

Example 4.6. The distance between adjacent bases of a baseball diamond is 90 ft. What is the distance between home plate and second base?

This distance is the hypotenuse of a 45° right triangle with equal legs of 90 ft. According to Table 4.3, the hypotenuse is $\sqrt{2} \times$ leg. Therefore,

$$d = \sqrt{2} \times 90 = 1.414 \times 90 = 127.3 \text{ ft}$$

Example 4.7. In a resistance-capacitance, ac (alternating current) circuit, the impedance vector Z, the reactance vector X_c, and the resistance vector R form a right triangle, as shown in Fig. 4.8. If R = 4.1 ohms and Z = 6.3 ohms, what is the value of X?

According to the Pythagorean theorem (Table 4.3), we have $Z^2 = R^2 + X_c^2$. Upon solving for X_c we obtain

$$X_c = \sqrt{Z^2 - R^2} = \sqrt{6.2^2 - 4.1^2} = 4.7 \text{ ohms}$$

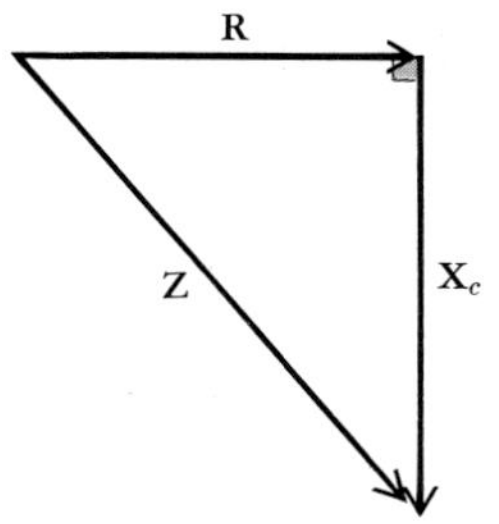

Fig. 4.8 Example 4.7. Impedance vectors.

Example 4.8. A 100-ft vertical antenna is stabilized by guy wires of equal length and in the same vertical plane as the antenna. See Fig. 4.9. If the distance between anchor points equals the length of each wire, (a) what is the length of each wire, and (b) what is the distance from the foot of the antenna to each anchor point?

Solution. The two wires and the ground form an equilateral triangle in which the antenna is an altitude. According to Table 4.3, the altitude is

TABLE 4.3. SPECIAL TRIANGLES AND PROPERTIES

Isosceles Triangle

a. The two legs are equal: $AB = BC = s$

b. The base angles are equal: $\angle A = \angle C = \theta$

c. The altitude on the base bisects the base and bisects the vertex angle: $AD = CD$

$$\angle ABD = \angle CBD = \frac{1}{2} \angle ABC = 90° - \theta$$

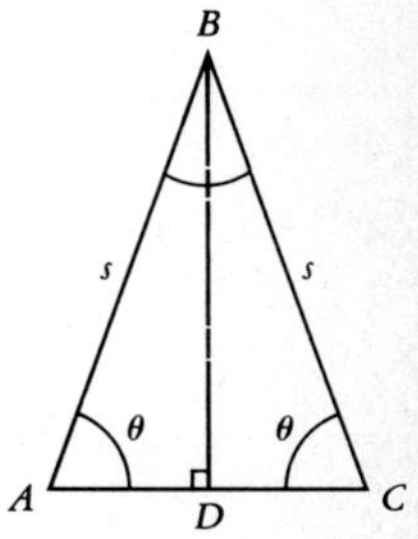

Equilateral Triangle

a. All three sides are equal.

b. Each angle is 60°.

c. Each altitude bisects its base.

d. $h = (\sqrt{3}/2)s$

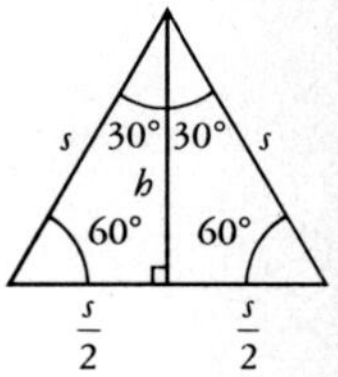

Right Triangle (with hypotenuse c)

a. $\alpha + \beta = 90°$.

b. Pythagorean theorem: $c^2 = a^2 + b^2$

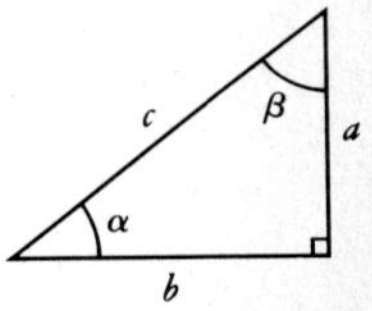

c. 45° right triangle: $a:b:c = 1:1:\sqrt{2}$

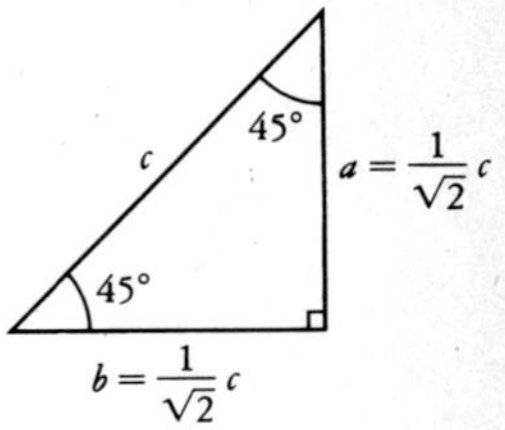

d. 30-60° right triangle: $a:b:c = \frac{\sqrt{3}}{2}:\frac{1}{2}:1$

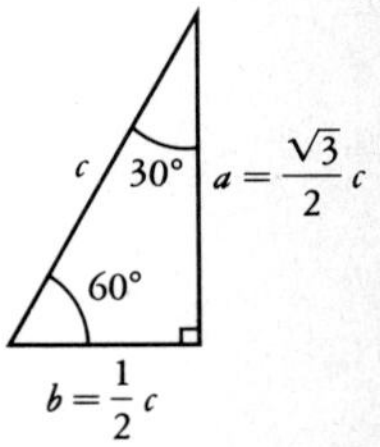

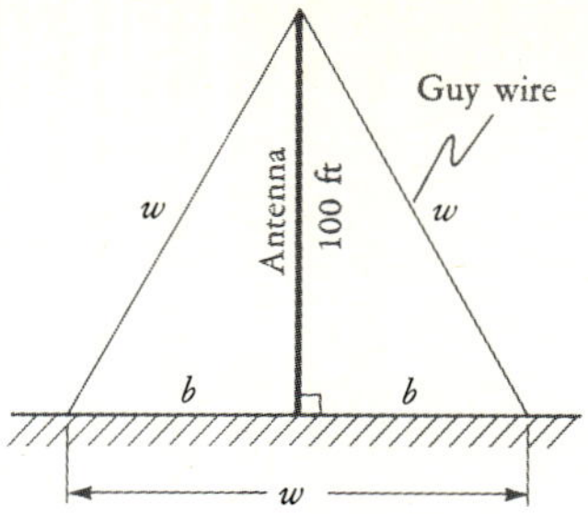

Fig. 4.9 Example 4.8. Antenna and guy wires.

$(\sqrt{3}/2)w$, where w is the length of each wire (also the distance between anchor points). Solving for w, we have

$$w = \frac{2}{\sqrt{3}} \times 100 = \frac{200}{1.732} = 115.5 \text{ ft}$$

Since the altitude of an equilateral triangle bisects the base, the distance from the antenna to each anchor is $(1/2)w = 57.8$ ft.

Example 4.9. As shown in Fig. 4.10, a camera lens projects a circular cone of light, whose base must at least circumscribe the picture area. What is the minimum diameter d for the 35-mm film shown?

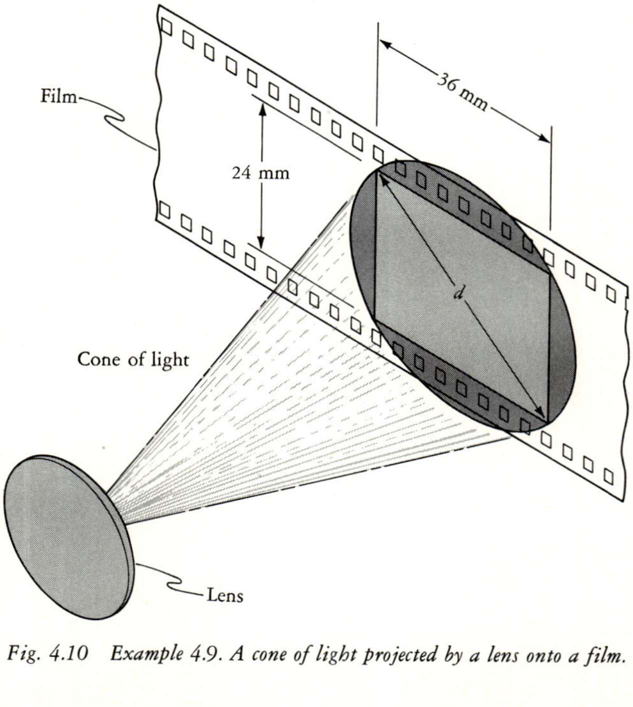

Fig. 4.10 Example 4.9. A cone of light projected by a lens onto a film.

Solution. The minimum diameter is the diagonal of the picture rectangle; therefore it is the hypotenuse of a right triangle having legs of 24 mm and 36 mm. From the Pythagorean theorem we have

$$d = \sqrt{24^2 + 36^2} = \sqrt{1{,}872} = 43 \text{ mm}$$

EXERCISE 4.3

In Probs. 1 to 12, determine the missing sides a, b, and c and the missing angles α, β, and γ, as indicated on each triangle (not necessarily drawn to scale).

1.
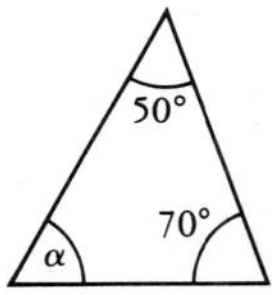

2.
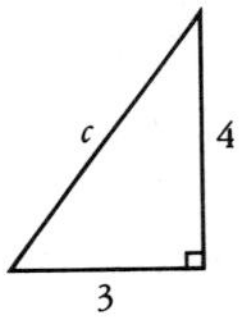

3.
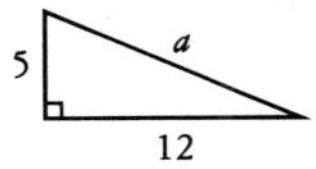

4.
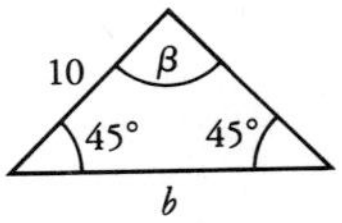

5.
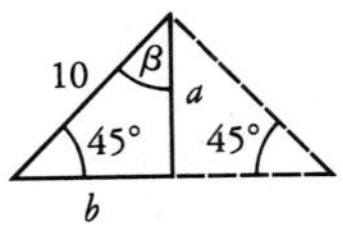

6.
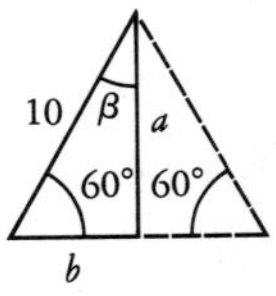

7.
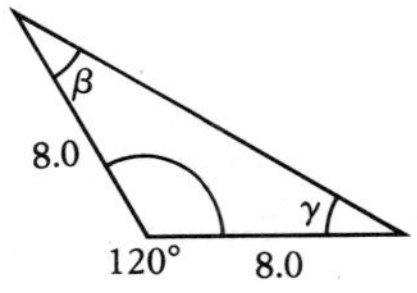

8.
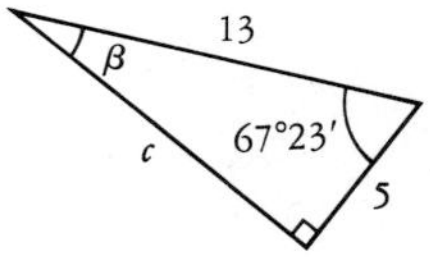

9.
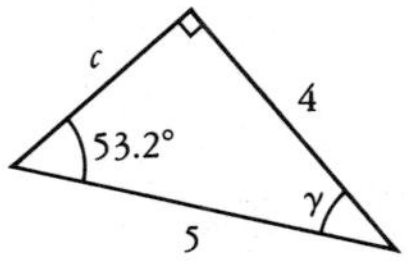

10.
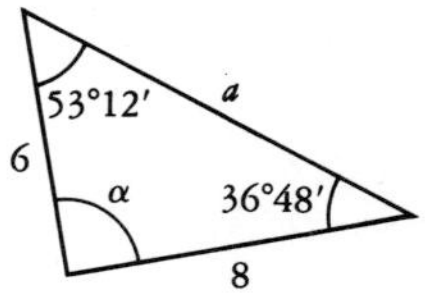

11.
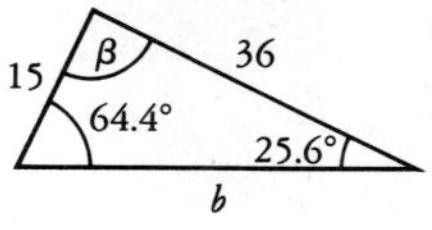

12.
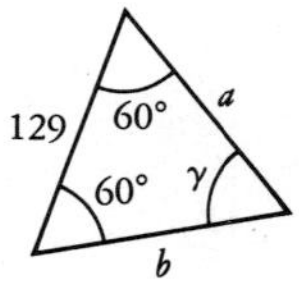

13. If the long leg of a 30-60° drafting triangle is 10 in., how long are the short leg and the hypotenuse?
14. If the hypotenuse of a 45° drafting triangle is 9 in., how long are the legs?
15. If the coordinates of a point in a rectangular coordinate plane are (3,4), how far is the point from the origin? Repeat, with the coordinates (−3,4), (−3,−4), and (3,−4).
16. If a point is 13 units from the origin of a rectangular coordinate plane and the x coordinate is −5, what is the y coordinate? (*Hint:* There is more than one answer.)
17. A triangular sign is to be supported, as shown in Fig. 4.11, by a single post under the sign's center of gravity. How far from the 90° corner should the post be located?

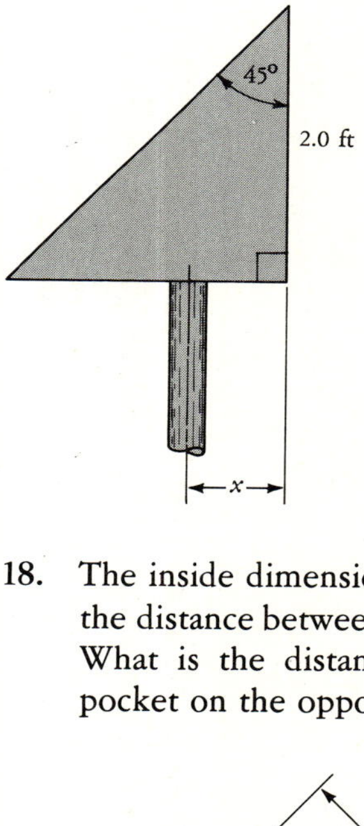

Fig. 4.11 Prob. 17. A triangular sign.

18. The inside dimensions of a pool table are 3½ ft by 7 ft. What is the distance between the pockets at diagonally opposite corners? What is the distance between a center pocket and a corner pocket on the opposite side?

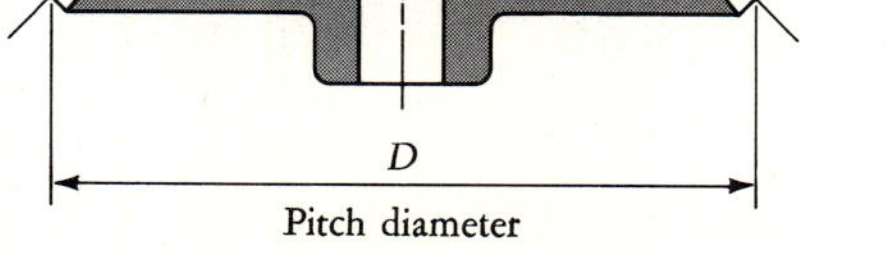

Fig. 4.12 Prob. 22. A bevel gear.

19. If **R** of Fig. 4.8 is 2.2 ohms, and $\mathbf{X}_c$ is 5.3 ohms, what is the impedance **Z** of the circuit?
20. If **Z** of Fig. 4.8 is 20 ohms and $\mathbf{X}_c$ is 12 ohms, what is the resistance **R** of the circuit?
21. A certain cofferdam has the form of an open rectangular box 9 ft wide by 24 ft long by 15 ft deep. What is the minimum length of a timber needed to make a diagonal cross brace from top to bottom, through the center of the box?
22. What is the pitch diameter D of the 90° bevel gear shown in Fig. 4.12 if the cone distance A_0 is 3.52 in.?
23. As shown in Fig. 4.13, six bolts are equally spaced on a 5-in.-diameter bolt circle. What is the center distance between adjacent bolts? (*Hint:* Draw lines between the bolts to form equilateral triangles.)

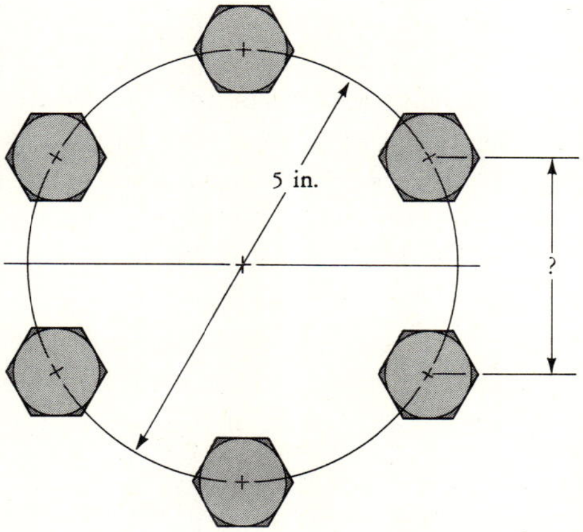

Fig. 4.13 Prob. 23. Six hex-head bolts equally spaced on a 5-in.-diameter bolt circle.

24. Repeat Prob. 23 for four equally spaced bolts.
25. The head of a hex-head bolt is a regular hexagon circumscribed around a circle with a diameter $1\frac{1}{2}$ times the diameter of the bolt. See Fig. 4.14. What is the distance across the corners of the head of a $\frac{9}{16}$-in. bolt? (*Hint:* Divide the hexagon into six equilateral triangles.)

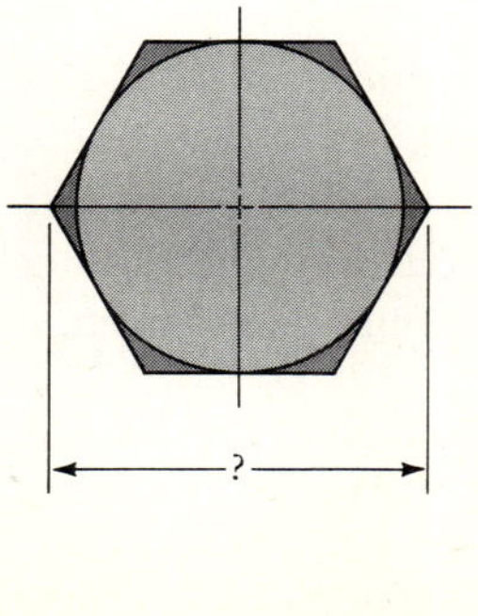

Fig. 4.14 Prob. 25. The hexagonal head of a bolt.

26. What is the depth h of the hole left by the point of a 5/8-in.-diameter drill with a 120° point? See Fig. 4.15.

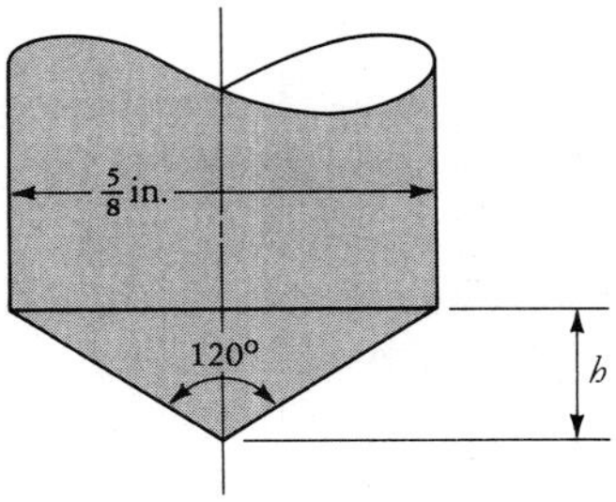

Fig. 4.15 Prob. 26. The hole formed by a drill.

27. Repeat Prob. 26 for a 1/4-in. drill, with a 90° point.
28. What is the diameter of the base of the cone of Fig. 4.10 if the negative area is a 2 1/4- by 2 1/4-in. square?
29. Repeat Prob. 28 if the negative area is a 6- by 7-cm rectangle.
30. A cube 2 in. on a side is bisected by a cut along the diagonal of one face and perpendicular to that face. What is the area of the rectangular surface thus formed?
31. One corner of a 2-in. cubic block is removed by an oblique cut along the diagonals of three mutually adjacent faces of the cube. What is the area of each of the four triangular surfaces thus formed on the block?
32. Figure 4.16 shows a pyramid with a square base and four equilateral triangular faces. If the vertical height h is 100 ft, how long is each side of the base?

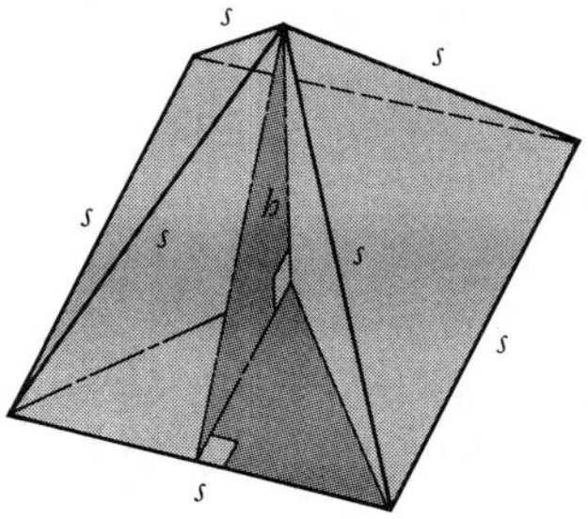

Fig. 4.16 Prob. 32. A pyramid with a square base and four equilateral triangles for sides.

33. A certain pyramid is an equilateral tetrahedron (an equilateral triangular base, with three equilateral triangular faces). See Fig. 4.17. If the vertical height is 100 ft, how long is each side of the base? (*Hint:* The altitude from the vertex to the base touches the base at its centroid.)

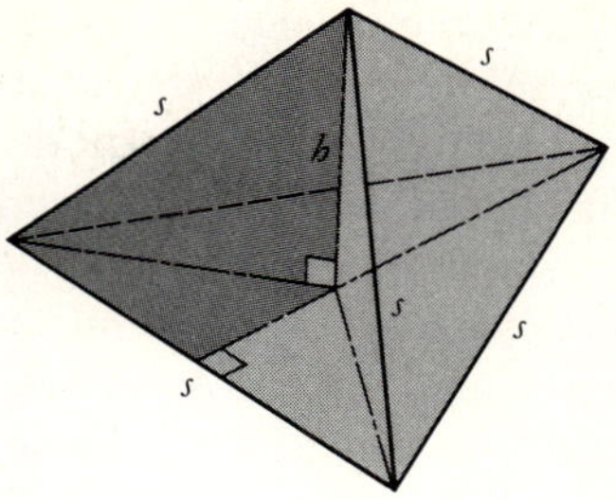

Fig. 4.17 Prob. 33. An equilateral tetrahedron—a pyramid consisting of four congruent, equilateral triangles.

B. SIMILAR TRIANGLES ~

Similar triangles are triangles of the same shape (but not necessarily of the same size). Therefore by definition (see Fig. 4.18):

> Similar triangles are triangles for which:
>
> *a.* All three angles of one triangle are respectively equal to the angles of the other triangle.
> *b.* Corresponding sides are in proportion.

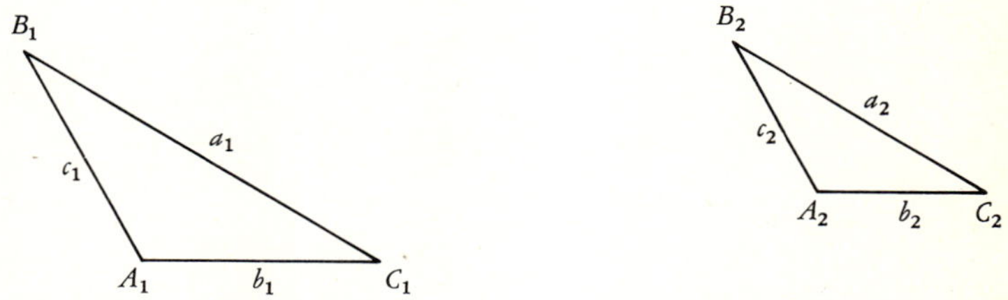

(*a*) Corresponding angles are equal:

$$A_1 = A_2$$
$$B_1 = B_2$$
$$C_1 = C_2$$

(*b*) Ratios of corresponding sides are equal:

$$\frac{a_1}{a_2} = \frac{b_1}{b_2} = \frac{c_1}{c_2}$$

Fig. 4.18 Two similar triangles and the properties of similar triangles.

The expression $\triangle A_1B_1C_1 \sim \triangle A_2B_2C_2$ reads "triangle $A_1B_1C_1$ is similar to triangle $A_2B_2C_2$."

Principles of Similar Triangles

The basic definitions of similar triangles can be combined with various properties of angles and triangles to establish several principles. For example, if only two angles of one triangle equal two angles of another triangle, the triangles are similar. This principle follows from the fact that the sum of the angles of any triangle is 180°. The following list summarizes some of the more important principles; from these, many other more specialized principles can be established.

1. If two angles of one triangle equal two angles of another triangle, the triangles are similar. (A special case: If an acute angle of one right triangle equals an acute angle of another right triangle, the triangles are similar. Why?)
2. If the three sides of one triangle are, respectively, parallel to or perpendicular to the three sides of another triangle, the triangles are similar. In Fig. 4.19, $\triangle DBE \sim \triangle ABC$.

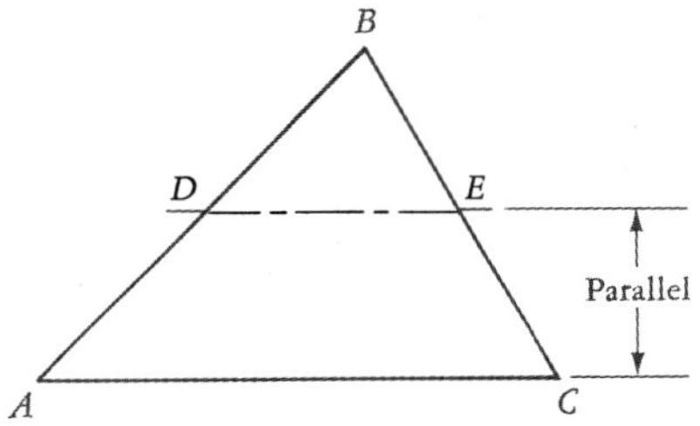

Fig. 4.19 Line DE ‖ Line AC forms similar triangles: $\triangle DBE \sim \triangle ABC$.

3. If one angle of a triangle equals one angle of another triangle and if the sides of these angles are in proportion, the triangles are similar. In Fig. 4.20, given that $\angle H_1 = \angle H_2$ and $f_2/f_1 = g_2/g_1$, then $\triangle F_2G_2H_2 \sim \triangle F_1G_1H_1$.

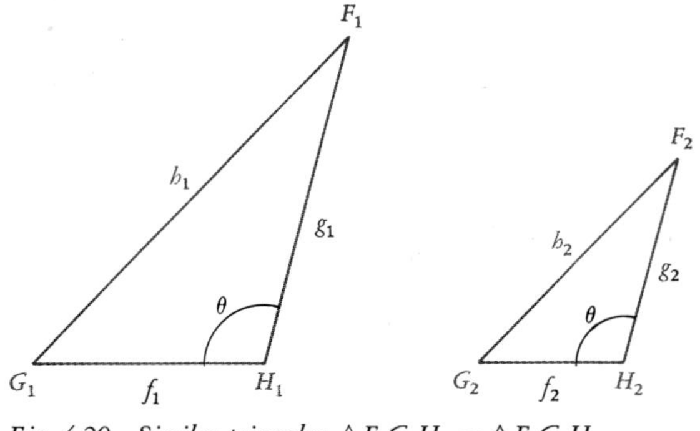

Fig. 4.20 Similar triangles: $\triangle F_2G_2H_2 \sim \triangle F_1G_1H_1$.

4. The altitude to the hypotenuse divides a right triangle into two right triangles that are similar both to each other and to the original triangle. In Fig. 4.21, $\triangle ABD \sim \triangle BCD \sim \triangle ABC$.

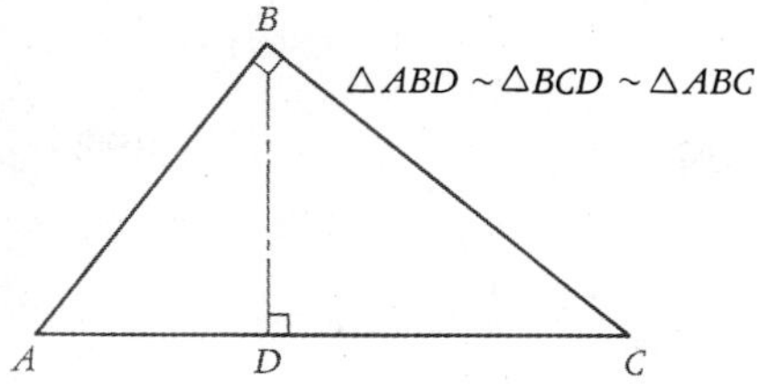

Fig. 4.21 Similar triangles: $\triangle ABD \sim \triangle BCD \sim \triangle ABC$.

Any one of these principles is sufficient to establish that two triangles are similar. And once we have similar triangles, any of the other characteristics can be employed to solve a triangle.

Example 4.10. **Determine the lengths of sides *e* and *k* of the triangles in Fig. 4.22.**

The triangles are similar because $\angle G = 45°$ and $\angle H = 120°$, which makes all three pairs of corresponding angles equal. Corresponding sides of similar triangles are in proportion. Therefore $f/h = e/j = g/k$, from which we find

$$e = \frac{f}{h} j = \frac{100}{75}(22.4) = 29.9 \text{ ft}$$

$$k = \frac{h}{f} g = \frac{75}{100}(81.6) = 61.2 \text{ ft}$$

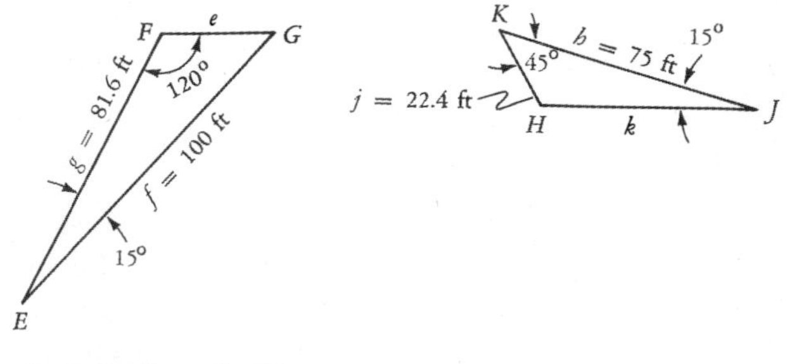

Fig. 4.22 Example 4.10.

Example 4.11. Figure 4.23 shows how we can determine the height of a tree by holding a yardstick parallel to the tree. According to Principle 2, the stick and its shadow form a triangle similar to that formed by the tree and its shadow. The ratios of corresponding sides of similar triangles are equal:

$$\frac{h}{3 \text{ ft}} = \frac{27 \text{ ft } 10 \text{ in.}}{5 \text{ ft } 3 \text{ in.}} = \frac{27.83 \text{ ft}}{5.25 \text{ ft}} \longrightarrow h = 3\left(\frac{27.83}{5.25}\right) = 15.9 \text{ ft}$$

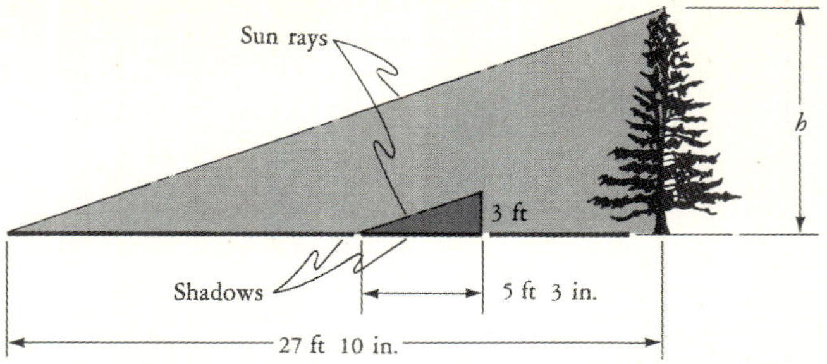

Fig. 4.23 Example 4.11. Determining the height of a tree by its shadow.

C. CONGRUENT TRIANGLES ≅

Congruent triangles are similar triangles with the corresponding sides *equal.* Thus congruent triangles are exactly alike: they have equal angles and equal sides. The symbol for congruency is "≅". Congruent triangles have all the properties of other similar triangles as well as the following additional properties.

Properties of Congruent Triangles

1. If two triangles are congruent, the corresponding parts are equal. See Fig. 4.24. Given $\triangle A_1B_1C_1 \cong \triangle A_2B_2C_2$, then corresponding angles are equal

$$\angle A_1 = \angle A_2 \quad \angle B_1 = \angle B_2 \quad \angle C_1 = \angle C_2$$

   and corresponding sides are equal

$$a_1 = a_2 \quad b_1 = b_2 \quad c_1 = c_2$$

2. SAS = SAS: *Side-Angle-Side equals Side-Angle-Side*
   If two sides and the included angle of one triangle equal two sides and the included angle of another triangle, the triangles are congruent. In Fig. 4.24, given

$$\angle A_1 = \angle A_2 \qquad b_1 = b_2 \qquad \text{and} \qquad c_1 = c_2$$

   then $$\triangle A_1B_1C_1 \cong \triangle A_2B_2C_2$$

   from which it follows that

$$a_1 = a_2 \qquad \angle B_1 = \angle B_2 \qquad \text{and} \qquad \angle C_1 = \angle C_2$$

3. ASA = ASA: *Angle-Side-Angle equals Angle-Side-Angle*
   If two angles and the included side of one triangle equal two angles and the included side of another triangle, the triangles are

congruent. In Fig. 4.24, given

$$\angle B_1 = \angle B_2 \qquad \angle C_1 = \angle C_2 \qquad \text{and} \qquad a_1 = a_2$$

then $$\triangle A_1B_1C_1 \cong \triangle A_2B_2C_2$$

4. SSS = SSS: *Side-Side-Side equals Side-Side-Side*
If the three sides of one triangle equal, respectively, the sides of another triangle, the triangles are congruent. In Fig. 4.24, given

$$a_1 = a_2 \qquad b_1 = b_2 \qquad \text{and} \qquad c_1 = c_2$$

then $$\triangle A_1B_1C_1 \cong \triangle A_2B_2C_2$$

from which it follows that

$$\angle A_1 = \angle A_2 \qquad \angle B_1 = \angle B_2 \qquad \text{and} \qquad \angle C_1 = \angle C_2$$

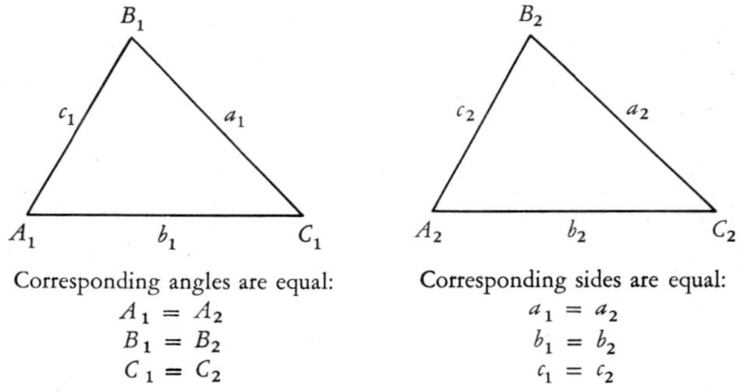

Fig. 4.24 Two congruent triangles and properties of congruent triangles.

Example 4.12. **Show that a diagonal of a parallelogram divides it into congruent triangles.**

Draw any parallelogram $ABCD$ and a diagonal AC as in Fig. 4.25. $\angle DAC = \angle BCA$ and $\angle DCA = \angle BAC$—alternate interior angles of parallel lines AD and BC.

side AC = side AC	by identity
$\triangle ADC \cong \triangle ABC$	ASA = ASA property

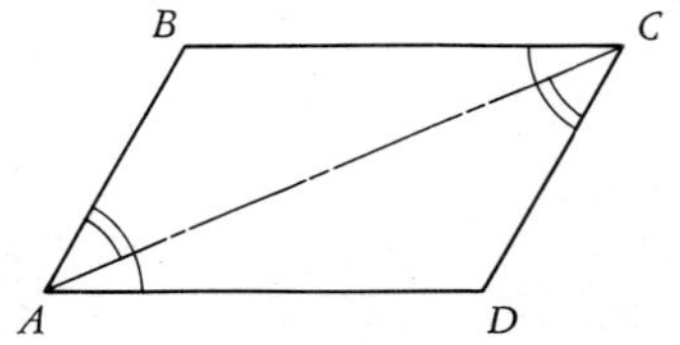

Fig. 4.25 Example 4.12. A parallelogram and one of its diagonals.

Example 4.13. Show that two right triangles, each having a side of the same square for a hypotenuse, are congruent.

For the two right triangles of Fig. 4.26 we can reason as follows:

$RQ = BQ$	Sides of a square are equal.
$\alpha + \theta = 90°$	Straight angle PQC: $\alpha + \theta + 90° = 180°$.
$\phi + \theta = 90°$	Acute angles of a right triangle are complementary.
$\alpha = \phi$	Complements of the same angle are equal.
$\beta = \theta$	Complements of equal angles are equal.
$\triangle PQR \cong \triangle ABC$	SAS = SAS property

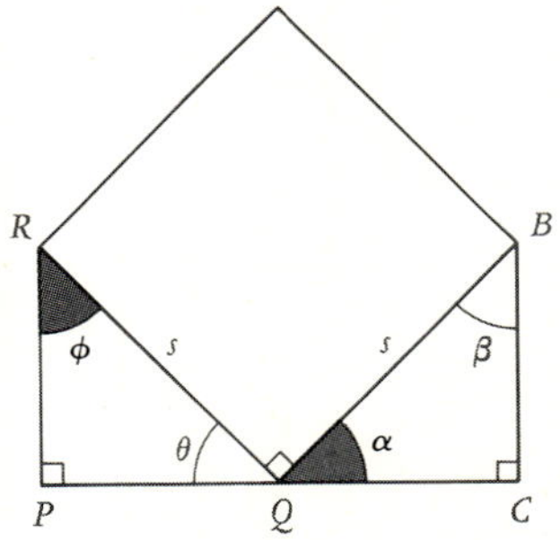

Fig. 4.26 Example 4.13.

EXERCISE 4.4

In Probs. 1 to 4, state the property that proves each pair of triangles similar, and determine the values of the missing parts x, y, α, and β, as indicated.

1. 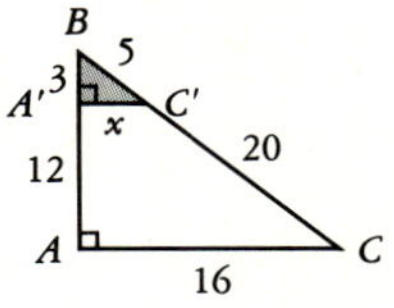

2. 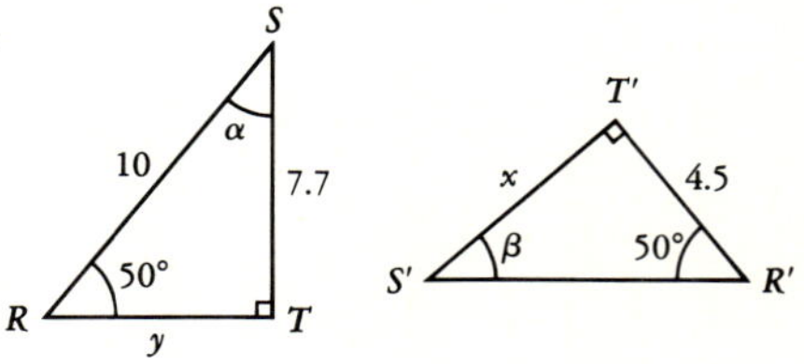

3. 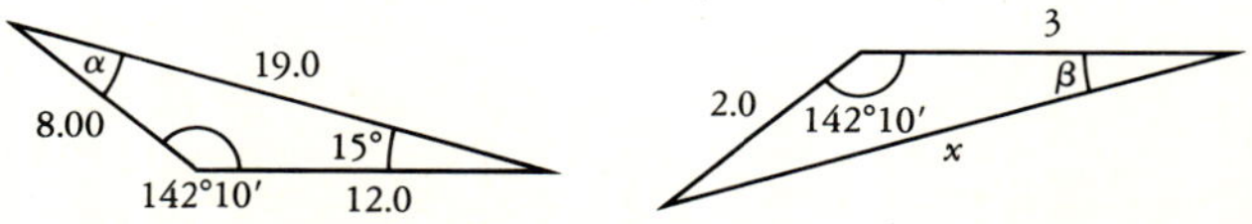

4. 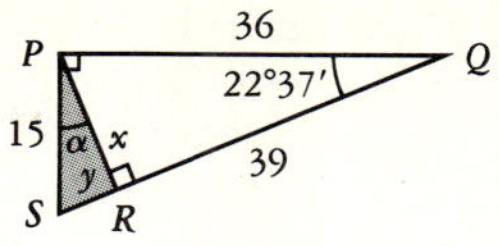

In Probs. 5 to 8, state the property that proves each pair of triangles congruent, and determine the missing parts, x, y, α, or β, as indicated. You may have to determine some of the missing parts before proving the triangles congruent.

5. 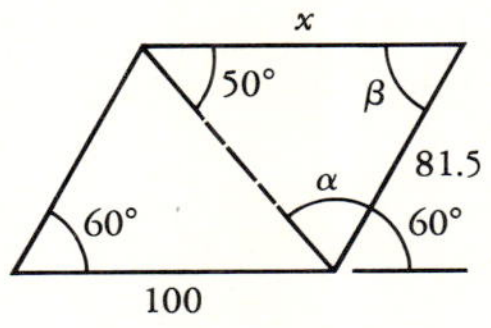

6. 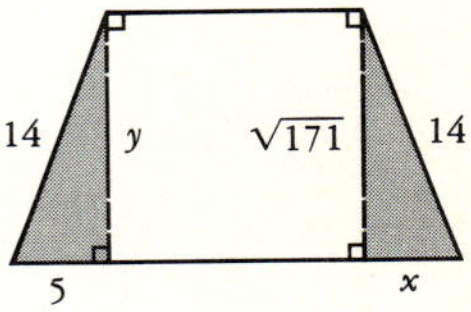

7. 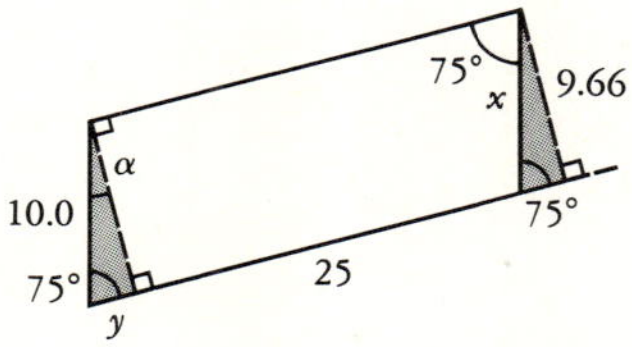

8. 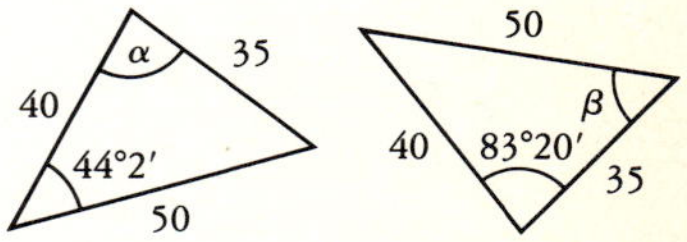

9. How tall is a flagpole if its shadow is 120 ft long and the shadow of a 5-ft stick held vertical is 15 ft long?
10. The coordinates of one point are $(-10, -\sqrt{44})$. What are the coordinates of a second point on the same line through the origin, if the second point is 18 units from the origin?

4.4 RADIAN MEASURE AND CIRCULAR DISTANCES

Although the degree is the most common unit for measuring angles, another unit called the *radian* (abbreviated rad) is better suited for many applications. Just as inches and meters are two different units for measuring lengths, degrees and radians are different units for angles; both units are arbitrary, but we can easily convert from one to the other.

A. DEFINITION OF RADIAN UNITS

We can best understand the radian unit by placing the vertex of an angle at the center of a circle. In this position (shown in Fig. 4.27) the angle is called a *central angle* of the circle. The sides of the angle θ are radii of the circle, and they cut off, or *intercept,* a circular arc s

from the circumference. If the length of the arc s equals the radius of the circle, then we say that the size of angle θ is 1 rad. The formal definition of 1 rad is as follows:

RADIAN MEASURE	A central angle of one *radian* intercepts, on a circle, an arc whose length equals the radius of the circle.

In Fig. 4.27, if $\theta = 1$ rad, then $s = r$. One reason for this definition is the direct relationship between the arc, the radius, and the angle. If $\theta = 1$ rad, then $s = r(1)$; if $\theta = 2$ rad, then $s = r(2)$; and in general

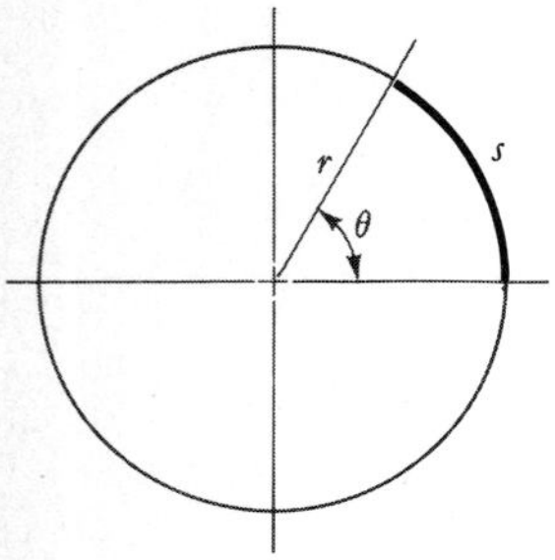

If $\theta = 1$ radian
Then $s = r$

Fig. 4.27 Definition of radian measure of an angle.

$$s = r\theta_R \tag{4.1}$$

$$\theta_R = \frac{s}{r} \tag{4.2}$$

where θ_R is in *radians*

Relationships between angles, arcs, and radii

Example 4.14. Arcs and angles

a. **In Fig. 4.27, if $r = 5.0$ in. and $\theta = 2.3$ rad, find the length of s. From Eq. (4.1) we have**

$$s = r\theta = 5.0(2.3) = 11.5 \text{ in.}$$

b. **If $s = 16.4$ ft and $r = 8.00$ ft, find the size of θ in radians. From Eq. (4.2) we have**

$$\theta = \frac{s}{r} = \frac{16.4}{8.00} = 2.05 \text{ rad}$$

Relationships between Degrees, Radians, and Revolutions

Frequently it becomes necessary to change angles from radians to degrees or revolutions, and vice versa. For this reason, and to gain a feeling for the size of a radian, we shall develop conversion formulas.

To find out how many radians are in 1 rev (abbreviation for revolution), we reason as follows. (See Fig. 4.28.) The circumference of a circle of radius r is $2\pi r$; an angle of 1 rad intercepts on the circumference an arc equal to the radius. Therefore 2π such arcs placed end to end would reach exactly around the circle, and 2π such angles placed side by side would exactly equal 1 rev: *1 rev contains 2π rad.* Having this relationship, we can write the following conversion formulas:

$$\theta_R = 2\pi n \tag{4.3}$$

$$n = \frac{1}{2\pi}\,\theta_R \tag{4.4}$$

where n is in *revolutions*
θ_R is in *radians*

Relationships between radians and revolutions

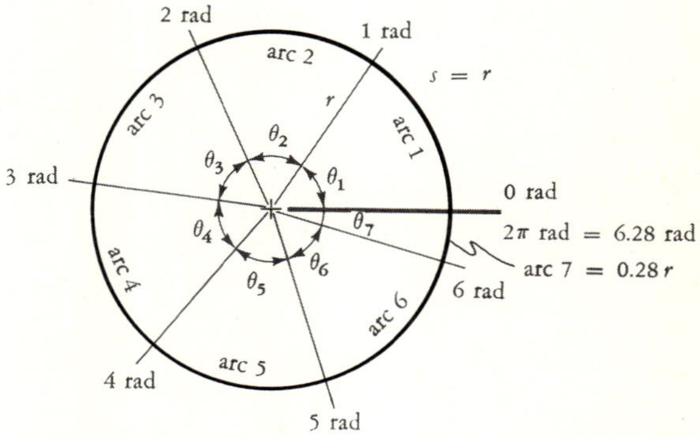

Fig. 4.28 One revolution contains 2π radians.

Since 1 rev contains 2π rad and also 360 degrees, we see that 360° is equivalent to 2π rad. Therefore 1 rad is equivalent to $360 \div 2\pi = 180/\pi$ degrees, and 1 degree is equivalent to $2\pi \div 360 = \pi/180$ rad.

$$1 \text{ rad} \longrightarrow \frac{180^\circ}{\pi} = 57.29\ldots^\circ \tag{4.5}$$

$$1^\circ \longrightarrow \frac{\pi}{180^\circ} = 0.01745\ldots \text{ rad} \tag{4.6}$$

The general conversion formulas are therefore:

$$\theta_D = \frac{180°}{\pi}\,\theta_R \tag{4.7}$$

$$\theta_R = \frac{\pi}{180°}\,\theta_D \tag{4.8}$$

where θ_D is an angle in *degrees*
θ_R is the same angle in *radians*

Relationships between degrees and radians

Example 4.15. Conversions involving degrees, radians, and revolutions

a. **Convert $\pi/3$ rad to revolutions and to degrees. Using Eqs. (4.4) and (4.7), we have**

$$n = \frac{1}{2\pi}\frac{\pi}{3} = \frac{1}{6}\text{ rev} \qquad \theta_D = \frac{180°}{\pi}\frac{\pi}{3} = 60°$$

b. **Convert 210° to radians in terms of π. From Eq. (4.8) we have**

$$\theta_R = \frac{\pi}{180°}210° = \frac{7\pi}{6}$$

c. **Convert 113.5° to radians. Using Eqs. (4.6) and (4.8), we have**

$$\theta_R = 0.01745(113.5°) = 1.981\text{ rad}$$

Comment: *Although many knowledgeable people say and write expressions such as "2π rad* ***equals*** *360°," or "180°* ***equals*** *π rad," this usage is not strictly correct. We contrasted equality and equivalency in Chap. 2. If we begin using expressions such as $\pi = 180°$* instead of *$\pi \rightarrow 180°$, then we may well become confused (as many students have) about where to use $\pi = 3.14\ldots$ and where to use $\pi = 180$. Equations (4.7) and (4.8) express correct equalities and help to avoid any misunderstanding about the value of π, which is always 3.14159. . . .*

EXERCISE 4.5

In Probs. 1 to 6, convert each angle in degrees to the equivalent angle in radians in terms of π.

1. $30°$	2. $75°$	3. $-90°$
4. $120°$	5. $570°$	6. $315°$

In Probs. 7 to 18, convert the angles in radians to degrees and to revolutions; draw the angles in standard position.

7. $3\pi/4$	8. $\pi/6$	9. $7\pi/3$
10. $-\pi$	11. $7\pi/6$	12. $7\pi/15$
13. 1.5	14. 5.80	15. 3.14
16. 7.62	17. -0.50	18. 0.03490

In Probs. 19 to 27, refer to Fig. 4.27. Determine the value of the missing part, r, s, or θ (in radians).

19. $r = 1$ cm, $\theta = 6$	20. $r = 12$ in., $\theta = 2\pi$
21. $r = 24$ in., $s = 3$ ft	22. $r = 44$ mm, $s = 88$ mm
23. $s = 100$ in., $\theta = 2$	24. $s = 2\pi$ cm, $\theta = 2\pi$
25. $r = 25$ ft, $\theta = 60°$	26. $r = 5$ m, $\theta = 215°$
27. $s = 0.750$ in., $\theta = 30°$	

B. APPLICATIONS

In several types of problems the circular arc relationships of Eqs. (4.1) and (4.2) are very useful. Their main value is the convenience and simplicity with which they enable us to set up an equation. Once the equation is written, we often have to convert to radians from degrees or from revolutions because of the widespread practice of specifying angles in those units.

Example 4.16. **In one day, how far does the earth travel in its solar orbit?**

Solution. **In Fig. 4.27, visualize the sun at the center of the circle; the radius r is the 93-million-mile distance to the earth, the angle θ is the angle turned in 1 day, and the arc s is the distance moved in 1 day. Using Eqs. (4.1) and (4.3) and the fact that the earth takes 365 days (approximately) per revolution, we have**

$$s = r\theta_R = (93 \times 10^6 \text{ miles}) \left(2\pi \frac{\text{rad}}{\text{rev}}\right)\left(\frac{1}{365} \text{ rev}\right) = 1{,}600{,}000 \text{ miles}$$

Linear Speeds on a Circular Path

We are familiar with the concept that average linear speed is distance divided by time. We can express this relationship with an equa-

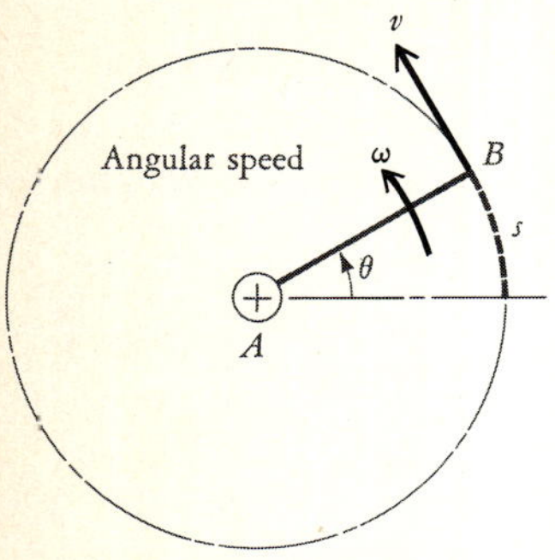

Fig. 4.29 Crank AB rotates with angular speed ω; point B moves with linear speed v.

tion such as $v = s/t$, where v, s, and t represent *speed, distance,* and *time,* respectively. A less familiar but parallel concept is that of average angular speed. *Average angular speed* is an angle divided by time. For example, if crank AB of Fig. 4.29 turns through 2 rad in 0.1 sec, then its average angular speed is $\theta/t = 2\text{ rad}/0.1\text{ sec} = 20\text{ rad/sec}$.

A typical equation for angular speed is

$$\omega = \frac{\theta}{t} \tag{4.9}$$

where ω is average angular speed
θ is an angle
t is time

A common practice is to use ω (omega) for angular speed in radians per unit time and N for revolutions per unit time. Then by using Eqs. (4.3) and (4.4) we have

$$\omega = 2\pi N \qquad \text{and} \qquad N = \frac{1}{2\pi}\,\omega \tag{4.10}$$

where ω is angular speed in radians per unit time
N is angular speed in revolutions per unit time

For example, if $\omega = 20$ rad/sec, then $N = 20/(2\pi) = 10/\pi$ rev/sec.

The distance that point B of Fig. 4.29 moves is the length of arc s. As we have seen, $s = r\theta$. Thus the average linear speed of point B is

$$\frac{s}{t} = \frac{r\theta}{t} = r\,\frac{\theta}{t}$$

The expression s/t is average linear speed v, and the expression θ/t is average angular speed ω. Therefore we have the following relationship between the linear speed of a point on a rotating arm and the angular speed of the arm:

$$v = r\omega \tag{4.11}$$

where v is linear speed of a point
r is the radius to the point
ω is the angular speed of the radius (in radians per unit time)

Linear and Angular Speeds

At any given instant, the direction of motion of a point (such as B of Fig. 4.29) on a revolving radius is at right angles to the radius. If point B were a rock being whirled on a string AB and the string broke, the rock would fly off on a straight line in the direction of the arrow v.

Example 4.17. **A certain earth satellite travels in a circular orbit at an altitude of 200 miles and completes one orbit in 1 hr 32 min (1.53 hr). If the radius of the earth is 4,000 miles, how fast is the satellite traveling?**

Solution.

$$v = r\omega = (4{,}000 + 200 \text{ miles}) \frac{1 \text{ rev}}{1.53 \text{ hr}} \left(2\pi \frac{\text{rad}}{\text{rev}}\right)$$

$$= 17{,}200 \text{ mph}$$

Area of a Circular Sector

As shown in Fig. 4.30, a sector of a circle is the figure formed by two radii and the included arc. The central angle of a sector represents some fraction of a revolution, and the area of the sector is the same fraction of the area of the circle. Thus for sector OPQ of Fig. 4.30 we have

$$\frac{\text{Area of sector}}{\text{Area of circle}} = \frac{\text{sector angle}}{\text{the angle in one revolution}} \qquad \text{or} \qquad \frac{A_S}{\pi r^2} = \frac{\theta_R}{2\pi}$$

We solve for A_S to obtain

$$A_S = \tfrac{1}{2} r^2 \theta_R = \tfrac{1}{2} rs \qquad (4.12)$$

where A_S is the sector area
r is the sector radius
s is the sector arc
θ_R is the sector angle in radians

Area of a circular sector

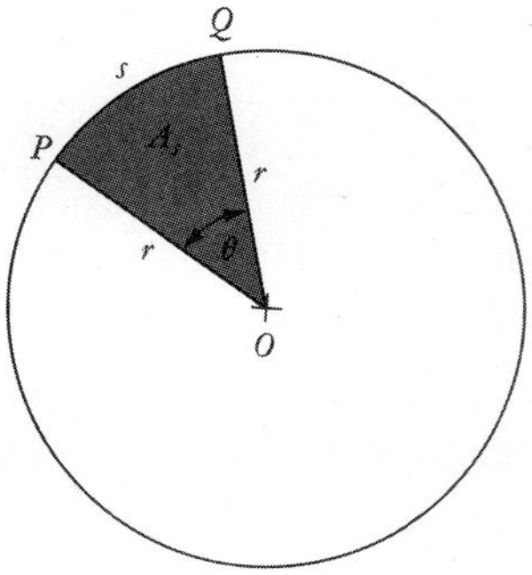

Fig. 4.30 Circular sector OPQ and its area A_s.

The extreme right side of the equation comes from substituting s for $r\theta$ in $r^2\theta$.

Example 4.18. **Figure 4.31*a* shows a sector gear with a pitch radius of 12.0 in. and an 8.00-in. arc of contact. The gear is made of ½-in.-thick steel (0.28 lb/in.³). What is the sector angle and how much does the gear weigh?**

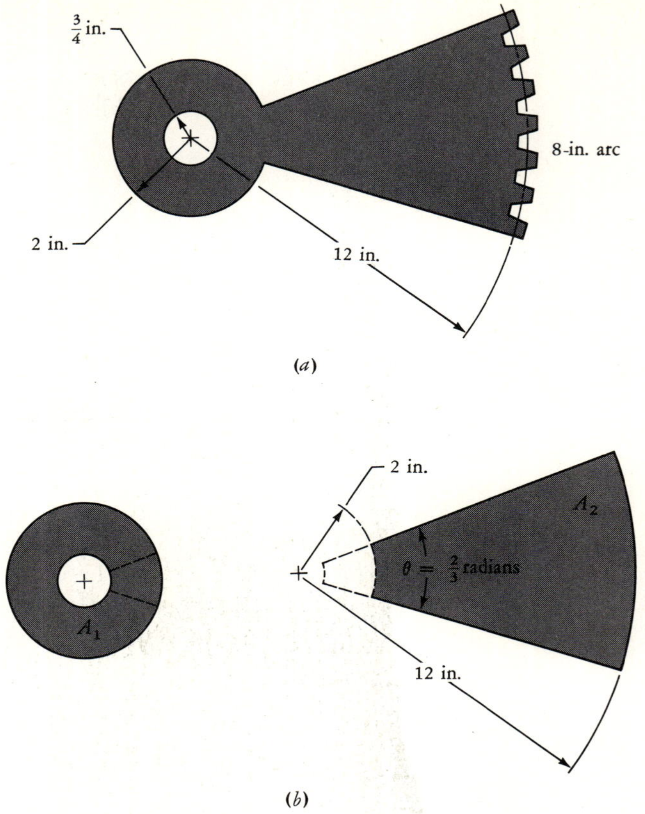

Fig. 4.31 Example 4.18. (a) Dimensions of a sector gear; (b) sector gear divided into two parts for area calculations.

Solution. From Eq. (4.2) we have

$$\theta = \frac{s}{r} = \frac{8.00}{12.0} = \frac{2}{3} \text{ rad}$$

To determine the weight, we divide the gear into circles and segments, as shown in Fig. 4.31*b*, and find the area of each part. The area between the circles is

$$A_1 = \pi(2^2) - \pi(3/4)^2 = \pi(4 - 9/16) = 10.8 \text{ in.}^2$$

The area between the sectors is [Eq. (4.12)]

$$A_2 = 1/2(12^2)(2/3) - 1/2(2^2)(2/3) = 46.7 \text{ in.}^2$$

The weight is area × thickness × density:

$$W = (A_1 + A_2)(\tfrac{1}{2}\text{ in.})(0.28\text{ lb/in.}^3) = 8.05\text{ lb}$$

EXERCISE 4.6

1. How many miles on the surface of the earth (radius = 4,000 miles) correspond to 1° of latitude?
2. The distance from Los Angeles to the equator is 2,380 miles. What is the latitude angle of Los Angeles?
3. West Palm Beach and Pittsburgh are almost exactly on the same longitude line of 80°. The latitude of West Palm Beach is N26.7° and that of Pittsburgh is N40.4°. What is the distance due north from West Palm Beach to Pittsburgh?
4. It takes an Apollo spacecraft approximately 68 hr to travel from the earth to the moon. Assuming that the moon travels in a circular orbit of radius 240,000 miles, how far does the moon move while the spacecraft is in transit? (*Hint:* How long does the moon take to complete 1 rev around the earth?)
5. How far does the earth move during the trip of an Apollo spacecraft to the moon? See Prob. 4 and Example 4.16.
6. A belt runs between pulleys of 6- and 3-in. radii that are mounted 18 in. apart. If the belt crosses between the pulleys, what is the length of the belt? [*Hints:* (1) Make a sketch. Notice that the belt comes tangent to each pulley, but not at the top or bottom. Note the similar right triangles with a common vertex where the belt crosses on the center line between pulleys. (2) To find the angle of contact, compare ratios of sides to those of a 30-60° right triangle (Table 4.3).]
7. A recommended cutting speed for turning aluminum on a lathe with a tungsten carbide tool is 1,500 ft/min. How fast should a lathe rotate in order to turn down the surface of a 6-in.-diameter cylinder of aluminum at the recommended speed?
8. How fast does the moon travel in its orbit around the earth? See Prob. 4 for data.
9. The centrifugal acceleration of a body revolving with angular speed ω rad per unit time around a circle of radius r is given by the formula

$$a = r\omega^2 \qquad \text{or} \qquad a = \frac{v^2}{r} \quad \text{where } v = r\omega$$

What is the centrifugal acceleration of an astronaut in the earth satellite of Example 4.17?

10. Since interstellar distances are measured in light-years, men would have to travel at or beyond the speed of light in order to explore other solar systems. Aside from the problems of attaining such speeds, a frequently overlooked obstacle to deep-space travel is the high centrifugal acceleration due to course corrections. How many g's of acceleration (1 g is 32.2 ft/sec^2 or 0.00610 miles/sec^2) would be felt by an astronaut traveling at the speed of light (186,000 miles/sec) if he turned along an arc of 1-million-mile radius? Use the formula given in Prob. 9.
11. How many seconds would be needed for the astronaut in Prob. 10 to make a course correction of 1°? (*Hint:* The change of direction equals the central angle of the arc on which the change takes place.)
12. At what linear speed does the rotor of an automobile distributor pass over the contacts if they are 2.00 in. from the center and the engine crankshaft is turning 5,000 rpm? Don't forget that each spark plug fires every other revolution of the crankshaft.
13. A typical disk drive mechanism is shown in Fig. 4.32. If the disk is a 33⅓-rpm turntable, the diameter of the drive wheel is 0.500 in., and the distance x is 4.00 in., what is the speed v of the point of contact? How many rpm must the drive wheel turn?

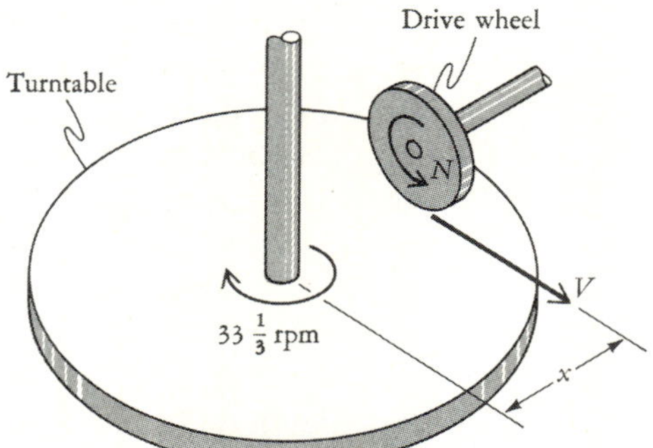

Fig. 4.32 Prob. 13. A disk drive.

14. The depth of oil in a horizontal fuel oil tank is 7½ in. The diameter of the tank is 30 in. How many gallons of fuel remain if the tank holds 100 gal when full? See Fig. 4.33.
15. What is the weight of the sector gear of Example 4.18 if the pitch radius is 0.75 in., the arc of contact is 1.96 in., the diameter of the center hole is 0.12 in., the hub diameter is 0.75 in., and the material is ¼-in.-thick aluminum (0.096 lb/in.^{3})?

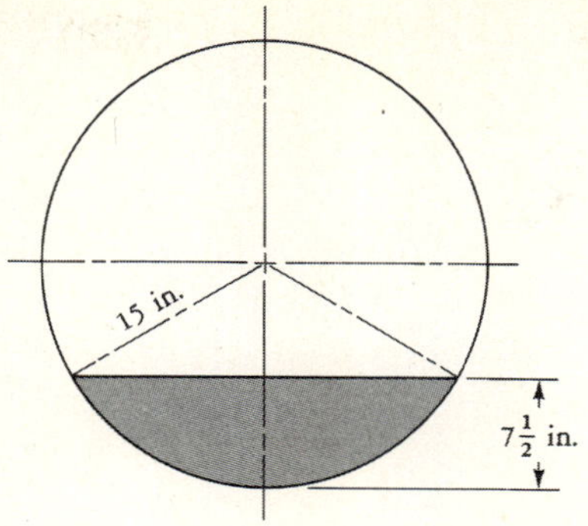

Fig. 4.33 Prob. 14. An oil tank.

16. The recommended cutting speed for a saw cutting hot metal is 25,000 ft/min. If the saw blade is 18 in. in diameter, at how many rpm should it turn?

TRIGONOMETRY AND RIGHT TRIANGLES

CHAPTER 5

5.1 DEFINITIONS OF THE TRIGONOMETRIC FUNCTIONS

Trigonometry is a branch of mathematics that has traditionally dealt with relationships between the angles and sides of triangles. *Trigonometry* literally means "triangle measurement." In modern times, however, other interpretations and applications have been discovered. In this chapter we shall (1) survey the general definitions of the trigonometric functions, and (2) emphasize the traditional, triangle-oriented applications of the subject. Later, in Unit 3, we shall explore other aspects in more detail.

We shall frequently use the word "function," as in *trigonometric function.* In mathematics, this word has a very specific meaning; we shall discuss it in Chap. 7. For our immediate purposes, we shall consider a function to be a relationship between two quantities. For example, the formula $A = s^2$ states a relationship between A, the area of a square, and s, the length of each side. We could say that A is a function of s, meaning that in some way A depends upon s or is related to s.

A. THE BASIS FOR THE TRIGONOMETRIC FUNCTIONS

Figure 5.1 shows an angle θ in standard position and a point P on the terminal side of θ. The distance from the origin to point P is called the *radius* or *radius vector* to the point. Thus r, the radius to point P, is the length of line segment OP. *A radius vector is always a positive distance,* no matter which way it points.

You should recognize the following simple geometric fact: the ratios ordinate/radius, abscissa/radius, and ordinate/abscissa of any point, such as P of Fig. 5.1, are related to the angle θ upon which P is located. If the angle changes, then these ratios change. For the configuration shown, the ratio y/r, for example, has a certain value.

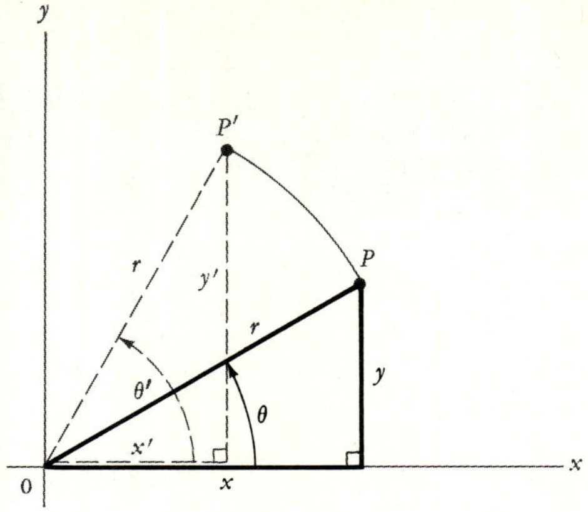

Fig. 5.1 Coordinates of two points at the same radius on different angles.

Suppose that the angle increases from θ to θ'. Then, although r remains the same, y increases to y', and the ratio y'/r is different from y/r. Thus, as we stated above, the values of these ratios are related to or depend upon the value of the angle. We say that y/r, x/r, and y/x are *functions* of θ. Also, the reciprocals, r/y, r/x, and x/y, are functions of θ.

Furthermore, these ratios depend upon only the angle θ and do not depend upon which point P is chosen on the terminal side of θ. In Fig. 5.2, the ordinate/radius ratio, for example, is the same for points P_1 and P_2, or for any point P in general, no matter how close or how far from the origin. Using the symbols of Fig. 5.2, we could say that

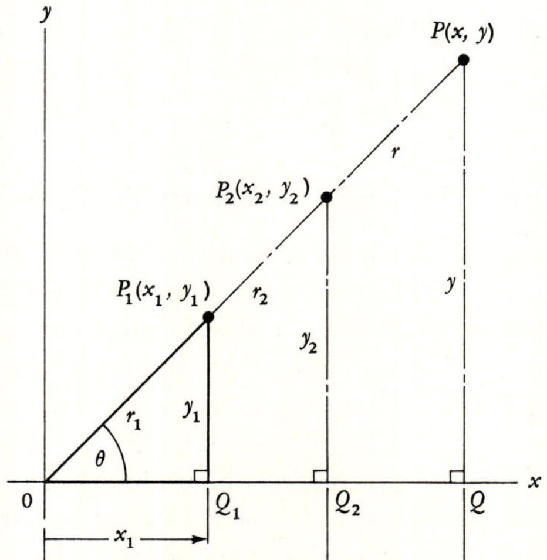

Fig. 5.2 Coordinates of three points on the same angle.

$$\frac{y_1}{r_1} = \frac{y_2}{r_2} = \frac{y}{r} \qquad \frac{x_1}{r_1} = \frac{x_2}{r_2} = \frac{x}{r} \qquad \frac{y_1}{x_1} = \frac{y_2}{x_2} = \frac{y}{x}$$

To demonstrate this fact, observe that the three right triangles OP_1Q_1, OP_2Q_2, and OPQ are similar triangles, and by definition, the ratios of corresponding sides are equal. (Recall the definitions and properties of similar triangles, Sec. 4.3B.)

To summarize, we recognize the existence of a relationship (function) between each of the six ratios y/r, x/r, y/x, r/y, r/x, and x/y, and the angle θ. Although we do not as yet know any equations or other means for expressing these functions, we recognize that the functions exist.

B. THE TRIGONOMETRIC FUNCTIONS

The six ratios referred to above are called the *trigonometric functions.* To avoid the awkwardness of referring to ordinate/radius, etc., mathematicians long ago gave names to the six ratios. Equations (5.1) to (5.6) on p. 92 give these names. These equations define the trigonometric functions.

The word "sine" and its abbreviation "sin" (no period) are both pronounced like the word "sign." Similarly, "cos" is pronounced co-sign.

Although the names "sine," "cosine," "tangent," and so forth, are strange looking and unfamiliar, the meanings they convey should not be mysterious. They are simply the names by which we identify certain ratios that are related to an angle. Thus if we wish to refer to the ordinate/radius ratio for any point on the terminal side of an angle θ, we can simply write sin θ.

Comment: *A frequent error of beginners in trigonometry is to interpret the symbolic notations of the functions as multiplications. The notation "sin θ" is read "the sine of angle θ," or more briefly "sine θ." It does* **not** *mean a multiplication such as sin times θ.*

An interesting point is that the cotangent, secant, and cosecant functions are reciprocals of the tangent, cosine, and sine functions, respectively. Notice, for example, that

$$\csc\theta = \frac{r}{y} = \frac{1}{y/r} \qquad \text{and} \qquad \sin\theta = \frac{y}{r}$$

Thus $$\csc\theta = \frac{1}{\sin\theta}$$

NAMES AND DEFINITIONS OF THE TRIGONOMETRIC FUNCTIONS OF AN ANGLE IN STANDARD POSITION	SYMBOLIC EQUATIONS

Sine: $\dfrac{\text{ordinate of any point on the terminal side}}{\text{radius to the point}}$

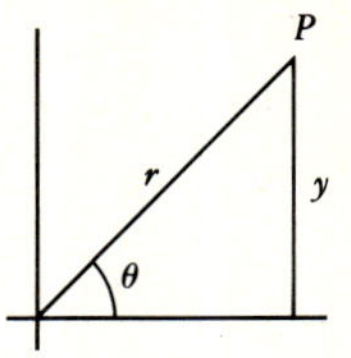

$$\sin \theta = \frac{y}{r} \tag{5.1}$$

Cosine: $\dfrac{\text{abscissa of any point on the terminal side}}{\text{radius to the point}}$

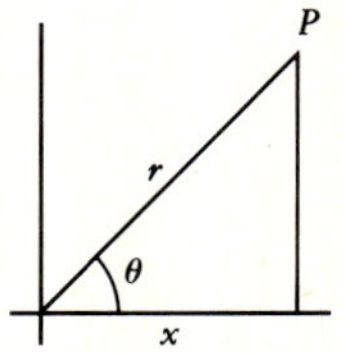

$$\cos \theta = \frac{x}{r} \tag{5.2}$$

Tangent: $\dfrac{\text{ordinate of any point on the terminal side}}{\text{abscissa of the point}}$

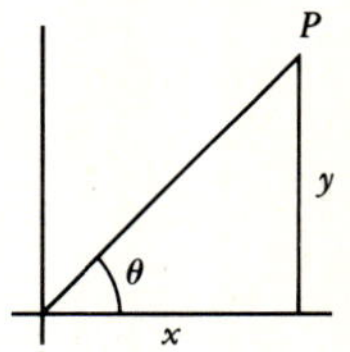

$$\tan \theta = \frac{y}{x} \tag{5.3}$$

Cotangent: $\dfrac{\text{abscissa of any point on the terminal side}}{\text{ordinate of the point}}$

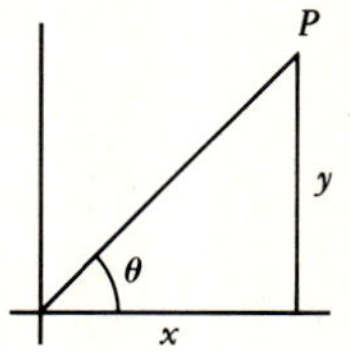

$$\cot \theta = \frac{x}{y} \tag{5.4}$$

Secant: $\dfrac{\text{radius to any point on the terminal side}}{\text{abscissa of the point}}$

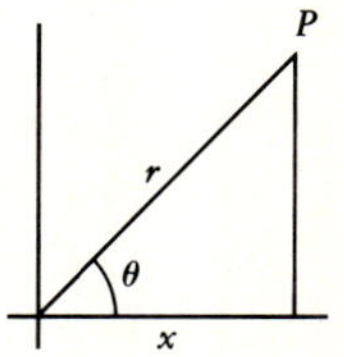

$$\sec \theta = \frac{r}{x} \tag{5.5}$$

Cosecant: $\dfrac{\text{radius to any point on the terminal side}}{\text{ordinate of the point}}$

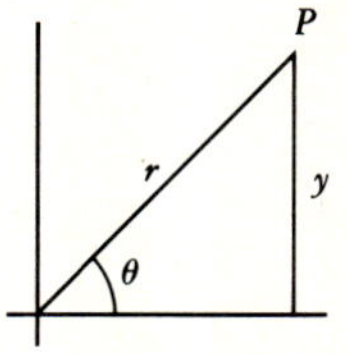

$$\csc \theta = \frac{r}{y} \tag{5.6}$$

Similarly we can show that

$$\cot \theta = \frac{1}{\tan \theta} \qquad \text{and} \qquad \sec \theta = \frac{1}{\cos \theta}$$

For example, if the value of $\tan \theta$ happens to be 2, then we immediately know that $\cot \theta = 1/2$.

Example 5.1. **For the point with coordinates $(x,y) = (2,5)$, determine the values of the six trigonometric functions.**

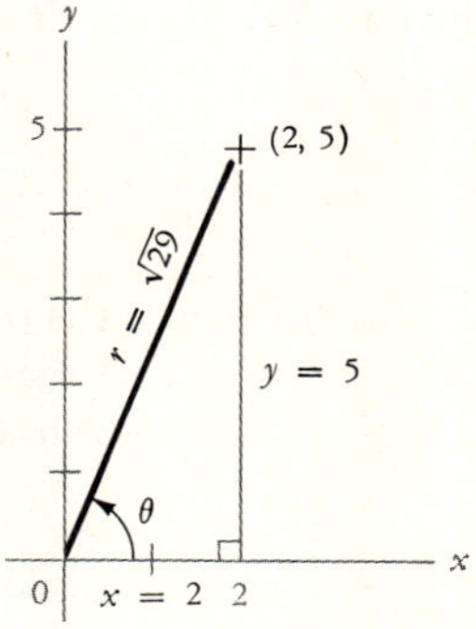

Fig. 5.3 Example 5.1.

First, make a diagram such as Fig. 5.3 to show the point, the angle, and the values of the coordinates (x, y). Notice that r is the hypotenuse of a right triangle with x and y as the legs. We compute r by using the Pythagorean formula:

$$r = \sqrt{x^2 + y^2} = \sqrt{2^2 + 5^2} = \sqrt{4 + 25} = \sqrt{29}$$

Remember that the radius is always *positive;* therefore we use only the positive root, $+\sqrt{29}$.

Then we use the definitions, Eqs. (5.1) to (5.6), to compute the values, as follows:

$$\sin \theta = \frac{y}{r} = \frac{5}{\sqrt{29}} \qquad \cos \theta = \frac{x}{r} = \frac{2}{\sqrt{29}} \qquad \tan \theta = \frac{y}{x} = \frac{5}{2}$$

$$\cot \theta = \frac{x}{y} = \frac{2}{5} \qquad \sec \theta = \frac{r}{x} = \frac{\sqrt{29}}{2} \qquad \csc \theta = \frac{r}{y} = \frac{\sqrt{29}}{5}$$

Example 5.2. **For a point in the second quadrant with a radius of 5 and an ordinate of 4, determine the values of the six trigonometric functions.**

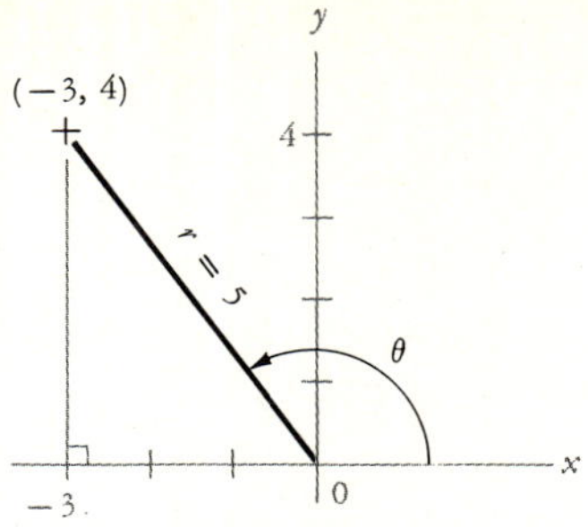

Fig. 5.4 Example 5.2.

As shown in Fig. 5.4, we have a right triangle with the hypotenuse r and one side y known. We use an alternate form of the Pythagorean formula to determine x as follows:

$$x = -\sqrt{r^2 - y^2} = -\sqrt{5^2 - 4^2} = -\sqrt{9} = -3$$

Notice that x is ***negative*** 3 because the point is stated to be in the second quadrant. (Recall Fig. 4.1 which defines the quadrants.)

Now we use Eqs. (5.1) to (5.6) to compute the values:

$$\sin\theta = \frac{y}{r} = \frac{4}{5} \qquad \cos\theta = \frac{x}{r} = \frac{-3}{5} = -\frac{3}{5}$$

$$\tan\theta = \frac{y}{x} = \frac{4}{-3} = -\frac{4}{3} \qquad \cot\theta = \frac{x}{y} = \frac{-3}{4} = -\frac{3}{4}$$

$$\sec\theta = \frac{r}{x} = \frac{5}{-3} = -\frac{5}{3} \qquad \csc\theta = \frac{r}{y} = \frac{5}{4}$$

Example 5.3. Given that $\cos\theta = -5/13$ and $180° < \theta < 270°$, determine the values of the other five functions of θ.

Solution. As shown in Fig. 5.5, locate a point in the third quadrant (where $180° < \theta < 270°$); label the radius 13 and the abscissa −5, thus making $\cos\theta = \text{abscissa/radius} = x/r = -5/13$ as specified. Remember that the radius is always positive; thus x has to be negative.

We can use the Pythagorean formula to determine y, or we might recognize the 5:12:13 right triangle, and see that $y = -12$. We should recognize that y has to be negative in the third quadrant.

From Eqs. (5.1) and (5.3) to (5.6) we have

$$\sin\theta = \frac{y}{r} = \frac{-12}{13} = -\frac{12}{13} \qquad \tan\theta = \frac{y}{x} = \frac{-12}{-5} = \frac{12}{5}$$

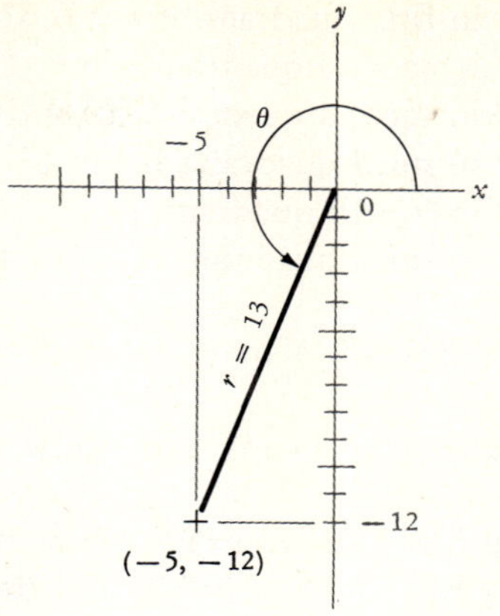

Fig. 5.5 Example 5.3.

$$\cot\theta = \frac{x}{y} = \frac{-5}{-12} = \frac{5}{12} \qquad \sec\theta = \frac{r}{x} = \frac{13}{-5} = -\frac{13}{5}$$

$$\csc\theta = \frac{r}{y} = \frac{13}{-12} = -\frac{13}{12}$$

EXERCISE 5.1

In Probs. 1 to 5, determine the values of the six trigonometric functions of the angle θ whose terminal side contains the given point $P(x,y)$.

1. (4,3)
2. (5,12)
3. (10,−6)
4. (−5,−5)
5. (−8,6)

In Probs. 6 to 11, determine the values of the other five functions. See Example 5.3.

6. $\tan\theta = 4/5,\quad 0° < \theta < 90°$
7. $\cot\theta = -0.6,\quad 270° < \theta < 360°$
8. $\sin\theta = 3/5,\quad 0° < \theta < 90°$
9. $\cos\theta = -4/5,\quad 90° < \theta < 180°$
10. $\sec\theta = -7/5,\quad 90° < \theta < 180°$
11. $\csc\theta = -13/12,\quad 180° < \theta < 270°$

In Probs. 12 to 17, determine the value of the indicated trigonometric function of the angle θ whose terminal side passes through the point located by the coordinates r, x, or y.

12. $\sin\theta$ in first quadrant, if $r = 6$, $x = 3$
13. $\tan\theta$ in second quadrant, if $r = 15$, $y = \sqrt{29}$
14. $\cos\theta$ in second quadrant, if $r = 10$, $y = 6$
15. $\csc\theta$ in third quadrant, if $r = 12$, $x = -\sqrt{23}$
16. $\cot\theta$ in fourth quadrant, if $r = \sqrt{50}$, $x = 5$
17. $\sec\theta$ in third quadrant, if $r = \sqrt{40}$, $y = -2$

18. Using Eqs. (5.1) to (5.4), show that $\tan\theta = (\sin\theta)/(\cos\theta) = 1/(\cot\theta)$.
19. Using Eqs. (5.1) to (5.4), show that $\cot\theta = (\cos\theta)/(\sin\theta) = 1/\tan\theta$.
20. Using Eqs. (5.2) and (5.5), show that $\sec\theta = 1/(\cos\theta)$.
21. Using Eqs. (5.1) and (5.2) and the Pythagorean theorem, show that $(\sin\theta)^2 + (\cos\theta)^2 = 1$.

5.2 VALUES OF THE TRIGONOMETRIC FUNCTIONS

The trigonometric functions would be of little use if their values were not known. In this section we shall examine means of determining numerical values of $\sin\theta$, $\cos\theta$, and the other functions, for any given value of θ. These values can then be used in conjunction with Eqs. (5.1) to (5.6) to find coordinates of points and to solve triangles.

A. SPECIAL ANGLES

The values of the trigonometric functions of the angles 0°, 30°, 45°, 60°, and 90° can be determined by geometry. These angles are used so frequently in engineering and science that you should learn the values derived below.

0° and 90°

The values of the functions of 0° and 90° can be readily determined from the definitions, Eqs. (5.1) to (5.6). For example, $\sin 0° = 0$, because by definition $\sin\theta$ = ordinate/radius, and for 0° the ordinate is zero. However, $\csc 0°$ is undefined, since $\csc\theta$ is defined as radius/ordinate, and for 0° the zero ordinate results in a division by zero. Values of all of the functions of 0° and 90° are listed in Table 5.1. You should be able to verify them by using the definitions of the functions as just demonstrated.

TABLE 5.1. TRIGONOMETRIC FUNCTIONS OF SPECIAL ANGLES

ANGLE θ		VALUES OF FUNCTIONS					
DEGREES	RADIANS	$\sin\theta$	$\cos\theta$	$\tan\theta$	$\cot\theta$	$\sec\theta$	$\csc\theta$
0°	0	0	1	0	undefined	1	undefined
30°	$\frac{\pi}{6}$	$\frac{1}{2}$	$\frac{\sqrt{3}}{2}$	$\frac{1}{\sqrt{3}}$	$\sqrt{3}$	$\frac{2}{\sqrt{3}}$	2
45°	$\frac{\pi}{4}$	$\frac{1}{\sqrt{2}}$	$\frac{1}{\sqrt{2}}$	1	1	$\sqrt{2}$	$\sqrt{2}$
60°	$\frac{\pi}{3}$	$\frac{\sqrt{3}}{2}$	$\frac{1}{2}$	$\sqrt{3}$	$\frac{1}{\sqrt{3}}$	2	$\frac{2}{\sqrt{3}}$
90°	$\frac{\pi}{2}$	1	0	undefined	0	undefined	1

30°, 45°, and 60°

Table 4.3 indicates that any altitude of an equilateral triangle divides it into two 30°-60° right triangles. Furthermore, two of the original equilateral sides become hypotenuses of the right triangles, the altitude becomes the side adjacent to the 30° angles and is $\sqrt{3}/2$ times the hypotenuse, and the sides opposite the 30° angles are one-half the hypotenuse.

We can use these facts to determine the values of the trigonometric functions of 30°. To do so, we place the 30° angle of a 30°-60° right triangle in standard position, as shown in Fig. 5.6. Then the sides of the triangle become the coordinates of a point $P(x,y)$ on the terminal side of the 30° angle. By using the relative lengths of the sides

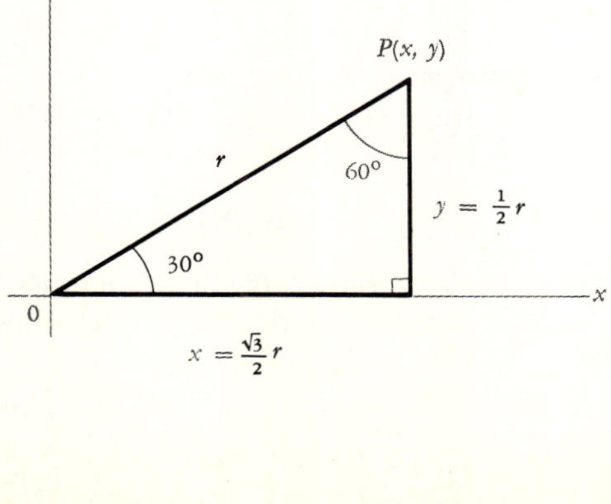

Fig. 5.6 A 30°–60° right triangle with the 30° angle in standard position.

as noted in Eqs. (5.1) to (5.6) we can compute exact values of the trigonometric functions of 30° as follows:

$$\sin 30° = \frac{y}{r} = \frac{(1/2)\cancel{r}}{\cancel{r}} = \frac{1}{2} \qquad \cot 30° = \frac{x}{y} = \frac{(\sqrt{3}/\cancel{2})\cancel{r}}{(1/\cancel{2})\cancel{r}} = \sqrt{3}$$

$$\cos 30° = \frac{x}{r} = \frac{(\sqrt{3}/2)\cancel{r}}{\cancel{r}} = \frac{\sqrt{3}}{2} \qquad \sec 30° = \frac{r}{x} = \frac{\cancel{r}}{(\sqrt{3}/2)\cancel{r}} = \frac{2}{\sqrt{3}}$$

$$\tan 30° = \frac{y}{x} = \frac{(1/\cancel{2})\cancel{r}}{(\sqrt{3}/\cancel{2})\cancel{r}} = \frac{1}{\sqrt{3}} \qquad \csc 30° = \frac{r}{y} = \frac{\cancel{r}}{(1/2)\cancel{r}} = 2$$

In a similar manner, by placing a 45° or 60° angle in standard position, we can derive values of the functions of those angles. Table 5.1 summarizes the trigonometric functions of these special angles. Notice that the table includes radian values of the angles as well as degrees. When we specify an angle, we are not limited to any one unit of measure, and we shall use both radian and degree units in the trigonometric functions.

EXERCISE 5.2

1. By using the proportions for the sides of a 30°-60° right triangle, show that
   a. $\sin 60° = \sqrt{3}/2$ b. $\cos 60° = 1/2$ c. $\tan 60° = \sqrt{3}$
2. By using the proportions for the sides of a 45° right triangle, show that
   a. $\sin 45° = 1/\sqrt{2}$ b. $\cos 45° = 1/\sqrt{2}$ c. $\tan 45° = 1$
3. By using the definitions of the functions, show that
   a. $\cos 0° = 1$ b. $\tan 0° = 0$ c. $\sec 0° = 1$
4. By using the definitions of the functions, show that
   a. $\sin 90° = 1$ b. $\cos 90° = 0$ c. $\cot 90° = 0$
5. Show, as in Prob. 3, that
   a. $\sin 180° = 0$ b. $\cos 180° = -1$ c. $\tan 180° = 0$
6. Show, as in Prob. 4, that
   a. $\sin 270° = -1$ b. $\cos 270° = 0$ c. $\cot 270° = 0$

B. TABLES OF TRIGONOMETRIC FUNCTIONS

Knowing the values of the functions of the special angles is not enough. We also need to know the values of the trigonometric functions of other angles. Mathematicians have discovered that the following formulas determine values of sin θ and cos θ, independently of the ordinate/radius and abscissa/radius ratios, respectively.

$$\sin\theta = \theta - \frac{\theta^3}{2\cdot 3} + \frac{\theta^5}{2\cdot 3\cdot 4\cdot 5} - \frac{\theta^7}{2\cdot 3\cdot 4\cdot 5\cdot 6\cdot 7} + \cdots \quad (5.7)$$

$$\cos\theta = 1 - \frac{\theta^2}{2} + \frac{\theta^4}{2\cdot 3\cdot 4} - \frac{\theta^6}{2\cdot 3\cdot 4\cdot 5\cdot 6} + \cdots \quad (5.8)$$

where θ is in radians.

These formulas are of a type called infinite power series. (An explanation of them is beyond the scope of this book.) Such formulas are awkward and time-consuming to use, so we shall use a table of values that has been constructed from the formulas. Table A5.4 in the Appendix is such a table.

Reading the Table

To find the value of one of the functions of any angle between 0 and 45°, (1) locate the angle in the left-hand column of Table A5.4, and (2) read across the row to the number under the desired function at the top of the page. Figure 5.7 illustrates the reading of cos 36°10′ or cos (0.6312 rad) equal to 0.8073.

ANGLE									
DEG-MIN	RADIANS	sine	tangent	secant	cosecant	cotangent	cosine		
36°00′	.6283	.5878	.7265	1.236	1.701	1.376	.8090	.9425	54°00′
10	.6312	.5901	.7310	1.239	1.695	1.368	.8073	.9396	50
20	.6341	.5925	.7355	1.241	1.688	1.360	.8056	.9367	40
30	.6370	.5948	.7400	1.244	1.681	1.351	.8039	.9338	30
40	.6400	.5972	.7445	1.247	1.675	1.343	.8021	.9308	20
50	.6429	.5995	.7490	1.249	1.668	1.335	.8004	.9279	10
37°00′	.6458	.6018	.7536	1.252	1.662	1.327	.7986	.9250	53°00′
10	.6487	.6041	.7581	1.255	1.655	1.319	.7969	.9221	50
20	.6516	.6065	.7627	1.258	1.649	1.311	.7951	.9192	40

Fig. 5.7 cos 36°10′ = cos (0.6312 rad) = 0.8073.

To find the value of a function of any angle between 45° and 90°, (1) locate the angle in the *right-hand* column, and (2) read across the row to the number above the desired function at the *bottom* of

secant	tangent	sine	ANGLE RADIANS	ANGLE DEG-MIN
3.906	3.776	.9667	1.3119	10
3.864	3.732	.9659	1.3090	75°00′
3.822	3.689	.9652	1.3061	50
3.782	3.647	.9644	1.3032	40
3.742	3.606	.9636	1.3003	30
3.703	3.566	.9628	1.2974	20
3.665	3.526	.9621	1.2945	10
3.628	3.487	.9613	1.2915	74°00′
3.592	3.450	.9605	1.2886	50
3.556	3.412	.9596	1.2857	40
3.521	3.376	.9588	1.2828	30

Fig. 5.8 *tan 74°40′ = tan (1.3032 rad) = 3.647.*

the page. Figure 5.8 illustrates the reading of tan 74°40′ or tan (1.3032 rad) equal to 3.647.

To find the angle when you know the value of one of the functions, look in the column for that function until you locate the value. If the value is located under the function at the top of a page, then read across the row to the angle in the left-hand column; or, if the value is located above the function at the bottom of a page, then read across the row to the angle in the right-hand column. For example, if we were given that cos $\theta = 0.8073$, we would locate 0.8073 under the *cosine* column, as shown in Fig. 5.7, and read $\theta = 36°10'$ or $\theta = 0.6132$ rad in the left-hand column. For tan $\theta = 3.647$, we would locate 3.647 *above* the *tangent* column, as shown in Fig. 5.8, and read $\theta = 74°40'$ or $\theta = 1.3032$ rad from the *right-hand* column.

EXERCISE 5.3

In Probs. 1 to 16, use Table A5.4 in the Appendix to find the value of each function.

1. sin 20°30′
2. cot 40°10′
3. sec 43°50′
4. cos 3°30′
5. tan 13°20′
6. csc 32°40′

7. cos 45°10′
8. sin 77°50′
9. cot 54°30′
10. sec 69°20′
11. tan 89°50′
12. csc 70°20′
13. sin (0.4451 rad)
14. tan (0.8261 rad)
15. cos (1.1839 rad)
16. cot (0.0640 rad)

In Probs. 17 to 28, use Table A5.4 to find θ in degrees and minutes and in radians for each of the given trigonometric functions.

17. $\tan\theta = 0.8342$
18. $\cos\theta = 0.9992$
19. $\csc\theta = 2.729$
20. $\sec\theta = 1.349$
21. $\cot\theta = 3.340$
22. $\sin\theta = 0.5495$
23. $\cos\theta = 0.6713$
24. $\sec\theta = 57.30$
25. $\cot\theta = 0.4628$
26. $\tan\theta = 1.303$
27. $\sin\theta = 0.9757$
28. $\csc\theta = 1.116$

Interpolation

As you might suspect, we frequently need to use values of angles or of functions that fall between those listed in the table. We can assume intermediate values between those tabulated. This procedure is called *interpolation.*

The basis for interpolation is as follows. Suppose we wish to know the value of cos 36°15′. 36°15′ is halfway between 36°10′ and 36°20′, the nearest two angles tabulated. Therefore we assume that cos 36°15′ is halfway between cos 36°10′ and cos 36°20′, which puts it halfway between 0.8073 and 0.8056. (See Fig. 5.7.) The value is

$$\begin{aligned}\cos 36°15' &= 0.8073 - \tfrac{1}{2}(0.8073 - 0.8056)\\ &= 0.8073 - \tfrac{1}{2}(0.0017)\\ &= 0.8073 - 0.0008\\ &= 0.8065 \qquad \text{(to four digits)}\end{aligned}$$

In general, if an angle is halfway between two tabulated angles, then we assume that the value of any function of that angle is halfway between the tabulated values for that function.

We can extend this idea to an angle that fits anywhere between two tabulated angles. For example, suppose we wish to find cos 36°13′. The difference between 36°13′ and 36°10′ is $\tfrac{3}{10}$ of the difference between 36°10′ and 36°20′. Thus we assume that the difference between cos 36°13′ and cos 36°10′ is $\tfrac{3}{10}$ of the difference between cos 36°10′ and cos 36°20′. Figure 5.9 shows an efficient scheme for this interpolation. First, list in a column the three angles involved. Then list in an adjacent column the values of cos 36°10′ and cos 36°20′

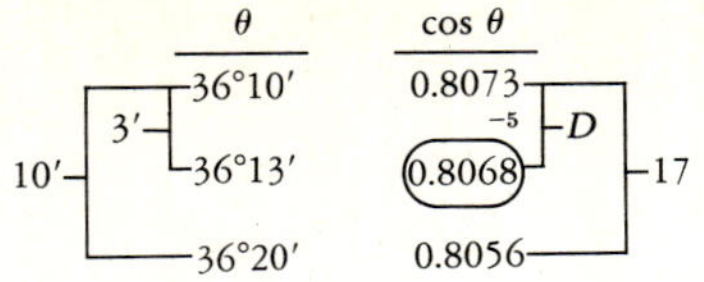

$$\frac{D}{17} = \frac{3}{10} \longrightarrow D = 5 \text{ parts, or } 0.0005$$

Therefore, $\cos 36°13' = 0.8073 - 0.0005 = 0.8068$

Fig. 5.9 A scheme for interpolating, for cos 36°13′.

obtained from Table A5.4. Indicate, as shown, the differences 3 and 10 minutes between the angles and the difference of 17 parts (meaning 0.0017) between the cosines. Compute the proportional difference D to be subtracted from $\cos 36°10'$ in order to obtain $\cos 36°13'$: $D = (3/10)17 = 5.1$ or 0.0005; we use only four digits. Then subtract this number from $\cos 36°10'$ to find $\cos 36°13'$: $\cos 36°13' = 0.8073 - 0.0005 = 0.8068$.

Comment: *Note that we* ***subtracted*** *from cos 36°10′ to find cos 36°13′. For another function we might have added. We add or subtract the proportional difference depending upon whether the function increases or decreases between the angles in question. We must always watch to see which way to go. It would have been foolish to add 0.0005 to cos 36°10′ in the example above, for we would have obtained cos 36°13′ = 0.8078, a number that is not between cos 36°10′ and cos 36°20′.*

We can interpolate in a similar way to find the angle when the value of the function is known. For example, suppose that $\sec \theta = 3.680$, and we want to find the value of θ. Figure 5.10 shows the book-

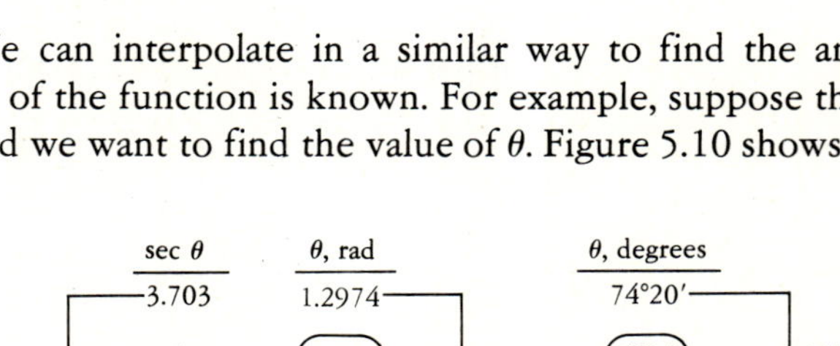

For θ in radians:

$$\frac{D}{29} = \frac{15}{38}$$

$$D = \frac{15}{38}(29) = 11.4$$

or

$$D = 0.0011$$

$$\theta = 1.2945 + 0.0011$$

$$\theta = 1.2956$$

For θ in degrees:

$$\frac{D}{10'} = \frac{15}{38}$$

$$D = \frac{15}{38}(10') = 3.9'$$

or

$$D = 4'$$

$$\theta = 74°10' + 4' = 74°14'$$

Fig. 5.10 Interpolations for θ, when sec θ = 3.680.

keeping format for this interpolation. We list the next-highest secant tabulated, 3.703, the given value 3.680, and the next lowest value tabulated, 3.665. These values are found in Table A5.4 (and also in Fig. 5.8). In an adjacent column, we list the angles 74°20′ and 74°10′ corresponding to sec $\theta = 3.703$ and sec $\theta = 3.665$, respectively. Figure 5.10 shows the angles in degrees and in radians. In this figure we indicate the differences 15 and 38 between the secants, and D and 10 minutes between the angles in degrees (or D and 29 between the angles in radians). The figure shows the computations which result in $\theta = 74°14'$ (or $\theta = 1.2956$ rad).

EXERCISE 5.3 (*Continued*)

In Probs. 29 to 36, use Table A5.4 and interpolate to find the value of each function to four digits.

29. sec 31°43′
30. sin 63°16′
31. tan 30°08′
32. cos 54°25′
33. sin (0.6910 rad)
34. cot (1.1215 rad)
35. csc (0.2170 rad)
36. tan (0.8160 rad)

In Probs. 37 to 44, use Table A5.4 and interpolate to find θ to the nearest minute and to the nearest 0.0001 rad, for each function.

37. $\sin \theta = 0.9057$
38. $\sec \theta = 1.279$
39. $\tan \theta = 0.2700$
40. $\cos \theta = 0.8592$
41. $\cot \theta = 0.6942$
42. $\csc \theta = 2.447$
43. $\tan \theta = 6.134$
44. $\sin \theta = 0.1630$

5.3 THE TRIGONOMETRY OF RIGHT TRIANGLES

The trigonometric functions are extremely useful for finding values of unknown sides and angles in right triangles. Because right triangles are widely used in science and technology, the trigonometric functions turn up in many diverse problems. Before we examine the applications, we shall modify the definitions of the trigonometric functions to better suit the terminology of right triangles.

A. SPECIAL DEFINITIONS OF THE TRIGONOMETRIC FUNCTIONS

Figure 5.11*a* shows a right triangle ABC placed on a rectangular coordinate plane so that angle α is in standard position. In this posi-

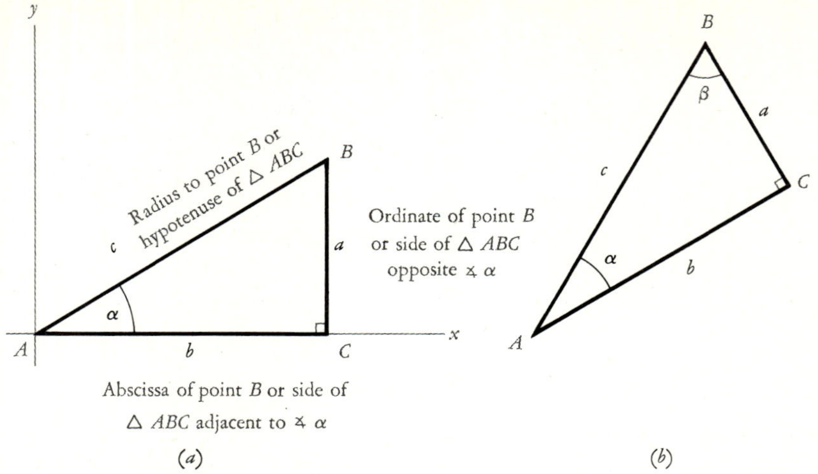

Fig. 5.11 Coordinates of a point stated in terms of the sides of a right triangle. (a) A right triangle ABC placed on the coordinate plane; (b) right triangle ABC moved to a random position off the coordinate plane.

tion, the sides of the triangle become the coordinates of point B. By substituting the words "hypotenuse" for "radius to the point," "side opposite α" for "ordinate of the point," and "side adjacent to α" for "abscissa of the point" in Eqs. (5.1) to (5.6), we can reword the definitions of the trigonometric functions as shown below.

These definitions are entirely consistent with the more general ones of Eqs. (5.1) to (5.6); the only differences are the changes of terminology. Because of the new terminology, the definitions are limited to the angles of a *right triangle,* but they are more appropriate and convenient to use when we discuss right triangles.

An advantage of these special definitions is their independence from any coordinate system. For example, we can move triangle ABC of Fig. 5.11a to a different position, as shown in Fig. 5.11b, and apply Eqs. (5.9) to (5.11) to angles α and β as follows:

$$\sin\alpha = \frac{a}{c} \qquad \sin\beta = \frac{\text{side opposite }\beta}{\text{hypotenuse}} = \frac{b}{c}$$

$$\cos\alpha = \frac{b}{c} \qquad \cos\beta = \frac{\text{side adjacent to }\beta}{\text{hypotenuse}} = \frac{a}{c}$$

$$\tan\alpha = \frac{a}{b} \qquad \tan\beta = \frac{\text{side opposite }\beta}{\text{side adjacent to }\beta} = \frac{b}{a}$$

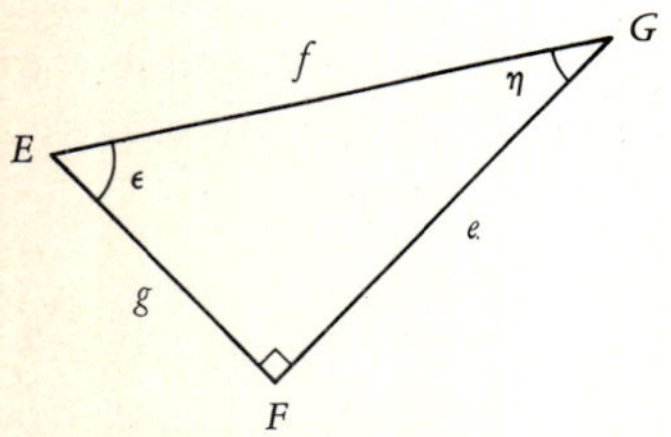

Fig. 5.12 A right triangle with different letters from those used to define the trigonometric functions.

The ratios (a/c etc.) do not change when we change the position of the triangle; thus the position has no bearing upon the special definitions or the values of the trigonometric functions.

	TRIGONOMETRIC FUNCTIONS OF ANGLES IN A RIGHT TRIANGLE		SYMBOLIC EQUATIONS	
Sine:	side opposite the angle / hypotenuse	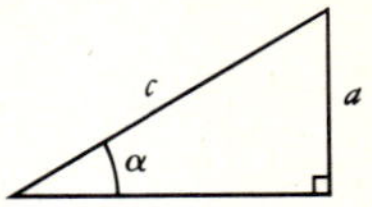	$\sin\alpha = \frac{a}{c}$	(5.9)
Cosine:	side adjacent to the angle / hypotenuse	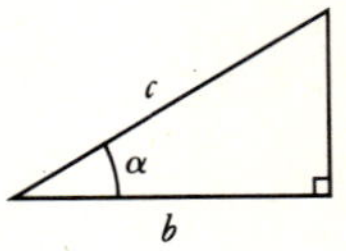	$\cos\alpha = \frac{b}{c}$	(5.10)
Tangent:	side opposite the angle / side adjacent to the angle	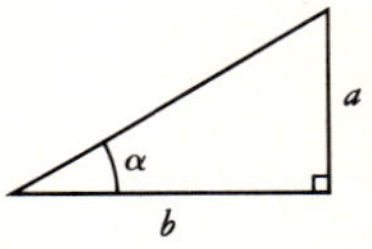	$\tan\alpha = \frac{a}{b}$	(5.11)
Cotangent:	side adjacent to the angle / side opposite the angle	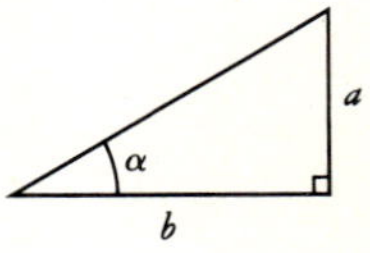	$\cot\alpha = \frac{b}{a}$	(5.12)
Secant:	hypotenuse / side adjacent to the angle	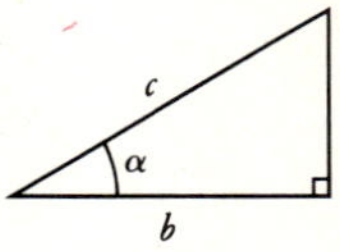	$\sec\alpha = \frac{c}{b}$	(5.13)
Cosecant:	hypotenuse / side opposite the angle	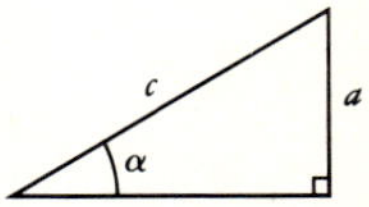	$\csc\alpha = \frac{c}{a}$	(5.14)

Comment: *It is important to learn definitions in terms of the* ***ideas they convey.*** *Thus you should learn the definitions of the trigonometric functions in terms of* ***words;*** *do not just memorize the symbolic equations. The symbols might be different for each problem you encounter, but the ideas are always the same. For example, we can apply the definition of the sine function to right triangle EFG of Fig. 5.12 to obtain sin* ϵ *= side opposite* ϵ*/hypotenuse =* e/f*. If we had memorized* $\sin\alpha = a/c$ *of Eq. (5.9) without learning the principle involved, we might have been lost—there is no* α*, a, or c in Fig. 5.12.*

Example 5.4. Evaluating trigonometric functions

a. **Evaluate** $\sin\alpha$**,** $\cos\alpha$**, and** $\tan\alpha$ **for angle** α **of Fig. 5.13.**

By applying Eqs. (5.9) to (5.11), we have

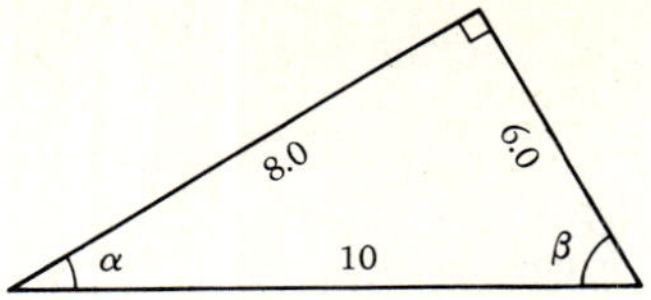

Fig. 5.13 Example 5.4a.

$$\sin \alpha = \frac{\text{side opposite } \alpha}{\text{hypotenuse}} = \frac{6.0}{10} = 0.60$$

$$\cos \alpha = \frac{\text{side adjacent to } \alpha}{\text{hypotenuse}} = \frac{8.0}{10} = 0.80$$

$$\tan \alpha = \frac{\text{side opposite } \alpha}{\text{side adjacent to } \alpha} = \frac{6.0}{8.0} = 0.75$$

b. Evaluate cot x, sec x, and csc x for angle x of Fig. 5.14.

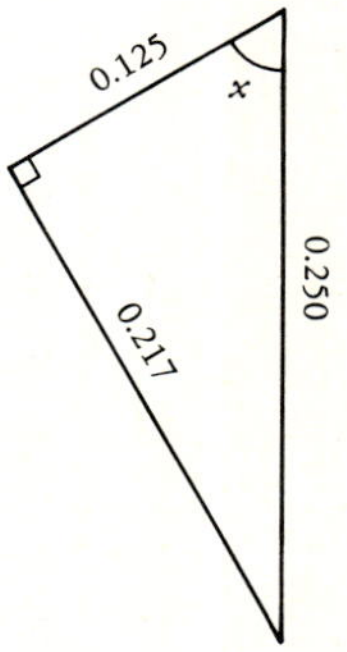

Fig. 5.14 Example 5.4b.

$$\cot x = \frac{\text{side adjacent to } x}{\text{side opposite } x} = \frac{0.125}{0.217} = 0.576$$

$$\sec x = \frac{\text{hypotenuse}}{\text{side adjacent to } x} = \frac{0.250}{0.125} = 2.00$$

$$\csc x = \frac{\text{hypotenuse}}{\text{side opposite } x} = \frac{0.250}{0.217} = 1.15$$

c. Use the results of Example 5.4*a* to determine the values of α and β to the nearest degree. Use Table A5.4 of the Appendix.

Using $\sin \alpha = 0.60$, we find in Table A5.4 that

$$\alpha = 37^\circ \qquad \beta = 90^\circ - \alpha = 53^\circ$$

(The acute angles of a right triangle are complementary.)

d. **Determine to the nearest 0.01 rad the value of x of part (*b*). Use Table A5.4.**

Using $\cot x = 0.576$, we find in Table A5.4 that $x = 1.05$ rad.

EXERCISE 5.4

In Probs. 1 to 6, use Fig. 5.15, and name the trigonometric functions of angles β and ϵ that correspond to each ratio listed.

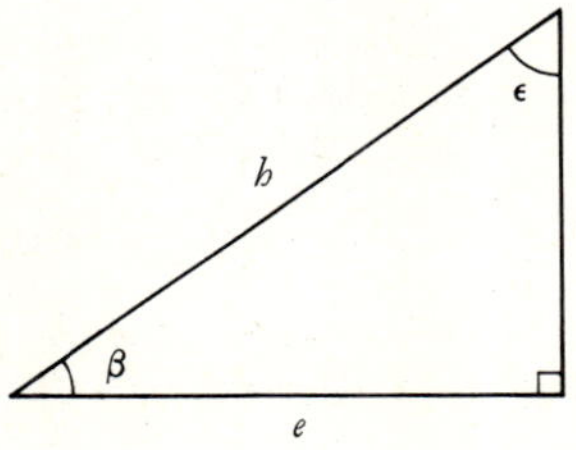

Fig. 5.15 Probs. 1 to 6.

1. e/b	2. h/e	3. b/h
4. b/e	5. e/h	6. h/b

7. In Fig. 5.15, let $h = 50$ and $e = 40$. Determine the values of the six trigonometric functions of angle β.
8. Repeat Prob. 7 for angle ϵ.
9. Determine to the nearest degree the values of β and ϵ of Prob. 7. Use Table A5.4.
10. Determine to the nearest 0.01 rad the values of β and ϵ of Prob. 7. Use Table A5.4.

B. SOLVING RIGHT TRIANGLES

To "solve a triangle" means to find the values of the unknown sides or angles. With the aid of the trigonometric tables, we can solve any right triangle if we know any two sides or any one side and either acute angle. We simply apply an appropriate function from one of Eqs. (5.9) to (5.14) and solve for the unknown part. The following examples show how to solve right triangles.

Example 5.5. **Solve $\triangle ABC$ of Fig. 5.16, given that $\alpha = 27°10'$ and $c = 10.0$.**

$$\beta = 90° - \alpha = 90° - 27°10' = 62°50'$$

$$\frac{a}{c} = \sin \alpha$$

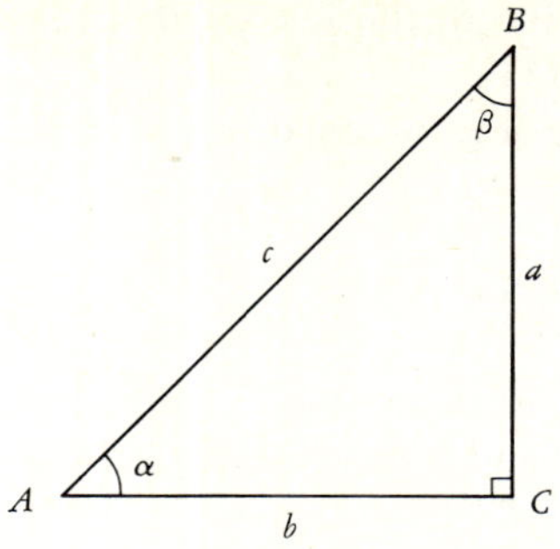

Fig. 5.16 Examples 5.5 and 5.6.

or, multiplying both sides by c, we have

$$\frac{a}{\not c}\not c = c \sin \alpha = 10.0(0.457)$$

$$a = 4.57$$

We use only three digits in sin 27°10′ because c is known to three digits.

$$\frac{b}{c} = \cos \alpha \longrightarrow b = c \cos \alpha = 10.0(0.890) = 8.90$$

$$\boxed{\beta = 62°50' \qquad a = 4.57 \qquad b = 8.90}$$

Example 5.6. Solve $\triangle ABC$ of Fig. 5.16 given that $a = 2{,}137$ and $b = 4{,}575$.

$$\tan \alpha = \frac{a}{b} = \frac{2{,}137}{4{,}575} = 0.4671 \longrightarrow \alpha = 25°02' \qquad \text{(from Table A5.4)}$$

$$\beta = 90° - \alpha = 90° - 25°02' = 64°58'$$

$$\frac{c}{a} = \csc \alpha \longrightarrow c = a \csc \alpha$$

$$= 2{,}137 \csc 25°02' = 2{,}137(2.363) = 5{,}050$$

$$\boxed{\alpha = 25°02' \qquad \beta = 64°58' \qquad c = 5{,}050}$$

Example 5.7. To determine the height of a tree, a surveyor stands 100 ft from it and measures the angle of elevation to the top as 41°20′. If the sighting point is 5 ft above the ground, how tall is the tree?

As shown in Fig. 5.17, the distance b' (from the horizontal line through the sighting point to the treetop) is the side opposite the 41°20′ angle in the right triangle. The side adjacent to the 41°20′ angle is 100 ft long.

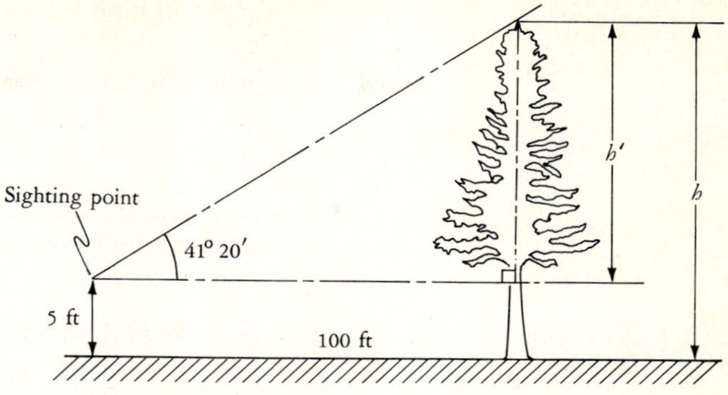

Fig. 5.17 Example 5.7.

Thus we choose the tangent function to solve for h'.

$$\frac{h'}{100} = \tan 41°20' \longrightarrow h' = 100 \tan 41°20' = 100(0.880) = 88.0 \text{ ft}$$

The height of the tree is h' plus the 5-ft elevation from the ground to the sighting point.

$$\boxed{h = h' + 5 = 93.0 \text{ ft}}$$

Example 5.8. An airplane flies at 450 mph on a straight-line course in the direction 72°35′ north of due east. After 2 hr, how far north and how far east is the airplane from its point of departure?

As shown in Fig. 5.18, the distance flown d is the hypotenuse of a right triangle, the northward distance n is the side opposite 72°35′, and the eastward distance e is the side adjacent to 72°35′.

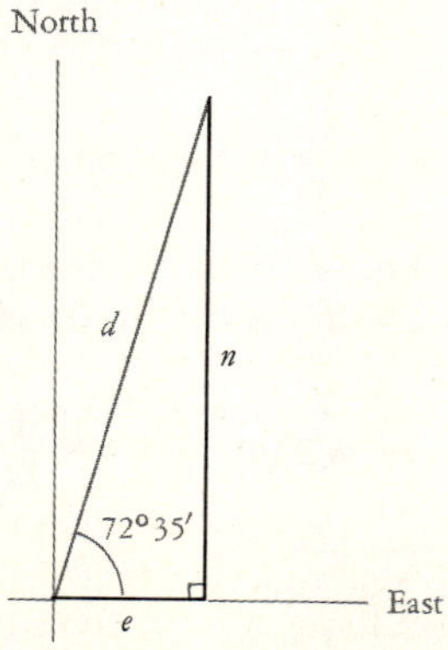

Fig. 5.18 Example 5.8.

$$d = 450 \text{ mph} \times 2 \text{ hr} = 900 \text{ miles}$$

$$n = d \sin 72°35' = 900(0.954) = 859 \text{ miles}$$

$$e = d \cos 72°35' = 900(0.299) = 269 \text{ miles}$$

$n = 859$ miles north	$e = 269$ miles east

Example 5.9. Chicago is at latitude N41.9°, and the radius of the earth is 4,000 miles (to the nearest 10 miles). How fast does Chicago move due to the rotation of the earth?

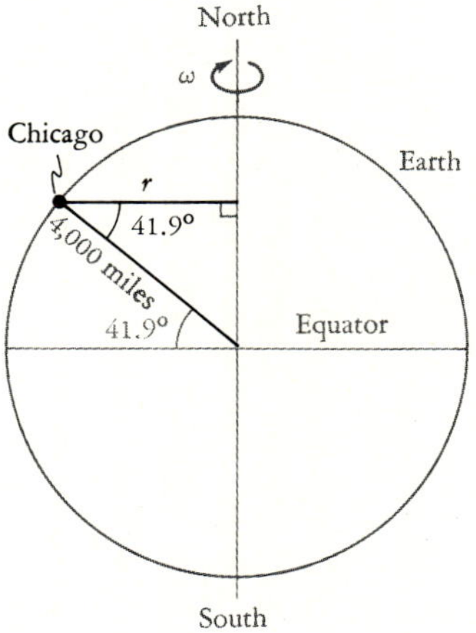

Fig. 5.19 Example 5.9.

As shown in Fig. 5.19, the radius of rotation is r, the perpendicular distance to the axis of rotation. By analyzing the right triangle with 4,000 miles as the hypotenuse, we have

$$\begin{aligned} r &= 4{,}000 \cos 41.9° \\ &= 4{,}000 \cos 41°54' \\ &= 4{,}000(0.7443) \end{aligned}$$

$$r = 2{,}980 \text{ miles}$$

Recall from Eq. (4.11) that $v = r\omega$, where ω is the angular speed in rad per unit time and v is the linear speed.

$$v = r\omega = 2{,}980 \text{ miles} \times \frac{1}{24} \frac{\text{rev}}{\text{hr}} \times 2\pi \frac{\text{rad}}{\text{rev}}$$

$$\boxed{v = 780 \text{ mph}}$$

Comment: 1. *An important part of solving right triangles is choosing the appropriate function. For example, if we know an angle and one side of a right triangle, we should choose the trigonometric function that uses the* ***known side and an unknown side.*** *Preferably, the unknown side should be in the numerator of the function. In Example 5.9, we chose the cosine function because it used the known side of 4,000 miles and the unknown side* ***r****; thus,* ***r*** *was in the numerator of cos 41°54′. Had we chosen the tangent function, we could not have solved the problem because the known side of 4,000 miles would not have been involved.*

A good procedure in working the exercises which follow is to make a list of the definitions, Eqs. (5.9) to (5.14). Refer to your list in order to choose the appropriate function. After a half-dozen or so exercises, practice choosing the function without the list. Eventually you should become so familiar with the trigonometric functions that you can use them instinctively.

2. *An important aspect of solving a stated problem that involves a right triangle is finding a suitable right triangle. An essential part of solving such problems is to draw a diagram based upon the problem statements. Then use your imagination and ingenuity to construct a suitable right triangle. Examples 5.7, 5.8, and 5.9 illustrate this process.*

EXERCISE 5.5

In Probs. 1 to 9, use Fig. 5.20. Assume that the listed parts of triangle *ABC* are known and solve for the remaining parts.

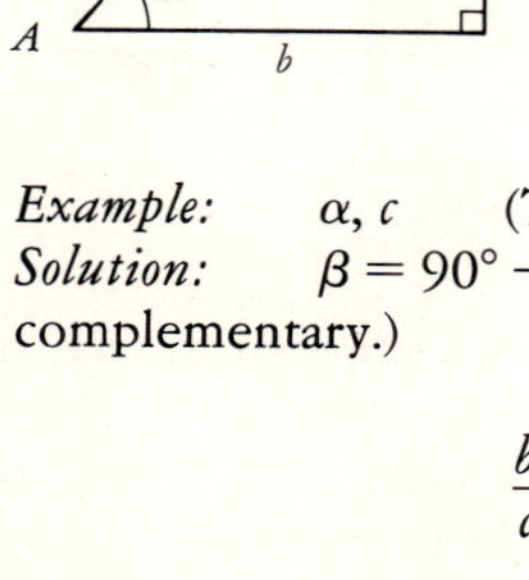

Fig. 5.20 Probs. 1 to 18.

Example: α, c (These are the known parts.)
Solution: $\beta = 90° - \alpha$ (Acute angles of a right triangle are complementary.)

$$\frac{b}{c} = \cos\alpha \longrightarrow b = c\cos\alpha$$

$$\frac{a}{c} = \sin\alpha \longrightarrow a = c\sin\alpha$$

1. β, c	2. α, b	3. a, c	4. β, a
5. α, c	6. b, c	7. β, b	8. α, a
9. a, b			

In Probs. 10 to 18, use Fig. 5.20; use the given values in each exercise to solve for the unknown parts.

10. $\alpha = 39°, c = 12$
11. $\beta = 73°, c = 45$
12. $a = 400, c = 500$
13. $\alpha = 11°16', a = 2.310$
14. $\beta = 67°30', a = 120$
15. $b = 500, c = 1{,}300$
16. $\alpha = 0°22', b = 1{,}000{,}000$
17. $\beta = 89°45', b = 123$
18. $a = 0.001375, b = 0.002750$

19. How tall is a tree if its shadow is 52 ft long when the sun is 32° above the horizon?
20. How high up a vertical wall will a 20-ft ladder reach if the angle between the ladder and the wall is 17°?
21. A cherry picker is positioned as shown in Fig. 5.21. How high is the basket?

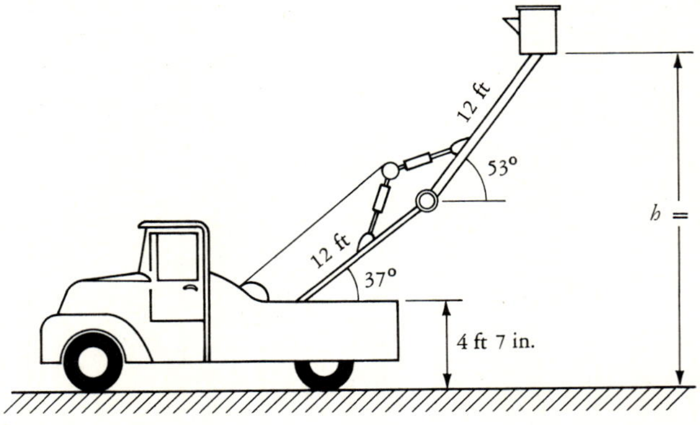

Fig. 5.21 Prob. 21.

22. Two buildings are face to face and 30 ft apart. If the distance between the floors is 10 ft, how long would a rope need to be in order to reach from a third-floor window to the fifth-floor window directly opposite?
23. The corner of a cast machine part is rounded with a circular arc of 3/4-in. radius, as shown in Fig. 5.22. What is the distance

x if the bend angle is 30°? (*Hint:* By analyzing angles of right triangles, determine θ; then solve the right triangle with θ and r.)

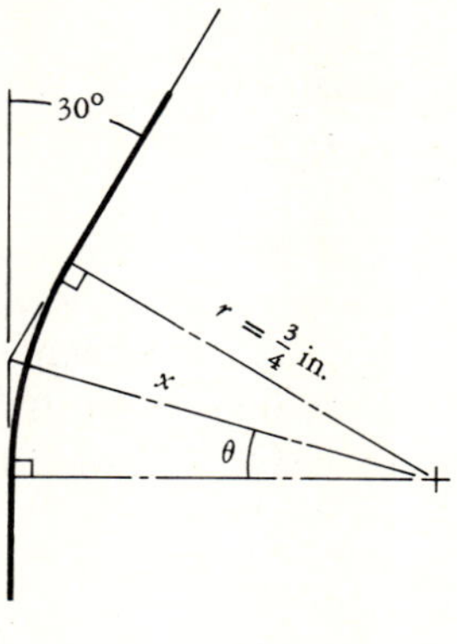

Fig. 5.22 Prob. 23.

24. An airplane is supposed to travel due west, but because of an error it travels 2°20′ north of west. After the plane has flown 225 miles, how far off course to the north is it?
25. The latitude of St. Petersburg, Florida, is approximately N27°50′. How fast does St. Petersburg travel due to the rotation of the earth (radius = 4,000 miles)?
26. In the mechanism shown in Fig. 5.23, what is the distance x?

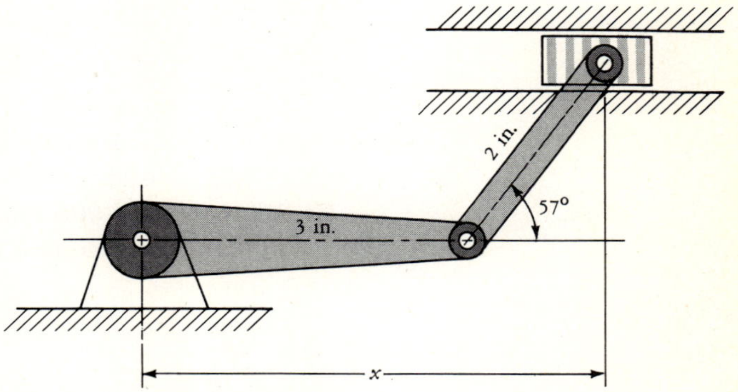

Fig. 5.23 Prob. 26.

27. In an electronic circuit with an inductive reactance X_L and a resistance R, the impedance Z is the hypotenuse of a right triangle with R and X_L being the legs. See Fig. 5.24.
    a. What is the phase angle θ if $X_L = 4.32$ ohms and $R = 19.5$ ohms?
    b. If $Z = 3.26$ ohms and $\theta = 37°$, what are R and X_L?
28. How high is the vertex of an A-frame if each leg is 10 ft long

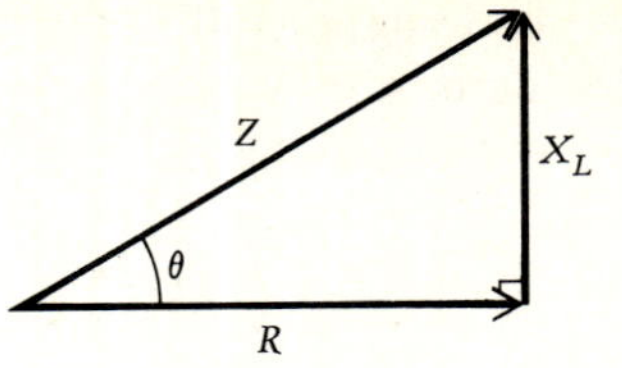

Fig. 5.24 Prob. 27.

and the angle between the legs is 20°? (*Hint:* The A-frame is an isosceles triangle. Draw the altitude on the base.)

29. As shown in Fig. 5.25, the diameter of the head of a 1/2-in. (diameter) flathead machine screw is 7/8 in. If the angle between the "sides" of the head is 82°, what is h, the height of the head?

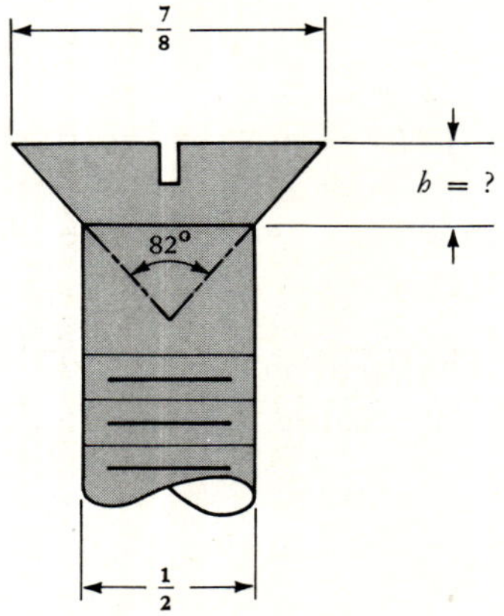

Fig. 5.25 Prob. 29. A flathead screw.

30. Suppose that an astronaut is in earth orbit at an altitude of 100 miles. What distance s over the earth's surface can he see? (Earth's radius = 4,000 miles.) How far, x, is the farthest point on the earth's surface that the astronaut can see? See Fig. 5.26.

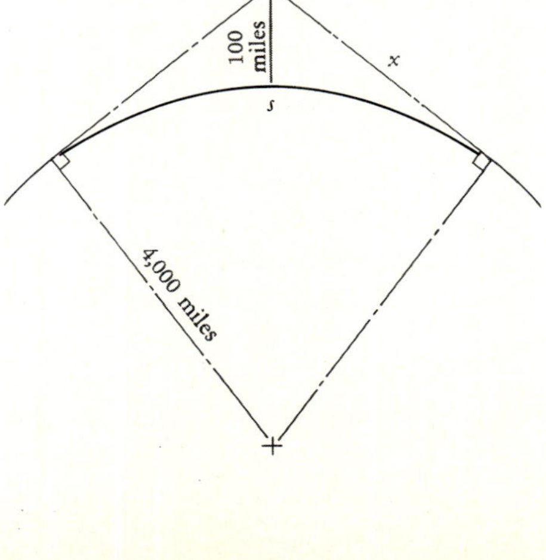

Fig. 5.26 Prob. 30. An astronaut 100 miles above the surface of the earth.

31. After a certain airplane climbs 20 ft off the runway, it can maintain a 27° climbing angle. If it is 100 ft from the end of the runway at this point, by how many feet will it clear a treetop that is 130 ft above and 150 ft beyond the end of the runway? See Fig. 5.27.

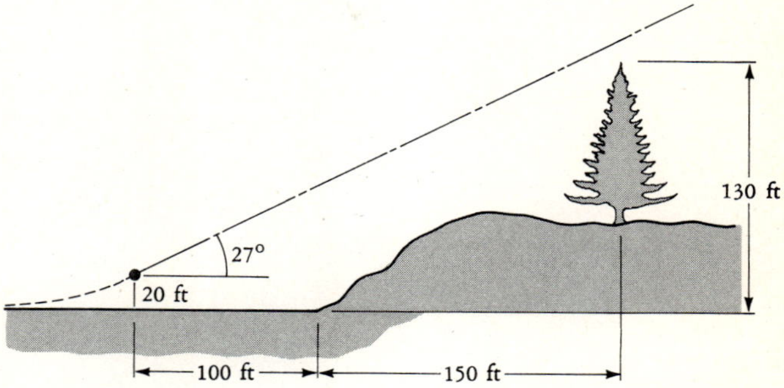

Fig. 5.27 Prob. 31.

32. A baseball "diamond" is a square, with 90 ft between the bases. The pitcher's mound is 60 ft 6 in. from home plate. What is the distance from the pitcher's mound to first base? (*Hint:* Draw lines from the pitcher's mound to home plate and first base; draw also a perpendicular to the first base line. Solve the two right triangles.)

33. An object such as the block shown in Fig. 5.28 becomes unstable and will fall over in the position where a vertical line through the center of gravity passes outside the pivot point. Through how many degrees can this block be tilted before it will become unstable?

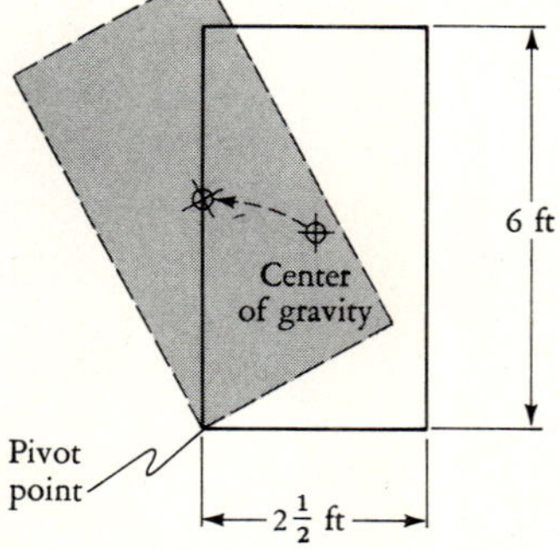

Fig. 5.28 Prob. 33. The block is tilted to an unstable position.

34. A modernistic fish tank is a truncated horizontal cylinder as shown in Fig. 5.29. If the tank is filled to the center as shown, how many degrees can it be tilted before water will spill?

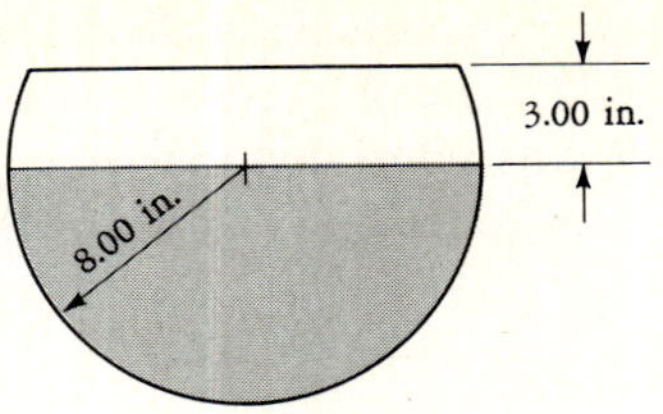

Fig. 5.29. Prob. 34. A cylindrical fish bowl.

35. The distance across the flats of a 1-in. hex-head cap screw is 1.875 in. See Fig. 5.30. The head of the screw is to be recessed into a cylindrical hole. What is the minimum diameter d of the hole needed for a wrench socket with a 0.062-in.-thick wall to be used?

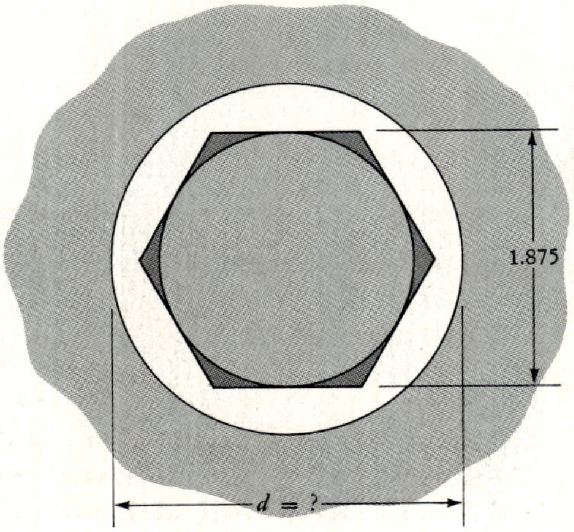

Fig. 5.30 Prob. 35. A hex-head cap screw.

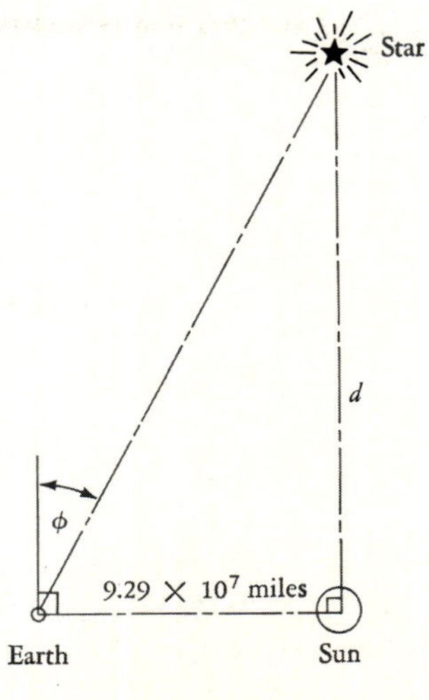

Fig. 5.31 Prob. 36. The parallax angle ϕ and distance d to a star.

36. A method that astronomers use to determine distances to nearby stars is illustrated in Fig. 5.31. When the earth is in just the right position, the parallax angle ϕ of a star is measured. Then, based upon the astronomical unit (the distance from the earth to the sun, 92.9 million miles), the distance d is determined by solving the right triangle. How far is Alpha Centauri, the nearest visible star, if the parallax angle is $0°00'0.741''$? (*Note:* For small angles such as this, the value of the sine function or the tangent function equals the measure of the angle in radians.)

What is this distance in light-years? (A *light-year* is the distance that a light beam travels in 1 year. The speed of light is 186,000 miles/sec.)

INTRODUCTION TO VECTOR QUANTITIES AND VECTORS

CHAPTER 6

Before we try to understand the general definition of a vector quantity, let us look at a specific example. Then we can better appreciate the general definition that follows.

6.1 BASIC IDEAS ABOUT VECTOR QUANTITIES

Suppose that the neighborhood kids are playing baseball in front of your house. One boy gets a solid hit, and crash! The ball smashes your new picture window! Oh, if only he hadn't hit the ball so *hard,* or if only it had traveled in a different *direction!*

The words "hard" and "direction" are keys to the concept of a vector quantity, for all vector quantities have a magnitude and a direction. The "hardness," meaning the *magnitude,* of the force applied to the ball was partly to blame for your broken window. Likewise, the *direction* of the ball's trajectory was partly to blame. Had it traveled in a slightly different direction, it would have missed the window.

This example illustrates that the effect of a force depends upon its magnitude (such as the number of pounds) and its direction in space. Any physical quantity that, like force, has a magnitude and a direction is called a *vector quantity.* Thus we are led to this definition:

VECTOR QUANTITY	Any physical quantity that has magnitude and direction

In contrast to vector quantities, those quantities that can be described with a magnitude only are called *scalar quantities.* For example, we specify temperature by measuring a number on a scale called a thermometer, and we describe time by referring to a number on a scale called a clock. Such a number tells us the magnitude (or size) of the entity. No direction is involved. Here is a list of some common scalar quantities.

Scalar Quantities:

Time	Speed
Temperature	Area
Humidity	Resistance
Distance	Voltage

Notice that none of these entities is dependent upon a direction in space. Your watch measures the same time no matter which way you point it; a thermometer indicates the same temperature whether you turn it up, down, or sideways; and a resistor maintains the same resistance whether it points north, south, east, or west.

As we have pointed out, vector quantities differ from scalar quantities in that they have *direction* as well as magnitude. Here is a list of some common vector quantities.

Vector Quantities:

Force	Velocity
Displacement	Acceleration

In the next few paragraphs you will find a definition and a brief description of each of these vector quantities.

A. FORCE

Force is defined as a directed push or pull on one body by another body. A force can cause a body to start, change direction, or stop its motion. As we have previously mentioned, the effect of a force depends both upon its magnitude and upon the direction in which it acts. Try to think of some actions that illustrate the *directional* characteristic of force. As a start, think about a billiard ball being struck by a cue stick or a space vehicle being lifted off its launch pad by the force of the booster engines.

B. DISPLACEMENT

The *displacement* between two points is the length and direction of the *straight line* between them. To understand displacement better, let's compare it to distance. The *distance* between two points is the length of any path that connects them, whether the path is a single straight line or a longer line with angles and curves. The displacement between two points, however, is always the length and direction of the *straight line* between them. Notice in Fig. 6.1 that three paths connect points A and B. The first path consists of two straight lines, one 3 miles long, due east, followed by a line 4 miles long, due north.

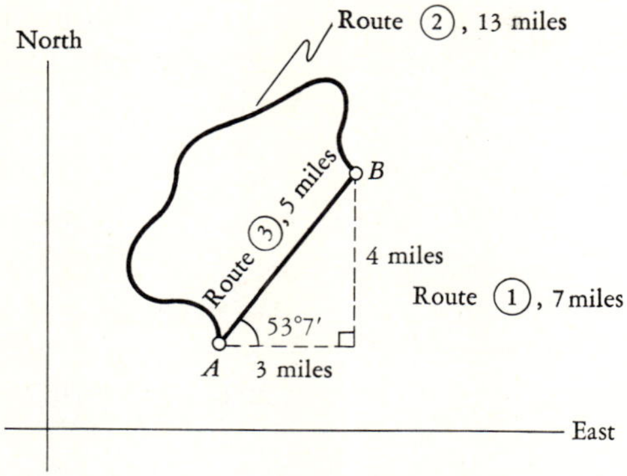

Fig. 6.1 Three different paths between points A and B.

The second route is an irregular, curved path 13 miles long. Thus we see that the *distance* between the points depends upon the path that connects them.

The third route is the *one and only one single straight line* that connects points A and B. The distance for this path is (using the Pythagorean formula)

$$\sqrt{3^2 + 4^2} = 5 \text{ miles}$$

It is this distance directed at 53°7′ north of east that we call the displacement from point A to point B. (The angle of 53°7′ comes about because $\tan 53°7' = 1.333 = 4/3$.) Thus displacement is sometimes called a *directed distance.* The direction of the line is just as important as its length. If we start at point A and walk 5 miles in any direction other than 53°7′, we will not reach point B. Displacement is a vector quantity, a directed distance.

The *distance* between two points depends upon the path, but *displacement* is independent of path. It is the same for every path, for no matter which path of Fig. 6.1 we take, we shall displace ourselves 5 miles in the direction 53°7′ north of east. When we talk about displacement, we are concerned only with the straight-line location of a terminal point relative to a starting point; we are not concerned about how we actually travel from one point to the other.

C. VELOCITY

Velocity is something like *speed.* You may be familiar with the idea that average speed is distance divided by time. For example, if you drive a car 120 miles in 3 hr, the average speed is 120 miles ÷ 3 hr = 40 mph. Likewise, *average velocity* is *displacement* divided by time. Thus velocity is straight-line speed in a specific direction—the direction of the displacement. Velocity is a vector quantity because it has magnitude (speed) and direction.

Suppose we travel route 1 of Fig. 6.1 in 2 hr. Then our average speed is 7 miles ÷ 2 hr = 3.5 mph. But our average velocity is displacement ÷ time = 5 miles ÷ 2 hr = 2.5 mph at 53°7′ north of east.

D. ACCELERATION

Suppose we travel from A toward B on route 3 of Fig. 6.1. Further, suppose that as we leave point A our velocity is 20 ft/sec and that we increase it until, 5 sec later, it is 80 ft/sec. On the average, how much has our velocity increased during each second?

$$\frac{(80-20)\text{ ft/sec}}{5\text{ sec}}=\frac{60\text{ ft/sec}}{5\text{ sec}}=\frac{12\text{ ft/sec}}{\text{sec}}\quad\text{at } 53^\circ 7' \text{ north of east}$$

The quantity we calculated is called *average acceleration.* Average acceleration is the *change* of velocity divided by time. It is a vector quantity having a direction equal to the direction of the change of velocity. The magnitude 12 (ft/sec)/sec means that on the average our velocity increased by 12 ft/sec for each second; thus the units are *feet per second, per second,* which is written (ft/sec)/sec or ft/sec^2.

If our velocity had decreased from 20 ft/sec to 5 ft/sec in 5 sec, then the average acceleration would have been

$$\frac{(5-20)\text{ ft/sec}}{5\text{ sec}}=\frac{-15\text{ ft/sec}}{5\text{ sec}}=-3\text{ ft/sec}^2$$

The negative sign indicates that the direction of acceleration is opposite to that of velocity; the acceleration is then called a *negative acceleration* or a *deceleration.*

6.2 GRAPHICAL REPRESENTATION OF VECTOR QUANTITIES AND GRAPHICAL VECTOR ALGEBRA

To represent a vector quantity on a drawing, it is customary to use an arrow. The length of the arrow represents the magnitude of the quantity, and the direction of the arrow is the direction of the vector quantity. Such an arrow is called a *vector.*

Letters used to name or represent vectors are distinguished

from ordinary algebraic (scalar) symbols by boldface type.* Thus **B** represents a vector, whereas *B* stands for a scalar quantity. We can use *B* to indicate the magnitude of vector **B**. In handwriting, it is customary to indicate a vector by placing a small arrow over its letter; thus $\vec{B}$ also means vector **B**.

A. VECTOR ADDITION

Because vectors involve direction, we cannot always add them by merely adding their magnitudes as we do for scalar quantities. The addition process must take the directions of the vectors into account. Therefore we add two vectors as follows (see Fig. 6.2):

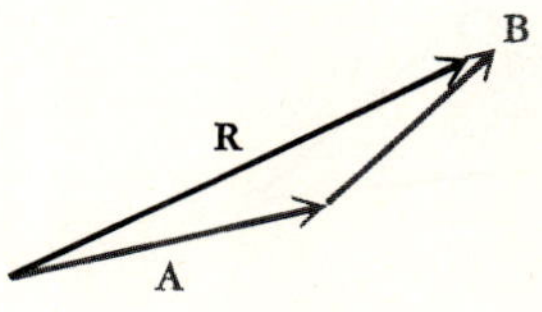

R = **A** + **B**

Fig. 6.2 Addition of two vectors—vector **R** *is the sum of vectors* **A** *and* **B**.

VECTOR ADDITION	To add two vectors, place the tail of the second arrow at the head of the first arrow. Their sum is the vector drawn from the tail of the first arrow to the head of the second arrow.

Notice that vector **B** is connected head to tail with vector **A**. The sum **A** + **B** is called the *resultant,* and is vector **R**.

This process of vector addition can be extended to three or more vectors. We simply place the vectors in a chainlike, head-to-tail arrangement. The resultant (vector sum) is the vector that reaches from the tail of the first vector to the head of the last vector. Figure 6.3 shows such an addition for the sum of four vectors. Because in this arrangement the vectors and their resultant form a closed polygon, this method is called the *polygon method* for adding vectors. The resultant is often called "the vector that closes the polygon."

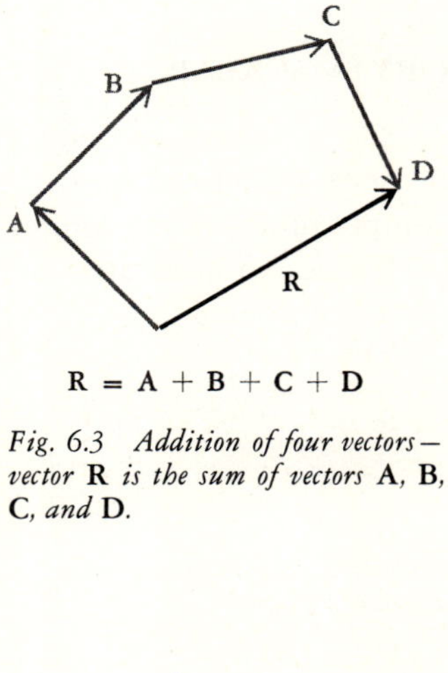

R = **A** + **B** + **C** + **D**

Fig. 6.3 Addition of four vectors—vector **R** *is the sum of vectors* **A**, **B**, **C**, *and* **D**.

Try rearranging the vectors in Figs. 6.2 and 6.3, and demonstrate that the order in which the vectors are added does not affect the resultant. In other words, vector addition is commutative: **A** + **B** = **B** + **A**.

B. VECTOR SUBTRACTION

We subtract one vector from another by adding the negative of the vector to be subtracted. In symbolic form we indicate this idea as follows:

* In earlier chapters, since examples were printed in **boldface type**, scalar symbols were also boldface. But in the examples of this chapter, the scalar symbols will be lightface and the vector symbols boldface (as they ordinarily appear). In the following chapters, if vectors are not involved, scalar symbols will again be boldface in the examples.

$$\mathbf{A} - \mathbf{B} = \mathbf{A} + (-\mathbf{B})$$

Thus we can subtract vectors by using exactly the same procedure as for addition. The *negative* of a *vector* is a vector of the same magnitude (length) that acts in the opposite direction. In Fig. 6.4*a*, $-\mathbf{U}$ is the negative of **U**, and **U** is the negative of $-\mathbf{U}$. Figure 6.4*b* illustrates vector subtraction.

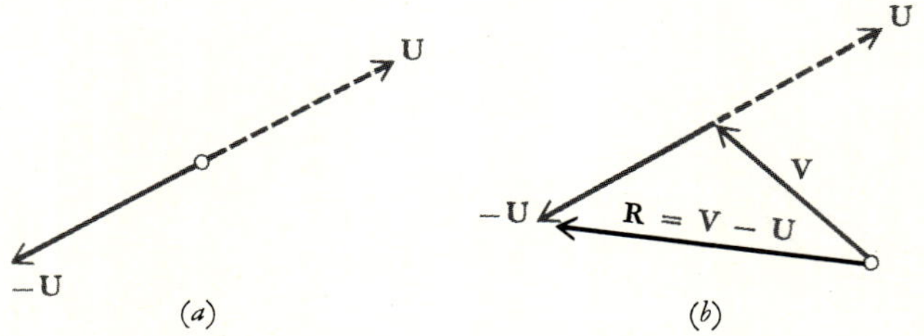

Fig. 6.4 Negative vectors and vector subtraction. (a) Vector $-\mathbf{U}$ *is the negative of* **U**, *and* **U** *is the negative of* $-\mathbf{U}$. *If two vectors are equal in length but opposite in direction, then each one is the negative of the other. (b) Subtraction of a vector by adding its negative. The resultant* **R** *is the sum of* **V** *and* $-\mathbf{U}$. *Thus* $\mathbf{R} = \mathbf{V} + (-\mathbf{U}) = \mathbf{V} - \mathbf{U}$.

C. MULTIPLICATION OF A VECTOR BY A SCALAR

An expression such as 3**B** signifies a vector that is 3 times as long as vector **B** and has the same direction as vector **B**. In general, to multiply a vector by any scalar number, we simply multiply the length of the vector. The result is another vector. Figure 6.5 illustrates the results of several vector multiplications.

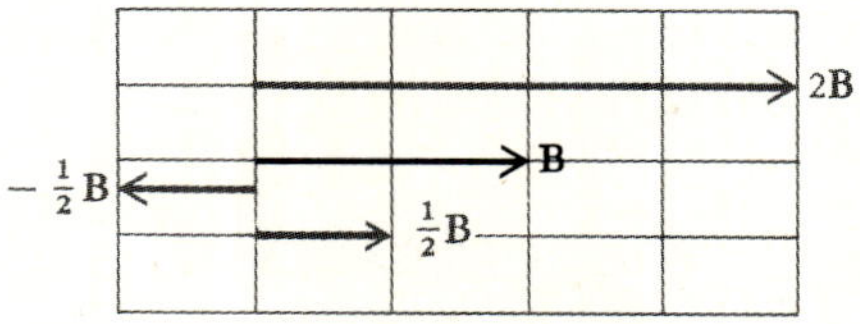

Fig. 6.5 Multiplication of a vector by a scalar.

Example 6.1. Find a resultant equal to $\mathbf{A} - 2\mathbf{B} + (1/3)\mathbf{C}$ for vectors **A**, **B**, and **C** of Fig. 6.6*a*.

Solution. The solution is vector **R** of Fig. 6.6*b*. Vectors **A**, $-2\mathbf{B}$, and $(1/3)\mathbf{C}$ are combined head to tail as shown. We used the grid lines to ensure the correct length and direction of each vector. For example, the number of horizontal grid lines between head and tail of $(1/3)\mathbf{C}$ is exactly 1/3 the number for **C**, and the number of vertical lines for $(1/3)\mathbf{C}$ is 1/3 the number for **C**.

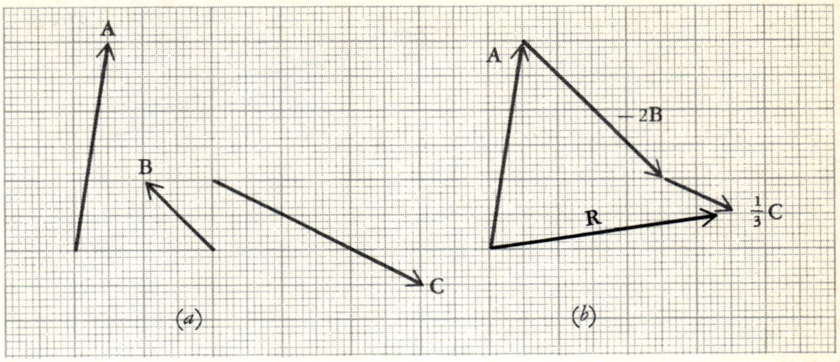

Fig. 6.6 Example 6.1. (a) The given vectors **A**, **B**, *and* **C**; *(b) the sum* $\mathbf{A} - 2\mathbf{B} + \frac{1}{3}\mathbf{C}$.

Example 6.2. An airplane flies 300 mph for 30 min in the direction 10° south of east, then turns 15° more toward the south, and flies 450 mph for 10 min. How far and in what direction is the plane from its starting point? What was the average velocity? If the plane required 30 sec to turn and make the change of speed, what was the average acceleration?

Solution. Figure 6.7 shows the solution. After drawing the two displacement vectors $\mathbf{D}_1 = 150$ miles $[(1/2)\ \text{hr} \times 300\ \text{mph}]$ and

$$\mathbf{D}_2 = 75 \text{ miles } [(1/6)\ \text{hr} \times 450 \text{ mph}]$$

with scale and protractor, we scale the resultant, $\mathbf{R} = \mathbf{D}_1 + \mathbf{D}_2$.

$$\mathbf{R} = 223 \text{ miles @ } 15° \text{ south of east}$$

The average velocity was

$$\mathbf{V} = \frac{\mathbf{R}}{t} = \frac{223 \text{ miles}}{(30 + 10) \text{ min}} = 5.58 \frac{\text{miles}}{\text{min}} \longrightarrow 335 \frac{\text{miles}}{\text{hr}} \text{ @ } 15° \text{ south of east}$$

The average acceleration was, from Fig. 6.7*b*,

$$\mathbf{A} = \frac{\mathbf{V}_2 - \mathbf{V}_1}{t} = \frac{177 \text{ mph}}{30 \text{ sec}} = 5.9 \frac{\text{mph}}{\text{sec}} \longrightarrow 21{,}240 \frac{\text{miles}}{\text{hr}^2} \text{ @ } 51° \text{ south of east}$$

Note that the acceleration involves the *change* of velocity, which is the vector difference $\mathbf{V}_2 - \mathbf{V}_1$. This change involves a change of direction as well as a change of speed. Figure 6.7*b* shows this vector difference, found by subtracting $\mathbf{V}_1$ from $\mathbf{V}_2$.

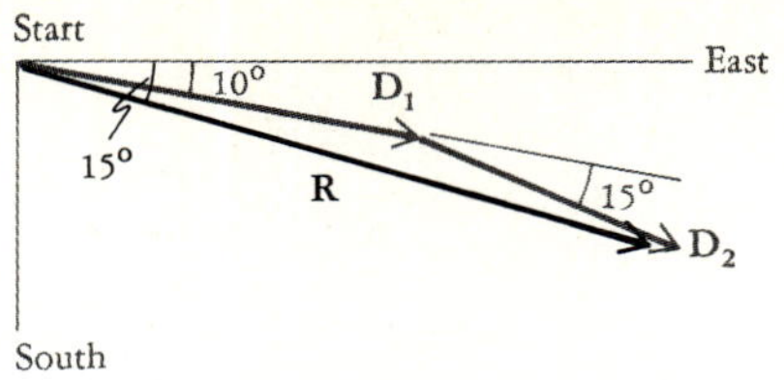

(*a*) Displacements

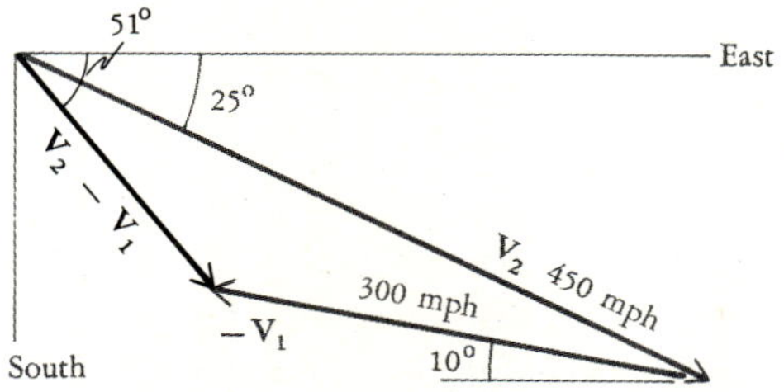

(*b*) Velocities

Fig. 6.7 Example 6.2.

EXERCISE 6.1

Use vectors **A**, **B**, **C**, **D**, and **E** below for Probs. 1 to 10. Use the grid lines to determine the magnitude and direction of each vector.

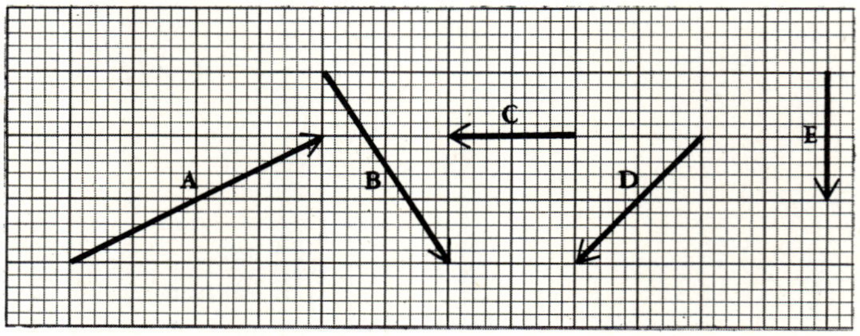

In Probs. 1 to 10 find the resultant for each vector sum.

1. **A** + **D**
2. **A** + **B**
3. **A** + **E** − 2**C**
4. **A** − 2**C** + **E**
5. **D** + **A** + **C** (What is the magnitude of the resultant?)
6. **B** + **D** − (1/2)**A**
7. 2**B** − **C** + **A** − 3**E**

8. **E** + **C** − **D** − **B** (How is the resultant related to one of the five given vectors?)
9. **A** − **B** + **C** − (3/2)**D** + **E**
10. **A** + **B** + (5/2)**C** + **D** + **E**

D. VECTOR COMPONENTS

Another way to add two vectors is to place them tail to tail as shown in Fig. 6.8. Then draw a parallelogram with the two vectors as adjacent sides. The resultant is the diagonal between the given vectors. This addition is called the *parallelogram method.* As you can see, it is hardly different from the previous method, but it works for only two vectors at a time.

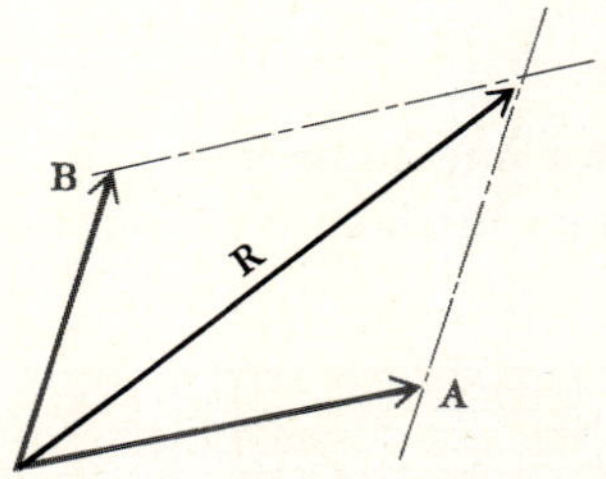

Fig. 6.8 The parallelogram method of vector addition: vector **R** *is the sum of vectors* **A** *and* **B**.

Now we shall examine the process of reversing vector addition. Suppose that we have a single resultant vector such as **R** of Fig. 6.8 and we wish to replace it by two vectors such as **A** and **B**, so that the sum of those two vectors will equal the given vector. The two vectors whose sum is the given vector are called *components* of that vector. The process of finding them is called *resolution.* Thus we want to resolve a vector into its components.

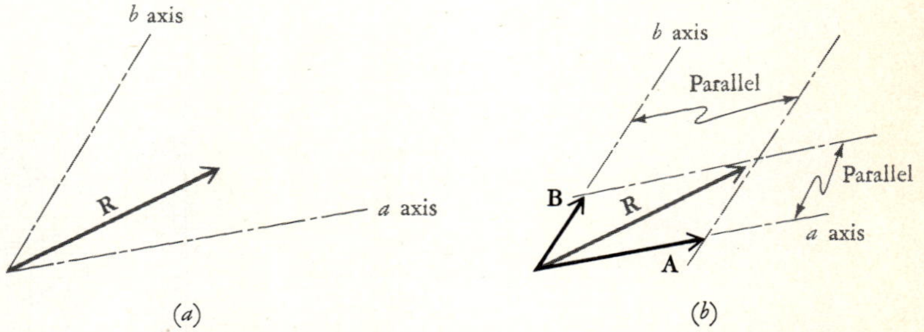

Fig. 6.9 Resolution of a vector into components. (a) Vector **R** *is to be resolved into components along the a and b axes. (b) Component* **A** *is found by constructing a line through the head of* **R** *and parallel to the b axis. The point where this line intersects the a axis is the head of* **A**. *Component* **B** *is found in a similar manner.*

Cover Fig. 6.9*b*, and study Fig. 6.9*a* for a minute; see if you can devise a basic procedure for resolving vector **R** into a pair of components along the *a* and *b* axes. Then look at Fig. 6.9*b* and the definition below.

Your procedure should go something like this:

VECTOR RESOLUTION	From the head of the given vector draw lines parallel to the given axes. The points where these lines intersect the axes determine the components of the given vector.

The procedure illustrated in Fig. 6.9*b* produces a unique solution; that is, there is one and only one pair of components along the given axes that will add to produce the given vector. In general there will be a different pair of components for each different vector and for each different pair of axes. There is no unique solution for components along three or more axes in the same plane.

EXERCISE 6.1 (*Continued*)

In Probs. 11 to 16, resolve each given vector into components parallel to the given axes. Use the parallelogram method as shown in Fig. 6.9*b*.

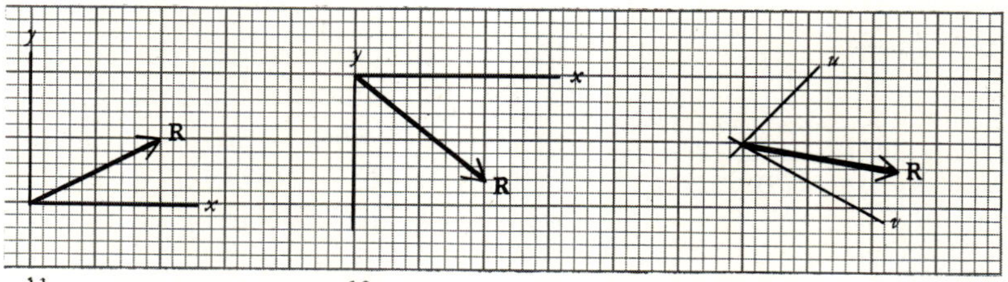

11. 12. 13.

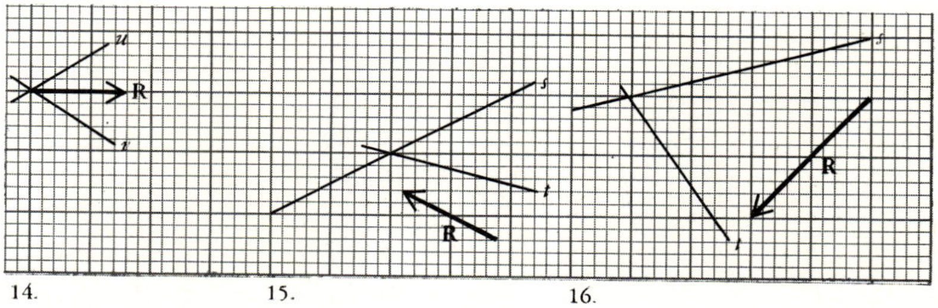

14. 15. 16.

Solve Probs. 17 to 20 graphically by using vector diagrams. Measure your diagrams with scale and protractor to obtain answers.

17. An airplane flies 200 miles in the direction 30° north of east, then turns 90° counterclockwise, flies 150 miles, and lands.
    a. How far and in what direction is the plane from its starting point?
    b. If the plane took 4 hr to make the trip, what was its average velocity?
    c. What was the direction for the second leg of the flight?
18. A 180-lb crate is pushed up the ramp, as shown in Fig. 6.10. The boy pushes horizontally with a force of 70 lb. The reaction of the ramp against the bottom of the crate is 191 lb. Determine the magnitude and direction of the resultant force on the crate. Neglect friction.

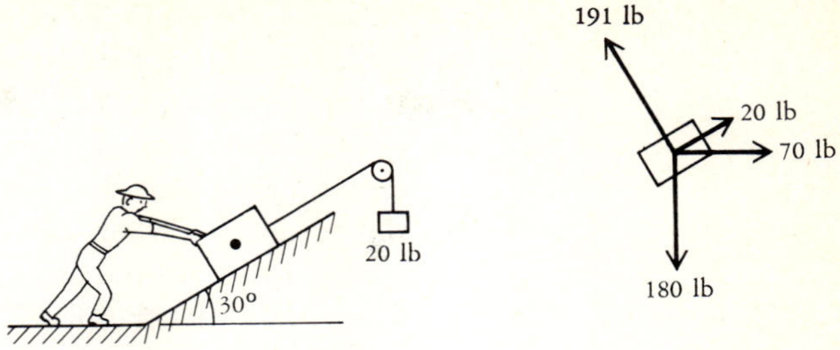

Fig. 6.10 Prob. 18.

19. A tourist is flying his own plane south for the winter. He becomes lost over north Florida, but uses his radio to "vector in" on the signal from the beacon at Tallahassee, and he reads 289°. (In aircraft navigation north is 0°00′, and all angles are read clockwise from north.) Then he picks up the Jacksonville beacon at 28°. See Fig. 6.11.
    a. If the plane is headed due south, what course correction must the pilot make, and on what vector must he fly, in order to fly straight to St. Petersburg? Assume no wind and assume no change of position between readings or during course correction. [*Hint:* To locate the plane, draw a 109° vector (289° − 180°) *from* Tallahassee and a 208° vector (180° + 28°) *from* Jacksonville. Find the point where these two vectors intersect.]
    b. If the plane is flying at 120 mph, how long will it take to reach St. Petersburg?

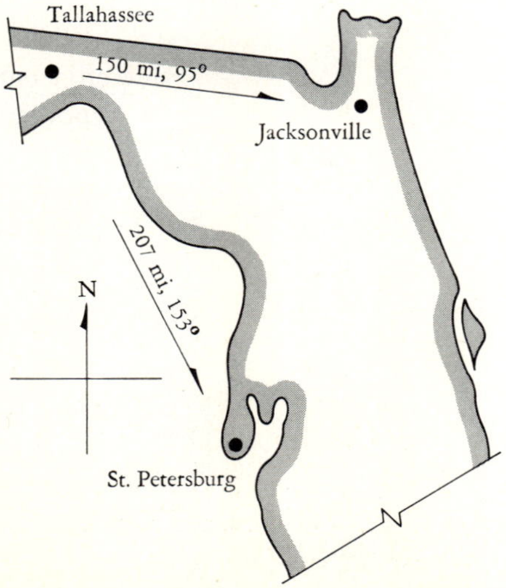

Fig. 6.11 Prob. 19.

20. Refer to Prob. 19. Suppose it takes the pilot 5 min from the time he tunes in the Tallahassee signal at 289° until the time he tunes in Jacksonville. Jacksonville now vectors in at 22½° due to the plane's change of position during the 5-min interval. Further suppose that the pilot needs an additional 5 min to figure where he was when he read Tallahassee. Now allow a 5-min countdown until the pilot corrects his course for St. Petersburg. Remember that the plane continues due south at 120 mph during these three 5-min intervals. What course should the pilot take when he changes course for St. Petersburg?

6.3 ADDITION OF VECTORS BY RIGHT-ANGLE COMPONENTS

We shall now develop a method for adding vectors mathematically rather than graphically. To do this we replace each vector by its right-angle components along two perpendicular axes such as the x and y axes of a rectangular coordinate system.

A. VECTOR COMPONENTS AND TRIGONOMETRIC FUNCTIONS

Mathematical relationships between a vector and its cartesian components are easily described by using trigonometric functions. Figure 6.12 shows an arbitrary vector $\mathbf{A}$ and its components $\mathbf{A}_x$ and $\mathbf{A}_y$ along the x and y axes, respectively. Notice that $\mathbf{A}$ is the hypotenuse of a right triangle and that $\mathbf{A}_x$ and $\mathbf{A}_y$ form the other two sides. From our previous study of right triangles we see that

$$\cos \theta_A = \frac{\text{side adjacent to } \theta_A}{\text{hypotenuse}} = \frac{A_x}{A}$$

$$\sin \theta_A = \frac{\text{side opposite } \theta_A}{\text{hypotenuse}} = \frac{A_y}{A}$$

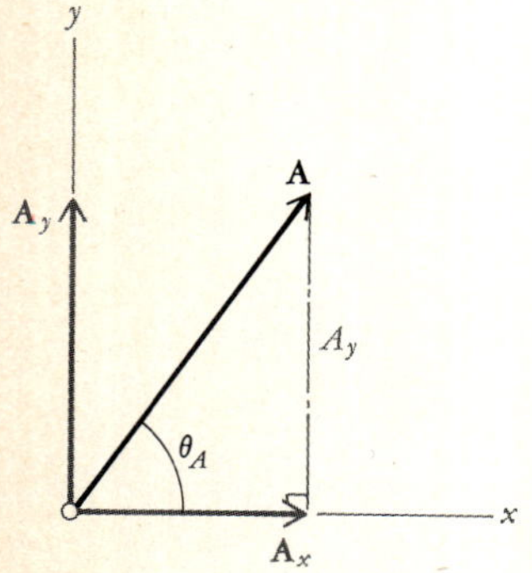

Fig. 6.12 Vector **A** *and its cartesian components,* $\mathbf{A}_x$ *and* $\mathbf{A}_y$.

Solving these equations for A_x and A_y, we have

$$\boxed{\begin{aligned} A_x &= A \cos \theta_A \\ A_y &= A \sin \theta_A \end{aligned}} \tag{6.1}$$

where A is the magnitude of $\mathbf{A}$
A_x is the magnitude of $\mathbf{A}_x$
A_y is the magnitude of $\mathbf{A}_y$

With these two equations, we can compute the magnitudes of the cartesian components of any vector. These are scalar equations; we cannot use vector symbols $\mathbf{A}$, $\mathbf{A}_x$, and $\mathbf{A}_y$ because the directions of $\mathbf{A}$, $\mathbf{A}_x$, and $\mathbf{A}_y$ are not the same. However, the x and y subscripts indicate the directions of A_x and A_y, respectively.

Example 6.3. Vector components

Suppose that you walk 5 miles in the direction 53°7′ north of east. How far east and how far north have you moved?

Let vector **A** of Fig. 6.12 represent your 5-mile displacement. Then $\mathbf{A}_x$ is your displacement toward the east and $\mathbf{A}_y$ is the northward displacement. Using Eq. (6.1), we obtain

$$A_x = A \cos \theta_A = 5 \cos 53°7' = 5(0.6) = 3 \text{ miles east}$$

$$A_y = A \sin \theta_A = 5 \sin 53°7' = 5(0.8) = 4 \text{ miles north}$$

As far as displacement from the starting point is concerned, the pair of vectors $\mathbf{A}_x$ and $\mathbf{A}_y$ is entirely equivalent to the resultant **A**. If you had actually walked 3 miles east, then turned and walked 4 miles north, you would have reached the same point as when you walked 5 miles along the straight line at 53°7′ north of east. The resultant is equivalent to, and can be replaced by, its components. This is true regardless of what the vectors represent—displacements, forces, velocities, etc. Thus we are justified in writing the **vector** equation $\mathbf{A} = \mathbf{A}_x + \mathbf{A}_y$. In this sense, 5 equals 3 plus 4. Note, however, that although your *displacements* are the same for both routes, the *distances* along the two routes are considerably different. This example illustrates the difference between the vector quantity *displacement* and the scalar quantity *distance.*

EXERCISE 6.2

In each exercise find the x and y components of the given vectors by using trigonometric functions. Assume a standard xy coordinate system with positive angles measured counterclockwise (ccw) from the positive x axis.

1. $\mathbf{A} = 6.3$ @ $35°$
2. $\mathbf{B} = 90$ @ $135°$
3. $\mathbf{C} = 278$ @ $-40°$
4. $\mathbf{D} = 0.048$ @ $240°$
5. $\mathbf{E} = 0.50$ @ 2 rad
6. $\mathbf{F} = 5.0$ @ $\pi/6$ rad

B. RESULTANTS AND TRIGONOMETRIC FUNCTIONS

Now we shall develop equations for finding a resultant in terms of its components. For example, if the components $\mathbf{A}_x$ and $\mathbf{A}_y$ of Fig. 6.12 were known, then we could determine the magnitude of their resultant **A** with the Pythagorean formula. Note that **A** is the hypotenuse of a right triangle, and $\mathbf{A}_x$ and $\mathbf{A}_y$ are the sides of the triangle. Thus we have

$$\boxed{A = \sqrt{A_x^2 + A_y^2}} \tag{6.2}$$

In the same right triangle, $\mathbf{A}_y$ is the side opposite angle θ_A, and $\mathbf{A}_x$ is the side adjacent to θ_A. Thus by the definition of the tangent function we have

$$\boxed{\tan\,\theta_A = \frac{A_y}{A_x}} \tag{6.3}$$

Example 6.4. Vector resultant

Suppose that you walk 3 miles due east, then turn and walk 4 miles due north. How far and in what direction are you from the starting point?

Let $\mathbf{A}_x$ of Fig. 6.12 be the 3-mile displacement eastward, and $\mathbf{A}_y$ the 4-mile displacement northward. Then A and θ_A are the desired distance and direction. Using Eqs. (6.2) and (6.3), we obtain

$$A = \sqrt{A_x^2 + A_y^2} = \sqrt{3^2 + 4^2} = \sqrt{25} = 5 \text{ miles}$$

and

$$\tan\,\theta_A = \frac{A_y}{A_x} = \frac{4}{3} = 1.33 \qquad \theta_A = 53°7'$$

EXERCISE 6.3

In each problem find the resultant magnitude and direction for the given right-angle components by using trigonometric functions or the Pythagorean formula.

1. $R_x = 22$
   $R_y = -5$
2. $R_x = -6$
   $R_y = 14$
3. $R_x = 125$
   $R_y = 240$
4. $R_x = -0.340$
   $R_y = -0.085$

C. ADDITION OF VECTORS

We are now in a position to add vectors mathematically. We shall make use of the ideas summarized in Eqs. (6.1) to (6.3). Suppose that we wish to add vectors **A** and **B** shown in Fig. 6.13*a*. First we resolve them into *x* and *y* components. The values of these components are computed using Eq. (6.1):

$$A_x = A \cos \theta_A \qquad A_y = A \sin \theta_A$$

$$B_x = -B \cos \theta_B \qquad B_y = B \sin \theta_B$$

Because $\mathbf{A}_x$ and $\mathbf{B}_x$ act along the same line (the *x* axis), they can be added as scalar quantities. $\mathbf{A}_y$ and $\mathbf{B}_y$ can also be added as scalar quantities. Thus by using pairs of right-angle components we reduce a vector addition to a pair of scalar additions. Figure 6.13*b* shows that R_x, the magnitude of the *x* component of the resultant, is merely the algebraic sum of A_x and B_x. Similarly, R_y is the algebraic sum of A_y and B_y. Be careful to notice that the algebraic sum of A_x and B_x becomes a *subtraction,* since they act in opposite directions. Thus we have

$$R_x = A_x + B_x = A \cos \theta_A - B \cos \theta_B$$

$$R_y = A_y + B_y = A \sin \theta_A + B \sin \theta_B$$

Fig. 6.13 Adding vectors by right-angle components. (a) **A** *and* **B** *are resolved into x and y components.*
(b) $\mathbf{R}_x = \mathbf{A}_x + \mathbf{B}_x$
$\mathbf{R}_y = \mathbf{A}_y + \mathbf{B}_y$
The x components add to give $\mathbf{R}_x$*. The y components add to give* $\mathbf{R}_y$*.*
(c) $\mathbf{R}_x$ *and* $\mathbf{R}_y$ *of part (b) combine to give* **R***, the resultant of* **A** *and* **B***.*

Figure 6.13*c* shows the resultant. To find the values of R and θ_R, we merely use the R_x and R_y components according to Eqs. (6.2) and (6.3):

$$R = \sqrt{R_x^2 + R_y^2} \qquad \tan \theta_R = \frac{R_y}{R_x}$$

Example 6.5. Vector addition

Suppose that you walk 6.3 miles in the direction 40° north of east and then turn and walk 9.8 miles at 25° north of west. How far and in what direction are you from the starting point?

Solution. First, sketch a vector diagram. Place the tails of both displacement vectors

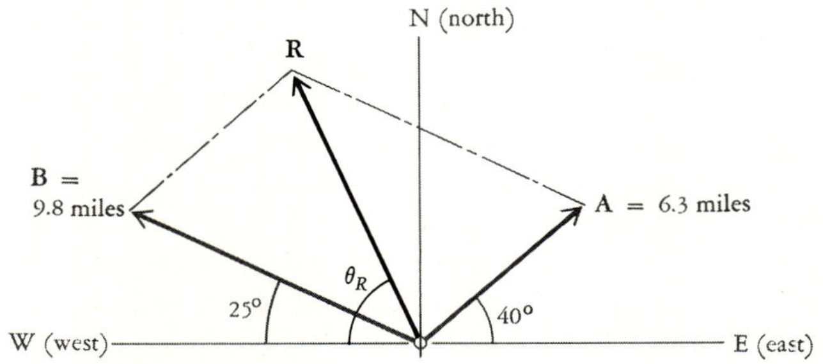

Fig. 6.14 Example 6.5.

at the origin as shown in Fig. 6.14. Then apply the procedures just developed, as follows:

Eastward:

$$\begin{aligned} R_e &= A_e + B_e \\ &= 6.3 \cos 40° - 9.8 \cos 25° \\ &= 6.3(0.766) - 9.8(0.906) \\ &= 4.83 - 8.88 \\ R_e &= -4.05 \text{ miles east, or } 4.05 \text{ miles west} \end{aligned}$$

Northward:

$$\begin{aligned} R_n &= A_n + B_n \\ &= 6.3 \sin 40° + 9.8 \sin 25° \\ &= 6.30(0.643) + 9.8(0.423) \\ &= 4.05 + 4.15 \\ R_n &= 8.20 \text{ miles north} \end{aligned}$$

Resultant:

$$R = \sqrt{R_w^2 + R_n^2} = \sqrt{4.05^2 + 8.20^2} = \sqrt{83.6}$$
$$R = 9.1 \text{ miles}$$

$$\tan \theta_R = \frac{R_n}{R_w} = \frac{8.20}{4.05} = 2.02 \longrightarrow \theta_R = 64°$$

Answer: **R** = 9.1 miles in the direction 64° north of west

Example 6.6. Given three concurrent forces as shown in Fig. 6.15, find the resultant force.

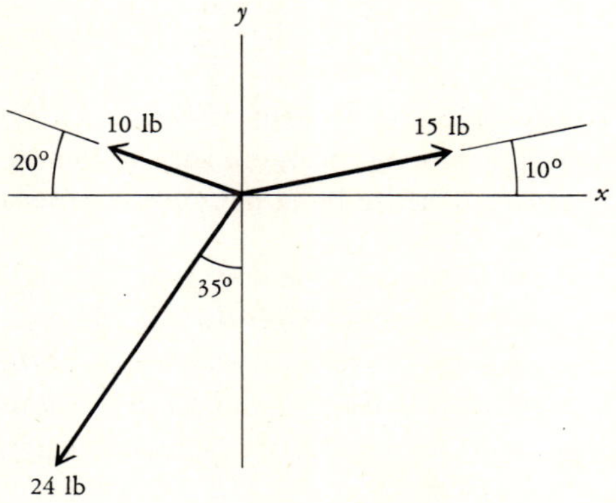

Fig. 6.15 Example 6.6.

Solution. As in the previous example, we break each vector down into components and then add the components.

X direction:

$$\begin{aligned} R_x &= 15 \cos 10° - 10 \cos 20° - 24 \sin 35° \\ &= 15(0.985) - 10(0.940) - 24(0.574) \\ &= 14.8 - 9.4 - 13.8 \\ R_x &= -8.4 \text{ lb} \end{aligned}$$

Y direction:

$$\begin{aligned} R_y &= 15 \sin 10° + 10 \sin 20° - 24 \cos 35° \\ &= 15(0.174) + 10(0.342) + 24(0.819) \\ &= 2.61 + 3.42 - 19.7 \\ R_y &= -13.7 \text{ lb} \end{aligned}$$

Resultant:

$$R = \sqrt{R_x^2 + R_y^2} = \sqrt{(-8.4)^2 + (-13.7)^2} = \sqrt{259}$$

$$R = 16.1 \text{ lb}$$

$$\tan \theta_R = \frac{R_y}{R_x} = \frac{-13.7}{-8.4} = 1.63$$

$$\theta_R = 58° \text{ below the negative } x \text{ axis}$$

R = 16.1 lb @ 238°

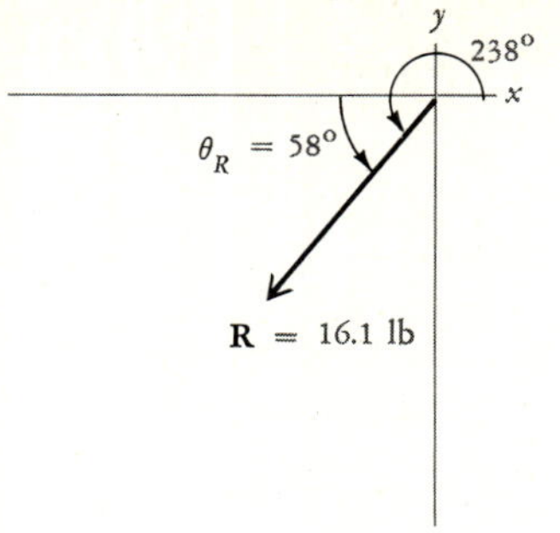

Fig. 6.16 The solution for Example 6.6.

Notice carefully the signs of R_x and R_y. You must use these signs to determine the position of θ_R on the coordinate axes. Since both R_x and R_y are negative, θ_R has to be in the *third quadrant*, as shown in Fig. 6.16.

Comment: *Did you wonder why we used 24 sin 35° instead of 24 cos 35° in the first line of Example 6.6? Well, notice that the 35° angle is measured from the y axis instead of the x axis. Therefore the x component of the 24-lb force is the side of the right triangle opposite the 35°, and the sine function is used. You have to be alert to such variations, analyze the problem, and apply the correct principles to solve it. Do not just memorize formulas. In technical work of this kind you should strive to develop an* ***understanding of principles.*** *Understanding principles is the heart of education. With a good understanding, the selection and use of formulas become a mechanical procedure. If you just memorize formulas, then you could be replaced by a computer.*

EXERCISE 6.4

1. Work Prob. 17 in Exercise 6.1 by using trigonometric functions.
2. Work Prob. 18 in Exercise 6.1 by using trigonometric functions.
3. You have walked 12 miles at 37° south of east, then 3.6 miles at 72° south of east. How far and in what direction are you from the starting point?
4. A ship was supposed to sail from A to B as shown in Fig. 6.17, but went from A to C instead. In what direction and how far must the ship now go in order to reach point B? (*Hint:* **AB** = **AC** + **CB** or **CB** = **AB** − **AC**.)

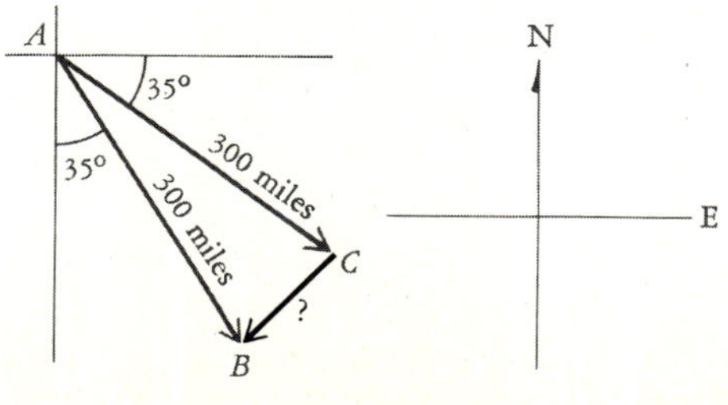

Fig. 6.17 Prob. 4.

5. John and Bob leave point P at the same time and walk for 2 hr in the directions shown in Fig. 6.18. If John stops, how far and in what direction should Bob walk to find him? (*Hint:* $\mathbf{B} + \mathbf{D} = \mathbf{J}$ or $\mathbf{D} = \mathbf{J} - \mathbf{B}$.)

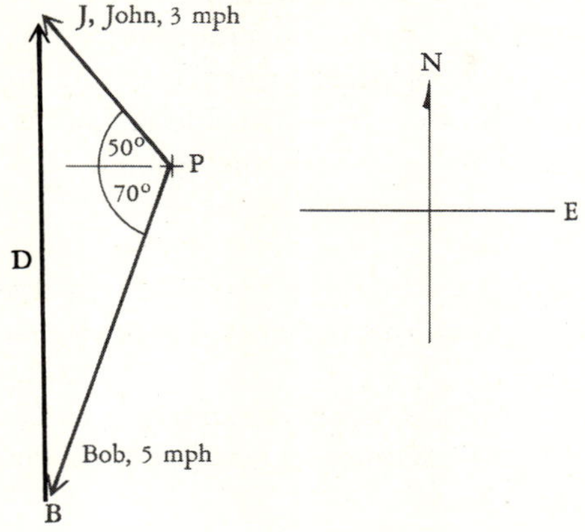

Fig. 6.18 Prob. 5.

6. Find the magnitude and direction of the resultant of the four forces shown (not to scale) in Fig. 6.19.

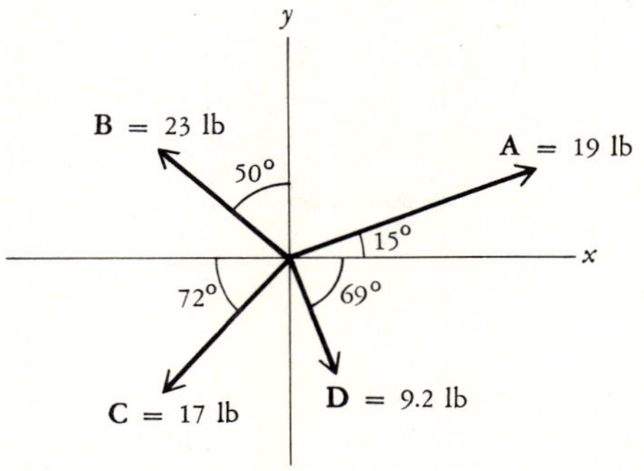

Fig. 6.19 Prob. 6.

7. Find the magnitude and direction of the resultant of the forces shown (not to scale) in Fig. 6.20.

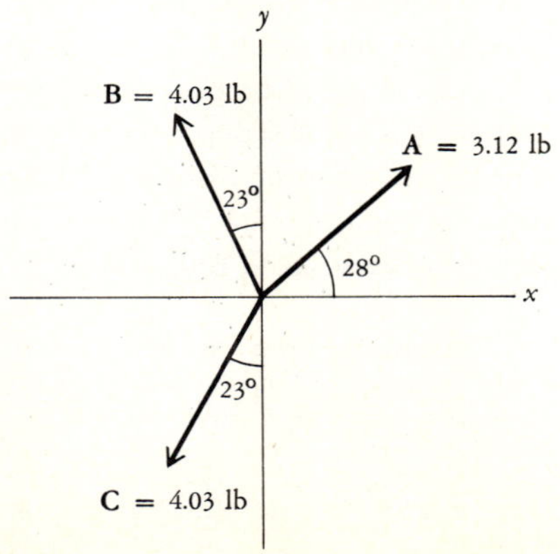

Fig. 6.20 Prob. 7.

6.4 FORCES IN EQUILIBRIUM

For reasons that we shall not discuss, forces are more complicated than other vector quantities. In order to avoid undue complications, we shall deal only with *concurrent forces,* that is, forces whose lines of action meet at a common point.

A physical principle called *Newton's first law** is frequently used to solve force problems involving bodies in equilibrium. This principle states that the resultant force on a body in equilibrium is zero. This means that the vector sum of all forces is zero if the body is at rest or is moving with constant velocity (constant speed and direction). In the examples that follow, we shall solve problems by combining this principle with the techniques of vector addition.

Example 6.7. A 400-lb sign is supported from a wall by a chain and a brace, as shown in Fig. 6.21*a*. What are the magnitudes of the force in the chain and of the force in the brace?

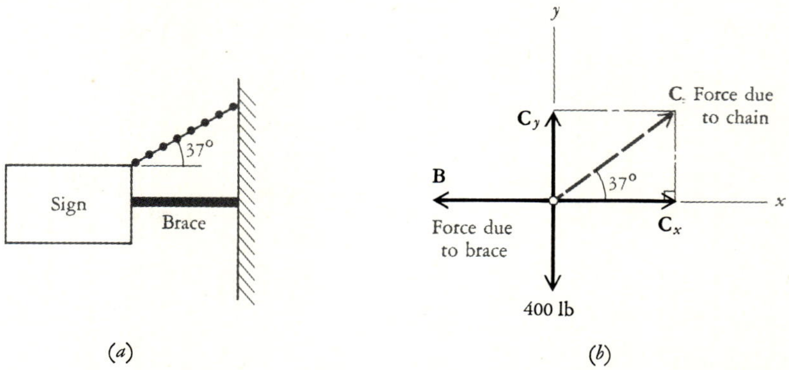

Fig. 6.21 Example 6.7. (a); (b) Vector diagram: all forces, and their components, that act on the sign.

Solution. To solve such problems, we draw a vector diagram that shows *all forces* that act *on the body*—the sign, in this case. We include the weight of the body itself. First, we assume a direction for each force. Notice that the direction of each force acting *on the body* is opposite to the direction of the corresponding force in the member (the chain or brace). This principle of equal and opposite reactions is *Newton's third law.* We assume the direction of force in a thin member such as a brace, strut, rod, etc., to be along the longitudinal axis of the

* This principle was discovered by the British mathematician/physicist Sir Isaac Newton (1642–1727). For an interesting biography of Newton and other famous mathematicians, see *Men of Mathematics* by E. T. Bell, Simon & Schuster, Inc., New York, 1961.

member. Furthermore, the force in a flexible member such as a chain, cable, etc., must place the member in tension. (Did you ever try to push something with a chain?)

Next we resolve the forces into cartesian components as shown in Fig. 6.21*b*. Since the sign is in equilibrium, the sum of the components along each of the perpendicular axes must be zero. We sum the forces as follows:

Y direction:

$$C_y - 400 = 0$$

$$C \sin 37° - 400 = 0$$ [Note here that we replace C_y by $C \sin 37°$.]

$$C \sin 37° = 400$$

$$C = \frac{400}{\sin 37°}$$

$\mathbf{C} = 664$ lb tension

(See Fig. 6.22.)

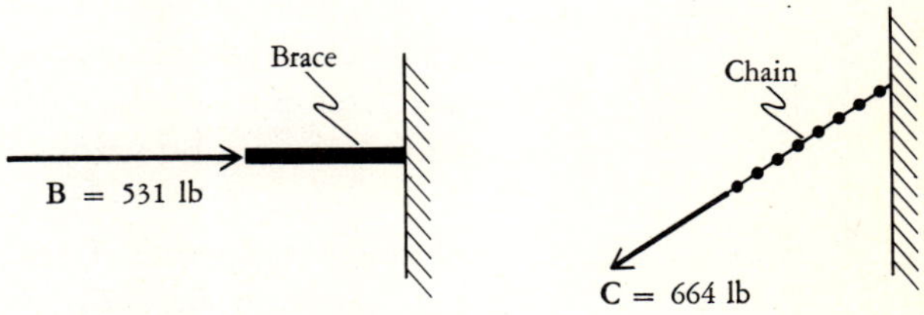

Fig. 6.22 Solutions for Example 6.7.

X direction:

$$C_x - B = 0$$

$$C \cos 37° - B = 0$$ [Replace C_x by $C \cos 37°$.]

$$B = C \cos 37°$$

$\mathbf{B} = 531$ lb compression

(See Fig. 6.22.)

The force in the chain is equal and opposite to the force of the chain on the sign. The same is true for the force in the brace.

Sign Conventions

The student should carefully observe how algebraic signs are treated in the equilibrium equations. First, we assume that each force is a positive quantity no matter which way it points in a vector diagram. Then if a component such as $\mathbf{C}_x$ of Example 6.7 acts in the positive x direction, we add it. But we subtract a component such as **B** that acts in the negative direction.

Notice also that we assumed directions for the unknown forces. We had the foresight to choose the correct directions in this example, and both forces **C** and **B** were positive. If we make an incorrect choice of direction, the answer for that force will be a negative number. The negative sign means that the force actually acts opposite to the direction shown in the vector diagram. For example, had we drawn vector **B** to the right in Fig. 6.21*b*, the equation for the forces in the x direction would have been

$$C_x + B = 0$$

$$B = -C_x = -531 \text{ lb}$$

The negative sign would tell us that **B** acts to the left.

We must always be extremely careful to observe the direction in which each component acts and to add or subtract accordingly. Reread Example 6.7 just to review how the vector directions and the additions or subtractions were correlated.

Example 6.8. A 400-lb sign is supported from a wall by a chain and from the ground by a brace, as shown in Fig. 6.23*a*. What are the forces in the chain and in the brace?

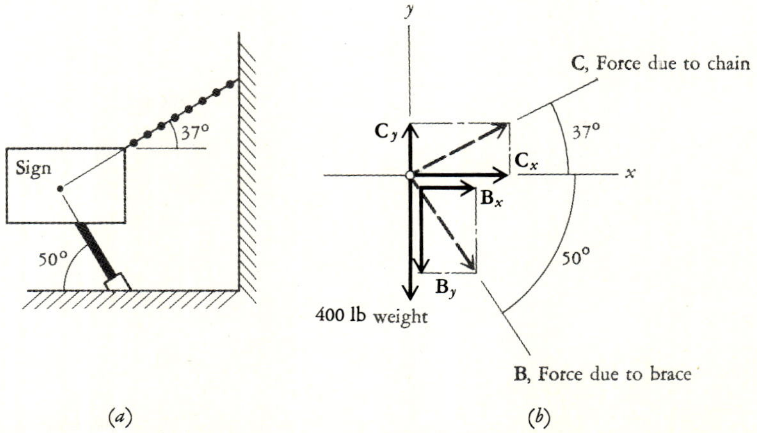

Fig. 6.23 Example 6.8. (a); (b) Vector diagram: all forces, and their components, that act on the sign.

Solution. Let **C** be the force due to the chain and **B** the force due to the brace; draw a vector diagram, as shown in Fig. 6.23*b*.

X direction:

$$C_x + B_x = 0$$

$$C_x = -B_x$$

$$C \cos 37° = -B \cos 50°$$

$$C = \frac{-B \cos 50°}{\cos 37°} = \frac{-B(0.643)}{0.800}$$

$$C = -0.804B$$

Y direction:

$$C_y - B_y - 400 = 0$$

$$C \sin 37° - B \sin 50° = 400$$

$$-0.804B(0.600) - B(0.766) = 400$$ [Note here that we replace C by $-0.804B$.]

$$-0.482B - 0.766B = 400$$

$$-1.248B = 400$$

$B = -320$ lb (opposite to the direction shown in Fig. 6.23).

or,

B = 320 lb compression in the brace (Fig. 6.24).

Then

$$C = -0.804B = -0.804(-320)$$

C = 257 lb tension in the chain (Fig. 6.24).

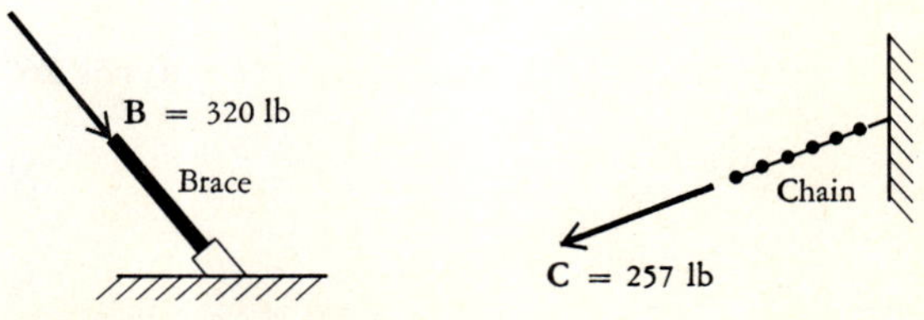

Fig. 6.24 Solutions for Example 6.8.

Example 6.9. A 732-lb block rests on an incline as shown in Fig. 6.25*a*.

a. What force **F** would a person have to exert to keep the block from sliding down the incline? Neglect friction.

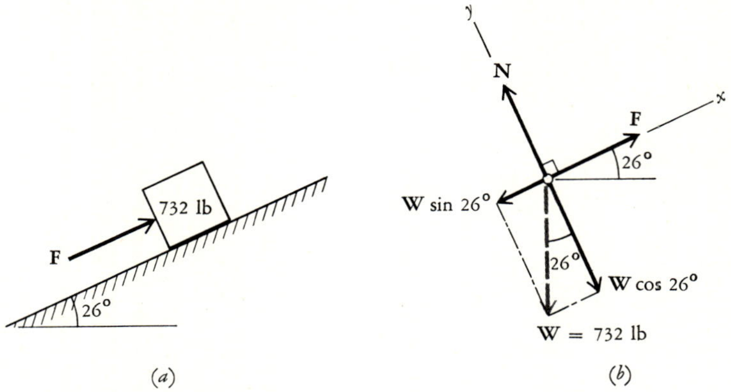

Fig. 6.25 (*a*) *Example 6.9;* (*b*) *vector diagram.*

b. What is the magnitude of the force that the incline exerts on the block?

Solution. This is a classic problem with many applications. The weight of the block acts straight down as shown in Fig. 6.25*b*. But only that component of the weight *parallel* to the incline will cause the block to slide down. Only the component of the weight that acts perpendicular to the incline will be the force on the incline. Thus we take our coordinate axes parallel to and perpendicular to the incline. See Fig. 6.25*b*.

X direction: $F - W \sin 26° = 0$

$$F = W \sin 26°$$

$$F = 732(0.438)$$

F = 321 lb Force needed to prevent block from sliding

Y direction: $N - W \cos 26° = 0$

$$N = W \cos 26°$$

$$N = 732(0.899)$$

N = 658 lb
Force that the incline exerts on the block

Notice how we used Newton's third law (see page 138). The force of the incline on the block is equal and opposite to the component of the weight of the block that acts on the incline.

EXERCISE 6.5

1. See Fig. 6.26. What is the force **F** needed to hold the 180-lb block in place on the inclined plane? Neglect friction.

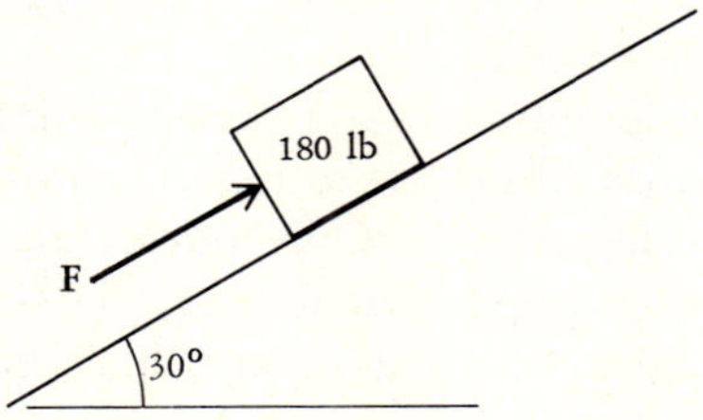

Fig. 6.26 Prob. 1.

2. Repeat Prob. 1 if force **F** acts horizontally instead of parallel to the incline.
3. What is the minimum force needed to hold a 3,000-lb car on a 3° incline?
4. See Fig. 6.27. What is the tension in each of the two ropes that support the swing and the 30-lb boy? With what horizontal force must the man pull in order to hold the boy in the position shown?

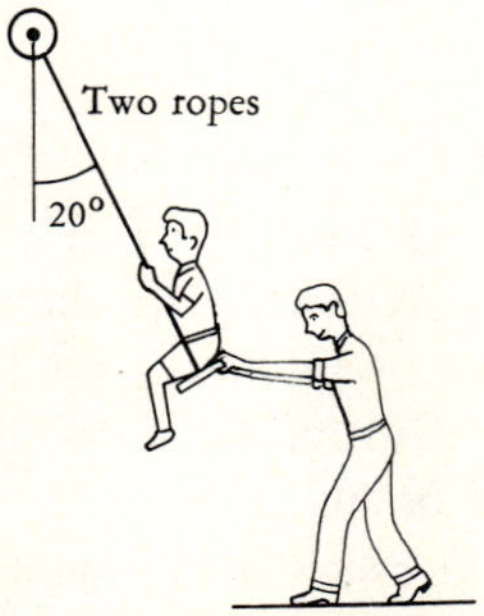

Fig. 6.27 Prob. 4.

5. What force must the man of Prob. 4 exert if he pulls at right angles to the ropes?
6. What are the tensions in cables A and B that support the 500-lb weight shown in Fig. 6.28?

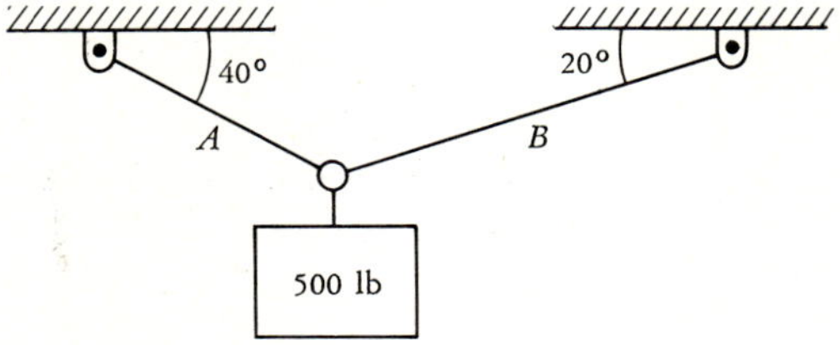

Fig. 6.28 Prob. 6.

7. A 183-lb tightrope walker stands in the middle of a rope that is fastened to poles 20 ft apart. His weight displaces the center of the rope down 5.4 in. What is the tension in the rope due to the man's weight?
8. To determine the size needed for a structural member, the stress analyst often has to resolve into components the forces acting on the member. One component is a "normal" component that acts perpendicular to a certain surface within the member; the other is a "shear" component that acts parallel to the surface. See Fig. 6.29.

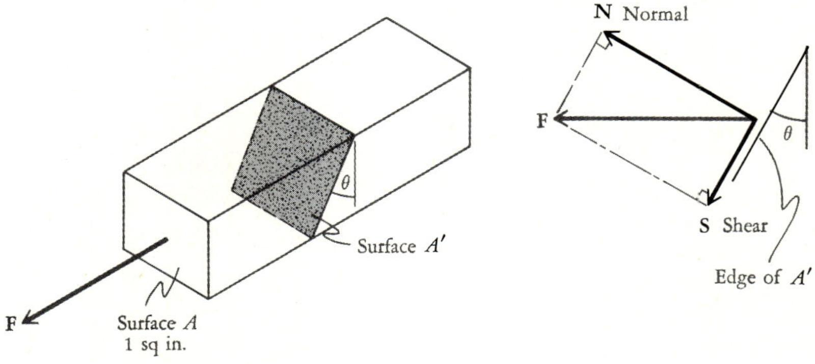

Fig. 6.29 Prob. 8.

The solid 1-in.² bar carries a load **F** along its axis. As a result of this force, what are the normal force **N** and the shear force **S** that act on plane A' tilted back at an angle θ as shown?
9. If the load **F** of Prob. 8 is 1,000 lb and angle θ is 30°, what are magnitudes of **N** and **S**?
10. The normal or tensile stress on a surface such as A' of Prob. 8 is defined as *normal force* ÷ *area.* Thus

$$S_n = \frac{N}{A'} \qquad \text{psi (pounds per square inch)}$$

where A' represents the area of surface A' (in in.^{2}). Similarly the shear stress is

$$S_s = \frac{S}{A'} \qquad \text{psi}$$

For surface A' of Prob. 8 determine S_n and S_s in terms of F and θ. Note that the area of A' is *not* equal to the area of A.

11. Determine S_n and S_s of Prob. 10 for the values of F and θ given in Prob. 9.

6.5 RELATIVE VELOCITY

Whenever we speak of the velocity of an object, we must establish a point of reference from which that velocity is measured. For example, suppose that a motorboat can travel 10 mph in still water. How fast can the boat travel downstream in a 6-mph current? The answer to this question depends upon the point of reference from which the speed is measured. If we stand on the ground and clock the speed, it will be $6 + 10 = 16$ mph, because the boat drifts at 6 mph with the current and also runs at 10 mph in addition to the current.

On the other hand, if we stand on a platform drifting along in the stream, then the boat passes us at 10 mph. The moving water is the "point of reference," and 10 mph is the boat's speed relative to the water. For the 16-mph speed, the ground is the point of reference; this speed is relative to the ground.

In general, the velocity of an object B as measured by an observer at a point A is called the *velocity of B relative to A.* We use the following special notation to indicate *relative velocity:*

RELATIVE VELOCITY NOTATION $\mathbf{V}_{b/a}$ is the velocity of B relative to A

Every velocity is relative to something; but if we wish to consider the point of reference as being stationary (that is, as having zero velocity), then any velocity relative to that point is an *absolute velocity.* In the notation above, if A is a stationary point, then we write $\mathbf{V}_b$ for the velocity of B, not $\mathbf{V}_{b/a}$. For earth-oriented objects, we consider the earth to be a stationary "point of reference." Thus $\mathbf{V}_b$ could represent the 16-mph absolute velocity of the boat (relative to the ground), and $\mathbf{V}_{b/w}$ could represent the 10-mph velocity of the boat relative to the water. If $\mathbf{V}_w$ represents the *absolute velocity* of the water, then the following equation relates these three velocities:

$$\mathbf{V}_b = \mathbf{V}_w + \mathbf{V}_{b/w}$$

This equation says that "the absolute velocity of the boat is the *vector sum* of the absolute velocity of the water plus the velocity of the boat relative to the water." If you think about this statement and relate it to the preceding discussion of the boat in the stream, it makes sense.

In general, if point A is a *moving* point of reference, then the absolute velocity of an object B can be stated as follows:

ABSOLUTE VELOCITY FORMULA $$\mathbf{V}_b = \mathbf{V}_a + \mathbf{V}_{b/a} \qquad (6.4)$$

where $\mathbf{V}_b$ is the absolute velocity of B
$\mathbf{V}_a$ is the absolute velocity of A
$\mathbf{V}_{b/a}$ is the velocity of B relative to A

Notice how this equation parallels the previous one above. This general velocity equation applies to any situation in which an object moves relative to a moving medium or carrier. Other examples are an airplane flying in a wind, a rock thrown from a moving automobile, a football thrown by a running quarterback, and a rocket fired from a moving airplane.

We can solve Eq. (6.4) for the relative velocity as follows:

RELATIVE VELOCITY FORMULA $$\mathbf{V}_{b/a} = \mathbf{V}_b - \mathbf{V}_a \qquad (6.5)$$

This equation tells us that a relative velocity is the *vector difference* between the absolute velocities.

Until now, our velocities have all been in the same direction. Vector Eqs. (6.4) and (6.5) could just as well have been scalar equations, but they do apply to vectors in different directions. The real power of these equations and their concepts lies in their application to velocities in different directions.

For example, let us find the absolute velocity of the boat (that would travel 10 mph in still water) if it heads across the stream at right angles to the 6-mph current.

The absolute velocity of the boat is the *vector sum* of the absolute velocity of the water and the velocity of the boat relative to the current. The vector equation is $\mathbf{V}_b = \mathbf{V}_w + \mathbf{V}_{b/w}$. Figure 6.30 illus-

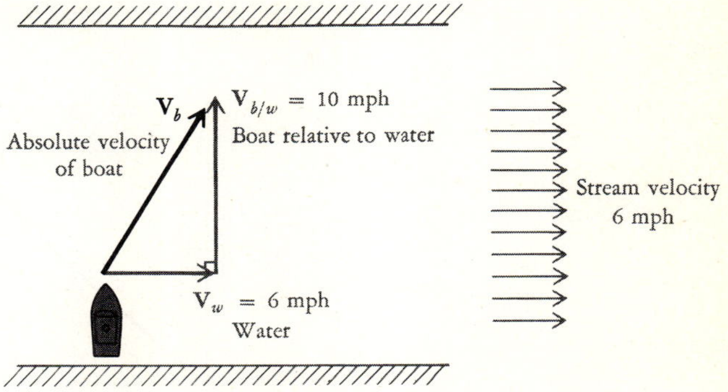

Fig. 6.30 The velocity of a boat in a stream.

trates the situation with vectors. $\mathbf{V}_b$ is the resultant of $\mathbf{V}_w$ and $\mathbf{V}_{b/w}$. We can find the magnitude of this resultant as follows [recall (Eq. (6.2)]:

$$V_b = \sqrt{V_w^2 + V_{b/w}^2} = \sqrt{6^2 + 10^2} = 11.7 \text{ mph}$$

The directional angle θ is found as follows:

$$\tan \theta = \frac{V_{b/w}}{V_w} = \frac{10}{6} = 1.67 \longrightarrow \theta = 59°10'$$

Thus $\mathbf{V}_b = 11.7$ mph @ $59°10'$ with the current. Notice that the boat always *points* across the stream, but it *moves* partly sideways.

The following examples illustrate various applications of the absolute-velocity and relative-velocity concepts.

Example 6.10. The wind is blowing due east at 60 mph. An airplane that would travel 100 mph in still air heads in the direction 65° north of east. What are the absolute speed and the track (direction of motion relative to the earth) of the airplane?

Solution. An airplane drifts with the wind just as a boat drifts in a stream. To find the resultant velocity, we resolve all velocities into eastward and northward components as shown in Fig. 6.31. Then we add the components as explained in Sec. 6.3.

Notation: Let $\mathbf{V}_w$ be the absolute velocity of the wind, $\mathbf{V}_g$ be the "groundspeed," or absolute velocity of the plane, $\mathbf{V}_{p/a}$ be the "airspeed," or velocity of the plane relative to the air, subscripts e refer to eastward components, and subscripts n refer to northward components.

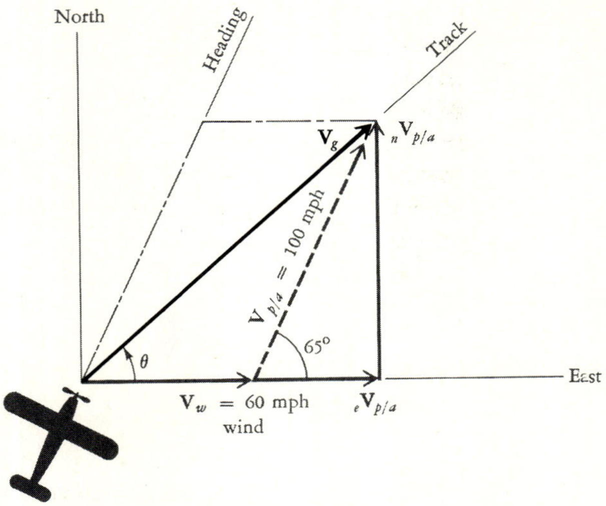

Fig. 6.31 Example 6.10. The velocity of an airplane in a wind.

Eastward components:
$$
\begin{aligned}
{}_eV_g &= {}_eV_w + {}_eV_{p/a} \\
&= 60 + 100 \cos 65^\circ \\
{}_eV_g &= 102 \text{ mph}
\end{aligned}
$$

Northward components:
$$
\begin{aligned}
{}_nV_g &= {}_nV_w + {}_nV_{p/a} \\
&= 0 + 100 \sin 65^\circ \\
{}_nV_g &= 91 \text{ mph}
\end{aligned}
$$

Resultant: $V_g = \sqrt{{}_eV_g^2 + {}_eV_g^2} = \sqrt{102^2 + 91^2} = 137$ mph (nearest mph)

$$\tan\theta = \frac{{}_nV_g}{{}_eV_g} = \frac{91}{102} = 0.89 \longrightarrow \theta = 42^\circ \text{ (nearest degree)}$$

Answer: $\boxed{\mathbf{V}_g = 137 \text{ mph @ } 42^\circ \text{ north of east}}$

Example 6.11. In what direction should the airplane of Example 6.10 head, in order that the track (the true direction of travel) be due north? If its destination is 260 miles due north, what is the flying time?

Solution. As shown in Fig. 6.32, the airplane must head into the wind so that its westward relative-velocity component just cancels the eastward wind velocity. Then the airplane will track due north. From the right triangle we can find angle θ as follows:

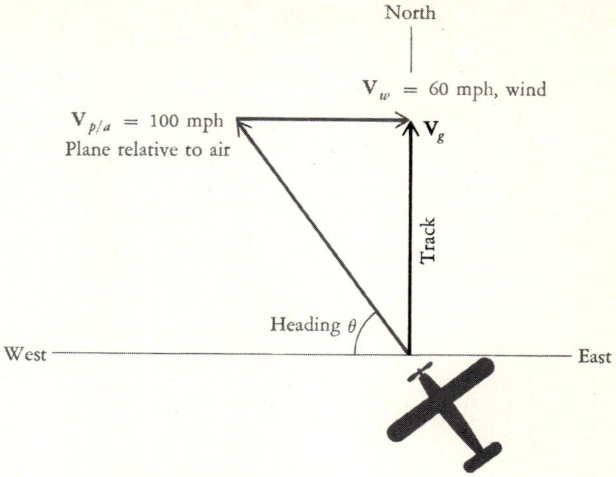

Fig. 6.32 Example 6.11. The heading of an airplane in a wind.

$$\cos\theta = \frac{V_w}{V_{p/a}} = \frac{60}{100} = 0.600$$

$$\boxed{\theta = 53^\circ \text{ north of west (nearest degree)}}$$

To find the flight time, we first determine, as we did before, the northward velocity component, which is the resultant, or absolute velocity. Note, however, that this time the resultant vector is *not* the hypotenuse.

$$V_g = \sqrt{V_{p/a}^2 - V_w{}^2} = \sqrt{100^2 - 60^2} = 80 \text{ mph}$$

$$\text{Time} = \frac{\text{distance}}{\text{speed}} = \frac{260 \text{ miles}}{80 \text{ mph}} = 3.25 \text{ hr}$$

$$\boxed{\text{Time} = 3 \text{ hr } 15 \text{ min}}$$

Example 6.12. A certain airline has a regularly scheduled flight from Kansas City to Chicago. The directional vector is 57°00′. (In aircraft navigation, due north is 0°, and all angles are clockwise from due north.) A 90-mph wind is blowing from 135° (that is, the wind is 45° toward the northwest). At what heading and with what airspeed should the pilot fly, if he wishes to maintain a groundspeed of 450 mph?

Solution. This time we need to find the relative velocity (or airspeed), and so we use Eq. (6.5). Figure 6.33 shows that we must subtract the cartesian components of the wind vector from those of the groundspeed vector in order to find

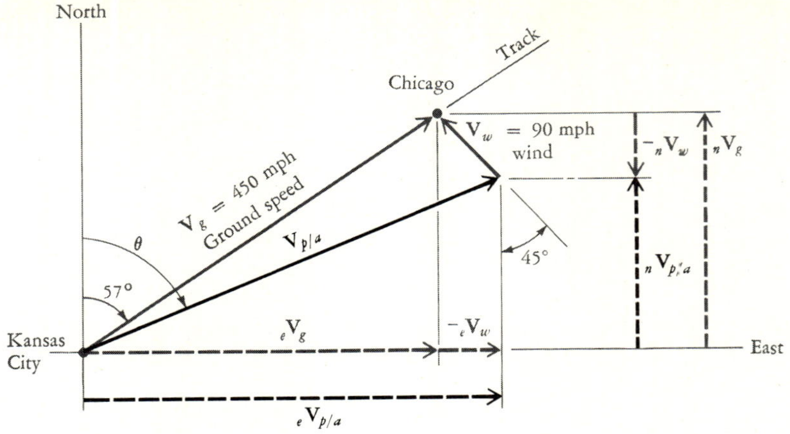

Fig. 6.33 Example 6.12.

components of the airspeed vector. From Eq. (6.5) we have

$$\mathbf{V}_{p/a} = \mathbf{V}_g - \mathbf{V}_w$$

which gives the following component equations.

Eastward components:

$$\begin{aligned} {}_eV_{p/a} &= {}_eV_g - {}_eV_w \\ &= 450 \sin 57° - (-90 \sin 45°) \\ &= 378 + 64 \\ {}_eV_{p/a} &= 442 \text{ mph} \end{aligned}$$

Northward components:

$$\begin{aligned} {}_nV_{p/a} &= {}_nV_g - {}_nV_w \\ &= 450 \cos 57° - 90 \cos 45° \\ &= 245 - 64 \\ {}_nV_{p/a} &= 181 \text{ mph} \end{aligned}$$

Resultant: $V_{p/a} = \sqrt{{}_eV_{p/a}^2 + {}_nV_{p/a}^2} = \sqrt{442^2 + 181^2} = 478$ mph

$$\tan\theta = \frac{{}_eV_{p/a}}{{}_nV_{p/a}} = \frac{442}{181} = 2.44 \longrightarrow \theta = 68° \text{ (nearest degree)}$$

Answer: $\boxed{\mathbf{V}_{p/a} = 478 \text{ mph @ } 68°}$

Comment: *Try to relate the following problems to the examples of this section. Some problems will be very similar to one of the examples; others will require variations of the examples. Then analyze the facts, reread the problem, make some sketches,* ***think,*** *and work out your own solution.*

When you do any problem, write it in a neat, orderly, step-by-step presentation. (Now is a good time to review Appendix 1.) Carefully define your notation (the letters that represent the various quantities of the problem). Make a neat sketch and label it. Clearly indicate each step of the solution and display the answer clearly. This approach will make problem solving easier for you because it forces you to clarify details and break problems into small, elementary steps. Example 6.10 is a good one to emulate.

A neat, readable solution will make more sense later when you review it for a test or final exam. Such a rare gem may even inspire your instructor to give you a better grade! But more important, you will be learning good techniques and habits. In industry, your work ***must*** *be correct, orderly, and intelligible. You* ***must*** *be able to communicate ideas and document your work.*

EXERCISE 6.6

1. A man stands on the rear platform of a railroad train going 20 mph. He throws a baseball straight back at 20 mph (relative to himself).
   a. Describe the motion of the ball from the viewpoint of the man on the train, but ignore the downward motion due to gravity.
   b. Describe the motion of the ball from the viewpoint of a man standing beside the tracks, but ignore the downward motion due to gravity.
2. Repeat Prob. 1 if the man throws the ball straight back at 15 mph.
3. Repeat Prob. 1 if the man throws the ball straight back at 25 mph.
4. If the man of Prob. 1 throws the ball at 50 mph, in what direction should he aim in order to hit a target that is on a line perpendicular to the tracks as he passes it?
5. A boy in a car going 45 mph due south throws a stone due west at 30 mph (relative to the car). With what speed and in what direction does the stone travel?
6. The boy of Prob. 5 throws the stone forward at an angle of 45° with the path of the car. What are the actual speed and direction of the stone?
7. A certain boat can go 16.4 mph in still water. It travels in a river that flows due south at 9.2 mph. If the boat heads due east at top speed, what are its actual speed and direction?
8. If the boat of Prob. 7 heads 13°10′ south of east, what are its actual speed and direction?
9. If the pilot of the boat of Prob. 7 wants to travel due east, in what direction should he point the boat? How long will he take to cross the river if it is 1.73 miles wide?

10. Repeat Prob. 7 if the river flows 58° south of west.
11. Repeat Prob. 8 if the river flows 58° south of west.
12. Repeat Prob. 9 if the river flows 58° south of west.
13. An airplane is headed 27° west of north at an airspeed of 326 mph. An 87-mph wind is blowing toward 13° north of east. What are the groundspeed and track (actual direction) of the airplane?
14. If the destination of the airplane of Prob. 13 is due north 723 miles away, in what direction should the airplane head? How long will the trip take?
15. A certain machine component consists of a slider within a disk, as shown in Fig. 6.34. If the disk rotates at 1,000 rpm and the outward speed of the slider is 80 ft/sec, what is the absolute speed of the slider in the position shown?

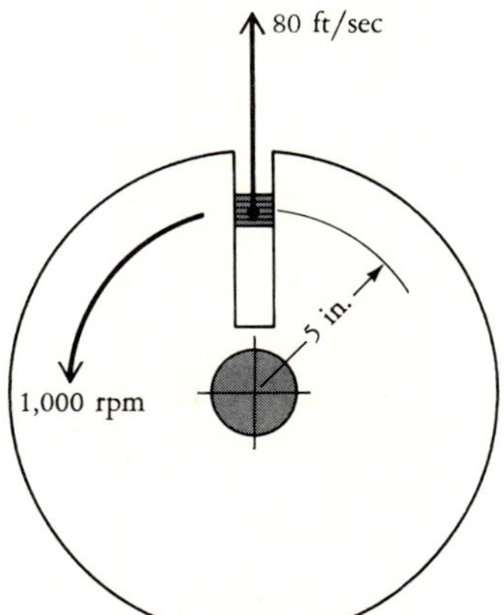

Fig. 6.34 Prob. 15.

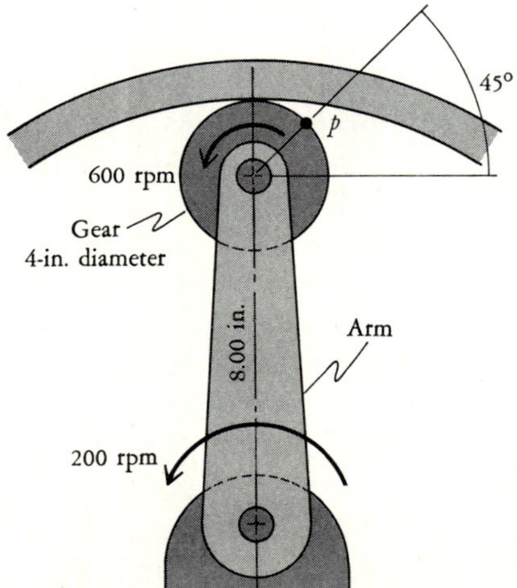

Fig. 6.35 Prob. 16.

16. An epicyclic gear train is shown in Fig. 6.35. The arm that carries the gear turns 200 rpm, and the gear turns 600 rpm on the pin at the end of the arm. What is the absolute velocity of point p on the rim of the gear? (*Hint:* The absolute velocity of point p equals the absolute velocity of the pin plus the velocity of point p relative to the pin.)
17. A car is traveling at 30 mph (44 ft/sec). If the diameter of each rear tire is 14 in., what is the velocity of the periphery of a rear tire
    a. at the point where the tire touches the ground?
    b. at the top of the tire?
    c. at both points on the periphery of the tire and halfway between the top and the bottom?

ADDITIONAL TOPICS IN ALGEBRA

UNIT 2

FUNCTIONS

CHAPTER 7

7.1 BASIC IDEAS ABOUT FUNCTIONS

In many aspects of our lives, we deal with relationships between quantities. Examples are the relationships between (1) postage and weight, (2) the area of a circle and its radius, (3) life expectancy and age, and (4) electric power consumption and the electric bill. Each such relationship consists of a rule that tells us how to find the value of one quantity (such as postage) when the value of the other quantity (weight) is known. Such relationships, or rules, are called *functions.* They can be expressed by equations, tables, graphs, or statements in words.

We customarily use phrases such as "the area of a circle is a *function* of its radius," meaning that area in some way depends upon radius. We can express the exact relationship by the formula, or rule, $A = \pi R^2$. In general, an equation such as $y = x^3 + 5$ is a rule that expresses a relationship between two sets of numbers, y and x. In this relationship, we consider y to be a *function* of x, because y is expressed in terms of x.

The literal quantities, such as x and y, in such relationships are called *variables,* because their values can change, or *vary.* Explicit numbers that cannot vary, such as 5, are called *constants.* Letters are sometimes used as constants also. For example, π stands for the irrational number 3.14159. . . . It is customary to use the letters a, b, c, d, etc., near the beginning of the alphabet for constants and the letters r, s, t, u, v, w, x, y, and z for variables. However, there are no ironclad rules about which letters are variables or constants.

In the equation $y = x^3 + 5$, the value of y depends upon the value of x, and so y is called the dependent variable, and x the independent variable. In general, a *dependent variable* is one whose value depends upon the value of some other variable. In an equation, the dependent variable is usually the one "solved for," just as y is "solved for" in terms of x in the equation $y = x^3 + 5$. An *independent variable* is one whose value is arbitrarily chosen. For example, we can arbitrarily choose any value for x in $y = x^3 + 5$; thus x is independent of y, but the value of y depends upon the value we assign to x.

In the strict mathematical sense, a function is a special kind of relationship: it is a relationship in which *exactly one* value of the dependent variable exists for any given value of the independent variable. Thus $y = x^3 + 5$ represents a function, but $y = \pm\sqrt{x}$ does not, because *two* values of y exist for *each* value of x. For example, if $x = 4$, then $y = +2$ or $y = -2$.

A good working definition of a function is as follows:

FUNCTION	A *function* is a rule that tells how to find a *unique* value of a dependent variable when a value of the independent variable is chosen.

The rule can be expressed by an equation. For example, the rule expressed by $y = x^3 + 5$ is: To find a value of y, cube the value of x and add 5.

In order to find one quantity as a function of another quantity, we state the rule (usually by writing an equation) that expresses the relationship between the quantities. The following examples illustrate the use of the term "function."

Example 7.1. **$A = s^2$ expresses A, the area of a square, as a function of s, the length of each side.**

Example 7.2. **$R = \sqrt{A/\pi}$ (the principal root only) expresses R, the radius of a circle, as a function of A, the area. To find this function, we start with the more familiar formula $A = \pi R^2$ and solve for R.**

Example 7.3. **To determine some functions, we often find a function of a function. For example, to find the area of a square as a function of its perimeter, we first write area as a function of the length of a side: $A = s^2$. Then we find s as a function of perimeter: $s = P/4$ (from $P = 4s$, the formula for perimeter of a square). Finally, we substitute the s function into the A function to obtain**

$$A = \left(\frac{P}{4}\right)^2 = \frac{P^2}{16}$$

Example 7.4. **Determine the volume of a cube as a function of its surface area.**

Solution. $V = s^3$ volume as a function of the length of a side

$A = 6s^2$

or $s = \sqrt{\dfrac{A}{6}}$ length of a side as a function of surface area A

from which

$$V = \left(\sqrt{\frac{A}{6}}\right)^3 = \left(\sqrt{\frac{A}{6}}\right)^2 \sqrt{\frac{A}{6}} = \frac{A}{6}\sqrt{\frac{A}{6}}$$

EXERCISE 7.1

1. Find the area of a circle as a function of its radius.
2. Find the area of a circle as a function of its diameter.
3. Find the circumference of a circle as a function of its diameter.
4. Find the area of a circle as a function of its circumference.
5. Find the circumference of a circle as a function of its area.
6. Find the area of a square as a function of the length d of its diagonal.
7. Find the volume of a cube as a function of its diagonal d.
8. Find the volume of a sphere as a function of its surface area s.
9. x ft of fencing are available to enclose a stock car track. Write an equation for the area that can be enclosed as a function of x if
   a. the fence forms a circle;
   b. the fence forms a square;
   c. the fence forms a rectangle with a 3:2 length:width ratio.
10. The cost of a certain carpet material is \$8.00 per yd^2. Determine the cost of circular carpets as a function of diameter, in *feet,* if
    a. no charge is made for waste material;
    b. waste material *is* charged and the carpets are cut from square stock sizes (assume any size square is available);
    c. waste material *is* charged, and the only stock width is 12 ft (any length can be cut from the 12-ft-wide roll, but the customer has to pay for the entire piece that is cut).
11. Repeat Prob. 10 if there is a \$0.10 per foot cutting charge for the circular part of the carpet.
12. In a simple series electric circuit, voltage E, as a function of current I and resistance R, is $E = IR$. The function for power (wattage) P is $P = EI$. Determine power as a function of current and resistance.
13. At liftoff, a Saturn V Apollo space vehicle weighs 6,424,000 lb. The first-stage engines burn fuel at a rate of 1,729,000 lb/min.

Find the weight of the vehicle as a function of time, *in seconds,* during the burn of the first-stage engines.

14. If an object is dropped in a vacuum, the number of feet s it falls as a function of time is $s = (1/2)gt^2$, where g is acceleration due to gravity in ft/sec^2 and t is time in seconds. To determine the value of g, experiments are devised to measure s and t; then g is computed. Determine a function by which g would be computed.
15. As a refinement of the experiments mentioned in Prob. 14, a ball is timed as it rolls a distance d down an inclined ramp. Determine g as a function of d, t, and the angle θ between the ramp and a horizontal plane. (*Hint:* Draw g as a downward, vertical acceleration vector and find the component of g that acts parallel to the ramp.)

7.2 FUNCTIONAL NOTATION

For convenience, we often refer to a certain function by using an abbreviated notation rather than an entire equation. One example of a function notation is the symbol "sin θ." We use "sin θ" to mean "the side of a right triangle opposite angle θ, divided by the hypotenuse." "sin θ" is a shorthand notation or name that calls to mind a specific function. Having defined this symbol, we have a convenient way of referring to the relationship without describing it in detail.

A. BASIC IDEAS

A notation that is used to stand for any function in general is of the form

FUNCTIONAL NOTATION $f(x)$ is the f function of x

The notation $f(x)$ is read "the f function of x," or, more briefly, "f of x."

Suppose we had a long, involved problem that used the expression $x^3 - 2x^2 + 3x - 9$ over and over again. To save time and space in writing this expression, we could define $f(x)$ to mean $x^3 - 2x^2 + 3x - 9$. Then we could write $f(x)$ instead of $x^3 - 2x^2 + 3x - 9$. We might refer to other functions by similar notations such as $g(x)$, $h(x)$, $Y(x)$, and so on. Such notations are like names of people, places, or objects. It is easier to say "Bobby Jones" than to say "the little boy who lives at 6226 Burlington Court Southeast."

Warning: *A functional notation such as $f(x)$ does* **not** *mean a multiplication. Thus the equation $y = f(x)$ means "y is the f function of x," not "y is f multiplied by x."*

B. EVALUATING FUNCTIONS

We can use the functional notation to indicate the value of a function for a specific value of the independent variable. For example, if

$$f(x) = x^3 - 2x^2 + 3x - 9$$

then we can indicate the value of $f(x)$ for $x = 2$, with the notation $f(2)$. Thus $f(2)$ means "the value of $f(x)$ when $x = 2$." To find $f(2)$, we replace x by 2, wherever x appears in the formula for $f(x)$. In this case,

$$f(2) = (2)^3 - 2(2)^2 + 3(2) - 9 = -3$$

In fact, we can replace x by any number, letter, or expression. Thus for the $f(x)$ above,

$$f(a) = a^3 - 2a^2 + 3a - 9 \qquad \begin{cases}\text{the value of} \\ f(x) \text{ when } x = a\end{cases}$$

$$\begin{aligned} f(5z) &= (5z)^3 - 2(5z)^2 + 3(5z) - 9 \\ &= 125z^3 - 50z^2 + 15z - 9 \qquad \begin{cases}\text{the value of} \\ f(x) \text{ when } x = 5z\end{cases}\end{aligned}$$

and

$$\begin{aligned} f(x^2) &= (x^2)^3 - 2(x^2)^2 + 3(x^2) - 9 \\ &= x^6 - 2x^4 + 3x^2 - 9 \qquad \begin{cases}\text{the value of} \\ f(x) \text{ when } x = x^2\end{cases}\end{aligned}$$

In this last example, we do not mean that x^2 *equals* x in the numerical sense; instead, we mean that x is to be *replaced by* x^2 in the formula for $f(x)$. Thus the equals sign can mean "is replaced by."

Which letter we decide to use to define a function is immaterial; it is a *dummy variable,* or place holder, and can be replaced by any expression. The *rule* or *pattern* defined by the equation is the important thing to watch, not the specific letter x, a, or whatever. We could just as well have defined the f function above by writing

$$f(y) = y^3 - 2(y)^2 + 3y - 9$$

or even

$$f(\quad) = (\quad)^3 - 2(\quad)^2 + 3(\quad) - 9$$

leaving a blank space where the value of the independent variable is to be inserted.

EXERCISE 7.2

1. Given that $f(x) = x^2 - 4x + 10$, find
   a. $f(0)$ b. $f(2)$ c. $f(-2)$ d. $f(a)$
2. Given that $h(z) = z^3 - 2z^2 + 4$, find
   a. $h(0)$ b. $h(5)$ c. $h(-3)$ d. $h(x^2)$
3. Given that $s(t) = \frac{1}{2}gt^2$, where g is 32.2 ft/sec², find
   a. $s(t+h)$ b. $s(t+h) - s(t)$
4. Area of a circle as a function of radius is given by $A = \pi R^2$. If the radius changes by a small amount ΔR (read "delta R," meaning a small change of R), then the new radius is $R + \Delta R$. The change of area is the new area $\pi(R + \Delta R)^2$ minus the old area πR^2. Determine a function for change of area in terms of R and ΔR; then find the change of area due to a 0.9-ft change of radius when the radius is 10 ft.
5. Given that $f(x) = x^2 + 2x - 5$ and $g(x) = x^2 - 1$, find $f[g(x)]$.

7.3 GRAPHS OF FUNCTIONS

The graph of a function is the graph of the equation, table, or statement that defines the function. To graph a function $f(x)$, we plot points with coordinates $[x,f(x)]$. For example, if $f(x) = x^2 - 5$, we would make a table of coordinates as follows:

x	± 5	± 4	± 3	± 2	± 1	0
x^2	25	16	9	4	1	0
$f(x) = x^2 - 5$	20	11	4	-1	-4	-5

Then we would plot the points (5,20), (4,11), and so on, that represent pairs of coordinates $[x,f(x)]$. Figure 7.1 shows the graph of this function. The graph is a smooth curve drawn through the plotted points. The standard practice is to use the abscissa for the independent variable and the ordinate for the dependent variable.

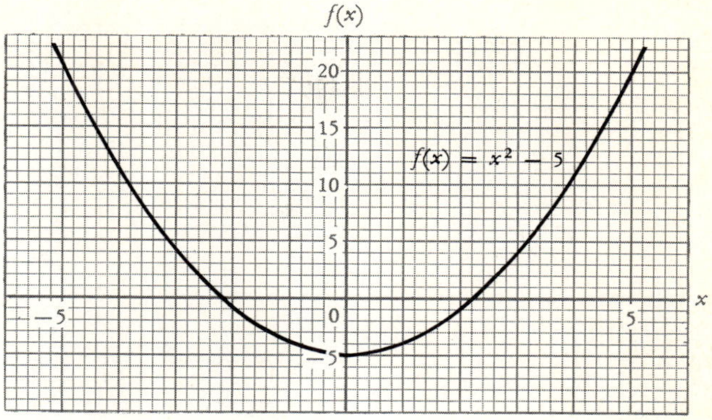

Fig. 7.1 A graph of the function $f(x) = x^2 - 5$.

Guidelines for Good Graphs

In Sec. 3.9, we mentioned some guidelines for constructing readable graphs. These few elementary rules are so important to the success of a graph that we shall expand upon them here.*

1. Identify the graph.
   Title the graph so that the reader can immediately identify it. The title should consist of a brief, descriptive phrase that tells what quantities are included in the graph and any special conditions that influence the graph. In addition, include in some prominent place the equation or other basis for the graph. See Figs. 7.2 and 7.3. If no physical quantities are involved, then at least identify the equation that the graph represents. See Fig. 7.1.
2. Identify the quantity associated with each axis.
   If you graph $f(x)$ vs. x, then place an "x" on the abscissa, and an "$f(x)$" on the ordinate. See Fig. 7.1. If you graph distance vs. time, label "Distance" on the ordinate, and "Time" on the abscissa as in Fig. 7.2. Such labels help the reader to identify the variables of the graph.
3. Identify the units of measurement.
   An abstract function such as $f(x) = x^2 - 5$ has no units, but often you will graph physical quantities. If you graph distance in feet vs. time in seconds, then write "Distance (ft)" on the ordinate, and "Time (sec)" on the abscissa. See Fig. 7.2*b*.
4. On each axis, label numerical values of the quantity represented.
   Without numbers on the axes, a graph cannot be read for specific

* Certain problems of publication, such as printing graphs small enough to fit the pages of a book, make it difficult to reproduce good graphs. Therefore even the graphs in this book may sometimes lack the features discussed here.

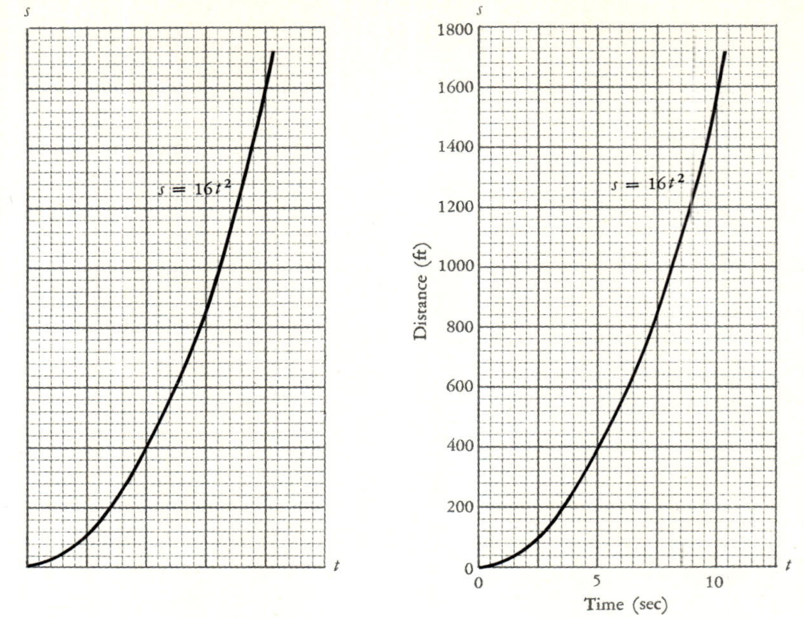

(*a*) Poor graphing practice: the variables are not identified, the units are not indicated, and the numerical values are not given.

(*b*) Good graphing practice: the axes are labeled with the quantities being graphed, their units, and the numerical values.

Fig. 7.2 Distance vs. time for a body falling under the influence of a 32 ft/sec² gravitational acceleration.

information; it is useless except to indicate a general relationship. Compare Figs. 7.2*a* and 7.2*b*.

5. Make each axis a direct-reading scale.

A *direct-reading scale* is one on which the value of each point (or each grid line on a graph) is obvious by inspection; no figuring is necessary.

A direct-reading graph axis is always formed if *the number of subdivisions within each major division is a factor of the value assigned to the first major division.* For example, if, as on the x axis of Fig. 7.3*b*, the first major division has a value of 10, then there should be 2, 5, or 10 subdivisions. Note that 2, 5, and 10 are factors of 10. If the value assigned to the first major division is 8, then there should be 2, 4, or 8 subdivisions. Figure 7.3*a* shows an awkward relationship between the number of subdivisions and the values placed on the major divisions. Compare Figs. 7.3*a* and *b* to see the relative ease of reading numbers such as 27 on the x axis and 165 on the $f(x)$ axis.

Since we frequently use the decimal system and since most standard graph papers are subdivided by 5's or 10's, a base 10 scale

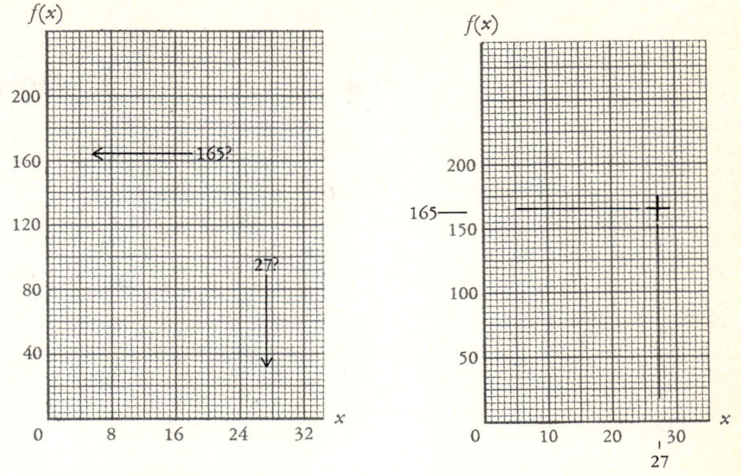

(*a*) Poor graphing practice: non-direct-reading scales

Although the major divisions are subdivided into ten parts, the axes are not base 10 scales, and it is difficult to locate most numbers.

(*b*) Good graphing practice: direct-reading scales

Each axis is a base 10 scale. Numbers can be located by inspection.

Fig. 7.3 A comparison of direct-reading and non-direct-reading scales.

is usually best for a direct-reading graph. On a direct-reading, base 10 scale, the value assigned to the first major division should be a factor of 10 (1, 2, 5, or 10) or a decimal multiple of one of these numbers (such as 0.01, 0.02, or 0.05; 0.1, 0.2, or 0.5; or 20, 50, or 100). The axes of Fig. 7.3*b* are good examples.

6. Utilize the graph paper efficiently.
Choose a scale on each axis so that the graph uses almost all the space available. Do not crowd the graph into a small space as in Fig. 7.4*a*; make it fill the available space without running off the paper. See Fig. 7.4*b*.

EXERCISE 7.3

In Probs. 1 to 6, graph each function between the limits +5 and −5 for the independent variable.

1. $f(x) = x^2 - 4x + 10$
2. $f(Y) = Y^2 + 4Y - 10$
3. $g(z) = z^3 - z^2$
4. $h(a) = 2 - \sqrt{25 - 5a}$
5. $T(t) = t^2 + 25$
6. $A(h) = h^3 - 5h^2 + 10h$

7. Graph the area of a circle as a function of circumference, if circumference varies from 0 to 25 in.

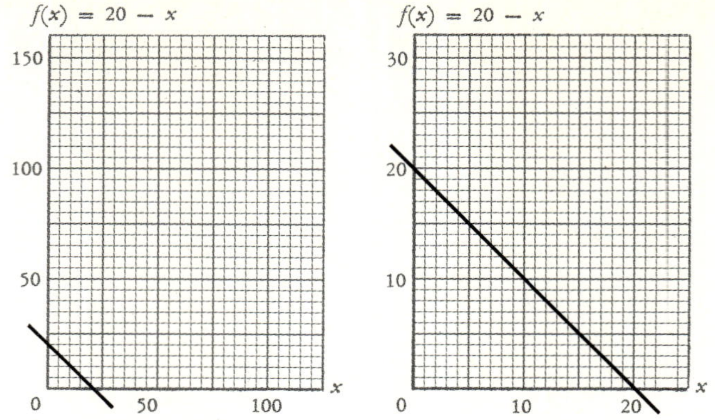

(*a*) Poor graphing practice: the graph is too small and does not use the space available.

(*b*) Good graphing practice: the graph utilizes the full space available.

Fig. 7.4 A comparison of good and poor utilization of space.

8. The earth's gravitational acceleration on a spaceship is given by the function $g = (5.05 \times 10^8)/r^2$ ft/sec^2, where r is the distance, in miles, from the center of the earth. Graph this function for a spaceship that travels from the surface of the earth (radius = 3,960 miles) to the moon 230,000 miles away. (Does the spaceship ever "escape" the earth's gravity?)
9. Graph the gravitational acceleration in Prob. 8 as a percentage of the surface gravity, 32.2 ft/sec^2, by graphing $g/32.2$.
10. Using Table A5.4 of the Appendix, graph sin θ, $0° \leq \theta \leq 90°$.
11. Using Table A5.4, graph cos θ, $0° \leq \theta \leq 90°$.
12. Make a graph of life expectancy as a function of age. A visit to your friendly neighborhood life insurance agent may help.
13. See Fig. 7.5. In photographic printing, the exposure time for any print is proportional to the projected area of the negative, and the area is proportional to the square of the distance d from the lens to the paper. Make a graph of t/t_0 vs. d/d_0, where d_0 is the minimum distance of the negative, t_0 is the exposure time needed for a certain print at a distance d_0, and t is the time needed at distance d.
14. If a 5- by 7-in. photographic print requires 23 sec for the best exposure, how much time is required for an 8- by 10-in. print from the same negative? Use the information of Prob. 13. Neglect the fact that 5 by 7 and 8 by 10 areas do not have lengths and widths in the same proportion.
15. Repeat Prob. 14, but assume that a 35-mm negative (actual dimensions 24 by 36 mm) is used. Take into account the fact that the 5 by 7 and 8 by 10 dimensions are *not* proportional to

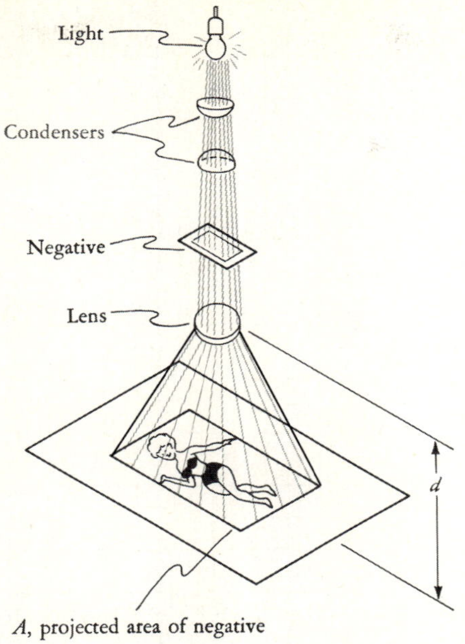

Fig. 7.5 Photographic printing system.

the 24 by 36 dimensions of the negative, and thus the projected area is greater than that of the print. For example, for the 5 by 7 print, the projected area has to be 5 by $7\frac{1}{2}$. What is the percent of error for Prob. 14?

7.4 SOLVING EQUATIONS GRAPHICALLY

We can use the graph of a function to solve an equation with a single variable. For example, we can solve the equation $x^3 - 5x^2 - 2x + 10 = 0$ by graphing the function $f(x) = x^3 - 5x^2 - 2x + 10$. The solutions to the equation are the values of x at points where $f(x) = 0$; thus $x^3 - 2x + 10 = 0$. These points are always the points where the graph of $f(x)$ crosses the x axis. Figure 7.6 shows that the solutions (to two significant digits) to $x^3 - 5x^2 - 2x + 10$ are $x = 5.0$, $x = 1.4$, and $x = -1.4$. If we desired greater accuracy, we could make enlarged graphs of the portions $-2.0 \leq x \leq -1.0$, $1.0 \leq x \leq 2.0$, and $4.5 \leq x \leq 5.5$.

Not all equations have real solutions. If no real solution exists, then the graph will never cross the x axis. Figure 7.7 shows that there is no real solution to the equation $x^2 + x + 5 = 0$.

In addition to solving single-variable equations with graphs, we can find conjunctions, or intersections, of two or more different functions. For example, suppose that the cost to produce a certain

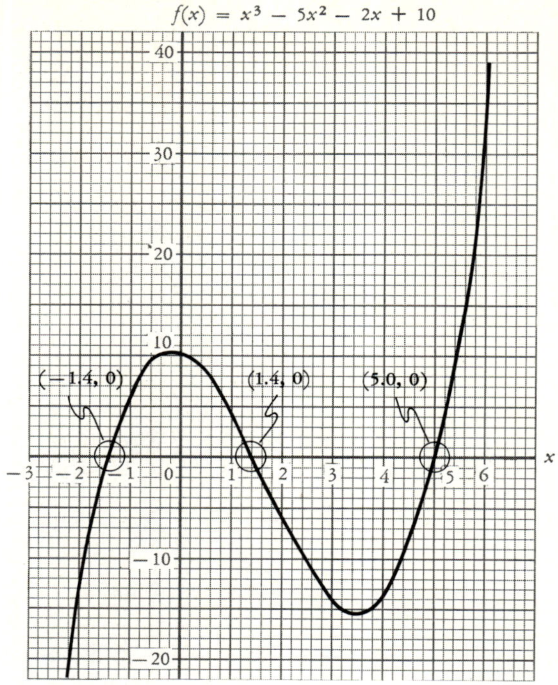

Fig. 7.6 Solutions of $x^3 - 5x^2 - 2x + 10 = 0$.

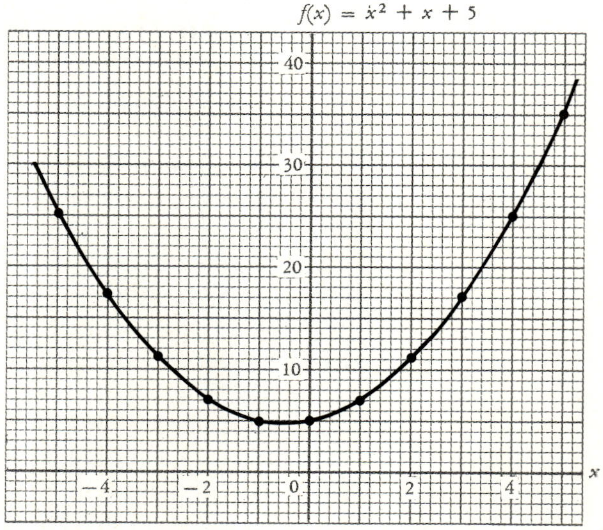

Fig. 7.7 The equation $x^2 + x + 5 = 0$ *has no real solution.*

part on a turret lathe is given by the function $C = \$50 + \$0.1275N$, where N is the number of parts. The cost to produce the same part on an automatic screw machine is $\$300 + \$0.0025N$. At what number of parts does the automatic screw machine begin to be more economical? In Fig. 7.8 we have graphed the two cost functions. The figure clearly shows that the two functions cross, and thus the costs are equal, where $N = 2{,}000$ parts. We can see that beyond this number the graph for the automatic machine is lower, and therefore on this machine the parts are less costly.

In general, to find a conjunction of two functions $f(x)$ and $g(x)$, we graph both functions. A *conjunction* is a point where the two

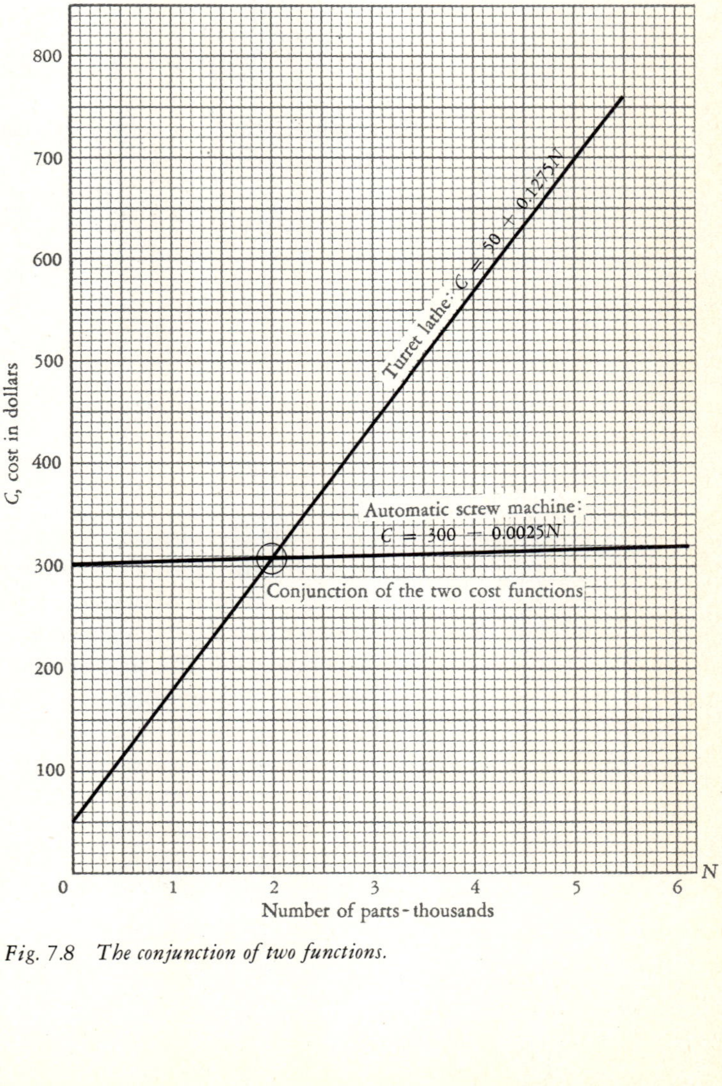

Fig. 7.8 The conjunction of two functions.

graphs intersect. At this point, $f(x) = g(x)$, and the value of x is the same in $f(x)$ and $g(x)$. If either function is of second degree or higher, there may be more than one conjunction.

EXERCISE 7.4

Solve the following equations graphically to two significant digits.

1. $2x - 5 = 0$
2. $5x + 16 = 0$
3. $x^2 - 9 = 0$
4. $x^2 + 9 = 0$
5. $x^2 - 9x = 0$
6. $x^2 + 9x = 0$
7. $2x^2 - 5x - 1 = 0$
8. $2x^2 + 5x - 1 = 0$
9. $3x^2 + x + 1 = 0$
10. $-x^2 + 3x + 18 = 0$
11. $x^3 + x + 1 = 0$
12. $x^2 - 10x - 20\sqrt{x} = 0$
13. $x^3 - 8x^2 - 60x + 90 = 0$
14. $x^3 + 3.7x^2 + x + 3.7 = 0$
15. $x^4 + x^3 - 24x^2 - 2x + 58 = 0$

By graphing, find the coordinates of the points where the two given functions intersect.

16. $f(x) = 2x - 5$
    $g(x) = x + 3$
17. $h(x) = x/2$
    $j(x) = 4 - x^2$
18. $Y(a) = a^3$
    $Z(a) = 2a^2 - 10$
19. $R(t) = \pm\sqrt{100 - t^2}, \quad -10 \leq t \leq 10$
    $S(t) = t$
    (*Hint:* For $R(t)$, after plotting 8 or 10 points, place a compass at the origin and see what you can do with it.)
20. A certain rectangular piece of land is 50 ft long. How wide must it be for the area to be numerically equal to the perimeter?
21. Acme Machine Works currently produces a certain part on a milling machine. The production costs are \$60 to set up the machine, \$50 for tooling, and \$0.10 per part to operate the machine (labor, including overhead and maintenance). If the part were made by die-casting, it would cost \$300 for molds and \$0.005 labor for each part. At what number of parts should Acme produce this item by die-casting?

Solve the following equations for θ by graphing each trigonometric function vs. θ, $0° \leq \theta \leq 90°$. Use Table A5.4 of the Appendix.

22. $\sin\theta = \cos\theta$
23. $\cos\theta = 2\sin\theta$
24. $\tan\theta = 3\cos\theta$

CHAPTER 8
LINEAR EQUATIONS AND DETERMINANTS

8.1 BASIC IDEAS ABOUT LINEAR EQUATIONS

A linear equation is an equation in which each variable is raised to the first power, no variables are multiplied together, no roots of variables occur, and no variable appears in a denominator. In general, a *linear equation* has the following form:

$$a_0 + a_1x_1 + a_2x_2 + a_3x_3 + \cdot \cdot \cdot + a_nx_n = 0$$

where the a's are constants and the x's are variables. Because all the variables are raised to the first power, a linear equation is also called a *first-degree equation.*

Examples 8.1. **Linear and nonlinear equations**

LINEAR EQUATIONS	NONLINEAR EQUATIONS	
a. $2x - 3y = 0$	$x^2 - 3y = 0$	x is not raised to the first power.
b. $x + 6 = y$	$xy + 6 = 0$	two variables are multiplied together.
c. $y = \sqrt{2}x + 3z + 5$	$y = \sqrt{2x} + 3z + 5$	A root of a variable occurs.
d. $z = x/2 - 3$	$z = 1/(2x) - 3$	A variable is in a denominator.

EXERCISE 8.1

Tell whether each equation is linear or nonlinear. Assume that each letter is a variable, unless otherwise noted.

1. $y^2 = x^2$
2. $y = x - 5$
3. $2y = -x + 2z$
4. $y = \sqrt{x}$
5. $y = \sqrt{x^2}$
6. $x = 2$

7. $I = PRT$ (P and R are constants)
8. $E = IR$ (R is a constant)
9. $y = -3$
10. $x/y = 2$
11. $y = |x|$
12. $u^3 = v^3$
13. $3r - 2s = t - r + 5$
14. $w + 2x - 2y = 8$
15. $xyz = 1$

A linear equation in two variables can always be written as follows in a form called the *slope-intercept form:*

$$y = mx + k$$

where m and k are constants.

We shall make, without proof, the following observations about linear equations in two variables.

1. The graph of a linear equation in *two variables* is always a straight line.
2. The constant k in $y = mx + k$ is the value of y where the line crosses the y axis. Thus the point $(0,k)$ is called the *y intercept.* For example, in Fig. 8.1 for $y = 3x + 5$, the y intercept is $(0,5)$.
3. On the graph of $y = mx + k$, y is proportional to x in the ratio $m{:}1$. In other words, y changes m units every time x changes 1 unit. For example, for $y = 3x + 5$ in Fig. 8.1, as x changes 1 unit from 0 to 1, y changes 3 units from 5 to 8.

   This ratio of the change of y to the change of x indicates the steepness of the line; thus m is called the *slope of the line*

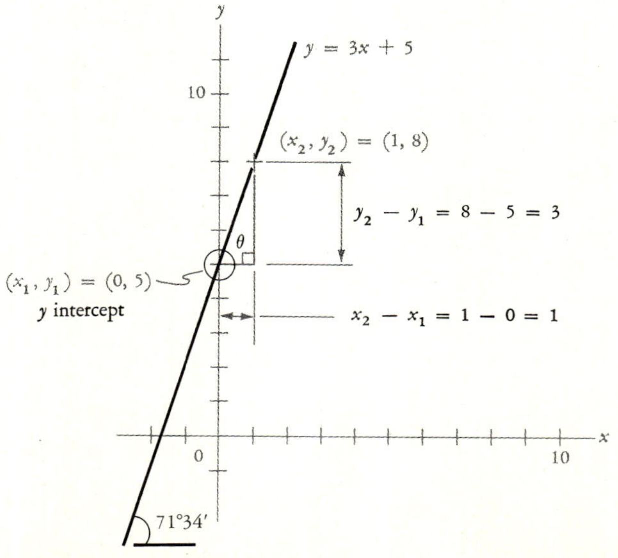

Fig. 8.1 A graph of the equation $y = 3x + 5$.

$y = mx + k$. On a graph of $y = mx + k$, if the scale factors for the x and y axes are different, then the slope of the graphed line will not be m; but the numerical slope is m.

4. The slope of a line is the tangent of the slope angle. The *slope angle* is the angle between the line and the positive abscissa (positive x axis). Notice in Fig. 8.1 that the slope angle is 71°34′ and that the slope of the line $y = 3x + 5$ is 3. Recall that $\tan \theta =$ side opposite ÷ side adjacent; for the right triangle of Fig. 8.1, $\tan 71°34' = 3 \div 1 = 3$.

    Here again, if the scale factors for the two axes are different, the slope angle of the graphed line will not be the numerical slope angle.

Example 8.2. Determine the y intercept, the slope, and the slope angle of the line described by the equation $3x - 2y - 8 = 0$.

Solution. First, we rewrite the equation in slope-intercept form. Solving for y we have

$$y = (3/2)x - 4$$

for which we see that (1) the y intercept is $(0,-4)$, (2) the slope is $3/2$, and (3) the slope angle is 56°20′ ($\tan 56°20' = 3/2$).

Example 8.3. Write the equation of the line for which the y intercept is (0,2) and the slope angle is −45°. See Fig. 8.2.

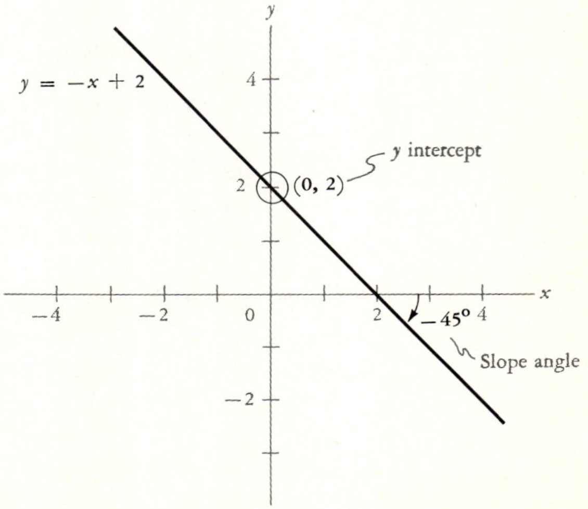

Fig. 8.2 Example 8.3. The line with a y intercept of 2 and a slope of −45°.

Since tan $(-45°) = -1$, the slope is -1, and thus the equation is

$$y = -x + 2$$

EXERCISE 8.1 (*Continued*)

For Probs. 16 to 27 determine the y intercept, the slope, and the slope angle of the line described by the equation.

16. $y = x$
17. $y = -x$
18. $2x + 2y = 0$
19. $y = x - 5$
20. $-y = 5 - x$
21. $x = y + 5$
22. $y = 6x + 10$
23. $y = 2(3x + 5)$
24. $y/2 = 3x + 5$
25. $2y = x - 4$
26. $3x + 5y + 20 = 0$
27. $4y + x - 3 = 0$

For Probs. 28 to 33, determine the equation of the line having the given slope angle and y intercept.

28. 45°, 0
29. −45°, 0
30. 30°, 5
31. 60°, −3
32. −60°, 3
33. −30°, −2

8.2 GRAPHS OF LINEAR EQUATIONS IN TWO VARIABLES

A. CONSTRUCTING THE GRAPH

Since the graph of a two-variable linear equation is always a straight line, we need only two points to construct it. We merely plot two points and draw the straight line through them. For the sake of accuracy, we should use two well-separated points, and to check for errors, we should plot a third point. The three points should lie on one straight line.

If the equation is in slope-intercept form, then we automatically have the point $(0,k)$ from the equation $y = mx + k$. To find a second point, we choose any value of x far from $x = 0$ and solve for y. To graph $10x - 2y - 7 = 0$ for example, we solve for y to obtain $y = 5x - 3.5$. Then $(0,-3.5)$ is one point on the graph. A second point is found by solving for $y = 21.5$ where $x = 5$. Figure 8.3 shows the line drawn through the points $(0,-3.5)$ and $(5,21.5)$. The point $(-5,-28.5)$ checks the others.

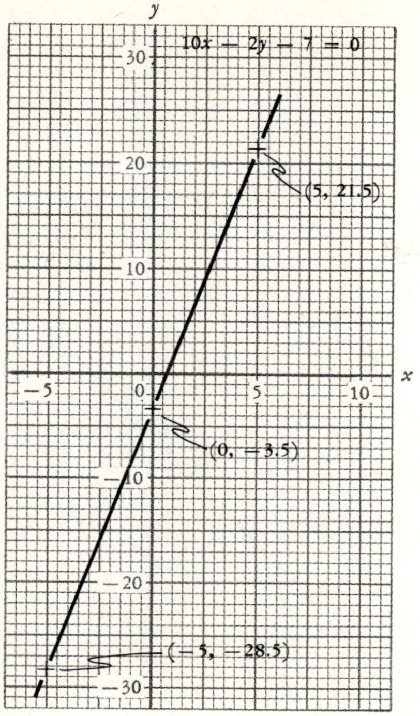

Fig. 8.3 A graph of $10x - 2y - 7 = 0$.

EXERCISE 8.2

For Probs. 1 to 7, graph the equations of Probs. 16, 17, 19, 22, and 25 to 27 of Exercise 8.1.

8. Change both of the following equations to slope-intercept form.
   a. $4x - 2y - 5 = 0$
   b. $6x - 3y + 9 = 0$

   Compare the slopes. How would the graphs look in relation to each other?
9. Repeat Prob. 8 for these equations.
   a. $3x - y + 2 = 0$
   b. $12x - 4y + 8 = 0$

B. SOLVING SIMULTANEOUS LINEAR EQUATIONS GRAPHICALLY

Figure 8.4 shows that the conjunction of the equations $y = 2x - 5$ and $y = x + 2$ is the point (7,9).

This conjunction has a special significance, for it is the one and only point that will solve *both* equations. Each of these equations

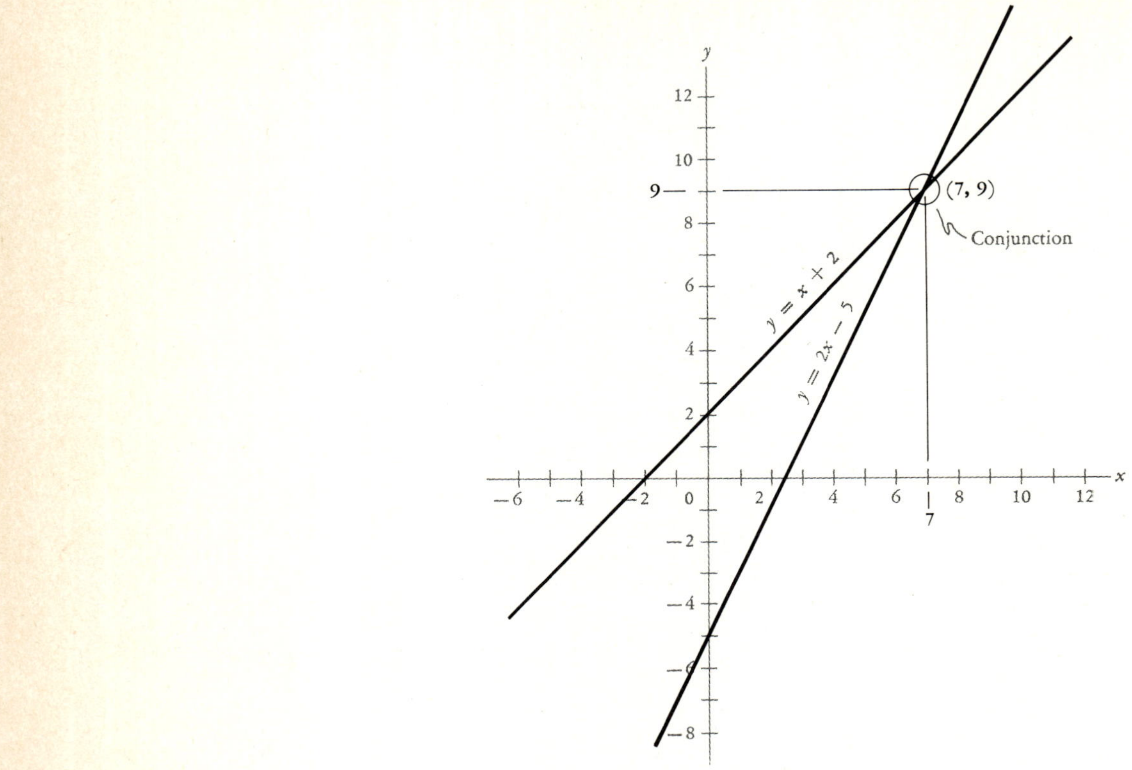

Fig. 8.4 The graphical solution of the simultaneous equations $y = x + 2$ *and* $y = 2x - 5$.

by itself has infinitely many solutions; for every value of x, there is a corresponding value for y. But only the pair of coordinates (7,9) will satisfy both equations. The coordinates of the intersection point are called the *simultaneous solution* of the two equations. As Fig. 8.4 shows, we can find the simultaneous solution of a pair of two-variable linear equations by graphing the equations.

Not every pair of linear equations in two variables has a unique simultaneous solution. If the lines are parallel, as in Prob. 8 above, they never intersect; no simultaneous solution is possible. If the lines coincide, as in Prob. 9, then every point that solves one equation also solves the other; still, no *unique* solution exists.

We can easily determine if a unique simultaneous solution exists by inspecting the equations. There are three simple tests:

1. If the slopes are *not equal,* a unique simultaneous solution exists. The two equations are called *independent equations.*
2. If the slopes *are equal* and the y intercepts are *not equal,* the lines are parallel. Such equations are called *inconsistent equations.*

3. If the slopes *are equal* and the y intercepts *are equal,* the lines coincide. Two such equations are called *dependent equations.*

The simultaneous solution of linear equations has many applications. The graphical solution often provides an insight that would be difficult to grasp otherwise. For an example, see Fig. 7.8. Here are some other illustrations.

Example 8.4. Car A leaves a certain point at 9:30 a.m. and averages 30 mph. Car B leaves the same point 1 hr 10 min later to overtake car A. If car B averages 40 mph, at what time will it overtake car A? How far will each car have traveled?

Solution. Figure 8.5 is a graph of distance vs. time for both cars. Both lines are based upon the fact that distance equals average speed multiplied by

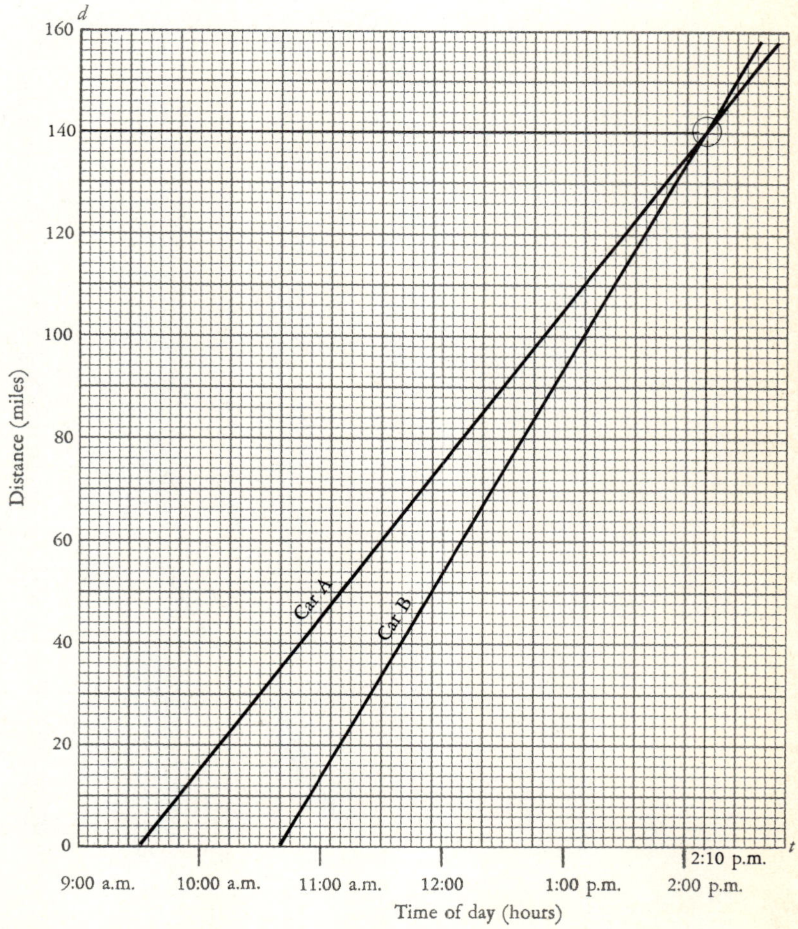

Fig. 8.5 Example 8.4.

time: $d = vt$. In this equation, v is the slope of the d-vs.-t line. At the intersection of the two lines in Fig. 8.5, the time is 2:10 p.m., and the distance for each car is 140 miles.

Example 8.5. In a certain sports car rally, the first instruction calls for your car to average 42 mph. Later, another instruction tells you to change your average speed to 24 mph. At that point, your corrected odometer reading is 10.3 miles and your elapsed time is 18 min. Are you on schedule? If not, what should you do to get on schedule?

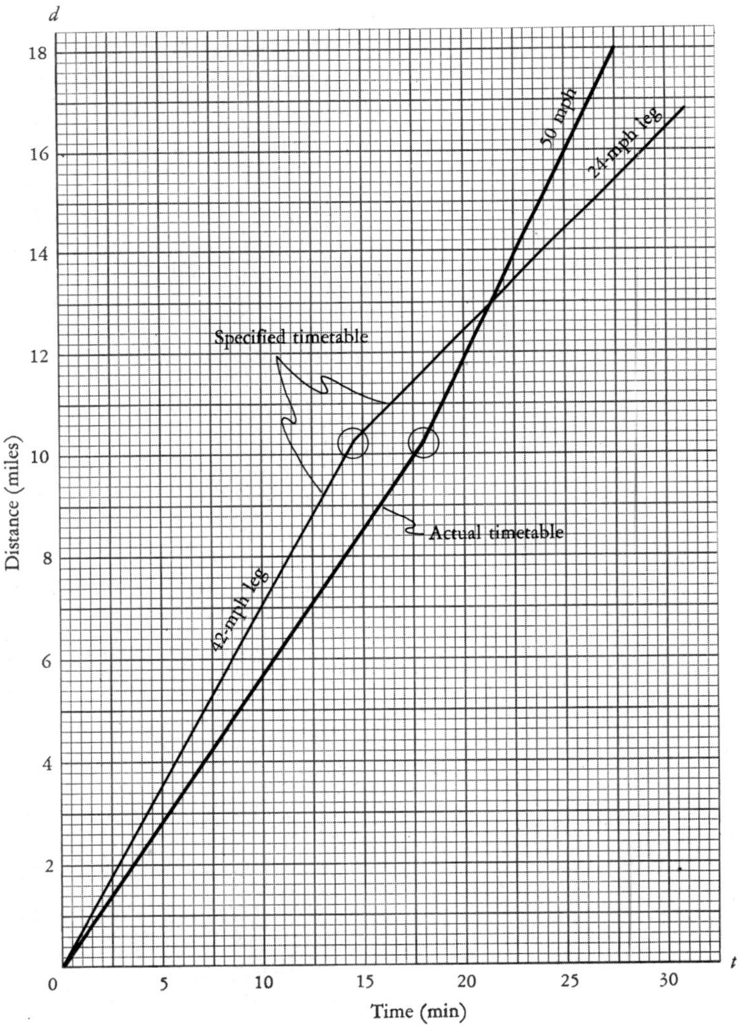

Fig. 8.6 Example 8.5.

Solution. Figure 8.6 is a graph of distance vs. time for the two legs of the rally. It shows that you are behind schedule. At a distance of 10.3 miles, the elapsed time should be 14.7 min. To get back on schedule, you should speed up (the faster the better, unless you acquire a speeding ticket or have an accident) until your distance and time coincide with any distance and time on the "24-mph leg" of Fig. 8.6. Then you should reduce speed and maintain a 24-mph average speed. The figure shows that by averaging 50 mph after the 10.3-mile point, you will be back on schedule when your corrected distance is 12.8 miles and the elapsed time is 21.0 min.

EXERCISE 8.3

Find the simultaneous solution to each pair of equations by graphing the equations.

1. $5x + 3y = 35$
   $3x - y = 7$
2. $y = 7x + 20$
   $y = -6x - 6$
3. $1.2u + 2.5v + 5.0 = 0$
   $3.2u - 0.50v - 9.6 = 0$
4. $r - 2s = 0$
   $2r + s = 250$
5. $4x - y = 0$
   $x + 2y = 81$
6. $6B + 4A = 26$
   $3A - (5/4)B + 5/2 = 0$
7. In a certain electric network, the currents through different branches are related by the formulas

$$I_1 + 5I_2 + 9 = 0$$

$$2I_1 + I_2 - 9 = 0$$

   Find the two currents by graphing I_2 vs. I_1.
8. The sum of two numbers is 24, and the difference between them is 43. What are the numbers? Solve by graphing.
9. A certain bronze alloy is 25 percent copper and 75 percent zinc, and another alloy is 40 percent copper and 60 percent zinc. How many pounds of each alloy should be melted and combined to produce 100 lb of an alloy with 31.2 percent copper and 68.8 percent zinc? (*Hint:* Let w_1 and w_2 represent the weights of the two alloys. Then for the copper, $0.25w_1 + 0.40w_2 = 31.2$ lb. Graph this equation; then graph a similar equation based on the weights of zinc.)
10. One airplane travels due south at an average speed of 250 mph. A second airplane leaves the same airport 45 min later and

travels due south at 462 mph. When will the second plane overtake the first one, and how far will they be from the airport? Find the solution graphically as in Example 8.4.

11. A certain bus line advertises 1½-hr travel time between two cities 75 miles apart. After leaving one of the cities, the driver notices that he has covered only 24 miles in 36 min. If he increases his average speed to 60 mph, how long will it take to get back on schedule? Find the solution graphically as in Example 8.5.

12. What is the minimum speed the bus of Prob. 11 could maintain after the first 36 min and still make the trip in the advertised time? (*Hint:* The average speed is the slope of the distance-vs.-time graph.)

8.3 ALGEBRAIC SOLUTIONS OF TWO SIMULTANEOUS LINEAR EQUATIONS

As an alternative to solving two simultaneous linear equations graphically, we can solve them algebraically. Algebraic solutions do not provide the insight of the graphical solution, but they are much faster and produce exact answers.

With algebraic solutions, the general procedure is to reduce the two equations and two variables to one equation and one variable. This reduction is accomplished by any one of several methods, but the basis for each method is elimination of one variable.

A. THE SUBSTITUTION METHOD

Given two independent linear equations $y = m_1x + k_1$ and $y = m_2x + k_2$, we know from graphing that one and only one pair of values (x, y) can satisfy both equations. Therefore, it is correct to equate the two expressions for y at *this one common point.* In effect we eliminate y by substituting $m_1x + k_1$ from the first equation for y in the second equation. The result is the following equation in x that applies only to the one point on both lines.

$$m_1x + k_1 = m_2x + k_2$$

We solve this equation for x, and having the value of x, we use it in either of the original equations to solve for y. The other equation can be used to check the solution.

Example 8.6. Find the simultaneous solution of the equations

$$y = 2x - 7$$

$$y = 6x + 5$$

We substitute $2x - 7$ from the first equation for y in the second equation as follows:

$$y = 6x + 5$$

$$2x - 7 = 6x + 5$$

Then we solve for x as follows:

$$6x - 2x = -7 - 5 = -12$$

$$4x = -12$$

$$\boxed{x = -3}$$

Now we substitute $x = -3$ in the first equation to find y:

$$\begin{aligned} y &= 2x - 7 \\ &= 2(-3) - 7 \\ &= -6 - 7 \end{aligned}$$

$$\boxed{y = -13}$$

To check, we substitute $x = -3$ and $y = -13$ in the second equation:

$$y = 6x + 5$$

$$-13 = 6(-3) + 5$$

$$-13 = -18 + 5$$

$$-13 = -13$$

If the original equations are not given in slope-intercept form, then at least one of them must be changed. Example 8.7 shows the procedure, and just to give a different viewpoint, it shows that we can solve for x in terms of y.

Example 8.7. Find the simultaneous solution of the equations

$$x - 2y + 6 = 0 \tag{8.1}$$

$$5x + 4y - 8 = 0 \tag{8.2}$$

First we solve Eq. (8.1) for x in terms of y.

$$x = 2y - 6 \tag{8.1a}$$

Then we substitute $2y - 6$ for x in Eq. (8.2) and solve for y:

$$5x + 4y - 8 = 0$$

$$5(2y - 6) + 4y - 8 = 0$$

$$10y - 30 + 4y - 8 = 0$$

$$14y - 38 = 0$$

$$\boxed{y = 19/7}$$

Finally, we substitute $y = 19/7$ into Eq. (8.1a), solve for x, and check in Eq. (8.2).

$$x = 2y - 6 \tag{8.1a}$$

$$x = 2(19/7) - 6 = 38/7 - 42/7$$

$$\boxed{x = -4/7}$$

Check:

$$5x + 4y - 8 = 0 \tag{8.2}$$

$$5(-4/7) + 4(19/7) - 8 = 0$$

$$-20/7 + 76/7 - 56/7 = 0$$

$$0 = 0$$

EXERCISE 8.4

Solve Probs. 1 to 10 of Exercise 8.3 by the substitution method.

B. THE MULTIPLICATION-ADDITION METHOD

Another way to eliminate one variable is illustrated by the following example. Given two equations $2x + 2y = 25$ and $4x - 2y = 17$, add them:

$$(2x + 4x) + (2y - 2y) = 25 + 17$$

$$6x + (0)y = 42$$

$$6x = 42$$

$$x = 7$$

The principle illustrated here is: *If in two linear equations the coefficients of one variable are equal in magnitude and opposite in sign, we can eliminate that variable by adding the two equations.*

If the coefficients of one variable are not equal or are not opposite in sign, then we *make* them so. We do this by multiplying *both* sides of one equation by a number that will result in the coefficients of one specific variable being equal but opposite. Example 8.8 illustrates this procedure.

Example 8.8. **Find the simultaneous solution of the equations**

$$7u + 3v - 5 = 0 \qquad (8.3)$$

$$2v + 3u + 20 = 0 \qquad (8.4)$$

First, rearrange the equations in a more orderly form so that like variables are aligned and the constants are on the right-hand side.

$$7u + 3v = \quad 5 \qquad (8.3)$$

$$3u + 2v = -20 \qquad (8.4)$$

Next, multiply *both sides* of Eq. (8.4) by $-3/2$ and add Eq. (8.3).

$$-(9/2)u - 3v = 30 \qquad (8.4a)$$

$$7u + 3v = \ 5 \qquad (8.3)$$

Add

$$(5/2)u \qquad = 35$$

$$\boxed{u = 14}$$

Substitute $u = 14$ into Eq. (8.4), solve for v, and check the results in Eq. (8.3).

$$3u + 2v = -20 \tag{8.4}$$

$$3(14) + 2v = -20$$

$$42 + 2v = -20$$

$$2v = -20 - 42 = -62$$

$$\boxed{v = -31}$$

Check:

$$7u + 3v = 5 \tag{8.3}$$

$$7(14) + 3(-31) = 5$$

$$98 - 93 = 5$$

$$5 = 5$$

An alternate scheme that avoids multiplying by a fraction is to multiply *both equations.* Example 8.9 repeats Example 8.8 but uses this alternate procedure.

Example 8.9. Find the simultaneous solution of the equations

$$7u + 3v = 5 \tag{8.5}$$

$$3u + 2v = -20 \tag{8.6}$$

Multiply Eq. (8.5) by 2 and Eq. (8.6) by -3; then add the results.

$$14u + 6v = 10 \tag{8.5a}$$

$$-9u - 6v = 60 \tag{8.6a}$$

Add

$$5u \qquad = 70$$

$$\boxed{u = 14}$$

Next substitute $u = 14$ into Eq. (8.6), solve for $v = -31$ as before, and check in Eq. (8.5).

EXERCISE 8.5

Solve Probs. 1 to 10 of Exercise 8.3 by the multiplication-addition method.

8.4 THE METHOD OF DETERMINANTS

A. BASIC IDEAS

We can derive a general simultaneous solution for two linear equations by applying either method of the preceding section. For example, given the general equations

$$a_1x + b_1y = c_1 \tag{8.7}$$

$$a_2x + b_2y = c_2 \tag{8.8}$$

we can solve for x by multiplying Eq. (8.7) by b_2 and Eq. (8.8) by $-b_1$ and then adding.

$$a_1b_2x + b_1b_2y = c_1b_2 \tag{8.7a}$$

$$-a_2b_1x - b_1b_2y = -c_2b_1 \tag{8.8a}$$

Add: $$(a_1b_2 - a_2b_1)x = c_1b_2 - c_2b_1$$

$$x = \frac{c_1b_2 - c_2b_1}{a_1b_2 - a_2b_1} \tag{8.9}$$

In a similar fashion we can show that

$$y = \frac{a_1c_2 - a_2c_1}{a_1b_2 - a_2b_1} \tag{8.10}$$

These equations contain some remarkable patterns and similarities. In the first place, the denominators are identical and are formed from the coefficients of x and y in Eqs. (8.7) and (8.8). In fact, the denominators can be formed by a crosswise pattern of multiplication and subtraction of the coefficients, as illustrated below.

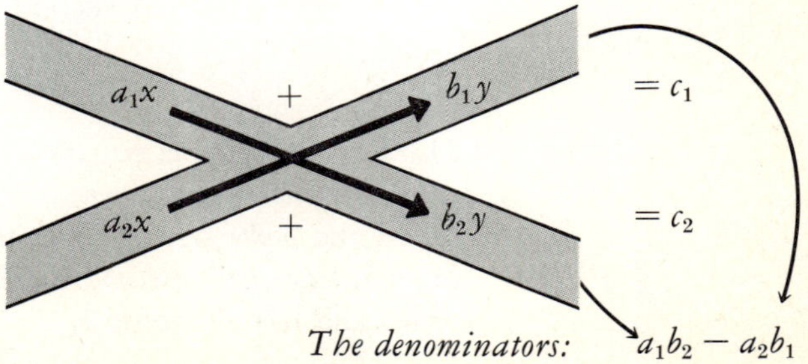

The denominators: $a_1b_2 - a_2b_1$ (8.11)

Also, the numerator for x in Eq. (8.9) has this same pattern but has the a coefficients of x replaced by the c terms on the right side of Eqs. (8.7) and (8.8). In a similar pattern, the numerator for y in Eq. (8.10) has c's where the denominator has b's.

As shown by the following example, we can use these patterns to find simultaneous solutions rapidly and without algebraic manipulations. The key is to write the equations in the same form as that of Eqs. (8.7) and (8.8) and to remember which numbers are the a's, b's, and c's.

Example 8.10. Simultaneous solution of linear equations

You are given

$$-x + 3y = 10$$

$$4x - 2y = 11$$

To solve, use the coefficients to form fractions according to Eqs. (8.9) and (8.10):

$$x = \frac{(10)(-2) - (11)(3)}{(-1)(-2) - (4)(3)} = \frac{-20 - 33}{2 - 12} = \frac{-53}{-10} = 5.3$$

$$y = \frac{(-1)(11) - (4)(10)}{(-1)(-2) - (4)(3)} = \frac{-11 - 40}{-10} = \frac{-51}{-10} = 5.1$$

B. THE DETERMINANT NOTATION

Mathematicians have devised a clever symbol called the determinant that enables us to describe crosswise patterns such as that of Expression (8.11) and to reconstruct Eqs. (8.9) and (8.10) from memory with remarkable ease. A *determinant* is an array of numbers in a square pattern [as illustrated on the left side of Eq. (8.12) below].

$$\begin{vmatrix} a_1 & b_1 \\ a_2 & b_2 \end{vmatrix} = a_1b_2 - a_2b_1 \qquad (8.12)$$

A second-order determinant and its expansion

The left side of this equation is the *determinant*, and the right side is the *expansion of the determinant*. The number of rows or columns is called the *order* of the determinant, and the numbers (or letters) in the array are called *elements*. Thus the determinant above is of second order and has elements a_1, a_2, b_1, and b_2.

Note that a second-order determinant is merely an algebraic symbol that represents a difference between two products. We can easily remember the expansion if we note that it is formed by multiplying the diagonal elements as indicated:

Expansion of a second-order determinant

$$\begin{vmatrix} a_1 & b_1 \\ a_2 & b_2 \end{vmatrix} = a_1b_2 - a_2b_1$$

The following examples show how to expand a second-order determinant.

Example 8.11.

$$\begin{vmatrix} 5 & 2 \\ 3 & 6 \end{vmatrix} = (5)(6) - (3)(2) = 30 - 6 = 24$$

Example 8.12.

$$\begin{vmatrix} 3 & 1 \\ 7 & -2 \end{vmatrix} = (3)(-2) - (7)(1) = -6 - 7 = -13$$

Example 8.13.

$$\begin{vmatrix} -12 & 16 \\ -2 & -3 \end{vmatrix} = (-12)(-3) - (-2)(16) = 36 + 32 = 68$$

Notice that the expansion of Eq. (8.12) is precisely the denominator of Eqs. (8.9) and (8.10). Recall that Expression (8.11) shows how to remember this denominator. Furthermore, if we replace the a's by c's, the resulting determinant is the numerator in Eq. (8.9); and if we replace the b's by c's, we have the numerator of Eq. (8.10). Thus we can write the simultaneous solutions for Eqs. (8.7) and (8.8) *in determinant form* as follows.

Given

$$a_1x + b_1y = c_1$$

$$a_2x + b_2y = c_2$$

the solutions are

$$x = \frac{\begin{vmatrix} c_1 & b_1 \\ c_2 & b_2 \end{vmatrix}}{\begin{vmatrix} a_1 & b_1 \\ a_2 & b_2 \end{vmatrix}} = \frac{c_1b_2 - c_2b_1}{a_1b_2 - a_2b_1} \qquad y = \frac{\begin{vmatrix} a_1 & c_1 \\ a_2 & c_2 \end{vmatrix}}{\begin{vmatrix} a_1 & b_1 \\ a_2 & b_2 \end{vmatrix}} = \frac{a_1c_2 - a_2c_1}{a_1b_2 - a_2b_1} \qquad (8.13)$$

Example 8.14. Simultaneous solutions of two linear equations by second-order determinants

Given

$$16x - 5y = 32$$

$$4x + 12y = -20$$

the solutions are

$$x = \frac{\begin{vmatrix} 32 & -5 \\ -20 & 12 \end{vmatrix}}{\begin{vmatrix} 16 & -5 \\ 4 & 12 \end{vmatrix}} = \frac{(32)(12) - (-20)(-5)}{(16)(12) - (4)(-5)} = \frac{384 - 100}{192 + 20} = \frac{284}{212} = \frac{71}{53}$$

$$y = \frac{\begin{vmatrix} 16 & 32 \\ 4 & -20 \end{vmatrix}}{212} = \frac{(16)(-20) - (4)(32)}{212} = \frac{-320 - 128}{212} = -\frac{448}{212} = -\frac{112}{53}$$

EXERCISE 8.6

Use second-order determinants to solve Probs. 1 to 10 of Exercise 8.3.

8.5 SYSTEMS OF THREE OR MORE LINEAR EQUATIONS

A linear equation with three variables such as

$$ax + by + cz = d$$

represents a plane, not just a line. Thus graphical solutions of a system of three-variable linear equations are very difficult, if not impossible. However, the substitution, addition, and determinant methods apply to systems of three or more linear equations.

The only condition needed to ensure a set of simultaneous solutions is that we have at least as many independent equations as we have variables (the term "independent equations" is defined in the next paragraph). In other words, if we have three variables, then we must have at least three independent equations; if four variables, then at least four independent equations; and so on. If we have more equations than variables, then some of the equations will be redundant. We must use only as many equations as we have variables, no more and no less.

Independent equations means equations in which the coefficients of corresponding variables are *not* all in the same ratio. For example,

$$2x - 5y + 3z = 1$$

and

$$6x - 15y + 9z = 5$$

are *not* independent because the coefficients of all three variables *are* in the same ratio. That is,

$$\frac{6x}{2x} = \frac{-15y}{-5y} = \frac{9z}{3z} = 3$$

If any one of the six coefficients were changed, then these equations would be independent of each other.

The following example illustrates simultaneous solutions to systems of linear equations with three or more variables.

A. THE SUBSTITUTION METHOD

Given the following three equations,

$$2x - 3y + 5z = 1 \tag{8.14}$$

$$x + 2y - z = 0 \tag{8.15}$$

$$8x + y + 2z = 3 \tag{8.16}$$

we solve one equation for x, y, or z and then eliminate that variable by substituting it into the other two equations. For example, we can solve Eq. (8.15) for z in terms of x and y, as follows:

$$z = x + 2y \tag{8.15a}$$

Then we substitute this expression into Eqs. (8.6) and (8.8) and combine similar terms, as follows:

$$2x - 3y + 5z = 1 \tag{8.14}$$

$$2x - 3y + 5(x + 2y) = 1$$

$$7x + 7y = 1 \tag{8.14a}$$

$$8x + y + 2z = 3$$

$$8x + y + 2(x + 2y) = 3$$

$$10x + 5y = 3 \tag{8.16a}$$

$$\left.\begin{aligned} 7x + 7y &= 1 \\ 10x + 5y &= 3 \end{aligned}\right.$$

Equations (8.14*a*) and (8.16*a*) constitute a system of two equations and two variables. We can solve them by any appropriate method. The results are

$$\boxed{x = {}^{16}\!/_{35} \qquad y = -{}^{11}\!/_{35}}$$

By substituting these values into Eq. (8.15*a*), we find

$$\boxed{z = -{}^{6}\!/_{35}}$$

If we have four equations and four variables, the procedure is much the same: (1) solve for one variable in terms of the others, (2) substitute this variable in all the other equations, and (3) repeat until only one equation with one variable remains.

B. THE MULTIPLICATION-ADDITION METHOD

In a certain electric network, the currents through different branches are related by the following four equations:

$$3I_1 + 3I_2 + 2I_3 + 6I_4 = 6 \tag{8.17}$$

$$2I_1 - 3I_2 + I_3 - 5I_4 = 5 \tag{8.18}$$

$$I_1 + I_2 - I_3 + 4I_4 = 0 \tag{8.19}$$

$$5I_1 - 2I_2 - 4I_3 - 6I_4 = 1 \tag{8.20}$$

Solve for the four currents.

The basic procedure is to eliminate the *same variable* 3 times by combining *different* pairs of equations 3 times. The result will be three equations with three variables. Then we eliminate one of these variables twice, thereby reducing the system to two equations and two variables.

Sometimes we can eliminate two variables at one time. A few minutes of study often pay dividends. In this problem, we shall start by eliminating I_1 and I_2 by multiplying Eq. (8.19) by -3 and adding the result to Eq. (8.17).

$$3I_1 + 3I_2 + 2I_3 + 6I_4 = 6 \tag{8.17}$$

$$-3I_1 - 3I_2 + 3I_3 - 12I_4 = 0 \tag{8.19a}$$

Add $$5I_3 - 6I_4 = 6 \tag{8.21}$$

Eliminate I_2 again by adding Eqs. (8.17) and (8.18).

$$3I_1 + 3I_2 + 2I_3 + 6I_4 = 6 \qquad (8.17)$$

$$\begin{array}{llr} & 2I_1 - 3I_2 + I_3 - 5I_4 = 5 & (8.18) \\ \text{Add} & 5I_1 \qquad + 3I_3 + I_4 = 11 & (8.22) \end{array}$$

Remember that we must keep on eliminating the *same variable.* Next we eliminate I_2 by multiplying Eq. (8.19) by 2 and adding to Eq. (8.20).

$$2I_1 + 2I_2 - 2I_3 + 8I_4 = 0 \qquad (8.19b)$$

$$\begin{array}{llr} & 5I_1 - 2I_2 - 4I_3 - 6I_4 = 1 & (8.20) \\ \text{Add} & 7I_1 \qquad - 6I_3 + 2I_4 = 1 & (8.23) \end{array}$$

Equations (8.21), (8.22), and (8.23) constitute a system of three linear equations and three variables.

$$5I_3 - 6I_4 = 6 \qquad (8.21)$$

$$5I_1 + 3I_3 + I_4 = 11 \qquad (8.22)$$

$$7I_1 - 6I_3 + 2I_4 = 1 \qquad (8.23)$$

Since Eq. (8.21) has no I_1, we shall eliminate I_1 by multiplying Eq. (8.22) by 7, multiplying Eq. (8.23) by -5, and adding.

$$35I_1 + 21I_3 + 7I_4 = 77 \qquad (8.22a)$$

$$\begin{array}{llr} & -35I_1 + 30I_3 - 10I_4 = -5 & (8.23a) \\ \text{Add} & 51I_3 - 3I_4 = 72 & \\ \div \text{ by } 3 & 17I_3 - I_4 = 24 & (8.24) \end{array}$$

Equations (8.21) and (8.24) now form a system of two linear equations with two variables.

$$5I_3 - 6I_4 = 6 \qquad (8.21)$$

$$17I_3 - I_4 = 24 \qquad (8.24)$$

To eliminate I_4, we multiply Eq. (8.24) by -6 and add.

$$5I_3 - 6I_4 = 6 \qquad (8.21)$$

$$\begin{array}{llr} & -102I_3 + 6I_4 = -144 & (8.24a) \\ \text{Add} & -97I_3 = -138 & \end{array}$$

$$\boxed{I_3 = {}^{138}/_{97}}$$

By successive substitutions into Eq. (8.21) then (8.22), and finally any one of (8.17) to (8.20), we find I_4, I_1, and I_2, respectively:

$$\boxed{I_4 = {}^{18}\!/_{97} \qquad I_1 = {}^{127}\!/_{97} \qquad I_2 = -{}^{61}\!/_{97}}$$

EXERCISE 8.7

Find the simultaneous solutions for each system of equations. Use one of the two methods illustrated in this section, or a combination of the methods.

1. $$\begin{aligned} x + y + z &= 1 \\ 2x - y + z &= 2 \\ x - 2y - z &= 0 \end{aligned}$$

2. $$\begin{aligned} 2a - b + 3c &= 9 \\ -a + 3b - 5c &= -10 \\ 7a - 2b + 2c &= 9 \end{aligned}$$

3. $$\begin{aligned} 8r + 2s + 4t &= 21 \\ 2r - s + t &= 3 \\ -r + 3s + 2t &= 1 \end{aligned}$$

4. $$\begin{aligned} 5x \qquad - z &= -23.5 \\ 3x - 2y + z &= 8.5 \\ 7x + y + 3z &= 18.2 \end{aligned}$$

5. $$\begin{aligned} a - b + c &= 30 \\ 5a + b \qquad &= 130 \\ 2a + 3b - c &= 90 \end{aligned}$$

6. $$\begin{aligned} 2u - v - 2w &= 0 \\ 5u + 3v + 3w &= 115 \\ 3u - 2v - 3w &= -25 \end{aligned}$$

7. $$\begin{aligned} s + t + u + v &= 4 \\ s - t + 2u + v &= 6 \\ 3s - 4t + 5u - 2v &= -1 \\ 2s \qquad - 3u - v &= -2 \end{aligned}$$

8. $$\begin{aligned} 3p - r + s - t &= 0 \\ -2p + 2r - 3s - 3t &= 2 \\ p + 3r + 2s + t &= -5 \\ 4p + r \qquad - t &= 3 \end{aligned}$$

9. The equations below are derived from the electric circuit shown. Solve for the three currents I_1, I_2, and I_3. Negative values are permissible.

$$3I_1 + 6I_2 = 9$$

$$6I_2 + 5I_3 = 6$$

$$I_1 - I_2 + I_3 = 0$$

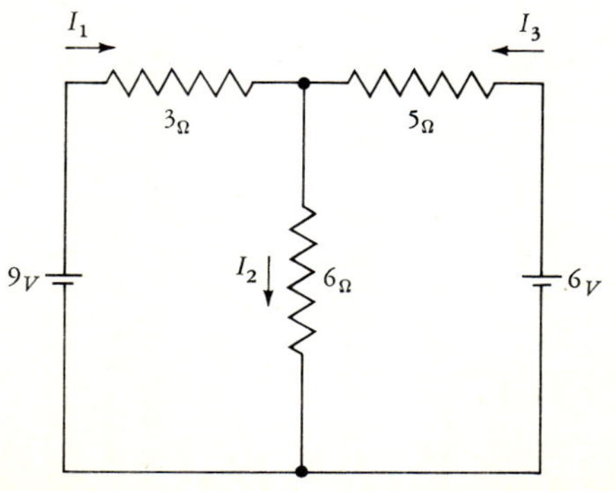

10. Three pieces of aluminum alloy are to be melted down and mixed to form a new alloy. Determine if this is possible, and if so, how many pounds of each alloy are needed for each pound of the mixture, if the percentages of the elements are as follows:

	FIRST ALLOY	SECOND ALLOY	THIRD ALLOY	NEW ALLOY
Percent aluminum	60	55	65	57
Percent nickel	25	35	10	22
Percent zinc	15	10	25	21

11. Repeat Prob. 10 if the percentages of the elements are as follows:

	FIRST ALLOY	SECOND ALLOY	THIRD ALLOY	NEW ALLOY
Percent aluminum	50	60	60	55
Percent nickel	25	20	30	25
Percent zinc	25	20	10	20

12. To find the number of teeth on each gear of the reverted gear train shown in Fig. 8.7, we must solve the following equations. Find the number of teeth on each gear.

$$T_A + T_D = 30 \qquad T_B + T_F = 30$$

$$\frac{T_A}{T_D} = 4 \qquad \frac{T_B}{T_F} = 5$$

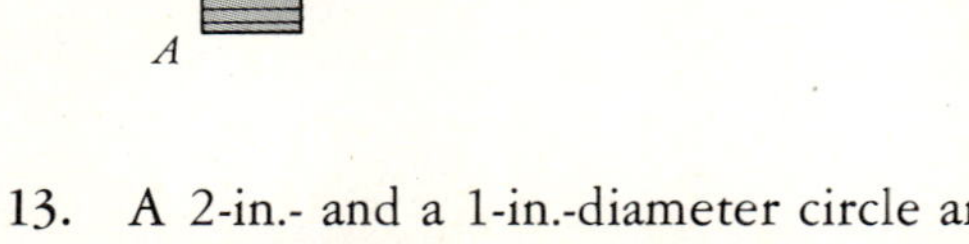

Fig. 8.7 Prob. 12. A reverted gear train with gears A, B, D, and F.

13. A 2-in.- and a 1-in.-diameter circle are inscribed in a 3-in.-diameter circle as shown in Fig. 8.8*a*. Find the radius and the coordinates of the center of the circle that is drawn tangent to the three given circles. (*Hint:* For each pair of circles, a right triangle can be formed as shown in Fig. 8.8*b*. By applying the Pythagorean formula to three such right triangles, three equations can be derived.)

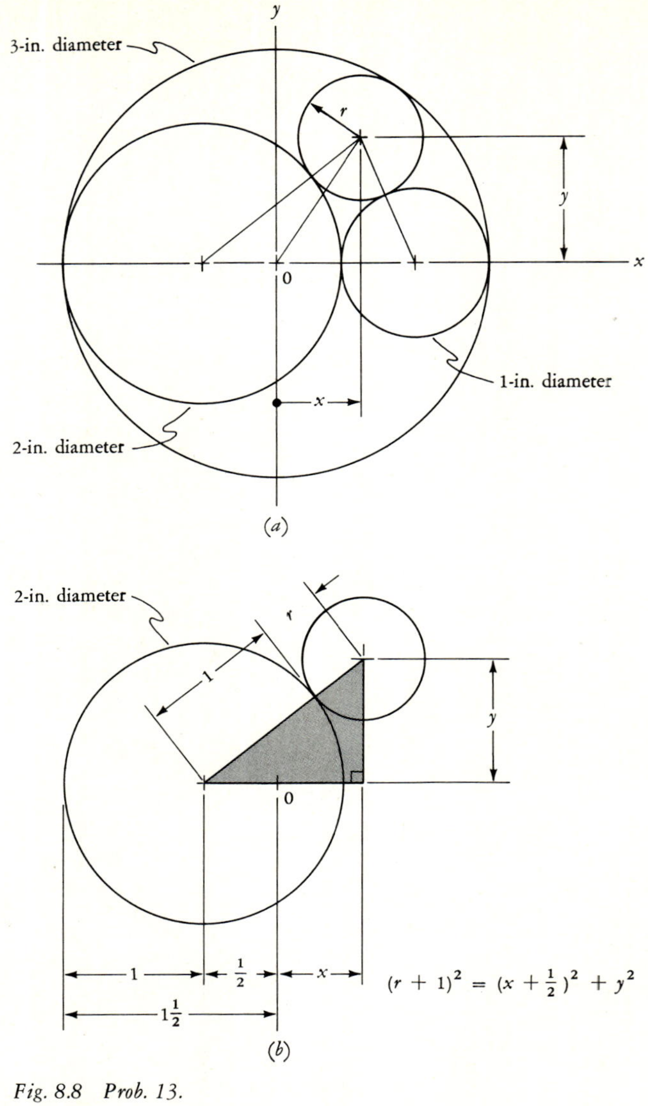

Fig. 8.8 Prob. 13.

8.6 DETERMINANTS

Before we can solve a system of three or more linear equations by determinants, we shall need to learn more about expanding determinants. We shall also examine some properties that greatly simplify the process.

A. EXPANSION BY MINORS

We shall illustrate the expansion of determinants in general by expanding the *third-order determinant* below.

$$\begin{vmatrix} a_1 & b_1 & c_1 \\ a_2 & b_2 & c_2 \\ a_3 & b_3 & c_3 \end{vmatrix}$$

One expansion of this determinant is defined in terms of second-order determinants as follows:

$$\begin{vmatrix} a_1 & b_1 & c_1 \\ a_2 & b_2 & c_2 \\ a_3 & b_3 & c_3 \end{vmatrix} = a_1 \begin{vmatrix} b_2 & c_2 \\ b_3 & c_3 \end{vmatrix} - a_2 \begin{vmatrix} b_1 & c_1 \\ b_3 & c_3 \end{vmatrix} + a_3 \begin{vmatrix} b_1 & c_1 \\ b_2 & c_2 \end{vmatrix} \qquad (8.25)$$

Each second-order determinant is called the *minor* of the element which is its coefficient. The sign preceding each coefficient a_1, a_2, or a_3 has a definite relationship to the position of that element in the determinant. The signs always alternate in a *checkerboard pattern* as illustrated in the diagrams below.

$$\begin{vmatrix} + & - & + \\ - & + & - \\ + & - & + \end{vmatrix} \qquad \begin{vmatrix} + & - & + & - \\ - & + & - & + \\ + & - & + & - \\ - & + & - & + \end{vmatrix}$$

Third order Fourth order

Signs corresponding to the elements of third-order and fourth-order determinants

Any expansion, as in Eq. (8.25), can easily be developed by the following procedure:

1. Choose *any* element and cross out the row and column containing it. This element is called the *pivotal element.* One term of the expansion is the product of the pivotal element and its minor—the determinant of the remaining elements. The diagram below illustrates this procedure, using a_1 as the pivotal element.

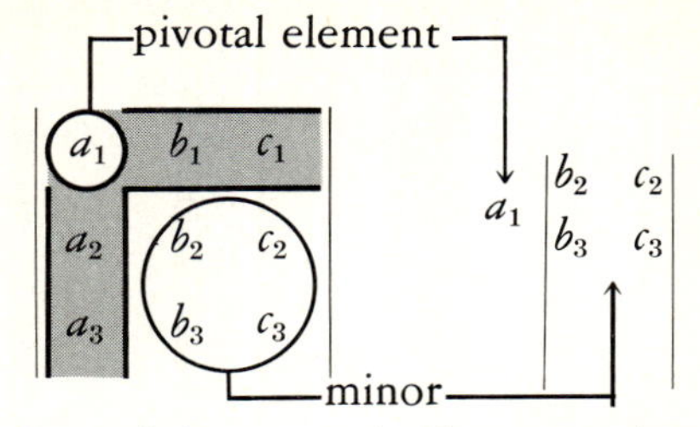

One of the terms in the expansion of a third-order determinant

2. Form the other terms of the expansion by repeating step 1 for each of the remaining elements *either* in the same row *or* in the same column as the pivotal element. A check of Eq. (8.25) will show how this procedure was used on elements a_2 and a_3.

The thoughtful reader may conclude that there are many ways to expand a determinant and that all expansions must produce the same result. For example, we could expand the determinant of Eq. (8.25) as follows, by using b_3 as the pivotal element:

$$\begin{vmatrix} a_1 & b_1 & c_1 \\ a_2 & b_2 & c_2 \\ a_3 & b_3 & c_3 \end{vmatrix} = -b_3 \begin{vmatrix} a_1 & c_1 \\ a_2 & c_2 \end{vmatrix} + a_3 \begin{vmatrix} b_1 & c_1 \\ b_2 & c_2 \end{vmatrix} + c_3 \begin{vmatrix} a_1 & b_1 \\ a_2 & b_2 \end{vmatrix} \qquad (8.26)$$

The expansions of Eqs. (8.25) and (8.26) do not appear equal, but we can expand the minors and arrange the elements in a, b, c order to show that both expansions—and all other expansions of this determinant—equal the following expression:

$$\begin{vmatrix} a_1 & b_1 & c_1 \\ a_2 & b_2 & c_2 \\ a_3 & b_3 & c_3 \end{vmatrix} = \begin{aligned} &a_1b_2c_3 + a_2b_3c_1 + a_3b_1c_2 - a_3b_2c_1 - a_2b_1c_3 \\ &- a_1b_3c_2 \end{aligned} \qquad (8.27)$$

General expansion of a third-order determinant

Example 8.15. **Find the value of the determinant** $\begin{vmatrix} 5 & 4 & -6 \\ -2 & 1 & -1 \\ -3 & 10 & 7 \end{vmatrix}$

Solution. **Using 5 as the pivotal element and expanding the first column as in Eq. (8.25), we have**

$$\begin{vmatrix} 5 & 4 & -6 \\ -2 & 1 & -1 \\ -3 & 10 & 7 \end{vmatrix} = 5\begin{vmatrix} 1 & -1 \\ 10 & 7 \end{vmatrix} - (-2)\begin{vmatrix} 4 & -6 \\ 10 & 7 \end{vmatrix} + (-3)\begin{vmatrix} 4 & -6 \\ 1 & -1 \end{vmatrix}$$

$$\begin{aligned} &= 5[(1)(7) - (10)(-1)] + 2[(4)(7) - (10)(-6)] \\ &\quad - 3[(4)(-1) - (1)(-6)] \\ &= 5[7 + 10] + 2[28 + 60] - 3[-4 + 6] \\ &= 5[17] + 2[88] - 3[2] \end{aligned}$$

$$\begin{vmatrix} 5 & 4 & -6 \\ -2 & 1 & -1 \\ -3 & 10 & 7 \end{vmatrix} = 255$$

Example 8.16. Find the value of the determinant

$$\begin{vmatrix} 1 & 4 & -3 & -1 \\ 3 & -2 & 1 & 5 \\ -2 & 0 & -4 & -7 \\ 1 & 5 & 6 & 2 \end{vmatrix}$$

Solution. Because the second column contains a zero, we shall use the elements of that column to expand the determinant. Note how we choose the sign for each term according to the "checkerboard" rule.

$$\begin{vmatrix} 1 & 4 & -3 & -1 \\ 3 & -2 & 1 & 5 \\ -2 & 0 & -4 & -7 \\ 1 & 5 & 6 & 2 \end{vmatrix} = -(4)\begin{vmatrix} 3 & 1 & 5 \\ -2 & -4 & -7 \\ 1 & 6 & 2 \end{vmatrix}$$

$$+ (-2)\begin{vmatrix} 1 & -3 & -1 \\ -2 & -4 & -7 \\ 1 & 6 & 2 \end{vmatrix}$$

$$-(0)\begin{vmatrix} 1 & -3 & -1 \\ 3 & 1 & 5 \\ 1 & 6 & 2 \end{vmatrix}$$

$$+ (5)\begin{vmatrix} 1 & -3 & -1 \\ 3 & 1 & 5 \\ -2 & -4 & -7 \end{vmatrix}$$

The result, after expanding each minor as in Example (8.15), is

$$\begin{vmatrix} 1 & 4 & -3 & -1 \\ 3 & -2 & 1 & 5 \\ -2 & 0 & -4 & -7 \\ 1 & 5 & 6 & 2 \end{vmatrix} = -388$$

EXERCISE 8.8

1. Show that the right-hand side of Eq. (8.25) equals the right-hand side of Eq. (8.27).
2. Show that the right-hand side of Eq. (8.26) equals the right-hand side of Eq. (8.27).

In Probs. 3 to 10, find the value of each determinant.

3. $\begin{vmatrix} 1 & 2 & -3 \\ -1 & 4 & -5 \\ 2 & 6 & 7 \end{vmatrix}$

4. $\begin{vmatrix} -2 & -1 & 5 \\ 3 & 10 & 7 \\ 5 & 9 & 0 \end{vmatrix}$

5. $\begin{vmatrix} 1 & 2 & 1 \\ 2 & 0 & 2 \\ 3 & 2 & 3 \end{vmatrix}$

6. $\begin{vmatrix} 100 & -27 & 225 \\ 0 & 0 & 0 \\ 98 & -49 & 53 \end{vmatrix}$

7. $\begin{vmatrix} -29 & 78 & 0 \\ 32 & -62 & 0 \\ 109 & -85 & 0 \end{vmatrix}$

8. $\begin{vmatrix} 5 & 3 & 2 \\ -9 & 0 & 0 \\ 6 & -4 & 8 \end{vmatrix}$

9. $\begin{vmatrix} 1 & -2 & -1 & 2 \\ 5 & 3 & 6 & 4 \\ 3 & 4 & 3 & -5 \\ 2 & 0 & 0 & 1 \end{vmatrix}$

10. $\begin{vmatrix} 2 & 1 & -2 & 5 & -3 \\ 0 & 3 & 4 & 0 & 2 \\ 0 & 5 & -6 & 0 & 1 \\ 0 & 7 & 8 & 0 & 5 \\ 0 & 9 & -10 & 1 & 6 \end{vmatrix}$

11. Show that

$$k\begin{vmatrix} a_1 & b_1 \\ a_2 & b_2 \end{vmatrix} = \begin{vmatrix} ka_1 & b_1 \\ ka_2 & b_2 \end{vmatrix} = \begin{vmatrix} ka_1 & kb_1 \\ a_2 & b_2 \end{vmatrix}$$

where k is any real number.

12. Show that

$$\begin{vmatrix} a_1 & b_1 \\ a_2 & b_2 \end{vmatrix} = \begin{vmatrix} a_1 & (b_1 + ka_1) \\ a_2 & (b_2 + ka_2) \end{vmatrix} = \begin{vmatrix} a_1 & b_1 \\ (a_2 + ka_1) & (b_2 + kb_1) \end{vmatrix}$$

where k is any real number.

B. PROPERTIES OF DETERMINANTS

Determinants have several properties that enable us to evaluate them very quickly. We shall not attempt to list every property, and we shall not attempt a rigorous proof of any property. But we shall examine some of the more useful ones.

Property 1. **If all elements of a row (or a column) are zero, the value of the determinant is zero. Problems 6 and 7 of the preceding set illustrate this property.**

Property 2. **If two rows (or two columns) are identical, the value of the determinant is zero. Problem 5 of the preceding set illustrates this property.**

Property 3. **If all elements in any one row (or in any one column) are multiplied by the same number, the result is the same as if the value of the determinant were multiplied by that number. Problem 11 of the preceding set illustrates this property.**

Property 4. **The value of a determinant is unchanged if every element of a row (or column) is multiplied by the same number and added to the corresponding element of another row (or column). Problem 12 of the preceding set illustrates this property.**

Property 5. **If all but one element in a row (or in a column) is zero, the value of the determinant is the product of the non-zero element and its minor. Problems 8 and 10 of the preceding set illustrate this property.**

We can use the last two properties in particular to simplify the expansion of a determinant. The idea is to form zeros for all but one element in some one row or column. We do this by successively applying Property 4 until all corresponding elements of the other rows (or columns) are zero. For example, in the determinant below, we can choose the element 1 as the pivotal element and multiply the first column by -3. Then when we add the result to the second column, the corresponding element in the second column is zero.

$$\begin{vmatrix} 2 & -5 & 2 \\ 1 & 3 & -6 \\ 3 & 7 & 5 \end{vmatrix} = \begin{vmatrix} 2 & (-6-5) & 2 \\ 1 & (-3+3) & -6 \\ 3 & (-9+7) & 5 \end{vmatrix} = \begin{vmatrix} 2 & -11 & 2 \\ 1 & 0 & -6 \\ 3 & -2 & 5 \end{vmatrix}$$

Multiply first column by -3 and add to second column.

The result is a zero in the second column.

To make all but the pivotal element zero, we repeat this step by multiplying the first column by +6 and adding the result to the third column. Then the corresponding element in the third column is zero.

$$\begin{vmatrix} 2 & -11 & 2 \\ 1 & 0 & -6 \\ 3 & -2 & 5 \end{vmatrix} = \begin{vmatrix} 2 & -11 & (12+2) \\ 1 & 0 & (6-6) \\ 3 & -2 & (18+5) \end{vmatrix} = \begin{vmatrix} 2 & -11 & 14 \\ 1 & 0 & 0 \\ 3 & -2 & 23 \end{vmatrix}$$

Multiply first column by +6 and add to third column.

The result is a zero in the third column.

Now the determinant is very easy to expand. By applying Properties 4 and 5 we have

$$\begin{vmatrix} 2 & -5 & 2 \\ 1 & 3 & -6 \\ 3 & 7 & 5 \end{vmatrix} = \begin{vmatrix} 2 & -11 & 14 \\ \textcircled{1} & 0 & 0 \\ 3 & -2 & 23 \end{vmatrix} = -(1) \begin{vmatrix} -11 & 14 \\ -2 & 23 \end{vmatrix}$$

$$= (-1)(-253 + 28) = (-1)(-225) = 225$$

The higher the order of the determinant, the more valuable is this technique; and the technique can be repeated over and over on the resulting minors, as the following example shows. Note the advantage of using 1 as a pivotal element if possible: we do not have to multiply by fractions.

Example 8.17. Expanding a determinant by zeros.

Evaluate the following determinant.

$$\begin{vmatrix} 3 & 2 & -8 & 9 \\ -2 & 5 & 2 & -11 \\ 4 & -3 & 4 & -3 \\ 3 & 6 & 12 & 9 \end{vmatrix}$$

First we simplify by applying Property 3. To do this we take a factor of 2 from each element of the third column and a factor of 3 from each element of the fourth row.

$$\begin{vmatrix} 3 & 2 & -8 & 9 \\ -2 & 5 & 2 & -11 \\ 4 & -3 & 4 & -3 \\ 3 & 6 & 12 & 9 \end{vmatrix} = \begin{vmatrix} 3 & 2 & 2(-4) & 9 \\ -2 & 5 & 2(1) & -11 \\ 4 & -3 & 2(2) & -3 \\ 3(1) & 3(2) & 2(3)(2) & 3(3) \end{vmatrix} = (2)(3) \begin{vmatrix} 3 & 2 & -4 & 9 \\ -2 & 5 & 1 & -11 \\ 4 & -3 & 2 & -3 \\ 1 & 2 & 2 & 3 \end{vmatrix}$$

Then we choose the 1 in the third column as the pivotal element, form zeros of the corresponding elements in the other rows, and repeat the process on the resulting minor.

Multiply second row as indicated and add to other rows.

$$(2)(3)\begin{vmatrix} 3 & 2 & -4 & 9 \\ -2 & 5 & 1 & -11 \\ 4 & -3 & 2 & -3 \\ 1 & 2 & 2 & 3 \end{vmatrix} \begin{matrix} \times (4) \longrightarrow \\ \\ \times (-2) \longrightarrow \\ \times (-2) \longrightarrow \end{matrix} = 6\begin{vmatrix} -5 & 22 & 0 & -35 \\ -2 & 5 & \textcircled{1} & -11 \\ 8 & -13 & 0 & 19 \\ 5 & -8 & 0 & 25 \end{vmatrix}$$

Multiply the first row as indicated and add to other rows.

$$= -(1)(6)\begin{vmatrix} -5 & 22 & -35 \\ 8 & -13 & 19 \\ 5 & -8 & 25 \end{vmatrix} \begin{matrix} \\ \times (8/5) \longrightarrow \\ \times (1) \longrightarrow \end{matrix} = (-6)\begin{vmatrix} \textcircled{-5} & 22 & -35 \\ 0 & 111/5 & -37 \\ 0 & 14 & -10 \end{vmatrix}$$

$$= (-5)(-6)\begin{vmatrix} 111/5 & -37 \\ 14 & -10 \end{vmatrix} = 30(-222 + 518) = 8{,}880$$

$$\begin{vmatrix} 3 & 2 & -8 & 9 \\ -2 & 5 & 2 & -11 \\ 4 & -3 & 4 & -3 \\ 3 & 6 & 12 & 9 \end{vmatrix} = 8{,}880$$

EXERCISE 8.9

Examine each determinant in light of Properties 1 through 5 of this section. Then use the appropriate properties to evaluate each determinant. State each property used.

1. $\begin{vmatrix} -1 & 5 & 7 \\ 2 & -3 & 6 \\ -1 & 5 & 7 \end{vmatrix}$

2. $\begin{vmatrix} 0 & 1 & -1 \\ 0 & -1 & 1 \\ 25 & 24 & 23 \end{vmatrix}$

3. $\begin{vmatrix} 2 & 0 & 8 \\ 4 & 0 & 10 \\ 6 & 0 & 12 \end{vmatrix}$

4. $\begin{vmatrix} 1 & -2 & 3 \\ -4 & 5 & -6 \\ 7 & -8 & 9 \end{vmatrix}$

5. $\begin{vmatrix} 3 & 4 & -5 \\ -2 & 6 & -4 \\ 2 & -2 & 6 \end{vmatrix}$

6. $\begin{vmatrix} 2 & 4 & 6 \\ 8 & 10 & 12 \\ 14 & 16 & 18 \end{vmatrix}$

7. $\begin{vmatrix} -2 & 7 & -4 \\ 1 & -5 & 2 \\ 3 & 8 & 6 \end{vmatrix}$

8. $\begin{vmatrix} 2 & -1 & 3 \\ 10 & 3 & 15 \\ -4 & 0 & -6 \end{vmatrix}$

9. $\begin{vmatrix} 5 & 8 & 17 \\ -2 & 9 & -10 \\ 3 & -11 & 4 \end{vmatrix}$

10. $\begin{vmatrix} 1/2 & 1/3 & -1/5 \\ -1/4 & 2 & 6 \\ 3/4 & 5 & -2/3 \end{vmatrix}$

11. $\begin{vmatrix} 0.6 & 0.5 & -2.4 \\ -0.2 & 1.5 & 0.3 \\ 1.4 & -2.0 & 0.6 \end{vmatrix}$

12. $\begin{vmatrix} 3 & -2 & 4 & 3 \\ 2 & 5 & -3 & 6 \\ -8 & 2 & 4 & 12 \\ 9 & -11 & -3 & 9 \end{vmatrix}$

13. $\begin{vmatrix} 6 & 3 & -9 & -3 \\ 0 & 4 & -2 & 1 \\ -4 & 5 & 5 & 0 \\ 8 & -3 & -3 & 4 \end{vmatrix}$

14. $\begin{vmatrix} 2 & 0 & -3 & -1 & 12 \\ -5 & 3 & 7 & 1 & -3 \\ 3 & -2 & 1 & 8 & 9 \\ 0 & -5 & 0 & 5 & 0 \\ -2 & 4 & 6 & -2 & -6 \end{vmatrix}$

15. $\begin{vmatrix} -10 & 4 & 0 & -2 & 6 \\ 5 & 0 & 3 & 7 & -3 \\ -5 & 1 & 0 & -5 & 9 \\ 15 & 2 & -2 & 10 & 3 \\ 10 & 0 & 4 & -1 & -6 \end{vmatrix}$

8.7 THE DETERMINANT METHOD FOR SOLUTIONS OF SIMULTANEOUS LINEAR EQUATIONS

In Sec. 8.4 we saw that second-order determinants could be used to solve two simultaneous linear equations. This was true because the expansions of the determinants exactly matched the general solutions, Eqs. (8.9) and (8.10). We shall now see that third-order determinants match the general solutions for three simultaneous linear equations, that fourth-order determinants match the solutions for four equations, and so on.

Given the three general linear equations

$$\begin{aligned} a_1x + b_1y + c_1z &= d_1 \\ a_2x + b_2y + c_2z &= d_2 \\ a_3x + b_3y + c_3z &= d_3 \end{aligned} \tag{8.28}$$

we can find the general solutions by the method of multiplication-addition. The detailed steps would be similar to those we used to derive Eq. (8.9). The results are as follows:

$$x = \frac{d_1b_2c_3 + d_2b_3c_1 + d_3b_1c_2 - d_3b_2c_1 - d_2b_1c_3 - d_1b_3c_2}{a_1b_2c_3 + a_2b_3c_1 + a_3b_1c_2 - a_3b_2c_1 - a_2b_1c_3 - a_1b_3c_2} \tag{8.29}$$

$$y = \frac{a_1d_2c_3 + a_2d_3c_1 + a_3d_1c_2 - a_3d_2c_1 - a_2d_1c_3 - a_1d_3c_2}{a_1b_2c_3 + a_2b_3c_1 + a_3b_1c_2 - a_3b_2c_1 - a_2b_1c_3 - a_1b_3c_2} \tag{8.30}$$

$$z = \frac{a_1b_2d_3 + a_2b_3d_1 + a_3b_1d_2 - a_3b_2d_1 - a_2b_1d_3 - a_1b_3d_2}{a_1b_2c_3 + a_2b_3c_1 + a_3b_1c_2 - a_3b_2c_1 - a_2b_1c_3 - a_1b_3c_2} \tag{8.31}$$

Simultaneous solutions for Eqs. (8.28)

Now let's reexamine Eq. (8.27), the expansion of a general third-order determinant, which is repeated here.

$$\begin{vmatrix} a_1 & b_1 & c_1 \\ a_2 & b_2 & c_2 \\ a_3 & b_3 & c_3 \end{vmatrix} = \begin{aligned} &a_1b_2c_3 + a_2b_3c_1 + a_3b_1c_2 \\ &- a_3b_2c_1 - a_2b_1c_3 - a_1b_3c_2 \end{aligned} \tag{8.27}$$

Observe that this expansion exactly matches the denominators of Eqs. (8.29), (8.30), and (8.31). Thus *the denominator part of the simultaneous solution for each variable is the determinant of the coefficients of the variables.* Compare the arrangement of the elements in this determinant to the arrangement of the coefficients in Eqs. (8.28).

Furthermore, the numerator determinant for each variable is almost identical to the denominator determinant. The only difference is that the coefficients of the variable are replaced by the corresponding constant terms on the right-hand side of the linear equations. Notice, for example, in Eq. (8.29) for x, that the numerator would be identical to the denominator except that each a is replaced by the corresponding d.

Thus we can write the simultaneous solutions for Eqs. (8.28) in determinant form as follows:

$$x = \frac{\begin{vmatrix} d_1 & b_1 & c_1 \\ d_2 & b_2 & c_2 \\ d_3 & b_3 & c_3 \end{vmatrix}}{\begin{vmatrix} a_1 & b_1 & c_1 \\ a_2 & b_2 & c_2 \\ a_3 & b_3 & c_3 \end{vmatrix}} \qquad y = \frac{\begin{vmatrix} a_1 & d_1 & c_1 \\ a_2 & d_2 & c_2 \\ a_3 & d_3 & c_3 \end{vmatrix}}{\begin{vmatrix} a_1 & b_1 & c_1 \\ a_2 & b_2 & c_2 \\ a_3 & b_3 & c_3 \end{vmatrix}} \qquad z = \frac{\begin{vmatrix} a_1 & b_1 & d_1 \\ a_2 & b_2 & d_2 \\ a_3 & b_3 & d_3 \end{vmatrix}}{\begin{vmatrix} a_1 & b_1 & c_1 \\ a_2 & b_2 & c_2 \\ a_3 & b_3 & c_3 \end{vmatrix}} \tag{8.32}$$

Simultaneous solutions for Eqs. (8.28)

The same idea applies to any number of simultaneous linear equations. The following example illustrates the simultaneous solutions by the method of determinants.

Example 8.18. Solution of four simultaneous linear equations by determinants

The currents through branches of a certain electric circuit are described by the following equations. Find the value of each current. Negative values are permissible.

$$3I_1 + 3I_2 \qquad + 6I_4 = 6 \tag{8.33}$$

$$2I_1 - 3I_2 + I_3 \qquad = 5 \tag{8.34}$$

$$I_2 - I_3 + 4I_4 = 0 \tag{8.35}$$

$$5I_1 - 2I_2 \qquad - 6I_4 = 1 \tag{8.36}$$

The denominator part of each solution is obtained below. First we add the second row to the third row; then we extract a factor of 3 from the first row, a factor of 2 from the third row, and a factor of 2 from the fourth column. The resulting third-order determinant is reduced in a similar fashion.

$$D = \begin{vmatrix} 3 & 3 & 0 & 6 \\ 2 & -3 & 1 & 0 \\ 0 & 1 & -1 & 4 \\ 5 & -2 & 0 & -6 \end{vmatrix} = \begin{vmatrix} 3 & 3 & 0 & 6 \\ 2 & -3 & 1 & 0 \\ 2 & -2 & 0 & 4 \\ 5 & -2 & 0 & -6 \end{vmatrix}$$

$$= -(1)\begin{vmatrix} 3 & 3 & 6 \\ 2 & -2 & 4 \\ 5 & -2 & -6 \end{vmatrix} = -(1)(3)(2)(2)\begin{vmatrix} 1 & 1 & 1 \\ 1 & -1 & 1 \\ 5 & -2 & -3 \end{vmatrix}$$

$$= -(12)\begin{vmatrix} 2 & 1 & 2 \\ 0 & -1 & 0 \\ 3 & -2 & -5 \end{vmatrix} = -(12)(-1)\begin{vmatrix} 2 & 2 \\ 3 & -5 \end{vmatrix}$$

$$D = (12)(-10 - 6) = -192$$

$$N_1 = \begin{vmatrix} 6 & 3 & 0 & 6 \\ 5 & -3 & 1 & 0 \\ 0 & 1 & -1 & 4 \\ 1 & -2 & 0 & -6 \end{vmatrix} = \begin{vmatrix} 6 & 3 & 0 & 6 \\ 5 & -3 & 1 & 0 \\ 5 & -2 & 0 & 4 \\ 1 & -2 & 0 & -6 \end{vmatrix}$$

$$= (3)(2)(-1)\begin{vmatrix} 2 & 1 & 1 \\ 5 & -2 & 2 \\ 1 & -2 & -3 \end{vmatrix} = -(6)\begin{vmatrix} 2 & 1 & 1 \\ 9 & 0 & 4 \\ 5 & 0 & -1 \end{vmatrix}$$

$$= -(6)(-1)\begin{vmatrix} 9 & 4 \\ 5 & -1 \end{vmatrix} = 6(-9 - 20) = -174$$

The first current is thus

$$I_1 = \frac{N_1}{D} = \frac{-174}{-192} = \frac{29}{32}$$

The other three currents are found to be

$$I_2 = -\frac{1}{4} \qquad I_3 = \frac{39}{16} \qquad I_4 = \frac{43}{64}$$

EXERCISE 8.10

Solve Probs. 1 to 11 of Exercise 8.7 by using determinants.

Find the simultaneous solutions of each of the following systems of equations. Use determinants.

12. $$\begin{aligned} 6x + 3y - 3z &= 10 \\ 3x - 4y \phantom{{}-3z} &= -5 \\ 2x + y - z &= 0 \end{aligned}$$

13. $$\begin{aligned} 4x - y - z &= -2 \\ 5y + 3z &= 6 \\ x + 2y + z &= 2 \end{aligned}$$

14. $$\begin{aligned} a \phantom{{}+3b} - c + 2d - e &= 0 \\ 2a + 3b + c \phantom{{}+2d} + 2e &= 5 \\ 2b - 2c - d \phantom{{}+e} &= -1 \\ -a - b \phantom{{}+c} + d + e &= -2 \\ 3a \phantom{{}+3b} + c \phantom{{}+2d} - 2e &= 0 \end{aligned}$$

FACTORING

CHAPTER 9

9.1 BASIC IDEAS ABOUT FACTORING

The process of factoring is the reverse of multiplication. We multiply two expressions by multiplying every term of one factor by every term of the other factor. We factor an expression by separating it into two or more factors so that their product equals the given expression. For example, to *multiply* $a(b + c)$, we multiply both terms in $(b + c)$ by a, and we get the following product:

$$a(b + c) = ab + ac$$

But to *factor* the expression $ab + ac$, we separate it into the factors a and $b + c$, as follows:

$$ab + ac = a(b + c)$$

Your ability to factor is greatly enhanced by some experience with multiplication. If you can *recognize the patterns* formed in various products, then you stand a good chance of recognizing the factors into which an expression can be separated. In other words, to factor, you have to work a multiplication problem backward. This is like the TV quiz problem in which the contestant is given an answer and asked to supply the question.

For example you might recognize that the factors of

$$a^2 + 2ab + b^2 \qquad \text{are} \qquad (a + b)(a + b)$$

because from past experience you might recall that

$$(a + b)(a + b) = a^2 + ab + ab + b^2 = a^2 + 2ab + b^2$$

In order to sharpen our ability to factor by recognizing the patterns in products, we shall first study some products that occur frequently in mathematics.

9.2 SPECIAL PRODUCTS

Each formula below represents a product that occurs frequently in mathematics. Do *not* memorize the formulas as such, but *learn the rule* or *idea* expressed by each formula. The notes below each formula will help you recognize the idea involved. You should verify each product by multiplying the factors on the left side.*

$$a(x + y) = ax + ay \tag{9.1}$$

The *product of a sum* is the sum of the products.

$$(ax + b)(cx + d) = acx^2 + (ad + bc)x + bd \tag{9.2}$$

The *product of two binomials* is a trinomial consisting of (1) the product of the first terms (ax and cx), plus (2) the product of the end terms (ax and d) added to the product of the middle terms (b and cx), plus (3) the product of the last terms (b and d).

$$(x + b)(x + d) = x^2 + (b + d)x + bd \tag{9.3}$$

This is a *special case* of Eq. (9.2).

$$(x + y)(x - y) = x^2 - y^2 \tag{9.4}$$

The *product of the sum of and the difference between two numbers* is the difference between the squares of those numbers.

$$(x \pm y)^2 = x^2 \pm 2xy + y^2 \tag{9.5}$$

The *square of a binomial* is a trinomial consisting of (1) the square of the first term x, plus (or minus) (2) twice the product of the two terms, twice xy, plus (3) the square of the second term y.

$$(x \pm y)^3 = x^3 \pm 3x^2y + 3xy^2 \pm y^3 \tag{9.6}$$

* Several of the formulas combine two formulas in one by using the $\pm$ symbol. You should always read such formulas by reading either the top sign or the bottom sign all the way through. For example $(x \pm y)^2 = x^2 \pm 2xy + y^2$ means $(x + y)^2 = x^2 + 2xy + y$ and $(x - y)^2 = x^2 - 2xy + y^2$.

The *cube of a binomial* is a quadrinomial with the following characteristics: (1) The exponents of the first term x are in descending order, 3, 2, 1, 0. (2) The exponents of the second term y are in ascending order, 0, 1, 2, 3. (3) The coefficients are 1, 3, 3, 1, in that order. (4) If the sign in the binomial is negative, then the signs in the product alternate +, −, +, −.

$$(x \pm y)(x^2 \mp xy + y^2) = x^3 \pm y^3 \tag{9.7}$$

To understand the rule, look closely at the exponents, the coefficients, and the signs. The right side is the *sum* (or *difference*) of *two cubes.*

In the following examples we shall multiply the expressions by (1) associating the given factors with those of one of the special products above, and (2) using the pattern of that formula to write the product. We shall *not* multiply any factors the long way, term by term. The key to this method is *pattern association,* that is, recognizing that a seemingly new expression forms the same pattern of letters as one of the special products. Sometimes we may have to rearrange an expression in order to make it match one of the special ones. The equation numbers to the right of the examples below refer to the special product being used.

Example 9.1. $2x(3x^2 - 5y) = 6x^3 - 10xy$ by (9.1)

Example 9.2. $A^n(2A^n - 1 + 2A) = 2A^{2n} - A^n + 2A^{1+n}$ by (9.1)

Example 9.3.
$$\begin{aligned}(2x + 3y)^2 &= (2x)^2 + 2(2x)(3y) + (3y)^2 \qquad \text{by (9.5)}\\ &= 4x^2 + 12xy + 9y^2\end{aligned}$$

Example 9.4.
$$\begin{aligned}(3x - 2y)^3 &= (3x)^3 - 3(3x)^2(2y) + 3(3x)(2y)^2 - (2y)^3 \qquad \text{by (9.6)}\\ &= 27x^3 - 54x^2y + 36xy^2 - 8y^3\end{aligned}$$

Example 9.5.
$$\begin{aligned}(2x - 5)(3 + x) &= (2x - 5)(x + 3)\\ &= 2x^2 + (6 - 5)x - 15 \qquad \text{by (9.2)}\\ &= 2x^2 + x - 15\end{aligned}$$

Example 9.6.
$$\begin{aligned}(3x - 5y)(5y + 3x) &= (3x - 5y)(3x + 5y)\\ &= (3x)^2 - (5y)^2 \qquad \text{by (9.4)}\\ &= 9x^2 - 25y^2\end{aligned}$$

Example 9.7.
$$\begin{aligned}(4x + 3y)(16x^2 + 9y^2 - 12xy) &= (4x + 3y)(16x^2 - 12xy + 9y^2)\\ &= (4x + 3y)[(4x)^2 - (4x)(3y) + (3y)^2]\\ &= (4x)^3 + (3y)^3 \qquad \text{by (9.7)}\\ &= 64x^3 + 27y^3\end{aligned}$$

Example 9.8.
$$\begin{aligned}(x + y - 3)(x + y + 3) &= [(x + y) - 3][(x + y) + 3] \\ &= (x + y)^2 - (3)^2 && \text{by (9.4)} \\ &= x^2 + 2xy + y^2 - 9 && \text{by (9.5)}\end{aligned}$$

EXERCISE 9.1

Multiply the factors by using one or more of the special products as in the examples of this section.

1. $3ab(2a^2 - 3b^2)$
2. $(2x - 3y)^2$
3. $(r - 1)^3$
4. $(ax - 2b)(ax + 2b)$
5. $(2x - 1)(x + 3)$
6. $x^{2n}(x^{2n} + x^n - 2)$
7. $(r + 2)(r + 3)$
8. $(2x + 3)^3$
9. $(x - 1)(x^2 + x + 1)$
10. $(a - 2b)(2a - b)$
11. $(5x + 2)^2$
12. $2x(x - 1)(x + 5)$
13. $(3x + 2)(3x - 2)$
14. $(x + y - 1)^2$
15. $(x + y - 1)^3$
16. $(a + b + 2)(a + b - 2)$
17. $(N^2 - 1)(N^2 + 1)$
18. $(a + 3)(a^2 - 3a + 9)$
19. $(x^3 - y^3)^2$
20. $(4 - 5a)^3$
21. $(A^2 + 9)(A - 3)(A + 3)$
22. $(A^3 + 8)(A - 2)(A^2 + 2A + 4)$
23. $(2k - 1)^4$
24. $[(3 - 2s)(3 + 2s)]^3$
25. $(2 - 3z + x)(x + 2 + 3z)$
26. $(2x + 2 - 2y)^3$
27. $(5y^3 - 3x^2 + 2)^2$
28. $R^2S^2(R^2 + S^2 - 2)^2$
29. $x^3(x^2 - 4)(x^2 + 3)$
30. $(N^4 + m^4)(N^2 + m^2)(N^2 - m^2)$

9.3 COMMON FACTORS

In this and the following sections we shall examine several standard types of factorable expressions. Try to learn each expression according to the pattern that it represents or the rule that you would use to describe that pattern. Do not attempt to memorize by rote each formula that we use to represent a rule symbolically. Instead, as you look at each formula, get its meaning—translate it into words. Having studied the special products of Sec. 9.2, you have some experience on which to base the coming work in factoring. Try to recall those special products and apply them to the problems of factoring.

If every term of an expression contains the same factor, this *common factor* can be removed. The basis for this factoring is the distributive property. See Eq. (9.1). Check the following examples by multiplying the factors on the right side.

Example 9.9. **Every term of $3x^3y^3 + 6x^2y^2 - 12xy$ contains the factors 3, x, and y; thus the common factor $3xy$ can be removed as follows:**

$$3x^3y^3 + 6x^2y^2 - 12xy = 3xy(x^2y^2) + 3xy(2xy) - 3xy(4)$$
$$= 3xy(x^2y^2 + 2xy - 4)$$

Example 9.10. $6A^{2n} - 2A^n + 4A^{n+1} = 2A^n(3A^n) - 2A^n + 2A^n(2A)$
$$= 2A^n(3A^n - 1 + 2A)$$

Example 9.11. $(a+b)(a-b) + (a+b)^2 - 3(a+b) = (a+b)(a-b+a+b-3)$
$$= (a+b)(2a-3)$$

Example 9.12. $\dfrac{2x^2y}{3w^2z^3} - \dfrac{10xy^2}{6w^3z^2} = \dfrac{2xy}{3w^2z^2}\left(\dfrac{x}{z} - \dfrac{5y}{2w}\right)$

Part of the factoring problem is to determine what remains after removing the common factors. We can always do this by dividing *each* original term by the factors removed. We repeat Example 9.10 to show such a division.

Example 9.10. (*repeated*) To factor $2A^n$ from $6A^{2n} - 2A^n + 4A^{n+1}$, write $2A^n$ in front as a factor, and divide *each* term by $2A^n$.

$$6A^{2n} - 2A^n + 4A^{n+1} = 2A^n\left(\frac{6A^{2n}}{2A^n} - \frac{2A^n}{2A^n} + \frac{4A^{n+1}}{2A^n}\right)$$
$$= 2A^n(3A^{2n-n} - 1 + 2A^{n+1-n})$$
$$= 2A^n(3A^n - 1 + 2A)$$

The preceding example shows that we can interpret factoring as a *multiplication and division by the same quantity.* Using this technique, we can mechanically factor any expression we please from any given expression. Example 9.13 illustrates such factoring where no common factors occur naturally.

Example 9.13. Factor $3x^3y^3$ from $5x^2 - 6y^3 + x^4y^4$.

$$5x^2 - 6y^3 + x^4y^4 = 3x^3y^3\left(\frac{5x^2}{3x^3y^3} - \frac{6y^3}{3x^3y^3} + \frac{x^4y^4}{3x^3y^3}\right)$$
$$= 3x^3y^3\left(\frac{5}{3xy^3} - \frac{2}{x^3} + \frac{xy}{3}\right)$$

EXERCISE 9.2

In Probs. 1 to 10, factor the given expressions by removing all common factors; then simplify the remaining expression, if possible.

1. $2Y^2 - 3XY$

2. $4a + 8b + 12c$
3. $30X^2Y^3Z^4 + 10X^3Y^2Z^4 - 15X^4Y^3Z^2$
4. $6C^3D + 10C^2D^3 - 2CD^2$
5. $3S^{k+1} - 6S^{2k}$
6. $(2x + y)(x - y)^3 - 3(x - y)(2x + y)(x^2 - xy) + (2x + y)^2(x - y)^2$
7. $\dfrac{2ad}{3b^2c} + \dfrac{4de}{9bc^2} - \dfrac{6ae}{15b^2c^2}$
8. $AX^{3n+2} - A^2X^{2n+3} + A^3X^{n+2}$
9. $\dfrac{2(3ax - 2bx)}{(3a + b)x} + \dfrac{2b - 3a}{3a + b}$
10. $\dfrac{m(2mx - 4nx)}{(4m + 4n)x} + \dfrac{n(2my - 4ny)}{4my + 4ny}$

In Probs. 11 to 16, factor the first expression from the second expression by the division method. See Example 9.13.

11. 2 from $4x + 5$
12. 6 from $3y - 2x$
13. AX^2 from $A^2X + AX^2$
14. -3 from $8a - 3b$
15. $\dfrac{2}{3}$ from $\dfrac{4a}{9} - \dfrac{8b}{15}$
16. $\dfrac{3M}{N}$ from $2MN - \dfrac{3M^2}{N} + \dfrac{N}{M}$

9.4 THE DIFFERENCE BETWEEN TWO SQUARES

Since the product $(a + b)(a - b)$ equals $a^2 - b^2$, the difference between two squares, then the difference between two squares can always be factored into the product of the sum of and the difference between those numbers that are squared. If common factors occur, they should be removed first.

Example 9.14. $9X^2 - 36Y^2 = 9(X^2 - 4Y^2) = 9[X^2 - (2Y)^2] = 9(X + 2Y)(X - 2Y)$

Example 9.15.
$$4A^{2n} - 16B^{4k} = 4(A^{2n} - 4B^{4k}) = 4[(A^n)^2 - (2B^{2k})^2]$$
$$= 4(A^n + 2B^{2k})(A^n - 2B^{2k})$$

Example 9.16.
$$(X + 1)^2 - (Y - 2)^2 = [(X + 1) + (Y - 2)][(X + 1) - (Y - 2)]$$
$$= (X + 1 + Y - 2)(X + 1 - Y + 2)$$
$$= (X + Y - 1)(X - Y + 3)$$

Example 9.17. $7x - 10 = (\sqrt{7x} + \sqrt{10})(\sqrt{7x} - \sqrt{10})$

This example shows that the terms within the factors need not be integers or perfect squares. However, we ordinarily intend *integral* factors when we speak of factoring an expression.

EXERCISE 9.3

In these problems, factor the expressions by factoring the difference between two squares. If common factors occur, remove them first.

1. $X^2 - 4Y^2$
2. $64S^2 - 25T^2$
3. $\frac{18B^2}{25} - \frac{32A^2}{49}$
4. $\frac{A^3B}{16} - \frac{AB^3}{9}$
5. $27M^{6k} - 12N^{8p}$
6. $\frac{2X^2}{3} - \frac{8Y^2}{3}$
7. $(X^2 - 4)^2 - (X^4 - 16)$
8. $4M^{2a+2} - N^{2b+2}$
9. $A^4 - B^4$
10. $(3x + 2y)^2 - (3x - 2y)^2$
11. $(s + 1)^2 - 25t^2$
12. $16H^6 - 81K^8$

9.5 FACTORING BY GROUPS

Certain quadrinomial expressions can be partially factored into groups. These groups can sometimes be factored further until complete factoring of the original expression is achieved. Such expressions have the general form:

FACTORING BY GROUPS

$$AB + BC + AD + CD = B(A + C) + D(A + C) = (A + C)(B + D)$$

This factoring is a result of the special product of Eq. (9.2).

Example 9.18. $2X^2 + 4XY - 3XZ - 6YZ = 2X(X + 2Y) - 3Z(X + 2Y) = (X + 2Y)(2X - 3Z)$

Example 9.19. $8 - S^7 + S^3 - 8S^4 = 8 + S^3 - S^4(8 + S^3) = (8 + S^3)(1 - S^4)$

Note that further factoring is possible here. See if you can complete the problem. [*Hint:* See Eqs. (9.4) and (9.7).]

EXERCISE 9.4

Factor the following quadrinomials. Look for common factors and the difference between squares that could be removed first.

1. $4x + 2xy + 2z + yz$
2. $3A^2 + 6AB - 3AC - 6BC$
3. $24 + 6MN - 6M - 24N$
4. $ab - 6 + 2b - 3a$

5. $1 - 3S^n + 3S^{4n} - S^{6n}$
[*Hint:* $S^{6n} = (S^{3n})^2$
$S^{4n} = S^n \cdot S^{3n}$]

6. $s - 4t^2 + s^2 + 2t$

9.6 QUADRATIC TRINOMIALS

A *quadratic trinomial* is an expression like

$$12x^2 + x - 35$$

It contains a second-degree term ($12x^2$), a first-degree term (x), and a constant term (-35). The general form of a quadratic trinomial is as follows:

$$Mx^2 + Nx + P$$

where M, N, and $P \neq 0$

Not all quadratic trinomials are factorable. To be factorable, a quadratic trinomial must correspond to one of the special products of Eqs. (9.2), (9.3), or (9.5). For example,

$$12x^2 + x - 35 = (3x - 5)(4x + 7)$$

This is an example of Eq. (9.2). But $12x^2 + x + 35$ is *not* factorable.

A. THE TRIAL AND ERROR METHOD

How can we tell if a quadratic trinomial is factorable, and if so, what the factors are? One method is an educated trial and error, and the characteristic features illustrated by Eqs. (9.2), (9.3), and (9.5) can be used to guide the trials. The following examples illustrate several key points that you should observe in deriving the factors.

1. Signs

a. $x^2 + 5x \quad + 6 = (x + 2)(x + 3)$

"+" here and "+" here mean these signs are *alike* and *positive.*

b. $x^2 - 5x \quad + 6 \quad = (x - 2)(x - 3)$

"−" here and "+" here mean these signs are *alike* and *negative.*

c. $x^2 + 5x - 6 = (x + 6)(x - 1)$

"−" here means these signs are *unlike.*

2. Factors

a. $(3x - 5)(4x + 7) = 12x^2 + x - 35$

The product of these terms equals the *second-degree term.*

b. $(3x - 5)(4x + 7) = 12x^2 + x - 35$

The product of these terms equals the *constant term.*

$-20x$

c. $(3x - 5)(4x + 7) = 12x^2 + x - 35$

$+21x$

$21x - 20x = +x$

The end-terms product plus the middle-terms product equals the *first-degree term.*

Since there may be several possible factors, several trials may be necessary. For example, $12x^2$ is also $x(12x)$, $(2x)(6x)$, and $(6x)(2x)$. If the factors do not satisfy all conditions outlined above, then we must try different factors or a different arrangement of factors. If all possible factors fail, then the trinomial is not factorable.

EXERCISE 9.5

Factor each of the quadratic trinomials by using the six guidelines outlined above.

1. $x^2 + 2x + 1$
2. $x^2 - 2x + 1$
3. $x^2 + 2x - 1$
4. $N^2 - 2N - 3$
5. $N^2 + 2N - 3$
6. $N^2 + 2N + 3$
7. $2A^2 - A - 3$
8. $2A^2 + A - 3$

9. $2A^2 - A + 3$
10. $6x^2 - 5x - 6$
11. $32Y^2 - 76Y + 45$
12. $18z^2 - 57z + 42$
13. $12R^3 + 7R^2 - 12R$
14. $96P^4 + 100P^3 - 128P^2$
15. $24S^2 + 26S - 33$

B. THE GROUPING METHOD

You may have noticed that some quadratic trinomials are difficult to factor by trial and error. We shall now look at a method that eliminates all trial and error. This method is based upon factoring by groups and the format of Eq. (9.2), repeated here:

$$(ax + b)(cx + d) = acx^2 + (ad + bc)x + bd \qquad (9.2)$$

The key point here is that the coefficient of x, that is, $ad + bc$ on the right side, is the sum of certain of the factors of the end-terms product $(ac)(bd)$. Thus to find the factors, we proceed as follows:

1. Multiply the first and last terms:

$$(ac)(bc) = a \cdot c \cdot b \cdot d$$

2. Factor this number into all possible pairs of factors.

$$\begin{aligned} a \cdot c \cdot b \cdot d &= a \cdot (cbd) = b \cdot (acd) = c \cdot (abd) = d \cdot (abc) \\ &= (ab) \cdot (cd) = (ac) \cdot (bd) = (ad) \cdot (bc) \end{aligned}$$

3. Select the pair whose sum [or difference, if the third term bd in Eq. (9.2) is negative] equals the coefficient of x. In this case we select ad and bc.
4. Rewrite the quadratic trinomial using the selected factors as separate coefficients of x, and factor by grouping.

$$\begin{aligned} acx^2 + (ad + bd)x + bd &= acx^2 + adx + bcx + bd \\ &= ax(cx + d) + b(cx + d) \\ &= (ax + b)(cx + d) \end{aligned}$$

Observe how this method is implemented in the following examples.

Example 9.20. **Factor $12x^2 + 8x + 1$.**

First, we multiply the *ac* and *bd* coefficients and then factor their product, as follows:

$$12x^2 + 8x + 1 \qquad 2 + 6 = \textcircled{8}$$

$$(ac)(bd) = (12)(1) = (2)(6) = (3)(4)$$

Next, we choose the factors 2 and 6, whose sum equals the $ad + bc$ coefficient 8, and rewrite the expression as follows:

$$\begin{aligned} 12x^2 + 8x + 1 &= 12x^2 + 2x + 6x + 1 \\ &= 2x(6x + 1) + (6x + 1) \\ &= (2x + 1)(6x + 1) \end{aligned}$$

Example 9.21. Factor $6x^2 + 11x - 21$.

$$6x^2 + 11x - 21$$

$$\begin{aligned} (6)(-21) &= (-1)(126) = (-2)(63) = (-3)(42) \\ &= (-7)(18) = (-9)(14) \end{aligned}$$

$$-7 + 18 = \textcircled{11}$$

$$\begin{aligned} &6x^2 + 11x - 21 \\ &= 6x^2 - 7x + 18x - 21 \\ &= x(6x - 7) + 3(6x - 7) \\ &= (x + 3)(6x - 7) \end{aligned}$$

This method always works if the quadratic trinomial is factorable; and we can quickly tell if it is not factorable, because no suitable pair of factors will develop from the end-terms product.

The following examples illustrate several variations of quadratic trinomials. Study them carefully and apply the techniques of Secs. 9.6A and 9.6B.

Example 9.22.
a. $-3x^2 + 4x + 15 = -(3x^2 - 4x - 15) = -(3x + 5)(x - 3)$
b. $-3x^2 + 4x + 15 = 15 + 4x - 3x^2 = (5 + 3x)(3 - x)$
c. $-3x^2 + 4x + 15 = (-3x - 5)(x - 3)$

Verify that the factors of Examples 9.22*a*, *b*, and *c* are equivalent.

Example 9.23. $2A^4 + 7A^2 + 3 = 2(A^2)^2 + 7(A^2) + 3 = (2A^2 + 1)(A^2 + 3)$

Example 9.24. $S^{2n} + S^n - 2 = (S^n)^2 + (S^n) - 2 = (S^n + 2)(S^n - 1)$

Example 9.25. $$8M^{6k}N^2 - 6M^{3k}N - 14 = 2[4(M^{3k}N)^2 - 3(M^{3k}N) - 7]$$
$$= 2(4M^{3k}N - 7)(M^{3k}N + 1)$$

Example 9.26. $9x^2 - 24x + 16 = (3x)^2 - 2(3)(4)(x) + (-4)^2 = (3x - 4)^2$

C. THE PERFECT-SQUARE TRINOMIAL

Example 9.26 illustrates a special case called the *perfect-square trinomial.* It is an example of Eq. (9.5). Since this special form is easy to recognize [as indicated by Eq. (9.5) and the middle expression in Example 9.26], we should look for it first, before applying the more general procedures.

EXERCISE 9.6

Factor the quadratic trinomial expressions. Also remove any common factors or differences between two squares that may appear. You may wish to rearrange some of the expressions before factoring.

1. $X^2 + 10X + 25$
2. $9N^2 - 12N + 4$
3. $4Y^2 - 17Y + 4$
4. $3a^2 + 7a + 2$
5. $12A^2 + 8A - 15$
6. $-6M^2 + 41M - 13$
7. $32 - 68P + 35P^2$
8. $Y^2 - 8Y + 16$
9. $4x^2 - 4x + 1$
10. $x^2 + 2x + 2$
11. $9V^2 - 47V - 42$
12. $162R^2 + 468R + 348$
13. $Z^2 - Z - 1$
14. $12W^2 - 52W + 35$
15. $-5S^4 + 3S^2 + 2$
16. $15 - 44K - 20K^2$
17. $2X^2Y^4 + 12XY^2Z + 18Z^2$
18. $3P^2Q^2 - 36PQ + 108$
19. $-25A^x + 10 - 15A^{2x}$
20. $24x^2 + 5x - 36$
21. $16S^2 - 227S + 42$
22. $2R^8 - 31R^4S^4 - 16S^8$
23. $2X^2 + aX + 2a + 4X$ (*Hint:* Group the aX and $4X$ terms.)
24. $36Z^2 + 90Z + 36$
25. $21A^2 + 26A - 32$
26. $15n - 3Xn + 20X - 4X^2$

9.7 THE SUM OR DIFFERENCE OF TWO CUBES; THE PERFECT-CUBE QUADRINOMIAL

Equations (9.6) and (9.7) show how to factor expressions involving cubes. The following examples illustrate the use of these equations in factoring.

Example 9.27. $x^3+8=x^3+2^3=(x+2)(x^2-x\cdot 2+2^2)$ by (9.7)
$= (x+2)(x^2-2x+4)$

Example 9.28. $64Y^3-27=(4Y)^3-3^3=(4Y-3)[(4Y)^2+(4Y)(3)+3^2]$ by (9.7)
$=(4Y-3)(16Y^2+12Y+9)$

Example 9.29. $8Z^3-36Z^2+54Z-27=(2Z)^3-3(2Z)^2(3)+3(2Z)(3)^2-3^3$ by (9.6)
$=(2Z-3)^3$

A perfect-cube quadrinomial, such as that of Example 9.29, is sometimes difficult to recognize as such. But it can always be factored by grouping the two perfect-cube terms as illustrated in the next example. Therefore, whenever you encounter a quadrinomial with two perfect-cube terms, try this method.

Example 9.29. (*Repeated*) $8Z^3-36Z^2+54Z-27=(8Z^3-27)-(36Z^2-54Z)$
$=[(2Z-3)(4Z^2+6Z+9)]-[(18Z)(2Z-3)]$ by (9.7) and (9.1)
$=(2Z-3)(4Z^2+6Z+9-18Z)$ by (9.1)
$=(2Z-3)(4Z^2-12Z+9)$
$=(2Z-3)(2Z-3)^2=(2Z-3)^3$ by (9.5)

EXERCISE 9.7

The following problems involve all types of factorable expressions that you have studied in this chapter. Identify each expression as one of the standard types and factor accordingly. A good procedure is to remove common factors first, then examine the result for the possibility of further factoring.

1. X^3-8
2. $2aM^3+16aN^3$
3. $3X^2-27Y^2$
4. $2XY+12Y^2+X^2+5XY$
5. $A^3+6A^2+12A+8$
6. Z^2-6Z+9
7. $8t^3+27s^3$
8. $1-x+x^2-x^3$
9. $64X^3-1$
10. $2s-4st+4u-8ut$
11. $3a\pi-75a\pi^3$
12. $X^3/8-Y^3/27$
13. $3X^2/8-29X/2-20$
14. $a^{4n}+a^{2n}x-12x^2$
15. $27Y^6-54aY^4+36a^2Y^2-8a^3$ [*Hint:* $27Y^6=(3Y^2)^3$]
16. $E^{3x}-E^{9x}$
17. $192+375A^{3k}$
18. $2Z^{2k+1}X+4Z^{k+1}X+4XZ$
19. $25U^{2x}-30U^xV^y+9V^{2y}$
20. $540ST^2+125S^3+450S^2T+216T^3$
21. $2X^4-5X^2-12$
22. $5X^6+38X^3-16$
23. X^3+3X^2+3X+1
24. $8X^3+36X^2+54X+27$
25. $X^3+6X^2+12X-117$ (*Hint:* $-117=8-125$)
26. $16Y^4-1$

FRACTIONS

CHAPTER 10

10.1 SOME BASIC ASPECTS OF FRACTIONS

In this section we shall review some basic terminology and examine some basic properties of fractions. We have seen some of this material in earlier chapters, but other material may be new to you. You should be sure to master all of it now if you have not already done so.

A. TERMINOLOGY

The result of a division of two numbers is called a *quotient.* When the quotient of two numbers is implied by writing one expression above the other, the entire expression is called a *fraction.* For example,

$$\frac{1}{2} \qquad \frac{5}{10} \qquad \frac{-3}{5} \qquad \frac{10}{-3} \qquad \frac{3x^2}{Y} \qquad \text{and} \qquad \frac{(a-2x)(a+2x)}{(a+2x)^2}$$

are fractions.

In a division problem the expression to be divided is called the *dividend.* In a fraction this expression, the one on top of the dividing line, is called the *numerator.*

In a division problem the expression that is divided into the dividend is the *divisor.* In a fraction this expression, the one below the dividing line, is called the *denominator.* Thus in the fractions above, the expressions

$$1 \qquad 5 \qquad -3 \qquad 10 \qquad 3x^2 \qquad \text{and} \qquad (a-2x)(a+2x)$$

are numerators and

$$2 \qquad 10 \qquad 5 \qquad -3 \qquad Y \qquad \text{and} \qquad (a+2x)^2$$

are denominators.

If two or more fractions have the same denominator, we call it a *common denominator.* For example, the fractions

$$\frac{1}{4} \quad \frac{3}{4} \quad \text{and} \quad \frac{-7}{4}$$

have a common denominator, 4, and

$$\frac{2xy}{(a+2x)^2} \quad \text{and} \quad \frac{3x^2}{(a+2x)^2}$$

have a common denominator, $(a + 2x)^2$.

B. THE FUNDAMENTAL PRINCIPLE

The value of a fraction is unchanged if both its numerator and its denominator are multiplied or divided by the *same non-zero number.* This fact is called the *fundamental principle of fractions.* For example,

$$\frac{1}{2} = \frac{2 \cdot 1}{2 \cdot 2} = \frac{2}{4} \quad \text{and} \quad \frac{3}{-4} = \frac{(-1)(3)}{(-1)(-4)} = \frac{-3}{4}$$

We shall use this principle to simplify fractions, as follows:

$$\frac{3}{12} = \frac{3/3}{12/3} = \frac{1}{4} \qquad \frac{ax^2}{a^2x} = \frac{ax^2/(ax)}{a^2x/(ax)} = \frac{x}{a}$$

and

$$\frac{-1}{-3} = \frac{(-1)(-1)}{(-1)(-3)} = \frac{1}{3}$$

We shall also use the fundamental principle to change two different fractions to fractions which have a common denominator. For example, $1/2$ and $1/3$ can be made to have a common denominator of 6, as follows:

$$\frac{1}{2} = \frac{3 \cdot 1}{3 \cdot 2} = \frac{3}{6} \quad \text{and} \quad \frac{1}{3} = \frac{2 \cdot 1}{2 \cdot 3} = \frac{2}{6}$$

Likewise, $2x/a$ and $3x/b$ can be changed to have a common denominator of ab, as follows:

$$\frac{2x}{a} = \frac{b \cdot 2x}{b \cdot a} = \frac{2bx}{ab} \quad \text{and} \quad \frac{3x}{b} = \frac{a \cdot 3x}{a \cdot b} = \frac{3ax}{ab}$$

If two fractions have a common denominator, then we can add them by factoring the denominator, as follows:

$$\frac{1}{2}+\frac{1}{3}=\frac{3}{6}+\frac{2}{6}=\frac{1}{6}(3+2)=\frac{5}{6}$$

and

$$\frac{2x}{a}-\frac{3x}{b}=\frac{2bx}{ab}-\frac{3ax}{ab}=\frac{x}{ab}(2b-3a)$$

C. THE SIGNS OF A FRACTION

We consider that every fraction has three signs: the sign of its numerator, the sign of its denominator, and the sign of the entire fraction. *Any two signs of a fraction may be changed without changing the value of the fraction.* For example, the fraction $-2/3$ can be written four ways:

$$-\frac{2}{3}=+\frac{-2}{3}=-\frac{-2}{-3}=+\frac{2}{-3}$$

Notice that in each of the last three forms, two signs have been changed relative to the signs of the first form. But each of these fractions equals $-0.666\ldots$; thus the value of the original fraction is unchanged. This rule is based on the fundamental principle and the associative law. For example,

$$\frac{-2}{3}=\frac{(-1)(2)}{3}=(-1)\left(\frac{2}{3}\right)=-\frac{2}{3}$$

Other examples were given in Sec. 10.1B.

D. SIMPLEST FORM OF A FRACTION

The simplest form of a fraction is the form in which its numerator and its denominator have *no factors in common.* Thus to simplify a fraction, we factor the numerator and the denominator and divide out (or "cancel") all common factors. The following examples illustrate how to reduce fractions to simplest form.

Example 10.1. $$\frac{2}{4}=\frac{2\cdot 1}{2\cdot 2}=\left(\frac{\cancel{2}}{\cancel{2}}\right)\left(\frac{1}{2}\right)=\frac{1}{2}$$

Example 10.2. $$\frac{42}{18}=\frac{6\cdot 7}{2\cdot 9}=\frac{2\cdot 3\cdot 7}{2\cdot 3\cdot 3}=\left(\frac{\cancel{2}}{\cancel{2}}\right)\left(\frac{\cancel{3}}{\cancel{3}}\right)\left(\frac{7}{3}\right)=\frac{7}{3}$$

Example 10.3. $$\frac{9{,}450}{1{,}764}=\frac{2(3^3)(5^2)(7)}{2^2(3^2)(7^2)}=\left(\frac{\cancel{2}}{\cancel{2}}\right)\left(\frac{\cancel{3^2}}{\cancel{3^2}}\right)\left(\frac{\cancel{7}}{\cancel{7}}\right)\left(\frac{3\cdot 5^2}{2\cdot 7}\right)=\frac{75}{14}$$

Example 10.4. $$-\frac{4-2x}{6(x^2-4)}=-\frac{-2(-2+x)}{6(x+2)(x-2)}=\frac{\cancel{2}\cancel{(x-2)}}{\cancel{2}(3)(x+2)\cancel{(x-2)}}=\frac{1}{3(x+2)}$$

Comment: *A commonly used synonym for "simplest form" is the phrase "lowest terms." This phrase is something of a misnomer, since we divide out factors, not terms; and when a fraction is in lowest terms, it has the least possible number of factors.*

In fact, a common error in simplifying fractions is that of canceling terms. Remember: ***Cancel only factors: never cancel terms!*** *Canceling terms is one of the most common and flagrant errors that beginning students make. To see exactly what we mean, consider the following examples.*

Example 10.5. $\dfrac{\not{2}+\not{5}}{\not{2}(\not{5})} \neq \dfrac{1+1}{1} = 2$ *but* $\dfrac{2+5}{2(5)} = \dfrac{7}{10}$ or $\dfrac{2+5}{2(5)} = \dfrac{\not{2}}{\not{2}(5)} + \dfrac{\not{5}}{2(\not{5})}$

$= \dfrac{1}{5} + \dfrac{1}{2}$

Incorrect canceling — Correct procedure — Correct procedure

Example 10.6. $\dfrac{a^{\not{2}}+\not{2}b}{\not{2}\not{a}} \neq a + b$ *but* $\dfrac{a^2+2b}{2a} = \dfrac{a^{\not{2}}}{2\not{a}} + \dfrac{\not{2}b}{\not{2}a}$

$= \dfrac{a}{2} + \dfrac{b}{a}$

Canceling terms is an *incorrect* procedure.

Canceling factors is the *correct* procedure.

Remember that in division each and every term of the numerator must be divided by the *entire* denominator.

EXERCISE 10.1

In Probs. 1 to 9, use the fundamental principle to change the fraction to an equivalent one having the indicated denominator.

1. $\dfrac{2}{3} = \dfrac{?}{9}$
2. $\dfrac{3}{5} = \dfrac{?}{25}$
3. $\dfrac{-4}{6} = \dfrac{?}{3}$
4. $\dfrac{-1}{2} = \dfrac{?}{32}$
5. $\dfrac{9}{-7} = \dfrac{?}{28}$
6. $\dfrac{6}{32} = \dfrac{?}{16}$
7. $\dfrac{-50}{-125} = \dfrac{?}{5}$
8. $\dfrac{5}{4} = \dfrac{?}{64}$
9. $\dfrac{21}{49} = \dfrac{?}{7}$

In Probs. 10 to 24, change one or more fractions to equivalent fractions having the indicated new denominator as a common denominator. Then add the fractions by factoring the common denominator, as shown in Sec. 10.1B.

10. $\frac{1}{2}+\frac{1}{4}=\frac{?}{4}$

11. $\frac{1}{3}+\frac{1}{6}=\frac{?}{6}$

12. $\frac{1}{3}-\frac{1}{6}=\frac{?}{6}$

13. $\frac{1}{4}-\frac{1}{6}=\frac{?}{12}$

14. $\frac{2}{5}-\frac{1}{3}=\frac{?}{15}$

15. $\frac{3}{4}-\frac{9}{10}=\frac{?}{20}$

16. $\frac{3}{4}-\frac{1}{8}+\frac{2}{3}=\frac{?}{24}$

17. $\frac{2x}{5}+\frac{x}{8}+\frac{3x}{4}=\frac{?}{40}$

18. $\frac{2a-b}{3}+\frac{a+b}{4}=\frac{?}{12}$

19. $\frac{2}{x}-\frac{5}{y}=\frac{?}{xy}$

20. $\frac{18x}{AB^2}+\frac{12x}{A^2B}=\frac{?}{A^2B^2}$

21. $\frac{3}{a}-\frac{5}{a+b}=\frac{?}{a(a+b)}$

22. $\frac{5a}{a-b}+\frac{3ab}{(a-b)^2}=\frac{?}{(a-b)^2}$

23. $\frac{x+y}{x-y}+\frac{x-y}{x+y}=\frac{?}{x^2-y^2}$

24. $\frac{x+y}{x-y}-\frac{x-y}{x+y}=\frac{?}{x^2-y^2}$

In Probs. 25 to 42, reduce the fraction to simplest form. A good technique is to factor each numerator and denominator into its prime factors, as shown in Sec. 10.1D.

25. $^4/_{12}$

26. $^3/_6$

27. $-^6/_{24}$

28. $^{15}/_{10}$

29. $^{28}/_4$

30. $\frac{-50}{-36}$

31. $^{12}/_{18}$

32. $^{240}/_{112}$

33. $\frac{1,820}{2,275}$

34. $\frac{2,772}{-504}$

35. $\frac{25x^2}{50x}$

36. $\frac{(a-b)^2}{(a+b)(a-b)}$

37. $\frac{5N^2-10N}{10N^2}$

38. $\frac{5N^2-10N}{2-N}$

39. $\frac{3a^2+6a}{3a^2+12a+12}$

40. $\frac{x^3-8}{(x-2)^3}$

41. $\frac{18k^2-78k-60}{30-6k}$

42. $\frac{2x^2-11x+12}{2x^2+5x-12}$

10.2 MULTIPLICATION AND DIVISION OF FRACTIONS

To multiply two or more fractions, we merely multiply all numerators together and all denominators together. We express this rule with a general equation as follows.

$$\frac{A}{B}\cdot\frac{C}{D}=\frac{AC}{BD} \qquad (10.1)$$

Multiplication of fractions

Example 10.7. $$\frac{2}{3} \cdot \frac{5}{9} = \frac{(2)(5)}{(3)(9)} = \frac{10}{27}$$

Example 10.8. $$\frac{ax^3}{3y} \cdot \frac{a^3x}{2y^2} = \frac{(a)(a^3)(x^3)(x)}{(3)(2)(y)(y^2)} = \frac{a^4x^4}{6y^3}$$

If any parts of any fraction are factorable, factor first and if possible, reduce the fractions to lowest terms before multiplying them. Remember this! It makes multiplication easier! Don't forget: Factor and reduce before multiplying!

Example 10.9. $$\frac{25}{18} \cdot \frac{6}{10} = \frac{5 \cdot \cancel{5}}{2 \cdot 3 \cdot \cancel{3}} \cdot \frac{\cancel{2} \cdot \cancel{3}}{\cancel{2} \cdot \cancel{5}} = \frac{5}{6}$$

Example 10.10. $$\frac{2x + x^2}{x^2 + 5x + 6} \cdot \frac{2x + 6}{2x^2} = \frac{\cancel{x}\cancel{(2 + x)}}{\cancel{(x + 2)}\cancel{(x + 3)}} \cdot \frac{\cancel{2}\cancel{(x + 3)}}{\cancel{2}x \cdot \cancel{x}} = \frac{1}{x}$$

To divide one fraction by another, invert the divisor and multiply. This rule for division can be represented symbolically as follows.

$$\boxed{\frac{A/B}{C/D} = \frac{A}{B} \cdot \frac{D}{C} = \frac{AD}{BC}} \qquad (10.2)$$

Division of fractions

Why should this rule for division work? We can verify it by applying the fundamental principle. We multiply the numerator and the denominator of the left side of Eq. (10.2) by D/C as follows:

$$\frac{\frac{A}{B}}{\frac{C}{D}} = \frac{\frac{A}{B} \cdot \frac{D}{C}}{\frac{\cancel{C}}{\cancel{D}} \cdot \frac{\cancel{D}}{\cancel{C}}} = \frac{\frac{AD}{BC}}{1} = \frac{AD}{BC}$$

Also, from an intuitive point of view consider what happens when you divide by a fraction. For example, divide 1 by ½. Be careful to observe that "dividing by ½" is not the same thing as "dividing by 2" or "taking one-half," both of which mean multiplying by ½.

$$\frac{1}{1/2} = \frac{1}{0.5} = \frac{10}{5} = 2$$

Isn't this result equivalent to multiplying 1 by 2?

The following examples illustrate how to divide fractions.

Example 10.11. $\frac{4}{3} \div \frac{8}{9} = \frac{4}{3} \cdot \frac{9}{8} = \frac{\cancel{4}}{\cancel{3}} \cdot \frac{\cancel{3} \cdot 3}{\cancel{4} \cdot 2} = \frac{3}{2}$

Example 10.12. $\frac{3xy^2/(2ab)}{6x^2y/(4a^2b^2)} = \frac{3xy^2}{2ab} \cdot \frac{4a^2b^2}{6x^2y} = \frac{3}{6} \cdot \frac{4}{2} \cdot \frac{\cancel{a^2}}{\cancel{a}} \cdot \frac{\cancel{b^2}}{\cancel{b}} \cdot \frac{\cancel{x}}{\cancel{x^2}} \cdot \frac{\cancel{y^2}}{\cancel{y}} = \frac{aby}{x}$

Example 10.13. $\frac{2x + x^2}{x^2 - x - 6} \div \frac{2x^2}{2x - 6} = \frac{2x + x^2}{x^2 - x - 6} \cdot \frac{2x - 6}{2x^2} = \frac{\cancel{x}(\cancel{2 + x})}{(\cancel{x + 2})(\cancel{x - 3})} \cdot \frac{\cancel{2}(\cancel{x - 3})}{\cancel{2}x^{\cancel{2}}}$

$$= \frac{1}{x}$$

EXERCISE 10.2

Perform the indicated multiplications or divisions. If possible, simplify fractions first. Work all problems as fractions, even when decimal form might ordinarily be used.

1. $\frac{15}{8} \cdot \frac{2}{3}$
2. $\frac{49}{-25} \cdot \frac{-30}{21}$
3. $-\frac{25}{64} \div \frac{35}{-6}$
4. $\frac{98}{15} \div \frac{-14}{9}$
5. A boy mows lawns for $2\frac{1}{4}$ dollars per hour. For how many hours and minutes would he have to mow to earn $15?
6. If the boy in Prob. 5 mows for $3\frac{2}{3}$ hr, how much will he earn?
7. An automatic milling machine produces a certain part at the rate of $16\frac{2}{3}$ pieces per hour. What is the day's total production if the machine runs $15\frac{3}{5}$ hr/day?
8. How many hours and minutes will the machine of Prob. 7 need to operate to produce 1,000 pieces? If the machine operator's wage is $4\frac{3}{8}$ dollars per hour, what is the cost of labor per piece?
9. The formula relating voltage E to resistance R and current I in an electric circuit is $E = IR$. If the current is $6\frac{2}{3}$ amp and the resistance is $2\frac{2}{5}$ ohms, what is the voltage? If the voltage is $18\frac{3}{4}$ volts and the resistance $\frac{3}{8}$ ohms, what is the current?
10. The power dissipated by a resistor in an electric circuit is given by the formula $P = I^2R$. If the current I is $2\frac{1}{2}$ amp and the resistance $2\frac{2}{5}$ ohms, how much power is lost?
11. A 1-lb 2-oz can of powdered brand X film developer will make 1 gal of stock solution of liquid developer. It takes 9 oz of liquid developer to process one roll of 35-mm film in a small tank.

If you want to mix enough developer for just one roll of film, how much powder should you use?

12. If the liquid developer of Prob. 11 can be diluted 3:1 (3 parts water to 1 part developer), then how much powder would you need to develop one roll of film? If the can of powder costs 195 cents, what is the cost of the developer per roll of film?

13. $\dfrac{3uv}{w} \cdot \dfrac{w^2}{6uv^2}$

14. $\dfrac{(-5a)^3}{2b^3} \cdot \dfrac{6a}{5b^2} \cdot \dfrac{b^4}{3a^2}$

15. $\dfrac{3uv}{w} \div \dfrac{w^2}{6uv^2}$

16. $\dfrac{st}{ar} \div \dfrac{br}{at}$

17. $\dfrac{10a^2b^2}{21xy} \cdot \dfrac{x}{4b} \div \dfrac{5ax}{7by}$

18. $\dfrac{2M^{2k}}{3N^{k+1}} \cdot \dfrac{M^3x}{y} \div \dfrac{M^kx^2}{aN^{1+k}y^2}$

19. $\dfrac{h^2 - eh}{8h + 12e} \cdot \dfrac{6h + 9e}{eh - e^2}$

20. $\dfrac{4u + 2y}{st - t^2} \cdot \dfrac{us - ut}{4us + 2sv}$

21. $\dfrac{39ax^2}{13a^{n-1}y^3} \cdot \dfrac{a^{2n}bx^2y^3}{9b^nxy}$

22. $\dfrac{39ax^2}{13a^2y^3} \div \dfrac{a^{2n}bx^2y^3}{9b^nxy}$

23. $\dfrac{a^2 - x^2}{2a^2x^2} \cdot \dfrac{2x - 2a}{x^2 - 2x + ax - 2a}$

24. $\dfrac{x^2 + 5x + 6}{125 - 5x^2} \cdot \dfrac{10x - 50}{x^2 + 4x + 4}$

25. $\dfrac{Z^3 - 6Z^2 + 12Z - 8}{10 - 3Z + Z^2} \div \dfrac{Z^2 - 4Z + 4}{Z + 5}$

26. $\dfrac{3S^2 - 6ST}{4S + 2T} \cdot \dfrac{S^2 + 2ST + T^2}{4S^2 + 12ST + 9T^2} \div \dfrac{6ST - 12T^2}{2S + 3T}$

27. $\dfrac{Y^4 - 16}{16Y^2 + 16Y + 4} \div \dfrac{Y^2 + 2Y - 8}{4Y + 2}$

28. $\dfrac{A - 3}{A^3 + 9A^2 + 27A + 27} \cdot \dfrac{A^2 + 6A + 9}{6 - 2A}$

29. $\dfrac{3XY + 2Y + 15X + 10}{Y^2 + 3} \cdot \dfrac{2X - 1}{Y^2 + 10Y + 25}$

30. $\dfrac{A^3 + 27}{A^2 - 6A + 9} \div \left[\dfrac{2A^2 + 6A + 18}{A^2 - 9} \div (4A - 12)\right]$

31. The overall speed ratio of a train with three pairs of gears is the product of the tooth ratios of all the pairs, as follows:

$$N = \frac{T_D}{T_A} \cdot \frac{T_B}{T_C} \cdot \frac{T_E}{T_F}$$

Find the speed ratio if the numbers of teeth are as follows: $T_A = 12$, $T_B = 6$, $T_C = 15$, $T_D = 5$, $T_E = 4$, and $T_F = 20$.

32. A certain photographic print washer is an upright, 16-in.-diameter cylindrical tank. Water flows in the top at the rate of $1\frac{1}{2}$ gal/min. It drains out the bottom at $3\sqrt{d}/5$ gal/min, where d is the depth, in inches, of water in the tank. At what depth d will the water level stabilize?

10.3 ADDITION AND SUBTRACTION OF FRACTIONS

A. BASIC ASPECTS OF ADDITION

When we add two or more quantities, they must be the *same kind* of quantity. They must be alike in some sense or have some "common denominator" as a basis for comparison. Thus we can add X's to X's as in

$$2X + 3X = 5X$$

But we cannot add X's and Y's except to indicate the addition algebraically by writing $X + Y$.

As a specific example, suppose that you wish to add five nickels and three dimes. In a broad sense you can add these different items without changing them, by saying that you have eight *coins.* The "common denominator" here is "coins"; nickels and dimes are both members of this broad category of objects. In an even broader sense, you could add five of any item to three of any other item by saying that you have eight "things."

In mathematics we are interested in a more precise kind of addition—addition of items that are the same type in every way. In this respect you cannot add the nickels and dimes unless you change each one into a different set of coins *with an equivalent value.* These other sets of coins are, of course, pennies—a common denominator of sorts for nickels and dimes. Thus you add your five nickels and three dimes by converting them all to cents, saying that you have 55 cents.

Notice that a nickel is $1/20$ of a dollar, a dime is $1/10$ of a dollar, and a cent is $1/100$ of a dollar. And so to add the twentieths and tenths, you changed them into a larger denominator, hundredths, that could absorb them both.

We add all other fractions in a similar manner. The characteristic by which we compare fractions is the denominator. Fractions with the *same denominator* are the *same kind,* and we can add them. Thus we think of

$$\frac{5}{2} + \frac{3}{2}$$

as 5 halves plus 3 halves, which equals (5 + 3) halves, or 8 halves, or 4. Note that it is the *sameness of the denominators,* the "halves," that enables us to add these fractions. Thus

$$\frac{5}{2} + \frac{3}{2} = \frac{5+3}{2} = \frac{8}{2} = 4$$

But fractions with *unequal* denominators are "different," and we cannot add them. Thus we think of

$$\frac{5}{3}+\frac{3}{2}$$

as 5 thirds plus 3 halves. We cannot add these different fractions, just as we could not add nickels and dimes, unless we change both fractions to *equivalent* fractions which have a common denominator. Thus we realize that $5/3$ is equivalent to $10/6$, $3/2$ is equivalent to $9/6$, and the sum is $19/6$. We add them mathematically by using the fundamental principle to change both fractions to *equivalent* ones with the common denominator, sixths, as follows:

$$\frac{5}{3}+\frac{3}{2}=\frac{5}{3}\cdot\frac{2}{2}+\frac{3}{2}\cdot\frac{3}{3}=\frac{10}{6}+\frac{9}{6}=\frac{10+9}{6}=\frac{19}{6}$$

B. FRACTIONS WITH THE SAME DENOMINATOR

If two or more fractions have the same denominator, then we add them by adding their numerators. To subtract them, we use the rule of signs to change signs and proceed as in addition. In effect we just subtract numerators.

Example 10.14.

a. $\dfrac{x}{3}+\dfrac{y}{3}=\dfrac{x+y}{3}$

b. $\dfrac{x}{3}-\dfrac{y}{3}=\dfrac{x}{3}+\dfrac{-y}{3}=\dfrac{x-y}{3}$

Example 10.15.

a. $\dfrac{3}{x}+\dfrac{2}{x}=\dfrac{3+2}{x}=\dfrac{5}{x}$

b. $\dfrac{3}{x}-\dfrac{2}{x}=\dfrac{3}{x}+\dfrac{-2}{x}=\dfrac{3-2}{x}=\dfrac{1}{x}$

Example 10.16.

$$\frac{5x+2}{ax^2+b}-\frac{3x-7}{ax^2+b}=\frac{5x+2-(3x-7)}{ax^2+b}=\frac{5x+2-3x+7}{ax^2+b}=\frac{2x+9}{ax^2+b}$$

Observe very carefully how the signs of $(3x-7)$ were handled in this example.

Example 10.17.

$$\frac{6X^2}{3X^2-6XY}+\frac{5X^2}{3X^2-6XY}=\frac{6X^{\cancel{2}}}{3\cancel{X}(X-2Y)}+\frac{5X^{\cancel{2}}}{3\cancel{X}(X-2Y)}$$

$$=\frac{6X+5X}{3(X-2Y)}=\frac{11X}{3(X-2Y)}$$

This example shows that we factor and simplify before adding; but we did not divide the 3 into the 6, because the resulting denominators would have been different.

Example 10.18. $\frac{11a}{3xy}+\frac{8a}{3xy}-\frac{a}{3xy}=\frac{11a+8a-a}{3xy}=\frac{18a}{3xy}=\frac{6a}{xy}$

This example shows that we can sometimes simplify after adding.

EXERCISE 10.3

In Probs. 1 to 7, combine the given quantities into a single quantity (integer or fraction), first in terms of the first type of unit, then in terms of the other type(s) of unit.

1. 5 ft + 2 yd (Find how many feet, then how many yards.)
2. 9 in. + 4 ft
3. $^{13}/_{100}$ in. + $^{450}/_{1,000}$ in.
4. 10 oz + 3 lb
5. 7 pennies + 4 nickels + 2 dimes
6. 18 min + 2 hr
7. 2 miles + 5 hr
8. Suppose apples sell for 60¢ a dozen and oranges for $1.00 a dozen.
   a. How many apples is 1 orange worth?
   b. How many oranges is 1 apple worth?
   c. What would be a fair trade in apples for 9 oranges?
   d. What would be a fair trade in oranges for 10 apples?
   e. How much are 4 apples and 3 oranges worth?
9. How many apples and oranges of Prob. 8 can you buy if you spend exactly $.80 and buy at least one of each kind? No parts of fruit are allowed.

In Probs. 10 to 26, add or subtract the fractions as indicated and reduce the result to simplest form.

10. $^1/_5 + ^7/_5$
11. $^7/_6 + ^3/_6$
12. $^{19}/_8 - ^3/_8$
13. $\frac{3a}{2x^2}-\frac{5a}{2x^2}$
14. $\frac{3a}{3ax}-\frac{6b}{3ax}$
15. $\frac{a}{3}-\frac{a+3b}{3}$
16. $\frac{x^2}{6}+\frac{2+x^2}{6}$
17. $\frac{a+b}{4x}-\frac{a-b}{4x}$
18. $\frac{a+b}{a-b}+\frac{a-3b}{a-b}$
19. $^2/_3$ gal + $^3/_5$ gal
20. $^1/_2$ volt + $^3/_4$ volt − $^5/_8$ volt
21. $\frac{2}{a}+\frac{3}{b}+\frac{4}{c}$
22. $\frac{1}{x}+\frac{3}{y}-\frac{5}{z}$
23. $\frac{3b}{a}+\frac{2}{ab}$

24. $\dfrac{3}{ab} + \dfrac{2a}{b}$ 25. $\dfrac{2x}{3} + 1$ 26. $\dfrac{a+b}{a-b} - 1$

C. FRACTIONS WITH DIFFERENT DENOMINATORS

As we discussed earlier, to add fractions with different denominators, we must change them to equivalent fractions that have a common denominator. We have already done this intuitively with simple fractions, but for more complicated fractions we shall need a more systematic method. The difficulty lies in finding the smallest possible common denominator and in changing each fraction to an equivalent one containing this common denominator.

The following example illustrates a step-by-step procedure that will always produce a common denominator and will change each fraction to the desired equivalent one. Furthermore, the common denominator will be the smallest one possible. We call it the *lowest common denominator* (*LCD*), the smallest number (absolute value) that is evenly divisible by all denominators in a set of fractions. Such a number is also called the *lowest common multiple* (*LCM*) of the denominators.

This example outlines a procedure for finding the LCD and adding fractions.

Example 10.19.

$$\underbrace{\frac{5}{12} + \frac{7}{40}}_{(1)} = \underbrace{\frac{5}{2^2 \cdot 3} + \frac{7}{2^3 \cdot 5}}_{(2)} = \underbrace{\frac{5 \cdot (2 \cdot 5)}{2^2 \cdot 3 \cdot (2 \cdot 5)} + \frac{7 \cdot (3)}{2^3 \cdot 5 \cdot (3)}}_{(3)}$$

1. **Factor denominators into prime factors.**
2. **Form LCD ($2^3 \cdot 3 \cdot 5$).**
3. **Multiply and divide each fraction by all factors of LCD *not* in that denominator.**

$$\frac{5}{12} + \frac{7}{40} = \underbrace{\frac{50}{2^3 \cdot 3 \cdot 5} + \frac{21}{2^3 \cdot 3 \cdot 5}}_{(4)} = \underbrace{\frac{71}{120}}_{(4)}$$

4. **The new denominators are now equal; thus add the new numerators.**

Now let's translate the ideas of Example 10.19 into more general terms.

First, the *prime factors* of a number are factors that are *prime numbers*. For example, $60 = 2^2 \cdot 3 \cdot 5$; thus the prime factors of 60 are 2^2, 3, and 5. In the case of algebraic expressions such as

$$x^4 + 5x^3 + 6x^2 = x^2(x + 2)(x + 3)$$

the factors are *relatively prime* if they have no factors in common. Thus x^2, $x + 2$, and $x + 3$ are relatively prime factors. But x and $x^2 + 2x$ are *not* relatively prime factors because each has x as a common factor.

Keeping these ideas in mind, we can define the LCM of a set of numbers to be the product of all the different prime factors, each factor taken to its highest power. To find the LCM of 24, 60, and 72 for example, we factor each one into prime factors:

$$24 = 2^3 \cdot 3 \qquad 60 = 2^2 \cdot 3 \cdot 5 \qquad \text{and} \qquad 72 = 2^3 \cdot 3^2$$

and we multiply the different factors 2, 3, and 5, each one taken to its highest power. Thus we have

$$\text{LCM of } 24, 60, 72 = 2^3 \cdot 3^2 \cdot 5 = 360$$

Think and experiment to see why this procedure makes 360 a multiple of 24, 60, and 72, and why it is the smallest multiple.

The LCD of a set of denominators is the LCM of those denominators. Thus we can state the following definition:

LOWEST COMMON DENOMINATOR	The LCD of a set of fractions is the product of all prime factors of the denominators, each factor taken to the highest power to which it occurs.

To add fractions:

1. Find the LCD by applying the definition above.
2. Change each fraction to an equivalent one having the LCD as its denominator. To do this, multiply the numerator *and denominator* by all factors of the LCD which are *not* found in the denominator.
3. Add the resulting fractions having the LCD as a common denominator by adding the numerators and placing the sum over the LCD.

Reread Example 10.19 to see how these steps were applied.

Also experiment with the denominators to see just why this procedure produces the lowest common denominator and why step 2 works.

Observe closely to see how the detailed steps above are applied to the following examples.

Example 10.20.

$$\frac{8}{35}-\frac{11}{75}=\frac{8}{5\cdot 7}-\frac{11}{3\cdot 5^2}=\frac{8(3\cdot 5)}{5\cdot 7\cdot(3\cdot 5)}-\frac{11(7)}{3\cdot 5^2(7)}$$

$$=\frac{120}{3\cdot 5^2\cdot 7}-\frac{77}{3\cdot 5^2\cdot 7}=\frac{43}{525}$$

LCD $= 3\cdot 5^2\cdot 7$

Example 10.21.

$$\frac{2}{x^2y}+\frac{3}{xy^2}=\frac{2\cdot y}{x^2y\cdot y}+\frac{3\cdot x}{xy^2\cdot x}=\frac{2y+3x}{x^2y^2}$$

LCD $= x^2y^2$

Example 10.22.

$$\frac{3}{x^3-2x^2}-\frac{5}{x^3-4x}=\frac{3}{x^2(x-2)}-\frac{5}{x(x-2)(x+2)}$$

LCD $= x^2(x-2)(x+2)$

$$=\frac{3\cdot(x+2)}{x^2(x-2)\cdot(x+2)}-\frac{5\cdot x}{x(x-2)(x+2)\cdot(x)}$$

$$=\frac{3x+6}{x^2(x-2)(x+2)}-\frac{5x}{x^2(x-2)(x+2)}$$

$$=\frac{2(3-x)}{x^2(x-2)(x+2)}$$

EXERCISE 10.4

In Probs. 1 to 20, add or subtract the fractions as indicated and reduce the result to simplest form.

1. $1/7 - 1/21$
2. $1/11 + 2/33$
3. $\dfrac{5}{18a}+\dfrac{7}{12a}$
4. $\dfrac{2x}{15}-\dfrac{3x}{25}$
5. $4/75 + 7/45$
6. $5/56 - 3/98$
7. $2/7 + 5/12 - 13/21$
8. $7/12 - 11/60 + 14/225$
9. $\dfrac{3}{4x}+\dfrac{5}{2x^2}$
10. $\dfrac{2}{3s^2t}+\dfrac{5}{6st^2}$

11. $\dfrac{1}{ab}-\dfrac{2}{bc}+\dfrac{3}{ac}$

12. $\dfrac{1}{15}+\dfrac{3}{5x}-\dfrac{5}{3y}$

13. $\dfrac{X^2+2}{X^2-X}-\dfrac{1+2X}{4X-2}$

14. $\dfrac{a-2}{a^2-4}+\dfrac{1-a}{a^2+2a}$

15. $\dfrac{X^2+2X}{X^2-3X-10}+\dfrac{X}{3-X}$

16. $\dfrac{2X^2+6X}{2X^3-3X^2-9X}-\dfrac{3X}{X^2-9}$

17. $\dfrac{2S-4}{S^3-6S^2+12S-8}+\dfrac{S-3}{S^2-4S+4}-\dfrac{S+3}{S-4}$

18. $\dfrac{4X^2-9}{8X^3+36X^2+54X+27}-\dfrac{2X^2-3}{8X^3+27}$

19. $\dfrac{Z-3}{Z^3+27}-\dfrac{Z+3}{Z^3-6Z^2+18Z-27}$

20. $\dfrac{2K^2+K-10}{2KM-15+5M-6K}+\dfrac{4K+8+KM-2M}{M^2-7M+12}$

In Probs. 21 to 28, add the fractions on the right side of the formula.

21. A formula for bending stress in a curved beam is

$$S=\frac{2M}{Ad}\left[1+\frac{1}{Z}\left(\frac{2x}{d+2x}\right)\right]$$

22. Three formulas for maximum stress due to an eccentric load on a column are as follows:

$$S_c=\frac{Fea}{I}+\frac{F}{A}$$

and

$$S_t=\frac{Fea}{I}-\frac{F}{A}$$

$$\frac{S_t}{F}=\frac{1}{wl}+\frac{6k}{wl^2}$$

23. A formula for maximum shear stress in a coil spring is

$$S_{\max}=\frac{8FD}{\pi d^3}+\frac{4F}{\pi d^2}$$

24. A formula for the torque transmitted by a gear tooth in terms of the load P and the radius R is

$$T=SFP\times R\left(1-\frac{P}{l}+\frac{P^2}{3l^2}\right)$$

25. A formula for the stress factor in a coil spring is

$$f = \frac{4k-1}{4k-4} + \frac{0.526}{k}$$

26. The formula for the equivalent of three electric resistors "in parallel," as shown in Fig. 10.1, is

$$\frac{1}{R} = \frac{1}{R_1} + \frac{1}{R_2} + \frac{1}{R_3}$$

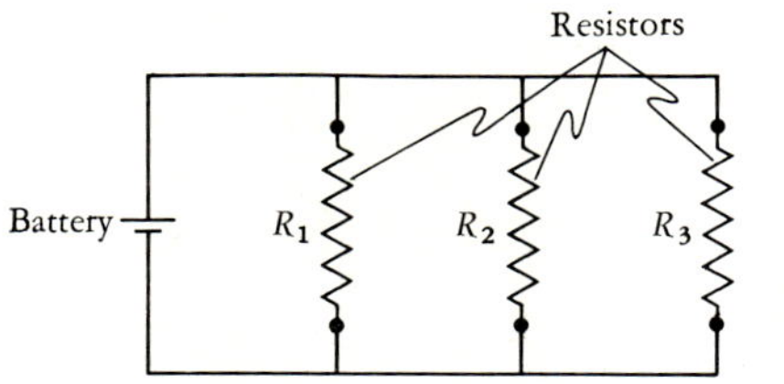

Fig. 10.1 Prob. 26. Three electrical resistors in parallel.

27. The equivalent stiffness factor for two springs connected "in series," as shown in Fig. 10.2, is

$$\frac{1}{k} = \frac{1}{k_1} + \frac{1}{k_2}$$

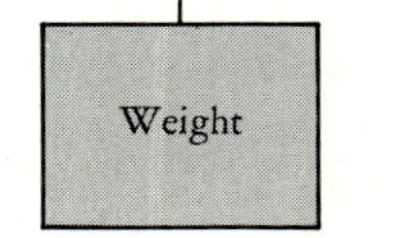

Fig. 10.2 Prob. 27. Two springs in series.

28. If a resistor, a capacitor, and an inductor are connected "in parallel" with an ac voltage source, the equivalent "impedance" is given as follows (see Fig. 10.3):

$$\frac{1}{Z^2} = \frac{1}{R^2} + \frac{1}{X_c^2} - \frac{2}{X_L X_c} + \frac{1}{X_L^2}$$

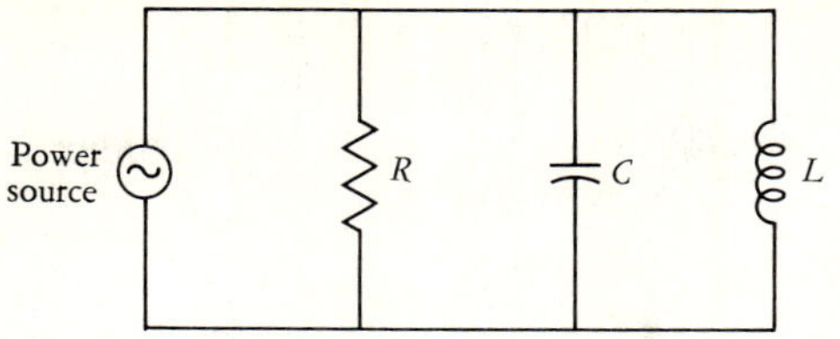

Fig. 10.3 Prob. 28. A resistor, a capacitor, and an inductor in parallel.

10.4 COMPOUND FRACTIONS

A *compound fraction* (sometimes called a *complex fraction*) is a fraction that has fractions within its numerator or its denominator, or both. To simplify a compound fraction, we change it to an equivalent fraction which has no fractions in either the numerator or denominator. The following examples show how to do this.

Example 10.23. **One method of simplifying a compound fraction is to add the fractions in the numerator and those in the denominator; then invert the denominator and multiply.**

$$\frac{2+5/2}{3-9/4}=\frac{(4+5)/2}{(12-9)/4}=\frac{4+5}{2}\cdot\frac{4}{12-9}=\frac{9}{2}\cdot\frac{4}{3}=6$$

An alternate method uses the fundamental principle and the LCD as follows:

$$\frac{2+5/2}{3-9/4}=\frac{(2+5/2)\cdot 4}{(3-9/4)\cdot 4}=\frac{8+10}{12-9}=\frac{18}{3}=6$$

Example 10.24.

$$\frac{(3x-y)/x}{x/y+3y/x}=\frac{(3x-y)/x}{(x^2+3y^2)/(xy)}=\left(\frac{3x-y}{x}\right)\left(\frac{xy}{x^2+3y^2}\right)$$
$$=\frac{y(3x-y)}{x^2+3y^2}$$

Using the alternate method, we would have

$$\frac{(3x-y)/x}{x/y+3y/x}=\frac{\left(\frac{3x-y}{x}\right)xy}{\left(\frac{x}{y}+\frac{3y}{x}\right)xy}=\frac{y(3x-y)}{x^2+3y^2}$$

EXERCISE 10.5

In Probs. 1 to 20, simplify the compound fractions.

1. $\dfrac{1}{1/2 - 1/4}$
2. $\dfrac{1}{1/3 - 1/5}$
3. $\dfrac{1}{1/2 + 2/3 - 3/5}$
4. $\dfrac{10}{1 + 2/3}$
5. $\dfrac{6/10}{2 - 7/5}$
6. $\dfrac{3 - 1/2}{2 + 1/2}$
7. $\dfrac{2 + 3/5}{3 - 2/5}$
8. $\dfrac{3 - 6/7}{2/3 + 2/7}$
9. $\dfrac{3/2 - 3/4}{3/8 - 9/16}$
10. $\dfrac{3/12 - 1/12}{1/6 - 5/12}$
11. $\dfrac{9/56 + 5/7}{3/8 - 13/16}$
12. $\dfrac{1 + (1/2 + 1/4)/(3/4)}{2 - 1/4}$
13. $\dfrac{1}{1/x - 1/y}$
14. $\dfrac{1/a}{2/a^2 + 1/a}$
15. $\dfrac{a/b - b/a}{a - b}$
16. $\dfrac{x^2 - x/a}{x - x/a}$
17. $\dfrac{a^2 - b^2}{(1 + a/b)(1 - b/a)}$
18. $\dfrac{(x^2 + 10x + 25)/(x - 5)}{1 + (x^2 + 7x + 20)/(x - 5)}$
19. $\dfrac{a^2 - ab + b^2 - (a^2b - ab^2)/(a - b)}{a^2/b - b^2/a}$
20. $\dfrac{1}{a + \dfrac{1}{a + \dfrac{1}{a + \dfrac{1}{a}}}}$

Simplify the compound fraction in each of the following formulas.

21. a. The equivalent resistance of two resistances in parallel is

$$R = \frac{1}{1/R_1 + 1/R_2}$$

b. The equivalent resistance of three resistances in parallel is

$$R = \frac{1}{1/R_1 + 1/R_2 + 1/R_3}$$

22. The lens-to-film distance in a camera with a focal length f is

$$d_i = \frac{1}{1/f - 1/d_0}$$

23. The Euler formula for the critical load on a column is

$$P = \frac{n\pi^2 AE}{(l/p)^2}$$

24. The half-width c of the contact band for two cylinders pressing against each other under a load F is

$$c = 2\sqrt{\frac{1-u^2}{\pi} \cdot \frac{F(1/E_1 + 1/E_2)}{h(1/R_1 + 1/R_2)}}$$

25. The transmissibility of force due to a vibrating load is

$$T = \frac{(p/p_n)^2}{1-(p/p_n)^2}$$

26. The amplitude of a forced vibration at a frequency w Hz (formerly cycles/sec) is

$$Z = \frac{F_0/k}{1-(w/w_n)^2} \sin(2\pi wt)$$

27. A formula for the radius of the end of a gear tooth is

$$r_0 = r\frac{1 - p/h + p^2/(3h^2)}{1 - p/(2h)}$$

28. Einstein's theory of relativity shows that time is a function of velocity as follows:

$$t = \frac{t_0}{\sqrt{1 - v^2/c^2}}$$

where t is the time interval on a stationary clock, corresponding to a time interval t_0 on a clock moving at speed v, and c is the speed of light.

29. According to Einstein's theory of relativity, the mass M of a body is a function of its velocity v, as follows:

$$M^2 = \frac{M_0{}^2}{1 - v^2/c^2}$$

where M_0 is the corresponding stationary mass and c is the speed of light.

30. The equivalent stiffness of two springs in series (see Fig. 10.2) is

$$K = \frac{1}{1/K_1 + 1/K_2}$$

10.5 EQUATIONS WITH FRACTIONS

Many problems in engineering and science involve equations with fractions. To solve such equations, we use the operations of the preceding sections. There are two general approaches for solving equations with fractions; we shall illustrate both methods in solving the important electrical engineering problem outlined below.

A. GENERAL METHODS FOR SOLVING EQUATIONS WITH FRACTIONS

Let R_1 and R_2 represent the resistances of two electric resistors in parallel branches of a circuit, as diagramed in Fig. 10.4. The equivalent single resistance that could replace R_1 and R_2 (as far as the total current and voltage drop between points 1 and 2 are concerned) is related by the following reciprocal formula:

$$\frac{1}{R} = \frac{1}{R_1} + \frac{1}{R_2} \tag{10.3}$$

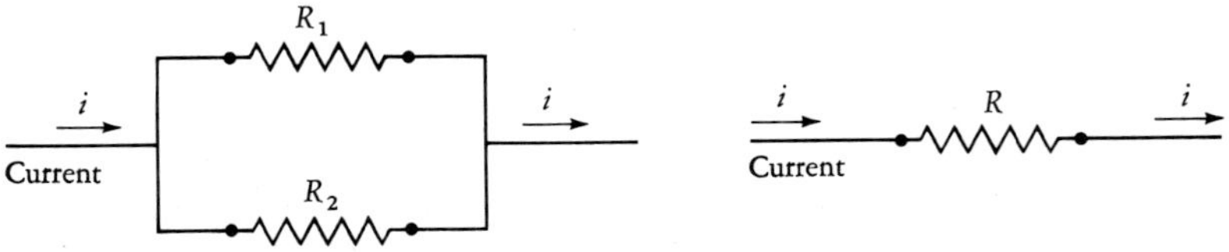

(*a*) Two parallel resistors

(*b*) The equivalent single resistor

Fig. 10.4

One method of solving such equations is to "clear fractions" by multiplying every term by the LCD of all the fractions involved. Thus we multiply every term by RR_1R_2, and cancel all denominators, as follows:

$$\frac{\cancel{R}R_1R_2}{\cancel{R}} = \frac{R\cancel{R_1}R_2}{\cancel{R_1}} + \frac{RR_1\cancel{R_2}}{\cancel{R_2}}$$

Notice how this procedure ensures that each original denominator will always divide into its new numerator, leaving an equation with no fractions. In this case we have

$$R_1R_2 = RR_2 + RR_1$$

Now we can solve for any one resistance in terms of the others. For example, we can factor the right side and solve for R as follows:

$$R_1R_2 = R(R_2 + R_1)$$

$$R = \frac{R_1R_2}{R_1 + R_2}$$

Note that although we used the LCD, we did *not* add fractions. In particular, we did *not* change each fraction to an equivalent one having the LCD as its denominator. Instead, we cleared fractions by multiplying both sides of the equation by the same quantity, the LCD. Let's not confuse these two different techniques, as many people tend to do.

The other method *does* make use of adding fractions. To solve for R, we could add the fractions on the right side of Eq. (10.3) as follows:

$$\frac{1}{R}=\frac{1}{R_1}+\frac{1}{R_2}=\frac{R_2}{R_1R_2}+\frac{R_1}{R_1R_2}=\frac{R_2+R_1}{R_1R_2}$$

Then we take the reciprocal of both sides to obtain

$$R=\frac{R_1R_2}{R_1+R_2}$$

Comment: *An important point here is that we* ***add fractions first, before we take the reciprocal.*** *A* ***common error*** *is to take the reciprocal of a sum by inverting each individual term. This is a rather illogical maneuver, as almost any example with simple numbers will show. For instance,* $1/2 + 1/8$ *is* $0.500 + 0.125 = 0.625$*. Thus the reciprocal of* $1/2 + 1/8$ *is unquestionably* $1/0.625 = 1.60$*. But if we take the reciprocal the* ***wrong*** *way, by inverting each term, we get* $2/1 + 8/1 = 2 + 8 = 10$*, which is clearly not 1.60. Remember, the reciprocal of any non-zero expression ( ) is 1/( ). Thus we can also interpret the reciprocal of* $(1/2 + 1/8)$ *as* $1/(1/2 + 1/8)$*, which we evaluate as follows:*

$$\frac{1}{1/2+1/8}=\frac{1\cdot 8}{(1/2+1/8)\cdot 8}=\frac{8}{4+1}=\frac{8}{5}=1.60$$

The following examples further illustrate the general procedure for solving equations with fractions.

Example 10.25. **Solve for a:**

$$\frac{2}{a^2+5a+6}-\frac{1}{a+3}=\frac{5}{3a+6}$$

Factor to find the LCD:

$$\frac{2}{(a+2)(a+3)}-\frac{1}{a+3}=\frac{5}{3(a+2)}$$

Multiply all terms by the LCD, $3(a + 2)(a + 3)$, and cancel denominators:

$$\frac{2 \cdot 3\cancel{(a+2)}\cancel{(a+3)}}{\cancel{(a+2)}\cancel{(a+3)}} - \frac{3(a+2)\cancel{(a+3)}}{\cancel{a+3}} = \frac{5 \cdot 3\cancel{(a+2)}(a+3)}{3\cancel{(a+2)}}$$

$$6 - 3a - 6 = 5a + 15$$

$$8a = -15$$

$$\boxed{a = -15/8}$$

Example 10.26. Solve for t:

$$\frac{1}{20} = \frac{1}{2/3 + 4/(t-2)}$$

$$\frac{1}{20} = \frac{1}{2/3 + 4/(t-2)} = \frac{1}{[2(t-2) + 3 \cdot 4]/3(t-2)} = \frac{3(t-2)}{2t+8}$$

$$\frac{20(2t+8)}{20} = \frac{3(t-2) \cdot 20(2t+8)}{(2t+8)}$$

$$2t + 8 = 60t - 120$$

$$58t = 128$$

$$\boxed{t = 64/29}$$

EXERCISE 10.6

Solve for the letters in Probs. 1 to 10.

1. $\dfrac{1}{R} = \dfrac{1}{2}$

2. $\dfrac{1}{R} = \dfrac{1}{2} + \dfrac{1}{4}$

3. $\dfrac{1}{R} + \dfrac{1}{4} = \dfrac{1}{2}$

4. $\dfrac{1}{R+4} = \dfrac{1}{2}$

5. $\dfrac{1}{a-2} = \dfrac{1}{3} - \dfrac{1}{5}$

6. $\dfrac{3b}{b-4} = 3 - \dfrac{4}{b}$

7. $\dfrac{2}{y^2-4} = \dfrac{3}{y+2} - \dfrac{5}{y-2}$

8. $\dfrac{1}{10} = \dfrac{1}{2/t + 3/t}$

9. $\dfrac{1}{t} = \dfrac{1}{1/2 - 2/3}$

10. $\dfrac{4x-3}{12x^2+7x-10} + \dfrac{1}{4x^2+5x} = \dfrac{1}{3x-2}$

11. The formula for the equivalent resistance R of three resistors R_1, R_2, and R_3 in parallel is (see Fig. 10.1)

$$\frac{1}{R} = \frac{1}{R_1} + \frac{1}{R_2} + \frac{1}{R_3}$$

a. Solve for R in terms of R_1, R_2, and R_3.
b. Solve for R_3 in terms of R, R_1, and R_2.

12. The relation between f, the focal length of a simple lens, and d_o and d_i, the distances from the lens to the object and from the lens to the image, respectively, is

$$\frac{1}{f} = \frac{1}{d_o} + \frac{1}{d_i}$$

If the focal length is 50 mm and the object is 10 in. away, what is the lens-to-image distance?

13. In photography the magnification, or ratio of image size to object size, equals the ratio of image distance to object distance, d_i/d_o. These distances are also related to the focal length of the lens as stated in Prob. 12. Thus, if $d_i = d_o$, a full-sized or 1:1 image is formed. For a 50-mm lens, what are the values of d_i and d_o for
a. a 2:1 (double-sized) image?
b. a 1:2 (half-sized) image?

14. Solve for P_i, the internal pressure, in the following formula for stress in a cylindrical pressure vessel:

$$S = \frac{P_i R_i}{t} + \frac{P_i}{2}$$

15. Solve the formula in Prob. 14 for t, the wall thickness of the vessel.

16. The efficiency of a screw thread is given by

$$e = \frac{\tan x\,(1 - F \sec y \tan x)}{\tan x + F \sec y}$$

Solve for sec y.

17. A formula for the speed ratio of an epicyclic-gear train is $n = (W_F - W_A)/(W_D - W_A)$. Solve for W_A, if $n = 30$, $W_F = 0$, and $W_D = 3{,}150$ rpm.

18. The stress factor in a thin-walled cylinder is given by

$$K = \frac{P_i R_i^2 - P_o R_o^2}{R_o^2 - R_i^2}$$

Solve for R_i, the inside radius, if $K = 2$.

19. The formula for the equivalent spring rate for two springs in series (see Fig. 10.2) is

$$\frac{1}{K} = \frac{1}{K_1} + \frac{1}{K_2}$$

If $K_1 = 3/4$ lb/in. and $K_2 = 5/8$ lb/in., what is K?

20. In Prob. 19, if $K = 1\frac{1}{2}$ lb/in. and $K_1 = 2\frac{1}{4}$ lb/in., what is K_2?
21. Solve for Z of Prob. 28 of Exercise 10.4 if $R = 3$ ohms, $X_c = 6$ ohms, and $X_L = 4$ ohms.
22. In Prob. 21, if $Z = 12/13$ ohm, $X_c = 6$ ohms, and $X_L = 12$ ohms, solve for R.
23. At what speed would an object have to travel in order that its mass double? See Prob. 29 of Exercise 10.5. The speed of light is 186,000 miles/sec.
24. A 30-year-old astronaut traveling at 99 percent of the speed of light starts a round trip to the star Arcturus, 40 light-years away from earth. How old will he be when he returns? See Prob. 28 of Exercise 10.5. (*Hint:* t, the earth time, is *distance* $\div v$, and t_0 is the "real time." As the real time slows down, so does the astronaut's aging; thus if the earth time were twice the real time, the astronaut would have aged half the normal amount.) How old would the astronaut's twin brother be if he stayed on earth?

B. COMBINED-RATE PROBLEMS

Many problems involve a combination of rates of doing something. The following example is typical of all such problems.

Example 10.27. Two cars are 1 mile apart and traveling toward each other. If one car maintains a steady speed of 40 mph and the other 50 mph, when will they pass?

The general relationship that applies to all rate problems is

$$\text{Amount} = \text{rate} \times \text{time}$$

or

$$\text{Time} = \frac{\text{amount}}{\text{rate}}$$

In this case *amount* represents the 1-mile combined distance traveled by the cars, or "amount of distance"; *rate* means average speed or "rate of travel." A special relationship of combined-rate problems is that *the rate is the sum of the rates of the individual participants,* in this case the two cars. In other words, a combined-rate problem always has two or more participants working simultaneously on a common "amount"; thus the equivalent rate of all participants working together is the sum of their individual rates. In this example, because the two cars work together to travel 1 mile, the combined rate is

$$r = r_1 + r_2 = 40 + 50 = 90 \text{ mph}$$

and the time to meet or pass is

$$t = \frac{1 \text{ mile}}{90 \text{ mph}} = \frac{1}{90} \text{ hr} \longrightarrow \frac{2}{3} \text{ min} \longrightarrow 40 \text{ sec}$$

Example 10.28. The difficulty in combined-rate problems comes when the rates are fractions. To illustrate, suppose that two boys are 1 mile apart and are walking toward each other. If one boy walks at 1/2 mph and the other at 3/4 mph, how long will they take to meet?

This example exactly parallels the "two cars" example above. The only differences are the values of the rates. Thus the time is as follows

$$\text{Time} = \frac{\text{amount}}{\text{rate}} = \frac{1 \text{ mile}}{r_1 + r_2} = \frac{1 \text{ mile}}{(1/2 + 3/4) \text{ mph}}$$

and we have a problem with a compound fraction. We must be careful to evaluate it correctly, according to the methods of the preceding section.

$$t = \frac{1}{1/2 + 3/4} = \frac{1}{2/4 + 3/4} = \frac{1}{5/4} = \frac{4}{5} \text{ hr}$$

or

$$t = 4/5 \text{ hr} \times 60 \text{ min/hr} = 48 \text{ min}$$

The following example illustrates that the "amount" in such problems need not be a distance and the "rate" need not be a speed in distance ÷ time, but the ideas and procedures are similar.

Example 10.29. One hose can fill a certain tank in 2 hr, and another hose can fill the same tank in 1 hr 20 min. If both hoses are used simultaneously, how many hours do they take to fill the tank?

Once again, the relationship is

$$\text{Amount} = \text{rate} \times \text{time}$$

But this time the amount is 1 tank and the rate is the number of tanks per hour. *It is important that you recognize the similarity between this and the preceding example.* Other than the differences in terminology, the two examples are identical, and the solutions are identical.

Since the first hose fills 1 tank in 2 hr, its rate is 1 tank ÷ 2 hr = $1/2$ tank/hr, and the rate of the second hose is 1 tank ÷ $4/3$ hr = $3/4$ tank/hr. Thus the time is

$$t = \frac{\text{amount}}{\text{rate}} = \frac{1\ \text{tank}}{(1/2 + 3/4)\ \text{tank/hr}} = \frac{1}{5/4} = \frac{4}{5}\ \text{hr}$$

Notice that we were given the individual *times* here, but we did *not* find the combined time as the sum of the individual times. Instead, we used the individual times to find the individual rates, and then we found the combined rate.

EXERCISE 10.7

1. Mr. Jones can mow his lawn in $2\frac{1}{2}$ hr, and his son can do it in 3 hr. How long does it take when they mow simultaneously with two mowers?
2. One crew can lay pipe at the rate of $2\frac{1}{4}$ miles/day, and another crew at $1\frac{3}{4}$ miles/day. How long will it take both crews together to lay 5 miles of pipe?
3. On the average, Bus 261 travels from Tampa to Washington, D.C., in 28 hr, and Bus 361 averages 24 hr from Washington to Tampa. If both buses leave at 6 p.m. on February 8 and use the same highways, at what time will they pass each other?
4. Two cars pass in opposite directions. If one car is going 35 mph and the other 45 mph, in how many seconds will they be 1 mile apart?
5. Repeat Prob. 4 if the cars travel in the same direction.
6. One pump can transfer Bunker C fuel oil from an oil tanker and fill a certain power plant storage tank in 14 hr 40 min. Another pump can fill the same tank in 18 hr 10 min. Assuming that both pumps can work simultaneously on the same tank, in how many hours can both pumps working together fill the tank?
7. The first pump of Prob. 6 transfers oil from the ship to the tank while the second pump simultaneously transfers the oil from that tank to another tank. How many hours will be needed to fill the first tank?
8. John and Bill have a TV repair shop. On the average, John can replace a TV picture tube in $1\frac{1}{4}$ hr. Bill averages 50 min for the same job. How long does it take both men together to re-

place 12 picture tubes? Assume that both work simultaneously on the last set.

9. Repeat Prob. 8 if only one man can work on the last set.
10. With a $3/8$-in.-diameter hose, Mr. Jones can fill his swimming pool in 12 hr. He can fill it in 6 hr 45 min with Mr. Brown's $1/2$-in. hose and in 3 hr with Mr. Smith's $3/4$-in. hose. How long will it take to fill the pool if all three hoses are used simultaneously?
11. If Mr. Smith's hose of Prob. 10 splits after 20 min, how long do the other two hoses take to complete the job?

QUADRATIC EQUATIONS

CHAPTER 11

In Chap. 9 we learned that some quadratic trinomials of the form $ax^2 + bx + c$ can be factored. We shall now solve, by factoring, equations of the form $ax^2 + bx + c = 0$, called *quadratic equations in one variable.* Furthermore, we shall develop a method for solving such equations even when the quadratic trinomial is not factorable in the usual sense. And we shall see that quadratic equations, like linear equations and fractional equations, have many applications in science and technology.

11.1 BASIC IDEAS ABOUT QUADRATIC EQUATIONS

A quadratic equation in one variable must be of the form $ax^2 + bx + c = 0$, where a, b, and c are constants and $a \neq 0$. Because of the second-degree term, quadratic equations are also called *second-degree equations.* We should be able to recognize such an equation from this general form. For example, $5Z^3 + 2Z + 7 = 0$ is *not* a quadratic, or second-degree, equation, and $x^2 + 1/x - 3 = 0$ is *not* a quadratic equation. But $2y^2 + y - 2 = y^2 + 2y + 3$ and $(2x - 3)^2 + x^2 - x = 0$ can be put into quadratic-equation form by combining the similar terms.

The following basic principle of algebra is used repeatedly to solve quadratic equations. It is presented now so that we can use it without pause later.

MULTIPLICATION		$a \cdot 0 = 0$
PROPERTY	Thus if	$A \cdot B = 0$
OF ZERO	then	$A = 0$
	or	$B = 0$

The idea here is that if the product of two numbers is zero, then one of the numbers has to be zero. For example, if

$$(x-3)(x-4)=0$$

then $\quad x-3=0 \quad$ or $\quad x-4=0$

which means that $x=3$ and $x=4$ are both solutions of the original equation.

Comment: *A fairly common error among beginning math students is to try to extend this principle to products other than zero. For example, if $(x-3)(x-4)=1$, we* ***cannot*** *say that either $x-3=1$ or $x-4=1$.*

11.2 SPECIAL CASES

We can easily solve the two special cases $ax^2+bx=0$ and $ax^2+c=0$, and also the general equation $ax^2+bx+c=0$, if the left side is factorable.

A. $ax^2+bx=0$

If the constant term is zero ($c=0$), we can always factor x and use the multiplication property of zero as follows:

$$ax^2+bx=0$$

$$x(ax+b)=0$$

$$\left.\begin{aligned} x&=0 \\ ax+b&=0 \end{aligned}\right] \rightarrow \boxed{\begin{aligned} x&=0 \\ x&=-b/a \end{aligned}}$$

Example 11.1. $2x^2+3x=0$

$$x(2x+3)=0$$

$$x=0 \quad \text{or} \quad 2x+3=0$$

$$\boxed{x=0,\ -3/2}$$

Example 11.2. $35x-5x^2=0$

$$5x(7 - x) = 0$$

$$5x = 0 \quad \text{or} \quad 7 - x = 0$$

$$\boxed{x = 0, 7}$$

B. $ax^2 + c = 0$

If the first-degree term is zero ($b = 0$), we can subtract c from both sides, divide by a, and take both square roots as follows:

$$ax^2 + c = 0$$

$$ax^2 = -c$$

$$x^2 = -c/a$$

$$\boxed{x = \pm\sqrt{\frac{-c}{a}}}$$

Of course either a or c (but not both) must be a negative number; otherwise the solutions are imaginary (not real numbers).

Example 11.3. $2x^2 - 50 = 0$

$$2x^2 = 50$$

$$x^2 = 25$$

$$\boxed{x = \pm 5}$$

Example 11.4. $36 - 2x^2 = 0$

$$-2x^2 = -36$$

$$x^2 = 18$$

$$\boxed{x = \pm 3\sqrt{2}}$$

You should check each solution in these four examples by substituting it into the original equation.

C. FACTORABLE QUADRATIC TRINOMIALS, $ax^2 + bx + c = 0$

The following examples show how to solve a quadratic equation with a *factorable* quadratic trinomial.

Example 11.5. $x^2 - 7x + 12 = 0$

$$(x - 3)(x - 4) = 0$$

$$x - 3 = 0 \quad \text{or} \quad x - 4 = 0$$

$$\boxed{x = 3, 4}$$

Example 11.6. $2x^2 - 9x - 35 = 0$

$$(2x + 5)(x - 7) = 0$$

$$2x + 5 = 0 \quad \text{or} \quad x - 7 = 0$$

$$\boxed{x = -5/2, 7}$$

Example 11.7. $(x - a)^2 + 1 = 2a - 2x$

$$(x - a)^2 + 2(x - a) + 1 = 0$$

$$[(x - a) + 1][(x - a) + 1] = 0$$

$$\left.\begin{array}{l} x - a + 1 = 0 \\ x - a + 1 = 0 \end{array}\right] \rightarrow \boxed{x = a - 1}$$

Perhaps you have noticed that until Example 11.7 every quadratic equation had *two solutions.* This is generally true. But in the case of a perfect-square trinomial, as in Example 11.7, the solutions are identical. This solution, or pair of identical solutions, is called a *double root.*

Example 11.8. A farmer has 100 ft of fencing to enclose a rectangular chicken yard. If he uses the 70-ft side of his barn as one side of the chicken yard, uses the entire 100 ft of fencing, and wants to enclose 1,200 sq ft, what are the dimensions of the chicken yard?

First, make a diagram as shown in Fig. 11.1.

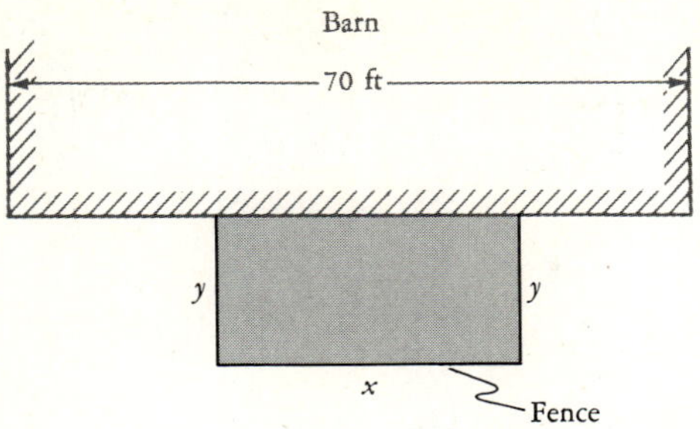

Fig. 11.1 Example 11.8.

Notation:
Let x = length of the parallel side
y = length of each perpendicular side

Equations:
Perimeter: $x + 2y = 100$ ft
Area: $xy = 1{,}200$ sq ft

Solution. By substituting $x = 100 - 2y$ into $xy = 1{,}200$, we have

$$(100 - 2y)y = 1{,}200$$

$$2y^2 - 100y + 1{,}200 = 0$$

$$y^2 - 50y + 600 = 0$$

$$(y - 20)(y - 30) = 0$$

$$y - 20 = 0 \quad \rightarrow \quad y = 20,\ x = 60$$

$$y - 30 = 0 \quad \rightarrow \quad y = 30,\ x = 40$$

Answer:
The chicken yard can be either 20 ft by 60 ft or 30 ft by 40 ft.

Question: What if the barn were only 50 ft long?

EXERCISE 11.1

Solve each equation by one of the methods of this section.

1. $x^2 - 16 = 0$
2. $2Z^2 - 24 = 0$
3. $y^2 + 5y = 0$
4. $a^2 + 4a + 4 = 0$
5. $3b^2 = 75$
6. $2x^2 - 3x - 5 = 0$

7. $5x^2 + 3x - 2 = 0$
8. $\frac{3}{2}B^2 = \frac{1}{4}B$
9. $aR^2 - bR = 0$
10. $mS^2 - n = 0$
11. $t^2 - 2 = t$
12. $2x^2 + x - 2 = x^2 - 3x + 3$
13. $(2x - 3)^2 = 2x - x^2$
14. $\frac{1}{4}U^2 - \frac{1}{3}U + \frac{1}{9} = 0$
15. $(x - 2)^2 + 2(x - 2) + 1 = 0$ (*Hint:* Let $x - 2 = U$, and solve for U.)
16. $\dfrac{6}{x^2} - \dfrac{23}{x} - 55 = 0$ (*Hint:* Let $1/x = U$ and solve for U.)
17. $acx^2 - adx = bd - bcx$
18. The sum of two numbers is 12, and their product is 11. What are the numbers?
19. The sum of two numbers is -30, and their product is 176. What are the numbers?
20. A certain rectangular field has a perimeter of 300 yd and an area of 5,000 yd^2. What are its dimensions?
21. A very thin board 12 ft long is to be cut to make a right triangle having a hypotenuse of 60 in. How should the board be cut?
22. If the board of Prob. 21 is 2 in. wide, make sketches to show the exact shape and size of each piece. Assume that all pieces are in the same plane.
23. The maximum tangential stress in a thick-walled cylindrical pressure vessel is $S_{t\text{-max}} = P_i(r_o^2 + r_i^2)/(r_o^2 - r_i^2)$, where r_o and r_i are the outside and inside radii, respectively, and P_i is the inside pressure. What is the inside radius if the outside radius is 20 in., the pressure is 100 psi, and the maximum allowable stress is 14,900 psi?
24. If air friction is neglected, the elevation, in feet, of a certain projectile as a function of time, in seconds, is given by the formula $S = 116t - 16t^2$.
    a. At what times is the elevation 100 ft going up and coming down?
    b. How long after reaching 100 ft coming down does the projectile strike the ground?
25. The retailer's percentage markup on the wholesale price of TV sets is 3 times the wholesaler's percentage markup on the manufacturer's price of \$100. If the retail price is \$192, what is the wholesale price? [*Hints:* (1) First find the percentages of markup. (2) Manufacturing price + wholesale markup + retail markup = retail price.]
26. A certain electric circuit involving two resistors requires that one resistance be 4 ohms less than the other and that the equiv-

alent parallel resistance be $1\frac{1}{2}$ ohms. See Eq. (10.3). What are the resistances of the two resistors?

27. A certain automatic transmission can be locked by shifting three gears so that each one is externally tangent to the other two and so that the lines between the centers form a right triangle. The center distance between the two largest gears is 6 in., and the distance between the two smallest is $1\frac{1}{5}$ in. less than the distance between the smallest and the largest. What is the radius of each gear?
28. Repeat Prob. 27 if the largest gear is a ring gear (internal teeth) with the other two gears inside the ring. Now the center distance between the largest and smallest is the 6 in., and the distance between the two smallest gears is $1\frac{1}{5}$ in. less than the distance between the two largest gears.
29. A rectangular air-conditioning duct is to be formed by bending a 68-in.-wide piece of sheet metal into a rectangle. The cross-sectional area of the duct has to be 2 ft^2. What are the height and width of the duct?
30. The net power output of an electric generator is the gross power EI produced, less the power loss I^2R due to internal resistance. What is the current I in a generator that gives a net output of 3,600 watts at $E = 120$ volts, if the internal resistance R is 1 ohm?
31. When two electric resistors are connected in series, the equivalent resistance is the sum of the individual resistances: $R_s = R_1 + R_2$. When they are connected in parallel, the equivalent resistance is given by Eq. (10.3). If the parallel resistance of two resistors is 3 ohms and the series resistance 12 ohms, what are the individual resistances?

11.3 THE QUADRATIC FORMULA

Now we shall develop a method for solving all kinds of quadratic equations in one variable, including those with nonfactorable quadratic trinomials. The method is called *completing the square,* and it will lead us to a general formula for the solution.

A. THE METHOD OF COMPLETING THE SQUARE

This method is based upon the factoring of a perfect-square quadratic trinomial. If we have a perfect square, such as $(x + n)^2$, on one side, we can always solve the quadratic equation by taking the square root of both sides. For example, to solve $x^2 + 6x + 9 = 4$, we factor the left side into a perfect square and take the square root of both sides as follows:

$$x^2 + 6x + 9 = 4$$

$$(x + 3)^2 = 4$$

$$x + 3 = \pm\sqrt{4} = \pm 2$$

$$x = -3 \pm 2$$

$$\boxed{x = -1, -5}$$

If we do not happen to have a perfect-square trinomial, we *make one by rearranging the constant term,* as follows:

$$x^2 + 6x + 7 = 0$$

$$x^2 + 6x + (9 - 2) = 0$$

$$x^2 + 6x + 9 = 2$$

$$(x + 3)^2 = 2$$

Then we take the square root of both sides and solve for x as before:

$$x + 3 = \pm\sqrt{2}$$

$$\boxed{x = -3 \pm \sqrt{2}}$$

How did we know to rewrite 7 as $9 - 2$? We did it because our amazing powers of foresight told us that we needed a 9 to make a perfect square with x^2 and $6x$. But just in case our foresight should fail, there is a general method for determining how to form the constant term.

In the first place, in $x^2 + 6x + 9$ notice that 9 is the square of half of 6: $[(1/2)6]^2 = 3^2 = 9$; and 9, of course, is the term that completes the square with x^2 and $6x$. This relationship between the constant term and the coefficient of the x term is always true *if the coefficient of x^2 is +1.* We can show this relationship even in the most general terms, as follows:

$$(mx + n)^2 = m^2x^2 + 2mnx + n^2$$

$$= m^2\left(x^2 + 2\,\frac{n}{m}\,x + \frac{n^2}{m^2}\right)$$

Notice that the *constant term n^2/m^2 is the square of one-half the coefficient of the first-degree term $2n/m$:*

$$\left[\frac{1}{2}\left(2\,\frac{n}{m}\right)\right]^2 = \left(\frac{n}{m}\right)^2 = \frac{n^2}{m^2}$$

Steps in Solving a Quadratic Equation

The relationship shown above is the *key point* that we shall apply to solve quadratic equations. We always perform five steps, as follows. We are given, say,

$$3x^2 - 5x - 13 = 0$$

1. Move the constant term to the right side:

$$3x^2 - 5x = 13$$

2. Divide both sides by the coefficient of the second-degree term:

$$x^2 - (5/3)x = 13/3$$

3. Complete the square by adding the square of one-half the coefficient of the first-degree term to both sides:

$$x^2 - (5/3)x + [1/2(-5/3)]^2 = 13/3 + [1/2(-5/3)]^2$$

4. Factor the left side and add the terms on the right side:

$$(x - 5/6)^2 = 181/36$$

5. Take the square root of both sides and solve for x:

$$x - \frac{5}{6} = \pm\frac{\sqrt{181}}{6}$$

$$x = \frac{5}{6} \pm \frac{\sqrt{181}}{6}$$

EXERCISE 11.2

Solve each equation by the method of completing the square.

1. $x^2 + 2x - 3 = 0$
2. $2y^2 - 4y - 6 = 0$
3. $2A^2 + 3A = 9$
4. $V^2 - 10V - 39 = 0$
5. $2Z^2 - 3Z - 6 = 0$
6. $U^2 + \frac{U}{2} = -\frac{1}{8}$
7. $5W^2 = 1 - W$
8. $1/3 - 2x - 3x^2 = 0$
9. $16y^2 - 16y + 1 = 0$
10. $x^2 - 2x - 1 = 0$
11. $x^2 - 2x + 3 = 0$
12. $x^2 - 6x + 12 = 0$
13. $ax^2 + bx + c = 0$
14. $mx^2 + nx + k = 0$

B. THE QUADRATIC FORMULA

Problem 13 of the preceding set shows that the method of completing the square can be applied to the generalized quadratic equation, $ax^2 + bx + c = 0$. The result is an equation called the *quadratic formula,* which is the general solution for any quadratic equation. Here is the derivation for the quadratic formula. We are given

$$ax^2 + bx + c = 0 \tag{11.1}$$

1. Move the constant term to the right side:

$$ax^2 + bx = -c$$

2. Divide both sides by the coefficient of the second-degree term:

$$x^2 + \frac{b}{a}x = -\frac{c}{a}$$

3. Complete the square:

$$x^2 + \frac{b}{a}x + \frac{b^2}{4a^2} = \frac{b^2}{4a^2} - \frac{c}{a}$$

4. Factor the left side and add the fractions on the right side:

$$\left(x + \frac{b}{2a}\right)^2 = \frac{b^2 - 4ac}{4a^2}$$

5. Take the square root of both sides and solve for x:

$$x + \frac{b}{2a} = \frac{\pm\sqrt{b^2 - 4ac}}{2a}$$

$$x = \frac{-b \pm \sqrt{b^2 - 4ac}}{2a} \tag{11.2}$$

The quadratic formula

Comment: *As in all other formulas, the objective here is to* ***learn the relationship*** *between the symbols of the solution, Eq. (11.2), and the symbols of the problem, Eq. (11.1). In other words, we are out to* ***learn the ideas*** *expressed by the formula, not to memorize the formula by rote. Thus we should translate Eq. (11.2) into words and get the gist of it. In this case the formula tells us that the solutions of the quadratic equation* $ax^2 + bx + c = 0$ *are (1) the negative of the coefficient of the first-degree term, plus or minus (2) the square root of the square of the coefficient of the first-degree term minus 4 times the coefficient of the second-degree term times the constant term, all divided by (3) twice the coefficient of the second-degree term.*

The following examples illustrate the solutions of quadratic equations by direct application of the quadratic formula. Watch closely to see how the ideas just expressed are employed.

Example 11.9. $2x^2 + 3x - 4 = 0$

$$x = \frac{-3 \pm \sqrt{3^2 - 4(2)(-4)}}{2(2)} = \frac{-3 \pm \sqrt{9 + 32}}{4}$$

$$\boxed{x = \frac{-3 \pm \sqrt{41}}{4}}$$

Decimal approximations are

$$x = \frac{-3 + \sqrt{41}}{4} = \frac{-3 + 6.403}{4} = 0.851$$

and

$$x = \frac{-3 - \sqrt{41}}{4} = \frac{-3 - 6.403}{4} = -2.351$$

Example 11.10. $8 - 15x^2 + 2x = 0$

$15x^2 - 2x - 8 = 0$

$$x = \frac{-(-2) \pm \sqrt{(-2)^2 - 4(15)(-8)}}{2(15)} = \frac{2 \pm \sqrt{4(1 + 120)}}{2(15)}$$

Note how the signs are handled.

Note how the radicand is made smaller if we can take out a perfect square.

$$x = \frac{2 \pm 2\sqrt{121}}{2(15)} = \frac{2(1 \pm 11)}{2(15)}$$

$$\boxed{x = \frac{-2}{3}, \frac{4}{5}}$$

The fact that the roots are rational indicates that the original equation could have been solved by factoring.

Example 11.11. $x^2 + x + 1 = 0$

$$x = \frac{-1 \pm \sqrt{1^2 - 4(1)(1)}}{2(1)} = \frac{-1 \pm \sqrt{1 - 4}}{2}$$

$$\boxed{x = \frac{-1 \pm \sqrt{-3}}{2}}$$

The roots are imaginary.

EXERCISE 11.3

For Probs. 1 to 14, use Probs. 1 to 14 of Exercise 11.2 and solve each equation by using the quadratic formula.

15. If two electric resistors are connected in series, the equivalent resistance is $R_s = R_1 + R_2$. If they are connected in parallel, the equivalent resistance is $R_p = R_1R_2/(R_1 + R_2)$. What two resistors should be used so that the series resistance is 2 ohms and the parallel resistance is $1/4$ ohm?
16. Repeat Prob. 15 if the series resistance is 3 ohms and the parallel resistance is 1 ohm.
17. If two springs are connected in parallel as shown in Fig. 11.2, the equivalent spring factor is $K_p = K_1 + K_2$. If they are connected in

series, the equivalent spring factor is $K_s = K_1K_2/(K_1 + K_2)$. What two springs should be used if the parallel spring factor is 2 lb/in. and the series spring factor is 1/4 lb/in.?

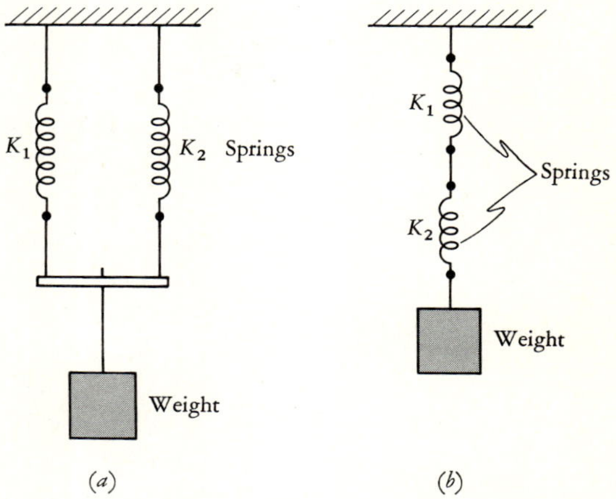

Fig. 11.2 Prob. 17. (a) Two springs in parallel, and (b) two springs in series.

18. Repeat Prob. 17 if the parallel spring factor is 3 lb/in. and the series spring factor is 1 lb/in.
19. In a certain machine, gear D drives two other gears A and B, as shown in Fig. 11.3. Two requirements of the machine are (1) that the diameter of B be 1 in. greater than that of A, and (2) that A turn 2/3 revolution more than B for each revolution of D. If D has a 4-in. diameter, what are the diameters of A and B? [*Hints:* The speed ratio of two gears is inversely proportional to their diameters. Let d_B be the diameter of gear B. Then $d_B - 1$ is the diameter of gear A. The speed ratio between B and D is $4/d_B$, and between A and D is $4/(d_B - 1)$.]

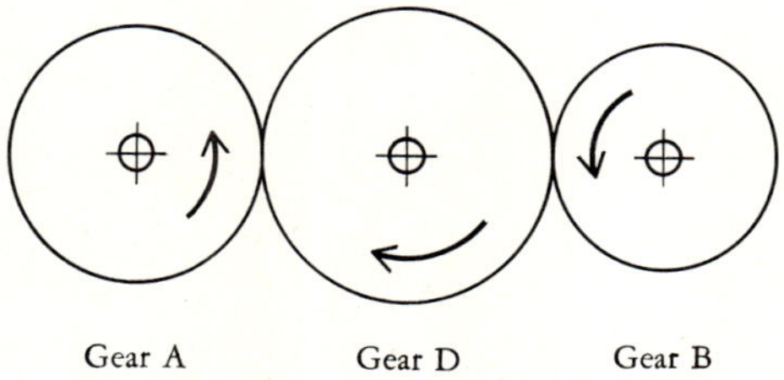

Fig. 11.3 Prob. 19. Three gears.

20. A resistor, a capacitor, and an inductor are connected in series as shown in Prob. 28 of Exercise 10.4. The circuit is to have an equivalent impedance of $Z = 1$ ohm, $R = 2$ ohms, and $X_L = 4$ ohms. What is X_c, the impedance of the capacitor?

21. A tube in the shape of a rectangle is to be formed by bending a 24-in.-wide piece of sheet metal. The area inside the rectangle is to be 10 in.^{2}. Find the width and height of the tube.
22. The winner of the 1971 Indianapolis 500-mile race averaged 18.97 mph faster than the winner of the 1960 race and took 26.00 min less time. What were the winning speed and time in 1971?
23. Two men with walkie-talkies started simultaneously at a right-angle corner and bicycled down the roads at right angles. One man rode 2.5 mph slower than the other. After 2 hr, they reached the 15-mile range of the walkie-talkies. How far did each man travel?
24. As shown in Fig. 11.4, the elongation e produced in a metal rod due to a falling weight is given by the formula $Wl/(AE) = e^2/[2(h+e)]$, where A is the cross-sectional area and E is the modulus of elasticity. If $W = 100$ lb, $l = 6$ in., $h = 3$ in., and $E = 30 \cdot 10^6$ psi, how much would a rod with a diameter of $2/\sqrt{\pi}$ in. elongate?

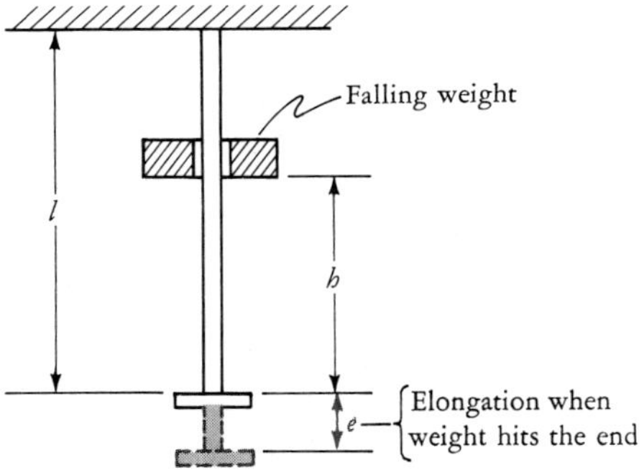

Fig. 11.4 Prob. 24. Elongation in a rod, due to a falling weight.

CHAPTER 12

EXPONENTS AND RADICALS

In Sec. 3.3 we limited our study of exponents to positive, integral exponents. We shall now extend our concepts to include zero, negative numbers, and fractions as exponents. And we shall see that radicals can be changed to exponents.

For reference, we repeat here the basic laws of exponents, but now we allow the exponents n and k to be any rational number, not just positive integers.

$$a^n \cdot a^k = a^{n+k} \tag{12.1}$$

$$\frac{a^n}{a^k} = a^{n-k} \qquad a \neq 0 \tag{12.2}$$

$$(a^n)^k = a^{nk} = a^{kn} = (a^k)^n \tag{12.3}$$

$$(ab)^n = a^n b^n \tag{12.4}$$

$$\left(\frac{a}{b}\right)^n = \frac{a^n}{b^n} \qquad b \neq 0 \tag{12.5}$$

12.1 ZERO AS AN EXPONENT

Previously, in equations such as (12.2) we restricted n to be larger than k so that $n - k$ was positive. Now we shall examine the possibility $n = k$, and show the meaning of zero as an exponent.

If we allow $n = k$ in the equation $a^n/a^k = a^{n-k}$, then we have

$$\frac{a^k}{a^k} = a^{k-k} = a^0$$

But we know that any non-zero number divided by itself is *unity* (+1). Therefore

$$\frac{a^k}{a^k} = 1 \qquad a \neq 0$$

By comparing these two results, we conclude that *any non-zero expression raised to the zero power is unity.*

ZERO EXPONENT	$a^0 = 1$	where $a \neq 0$	(12.6)

Example 12.1. $\dfrac{3^5}{3^5} = 3^{5-5} = 3^0 = 1$

Example 12.2. $\dfrac{(2x^2)^3}{2x^6} = \dfrac{2^3}{2} \cdot \dfrac{x^6}{x^6} = 2^2x^0 = 4(1) = 4$

Example 12.3. $[(-5)^3]^0 = (-5)^{3 \cdot 0} = (-5)^0 = 1$

Example 12.4. $(-x^7)^0 = (-1)^0(x^{7 \cdot 0}) = (1)(1) = 1 \qquad \text{for } x \neq 0$

Example 12.5. $(x^2 - y^2)^0 = 1 \qquad \text{for } x^2 \neq y^2$

Example 12.6. $a^2 \cdot a^0 = a^{2+0} = a^2 \qquad \text{or} \qquad a^2 \cdot a^0 = a^2 \cdot 1 = a^2$

12.2 NEGATIVE EXPONENTS

In the equation $a^n/a^k = a^{n-k}$, if we allow $k > n$, then we have a negative exponent. For example,

$$\frac{5^2}{5^4} = 5^{2-4} = 5^{-2} \qquad \text{or} \qquad \frac{x}{x^2} = x^{1-2} = x^{-1}$$

But we have previously seen that for $k > n$, $a^n/a^k = 1/a^{k-n}$. We demonstrated this fact by canceling factors as follows:

$$\frac{5^2}{5^4} = \frac{\cancel{5} \cdot \cancel{5}}{\cancel{5} \cdot \cancel{5} \cdot 5 \cdot 5} = \frac{1}{5^2}$$

Thus $$\frac{5^2}{5^4} = \frac{1}{5^{4-2}} = \frac{1}{5^2} \qquad \text{and} \qquad \frac{x}{x^2} = \frac{1}{x}$$

By comparing these examples, we see that

$$5^{-2} = \frac{1}{5^2} \qquad \text{and} \qquad x^{-1} = \frac{1}{x}$$

and in general

$$\frac{a^n}{a^k} = a^{n-k} = a^{-(k-n)} = \frac{1}{a^{k-n}}$$

We must conclude that a negative exponent implies a *reciprocal:* Any non-zero base with a negative exponent equals the reciprocal of that base with the corresponding positive exponent.

NEGATIVE EXPONENT	$a^{-n} = \frac{1}{a^n}$	and	$\frac{1}{a^{-n}} = a^n$	for $a \neq 0$	(12.7)

Example 12.7. $(10)^{-1} = \frac{1}{10} = 0.1$

Example 12.8. $0.50 = 5(1/10) = 5 \cdot 10^{-1}$

Example 12.9. $10^{-4} = \frac{1}{10^4} = \frac{1}{10{,}000} = 0.0001$

Example 12.10. $-0.00075 = -7.5(1/10{,}000) = -7.5 \cdot 10^{-4}$

Examples 12.8, 12.9, and 12.10 illustrate a form called *scientific notation:* The number is written in terms of 10 to some positive or negative power.

Example 12.11. $\frac{1}{2^{-1}} = \frac{1}{1/2} = 1 \cdot \frac{2}{1} = 2$

Example 12.12. $\frac{1}{(2x^{-1})^{-1}} = \frac{1}{2^{-1}x} = \frac{2}{x}$

Example 12.13. $\left(\frac{5x^2}{3y^3}\right)^{-3} = \left(\frac{3y^3}{5x^2}\right)^3 = \frac{27y^9}{125x^6}$

Example 12.14. $(x + y)^{-2} = \frac{1}{(x + y)^2} = \frac{1}{x^2 + 2xy + y^2}$

but

$$x^{-2} + y^{-2} = \frac{1}{x^2} + \frac{1}{y^2} = \frac{y^2 + x^2}{x^2y^2}$$

Study the two parts of Example 12.14 very closely. Notice that $(x + y)^{-2}$ is *not* equal to $x^{-2} + y^{-2}$.

Example 12.15. $$\frac{n + m^{-1}}{m - n^{-1}} = \frac{n + 1/m}{m - 1/n} = \frac{(mn + 1)/m}{(mn - 1)/n} = \frac{n(mn + 1)}{m(mn - 1)}$$

EXERCISE 12.1

In Probs. 1 to 21, remove the zero and negative exponents and carry out the indicated operations wherever possible.

1. 100^0
2. $(3/2)^0$
3. $3/2^0$
4. $3^0/2$
5. $(x^2 + y^2)^0$
6. $(x^0 + y^0)^2$
7. $(-2 \cdot 3)^0$
8. 3^{-2}
9. $(-3)^{-2}$
10. $(3^{-2})^0$
11. $(2 - x)^{-2}$
12. $2^{-2} - x^{-2}$
13. $\dfrac{a^3}{a^{-5}}$
14. $\left[\dfrac{3}{(x - y^0)^2}\right]^{-1}$
15. $\dfrac{2 + x^{-1}}{x - 2^{-1}}$
16. $(x^{-1} - y^{-1})^{-1}$
17. $\left(\dfrac{5x}{2y^2}\right)^{-3}\left(\dfrac{y}{5}\right)^{-5}$
18. $\dfrac{s^{-2} - r^{-2}}{s^{-1} - r^{-1}}$
19. $\left(\dfrac{s^{-2} - r^{-2}}{s^{-1} - r^{-1}}\right)^0$
20. $(x^2 - 2y)^2(x^2 - 2y)^{-1}$
21. $(n + 1)^{-1} + (n - 1)^{-1}$

In Probs. 22 to 27, change each number from scientific notation to ordinary decimal notation.

22. 3.2×10^{-1}
23. 3.2×10^1
24. -52×10^{-4}
25. 0.5×10^{-3}
26. -30×10^6
27. 1×10^{-6}

In Probs. 28 to 33, write each number in scientific notation.

28. 0.0004
29. −0.0032
30. 256
31. 5,000,000
32. 0.000001
33. −0.005

34. What power of 10 corresponds to
    a. one thousand? b. one thousandth?
    c. one millionth? d. one million?
35. The approximate distance from the earth to the sun is 93 million miles. Write this number in scientific notation.

36. One *micron* (abbreviated μm) is one millionth of a meter. Using scientific notation, change 3.5 μm to meters.
37. In electronics, the term *milliamp* means one thousandth of an ampere. Write 12 milliamps in scientific notation.
38. In several areas of technology, the Greek letter μ (mu) is used to indicate *millionths.* Write the number 5 μ in. (millionths of an inch) in scientific notation. Write the expression 120 *microvolts* (millionths of a volt) in scientific notation, in ordinary decimal notation, and in "μ" notation.
39. Young's modulus, or the modulus of elasticity, of most steels is about 30 million psi. Write this number in scientific notation.

In Probs. 40 to 48, solve for x. Watch out for the restriction $a \neq 0$ in Eq. (12.7)!

40. $x^{-2} = 16$
41. $x^{-2} = 1/16$
42. $x^{-1} = 0$
43. $2x^{-2} - x^{-1} = 0$
44. $2/x^{-2} - 1/x^{-1} = 0$
45. $x^{-2} + 6x^{-1} + 9 = 0$
46. $x^{-4} - 2x^{-2} = 0$
47. $2x^{-2} + x^0 = 3x^{-1} + 5$
48. $3x^{-1} - 2x = 6$

12.3 FRACTIONAL EXPONENTS

We shall now determine the significance of a fractional exponent, as in the expressions $a^{1/n}$ and $a^{k/n}$, where n and k are integers. For example, we want to learn the meaning of expressions such as $16^{1/2}$ and $x^{2/3}$.

In Eq. (12.3) for $(a^n)^k$, let's make a fractional exponent by replacing n by $1/n$. Then we would have $(a^{1/n})^k = a^{k/n}$. We can gain an insight into the meaning of $a^{k/n}$ by examining the following specific example. If $n = 2$ and $k = 2$, then we can interpret $a^{k/n}$ as follows:

$$(a^{1/2})^2 = a^{2/2} = a^1 = a$$

In effect, this equation says "$a^{1/2}$ squared equals a." Now think a minute: What expression involving the number a can you square and have the result be a itself? Would you believe $\sqrt{a}$? From past experience we should recall that

$$(\sqrt{a})^2 = a$$

By comparing this equation to the one above, we must conclude that

$$a^{1/2} = \sqrt{a}$$

In similar fashion we could show that $a^{1/3} = \sqrt[3]{a}$, $a^{1/4} = \sqrt[4]{a}$, and so on. Thus in general we must conclude that *a fractional exponent implies a root,* as follows:

FRACTIONAL EXPONENTS	$a^{1/n} = \sqrt[n]{a}$	(12.8)
	$a^{k/n} = (a^{1/n})^k = (\sqrt[n]{a})^k$	

Example 12.16. $16^{1/2} = \sqrt{16} = 4$

Example 12.17. $16^{-1/2} = \dfrac{1}{\sqrt{16}} = \dfrac{1}{4}$

Example 12.18. $16^{3/2} = (\sqrt{16})^3 = 4^3 = 64$

Example 12.19. $(16 + 9)^{1/2} = 25^{1/2} = \sqrt{25} = 5$

but

$$16^{1/2} + 9^{1/2} = \sqrt{16} + \sqrt{9} = 4 + 3 = 7$$

From Example 12.19 we see that $(a + b)^{1/n}$ is *not* the same as $a^{1/n} + b^{1/n}$.

Example 12.20. $\sqrt[-5]{32} = 32^{-1/5} = \dfrac{1}{32^{1/5}} = \dfrac{1}{2}$

Example 12.21. $\dfrac{A}{A^{31/32}} = A^{1-31/32} = A^{1/32}$

Example 12.22.

$$(x^{-3} - x^{-6})^{-1/6} = [x^{-3}(1 - x^{-3})]^{-1/6} = x^{-3/-6}\left(1 - \frac{1}{x^3}\right)^{-1/6}$$
$$= x^{1/2}\left(\frac{x^3 - 1}{x^3}\right)^{-1/6} = x^{1/2}\left(\frac{x^3}{x^3 - 1}\right)^{+1/6} = \frac{x}{(x^3 - 1)^{1/6}}$$

The inquisitive student might wonder if we can commute the factors k and $1/n$ in the middle expression of Eq. (12.8) and come up with the following formula:

$$a^{k/n} = (a^k)^{1/n} = \sqrt[n]{a^k}$$

In most cases we can do this; but for certain combinations of a, n, and k, the above equation is not true. The following example illustrates this point.

Example 12.23. $(-3)^{2/2} = (-3)^1 = -3$ and $(\sqrt{-3})^2 = -3$ — Therefore $(-3)^{2/2} = (\sqrt{-3})^2$.

$(\sqrt{-3})^2 = -3$ but $\sqrt{(-3)^2} = \sqrt{9} = 3$ — Therefore $(-3)^{2/2}$ is *not* $\sqrt{(-3)^2}$.

EXERCISE 12.2

Evaluate each expression in Probs. 1 to 16.

1. $36^{1/2}$
2. $36^{-1/2}$
3. $(8/27)^{1/3}$
4. $64^{2/3}$
5. $(1/64)^{-3/2}$
6. $(-125)^{4/3}$
7. $(-8)^{-1/3}$
8. $[(-2)^2]^{1/2}$
9. $[(-2)^{1/2}]^2$
10. $(2^4)^{3/2}$
11. $[(2/3)^6]^{-1/3}$
12. $(7^{3/2} \cdot 7^{1/2})/7$
13. $(16^{1/2})^{1/2}$
14. $(64^{1/3})^{1/2}$
15. $(54^{1/2})^{1/3}$
16. $[(-32)^{1/3}]^{1/5}$

In Probs. 17 to 28, simplify each expression by eliminating and combining exponents according to Eqs. (12.1) to (12.8).

17. $(4x^2)^{3/2}$
18. $(x^3 \cdot x^2)^{5/3}$
19. $(-27a^6b^0/c^2)^{-2/3}$
20. $(N^5)^{-0/5} - N^{-1}$
21. $\dfrac{K^{3/2} \cdot K^{1/3}}{K^{-1}}$
22. $\left(\dfrac{4x^{-4}}{9y^{-2}}\right)^{-1/2}$
23. $\left[\dfrac{(-a)^{-3}N^0M^3}{aM^{74/75}}\right]^{25/2}$
24. $(16x^4 - 32x^8)^{1/4}$
25. $1/[(s^2 + 8s + 16)^{-1/2}]$
26. $[x^{1/n}]^{1/m}$
27. $[(x^m)^{1/n}]^{1/m}$
28. $[(x^{1/m})^{1/n}]^m$
29. Does $(a^{1/n})^k$ always equal $(a^k)^{1/n}$?
30. What are the conditions under which $(a^{1/n})^k$ does not equal $(a^k)^{1/n}$?
31. Under what condition does $(x^2)^{1/2} = x$?
32. Under what condition does $(x^2)^{1/2} = -x$?
33. Determine the fallacy in the following "proof" that $-3 = 3$, where x is any real number.

$$(x^2 - 6x + 9)^{1/2} = (9 - 6x + x^2)^{1/2}$$

$$[(x-3)^2]^{1/2} = [(3-x)^2]^{1/2}$$

$$x - 3 = 3 - x$$

Since x can be any real number, let $x = 0$. Then

$$0 - 3 = 3 - 0$$

$$-3 = 3$$

12.4 SIMPLIFYING RADICALS

We can often simplify an expression involving radicals by performing any combination of the following three operations, which will be discussed in detail below:

1. Remove perfect nth powers from the nth root (Sec. 12.4A).
2. Rationalize the denominator (Sec. 12.4B).
3. Reduce the index (Sec. 12.4C).

A. REMOVING PERFECT nth POWERS

In the expression $\sqrt{4x}$, we say that 4 is a *perfect square* because $4 = 2^2$. We can remove any such perfect-square *factor* from a second-order radical by taking its square root. Thus $\sqrt{4x} = 2\sqrt{x}$. Similarly, we can write $\sqrt[3]{27a^4}$ as $\sqrt[3]{3^3a^3a}$ and remove the perfect-cube *factors* by taking the cube roots to obtain $3a\sqrt[3]{a}$. In general, *we can simplify an* n*th-order radical by removing any perfect* n*th factors from the radicand.*

We keep emphasizing *factor,* because *terms* cannot be removed in this way. Thus one step in simplifying a radical is to *factor the radicand* into prime factors and to remove the perfect nth powers. If the power of some factor, such as a^4 in the cube root above, is higher than a perfect nth power, then a perfect nth power can always be separated by using Eq. (12.1) as we did above. Example 12.26 below shows how to use Eq. (12.3) to form a perfect nth power.

Example 12.24. $\sqrt{32a^3} = \sqrt{16 \cdot a^2 \cdot 2a} = \sqrt{(4a)^2 \cdot 2a} = 4a\sqrt{2a}$ for $a \geq 0$

Example 12.25. $\sqrt[3]{32a^3} = \sqrt[3]{8a^3 \cdot 4} = \sqrt[3]{(2a)^3 \cdot 4} = 2a\sqrt[3]{4}$

Example 12.26. $\sqrt[5]{\dfrac{x^{13}}{32}} = \sqrt[5]{\left(\dfrac{x^2}{2}\right)^5 \cdot x^3} = \dfrac{x^2}{2}\sqrt[5]{x^3}$

Example 12.27. $\sqrt{4x^4 - 16x^2} = \sqrt{4x^2(x^2 - 4)} = 2x\sqrt{x^2 - 4} \quad \text{for } x \geqslant 0$

Example 12.27 shows that *we must factor first,* then remove perfect-square *factors.* We *cannot* remove the perfect-square *terms* $4x^4$ and $16x^2$ individually. Try it with $x = 3$ and see!

Example 12.28. $\sqrt{(-2x)^2} = \sqrt{(-1)^2(2x)^2} = \sqrt{(2x)^2} = 2x \quad \text{for } x \geqslant 0$

Notice how we must handle *negative* perfect squares. $\sqrt{(-2x)^2}$ is *not* $-2x$, unless x is negative.

Example 12.29. $\sqrt[n]{a^{2n}b^{3n}} = \sqrt[n]{(a^2b^3)^n} = a^2b^3$

A Special Definition for Square Roots

Example 12.28, along with the definition of a principal root (Chap. 3), leads us to make the following special definition for square roots.

SQUARE ROOTS $\quad \sqrt{x^2} = \begin{cases} x & \text{if } x \geqslant 0 \\ -x & \text{if } x < 0 \end{cases} \quad$ (12.9)

For example,

$$\sqrt{2^2} = 2 \qquad \text{and} \qquad \sqrt{(-2)^2} = -(-2) = 2$$

This is the reason why Example 12.24 has the restriction $a \geqslant 0$ and Example 12.27 has $x \geqslant 0$. How would these examples read if $a < 0$ and $x < 0$, respectively? Could we always allow negative values for a and x?

B. RATIONALIZING THE DENOMINATOR

An expression such as $\sqrt[3]{a/3}$ or $1/\sqrt{x}$ is made more simple if we remove the radical from the denominator. This process is called *rationalizing* (making rational) *the denominator.* To do this, we multiply the denominator *and the numerator* by an expression that will produce a perfect nth power in the denominator. For example, the denominator of $\sqrt[3]{a/3}$ needs to be made into a perfect cube. To do this, we ask ourselves: By what number should we multiply 3 to obtain 3^3? From experience we know immediately that $3 \cdot 3^2 = 3^3$; thus we multiply denominator and numerator by 3^2, as follows:

$$\sqrt[3]{\frac{a}{3}} = \sqrt[3]{\frac{a \cdot 3^2}{3 \cdot 3^2}} = \sqrt[3]{\frac{9a^2}{3^3}} = \frac{1}{3}\sqrt[3]{9a^2}$$

Similarly, to rationalize $1/\sqrt{x}$, we would multiply denominator and numerator by $\sqrt{x}$, as follows:

$$\frac{1}{\sqrt{x}} = \frac{1 \cdot \sqrt{x}}{\sqrt{x} \cdot \sqrt{x}} = \frac{\sqrt{x}}{(\sqrt{x})^2} = \frac{\sqrt{x}}{x}$$

Example 12.30. $\sqrt{\frac{5a}{2b}} = \sqrt{\frac{5a \cdot 2b}{2b \cdot 2b}} = \sqrt{\frac{10ab}{(2b)^2}} = \frac{1}{2b}\sqrt{10ab} \qquad \text{for } b > 0$

Example 12.31. $\sqrt{\frac{x^7}{-5}} = \sqrt{\frac{(x^3)^2 \cdot x(-5)}{(-5)(-5)}} = \frac{x^3}{5}\sqrt{-5x} \qquad \text{for } x > 0$

Example 12.32. $\sqrt[5]{\frac{3a^2}{4b^3}} = \sqrt[5]{\frac{3a^2 \cdot 2^3b^2}{2^2b^3 \cdot 2^3b^2}} = \sqrt[5]{\frac{24a^2b^2}{(2b)^5}} = \frac{1}{2b}\sqrt[5]{24a^2b^2}$

Example 12.33. $\sqrt[3]{\frac{M^{x+3}N^{x+1}}{N}} = \sqrt[3]{M^3 \cdot M^xN^{x+1-1}} = M\sqrt[3]{M^xN^x}$

Example 12.34. $\frac{1}{\sqrt{3}-1} = \frac{1(\sqrt{3}+1)}{(\sqrt{3}-1)(\sqrt{3}+1)} = \frac{\sqrt{3}+1}{(\sqrt{3})^2-(1)^2} = \frac{\sqrt{3}+1}{2}$

C. REDUCING THE INDEX

Sometimes the index of a radical can be changed to a smaller number. For example, we can simplify $\sqrt[4]{9x^2}$ by changing the index to a fractional exponent, as follows:

$$\sqrt[4]{9x^2} = (3^2x^2)^{1/4} = (3x)^{2/4} = (3x)^{1/2} = \sqrt{3x}$$

Writing the index as a fractional exponent is a helpful, though not necessary, step. As this example shows, *the basis for reducing the index is reducing the fractional exponent to simplest form.* It should be apparent, then, that we can reduce the index only if the index and the exponent of the radicand have some common factor. For example, the index of $\sqrt[4]{5^3x^2}$ cannot be reduced, because 4, 3, and 2 do not all have a common factor.

Example 12.35. $\sqrt[6]{x^3} = x^{3/6} = x^{1/2} = \sqrt{x}$

Example 12.36. $\sqrt[4]{64x^{10}} = \sqrt[4]{2^6x^8 \cdot x^2} = 2x^2\sqrt[4]{(2x)^2} = 2x^2\sqrt{2x}$

Example 12.37. $\sqrt[6]{25n^2 + 30mn + 9m^2} = (5n + 3m)^{2/6} = \sqrt[3]{5n + 3m}$

Example 12.38. $\sqrt[3]{\sqrt{64a^8}} = [(2^6a^8)^{1/2}]^{1/3} = [(2a)^6(a^2)]^{1/6}$
$= (2a)^{6/6}a^{2/6} = 2a\sqrt[3]{a}$

Example 12.39. $\sqrt[3n]{b^n x^{2n}} = [(bx^2)^n]^{1/(3n)} = (bx^2)^{1/3} = \sqrt[3]{bx^2}$

Example 12.40. $\sqrt[4]{x^2 + y^4}$ – cannot be simplified

EXERCISE 12.3

Simplify the radicals in Probs. 1 to 25. Remember to check each radical for three points, in order: (1) all perfect nth powers removed, (2) no radical in the denominator, and (3) no factors in common between the index and the exponent of the radicand.

1. $\sqrt[3]{54}$	2. $\sqrt{32}$	3. $\sqrt{2/5}$
4. $\sqrt[3]{256x^4}$	5. $\dfrac{1}{\sqrt{2}}$	6. $\dfrac{1}{\sqrt{3}}$
7. $\dfrac{3}{\sqrt{8}}$	8. $\sqrt[3]{-16}$	9. $\sqrt[4]{36x^2}$
10. $\sqrt[6]{9R^8S^6}$	11. $\sqrt[4]{64x^6/25}$	12. $\sqrt[3]{a^{3+n}b^{3n}/c^2}$
13. $\sqrt{1 \times 10^6}$	14. $\sqrt[3]{-8 \times 10^{-3}}$	15. $\sqrt[4]{4 \times 10^{-8}}$
16. $\sqrt{x^3 \times 10^3}$	17. $\sqrt[4]{16S^4 + S^8}$	18. $\sqrt{t^{-2} - 1}$
19. $\sqrt{\sqrt{16}}$	20. $\sqrt{25 - 10U + U^2}$	21. $\sqrt{(b-1)^3 \cdot 5^{-1}}$
22. $\sqrt[3]{\sqrt[4]{25x^2}}$	23. $\sqrt[6]{x^3 - x^2 + x - 1}$	24. $\dfrac{1}{\sqrt{2}+1}$
25. $\dfrac{1}{1-\sqrt{5}}$		

26. In Example 12.31, explain why it is not possible to have a real-number value for the expression $\sqrt{x^7/(-5)}$, $x > 0$.
27. In Example 12.30, explain why the restriction $b > 0$ is used. What would be the final expression if $b < 0$? Under what condition, for a, could b be negative and $\sqrt{10ab}/(2b)$ be a real number?
28. Does $\sqrt{x^2}$ always equal $(\sqrt{x})^2$? Explain.

12.5 ADDITION OF RADICALS

In Chap. 10 we discussed at length the fact that we can add only quantities of the same type. There are two characteristics that distinguish radicals as to type: the index and the radicand. Thus radicals with the *same index* and the *same radicand* are called *similar radicals.*

In other words, similar radicals can differ only in the coefficients outside the radical sign.

These examples illustrate similar and dissimilar radicals.

Example 12.41. $\sqrt[3]{x}$ and $\sqrt[3]{x^2}$ not similar—different radicands

Example 12.42. $\sqrt{2}$ and $\sqrt[3]{2}$ not similar—different indices

Example 12.43. $3\sqrt{2}$ and $5\sqrt{2}$ similar

Example 12.44. $a\sqrt[3]{x}$ and $b\sqrt[3]{x}$ similar

We can add only similar radicals, and we add them by factoring the similar radicals and adding the coefficients as in $3\sqrt{2}+5\sqrt{2}=(3+5)\sqrt{2}=8\sqrt{2}$.

If two radicals are not similar, we can sometimes change the form of one or both radicals to make them similar, as in Examples 12.46, 12.49, and 12.50.

The following examples illustrate how to add radicals.

Example 12.45. $a\sqrt[3]{x}+b\sqrt[3]{x}=(a+b)\sqrt[3]{x}$

Example 12.46. $2\sqrt{a}-3\sqrt{a^3}=2\sqrt{a}-3a\sqrt{a}=(2-3a)\sqrt{a}$

Example 12.47. $\sqrt{3}+\sqrt{2}$ cannot be added except as decimal approximations: $1.732+1.414=3.146$

Example 12.48. $(a+b)\sqrt{x}-(a-b)\sqrt{x}=(a+b-a+b)\sqrt{x}=2b\sqrt{x}$

Example 12.49.

$$\sqrt[3]{\frac{b^2}{a^4}}-\sqrt[3]{\frac{a^2}{b^4}}=\sqrt[3]{\frac{a^2b^2}{a^6}}-\sqrt[3]{\frac{a^2b^2}{b^6}}$$

$$=\left(\frac{1}{a^2}-\frac{1}{b^2}\right)\sqrt[3]{a^2b^2}=\frac{b^2-a^2}{a^2b^2}\sqrt[3]{a^2b^2}$$

Example 12.50.

$$\sqrt[5]{96a^4b^6}-\sqrt[10]{9a^{18}b^2}=2b\sqrt[5]{3a^4b}-a(3a^4b)^{2/10}$$

$$=2b\sqrt[5]{3a^4b}-a\sqrt[5]{3a^4b}$$

$$=(2b-a)\sqrt[5]{3a^4b}$$

EXERCISE 12.4

Add the radicals in Probs. 1 to 15.

1. $5\sqrt{6}-\sqrt{6}+3\sqrt{6}$
2. $3\sqrt{8}-\sqrt{2}+\sqrt{32}$
3. $a\sqrt[3]{2}-2b\sqrt[3]{2}$
4. $2\sqrt[3]{a^2}+3\sqrt[3]{a^5}$

5. $\dfrac{\sqrt{x^3}}{x+1} - \dfrac{2\sqrt{x^5}}{x+1} + \dfrac{3\sqrt{x}}{x+1}$
6. $\sqrt[5]{32a^4b^7} - \sqrt[5]{a^7b^4}$
7. $\sqrt{\dfrac{2a}{b}} - \sqrt{\dfrac{b}{2a}}$
8. $\sqrt[4]{\dfrac{3x^2}{y^3}} + \sqrt[4]{\dfrac{y}{27x^2}}$
9. $\dfrac{\sqrt{h(h+2)^2} - \sqrt{4h^3 + 16h^2 + 16h}}{h+2}$
10. $\sqrt{\dfrac{x+2}{x-2}} - \sqrt{\dfrac{x-2}{x+2}}$
11. $\sqrt[3]{(25 - 10x + x^2)^2} + (x-5)\sqrt[3]{5-x}$
12. $\sqrt{S - 6 + 9/S} + \sqrt{1/S}$
13. $\sqrt[4]{N^6} - \sqrt{N^3}$
14. $\sqrt[3]{5R^2} + \sqrt[6]{25R^4}$
15. $\sqrt{8x} - \sqrt[4]{x^2/4} + \sqrt[6]{8x^3}$

12.6 MULTIPLICATION OF RADICALS

If we use fractional exponents in Eq. (12.4), we have the basis for multiplying radicals.

$$a^{1/n}b^{1/n} = (ab)^{1/n}$$

which is equivalent to

$$\sqrt[n]{a}\sqrt[n]{b} = \sqrt[n]{ab}$$

For example,

$$3^{1/2} \cdot 5^{1/2} = (3 \cdot 5)^{1/2} = 15^{1/2}$$

which is equivalent to

$$\sqrt{3} \cdot \sqrt{5} = \sqrt{3 \cdot 5} = \sqrt{15}$$

The *key point* in multiplying radicals is that *both radicals must have the same index.* Thus we could not directly combine $\sqrt{3} \cdot \sqrt[3]{5}$ under the same radical sign.

To multiply two radicals with different indices, we must make the indices alike by changing the form of one or both. Such a change is usually easier to visualize if the radicals are written in exponential form. We could multiply $\sqrt{3} \cdot \sqrt[3]{5}$ as follows:

$$\sqrt{3} \cdot \sqrt[3]{5} = 3^{1/2} \cdot 5^{1/3} = 3^{3/6} \cdot 5^{2/6} = (3^3 \cdot 5^2)^{1/6}$$

$$\sqrt{3} \cdot \sqrt[3]{5} = \sqrt[6]{675}$$

With the possible exception of reducing the indices, we should simplify all radicals before attempting to multiply, and sometimes the resulting product can be simplified even though the factors could not.

The following examples illustrate how to multiply radicals.

Example 12.51. $\sqrt{2}\sqrt{6} = \sqrt{12} = 2\sqrt{3}$

Example 12.52. $\sqrt[3]{54}\sqrt[3]{4} = 3\sqrt[3]{2}\sqrt[3]{4} = 3\sqrt[3]{8} = 3(2) = 6$

Example 12.53. $\sqrt[4]{9}\sqrt[4]{8} = \sqrt[4]{72}$

Example 12.54.
$$(\sqrt{a} + \sqrt{b})(\sqrt{a} - \sqrt{b}) = (\sqrt{a})^2 + \sqrt{a}\sqrt{b} - \sqrt{a}\sqrt{b} - (\sqrt{b})^2$$
$$= a - b$$

Example 12.55.
$$(\sqrt{2} - 5\sqrt{3})(2\sqrt{2} + 3\sqrt{3}) = 2(\sqrt{2})^2 - 10\sqrt{2}\sqrt{3} + 3\sqrt{2}\sqrt{3} - 15(\sqrt{3})^2$$
$$= 2(2) - 10\sqrt{6} + 3\sqrt{6} - 15(3)$$
$$= -41 - 7\sqrt{6}$$

Example 12.56.
$$\sqrt[5]{x^2}\sqrt[3]{xy^5} = x^{2/5} \cdot x^{1/3}y^{5/3} = x^{6/15} \cdot x^{5/15} \cdot y^{10/15} \cdot y$$
$$= x^{11/15} \cdot y^{10/15} \cdot y = y\sqrt[15]{x^{11}y^{10}}$$

EXERCISE 12.5

In each problem, find the indicated product of radicals.

1. $\sqrt{2}\sqrt{5}\sqrt{10}$
2. $\sqrt{3}(-\sqrt{8})$
3. $\sqrt{x^3}\sqrt{x^5}$
4. $\sqrt[3]{2}\sqrt[3]{6}\sqrt[3]{18}$
5. $\sqrt{2/3}\sqrt{6/5}$
6. $\sqrt[3]{2a/b^2}\sqrt[3]{12b^2/a}$
7. $\sqrt[4]{4}\sqrt{8}$
8. $\sqrt[3]{2}\sqrt{2}$
9. $\sqrt[4]{16}\sqrt{3}$
10. $\sqrt{2}\sqrt[4]{20}$
11. $\sqrt[3]{x}\sqrt[6]{x^5}$
12. $\sqrt{3}\sqrt[5]{18}\sqrt[10]{24}$
13. $\sqrt{\sqrt{8x^3}\sqrt[3]{2x}}$
14. $(3\sqrt{2} + 2\sqrt{3})(3\sqrt{2} - 2\sqrt{3})$
15. $(\sqrt{2s} + \sqrt{3r})(\sqrt{2s} - \sqrt{3r})$
16. $(3 - \sqrt{5})(7 + \sqrt{5})$
17. $(\sqrt[3]{x} - \sqrt[3]{2})(\sqrt[3]{x^2} + \sqrt[3]{2}\sqrt[3]{x} + \sqrt[3]{4})$

12.7 DIVISION OF RADICALS

Equation (12.5) is the basis for dividing radicals. If the exponents in this equation are put into fractional form, we have

$$\frac{a^{1/n}}{b^{1/n}} = \left(\frac{a}{b}\right)^{1/n} \qquad \text{or} \qquad \frac{\sqrt[n]{a}}{\sqrt[n]{b}} = \sqrt[n]{\frac{a}{b}}$$

For example,

$$\frac{6^{1/3}}{10^{1/3}} = \left(\frac{6}{10}\right)^{1/3} = \left(\frac{3}{5}\right)^{1/3} = \left(\frac{3 \times 5^2}{5^3}\right)^{1/3} = \frac{75^{1/3}}{5}$$

or

$$\frac{\sqrt[3]{6}}{\sqrt[3]{10}} = \sqrt[3]{\frac{6}{10}} = \sqrt[3]{\frac{3}{5}} = \sqrt[3]{\frac{3 \times 5^2}{5^3}} = \frac{\sqrt[3]{75}}{5}$$

As this example shows, division by a radical amounts to rationalizing the denominator.

Here again, as in multiplying radicals, the only restriction is that *the radicals must be of the same order.* To divide two radicals with different indices, we must make the indices alike by changing the form of one or both. For example,

$$\frac{\sqrt{5}}{\sqrt[3]{10}} = \frac{5^{3/6}}{10^{2/6}} = \left(\frac{5^3}{10^2}\right)^{1/6} = \left(\frac{5^3}{2^2 \times 5^2}\right)^{1/6} = \left(\frac{5}{2^2}\right)^{1/6}$$
$$= \sqrt[6]{\frac{5 \times 2^4}{2^2 \times 2^4}} = \frac{1}{2}\sqrt[6]{80}$$

Of particular interest in electronics is the problem of dividing by a binomial with square roots. The expression $2/(\sqrt{2} + \sqrt{3})$ is an example. In electronics these problems involve imaginary numbers, but the technique of division is the same one that we shall show here for real numbers.

The basis for rationalizing the denominator of such an expression is to form the difference between two squares in the denominator. We know that a difference between two squares always results from the expression $(a + b)(a - b)$, the product of the sum of and difference between the same two numbers.

$$(a + b)(a - b) = a^2 - b^2$$

Likewise,

$$(\sqrt{2} + \sqrt{3})(\sqrt{2} - \sqrt{3}) = (\sqrt{2})^2 - (\sqrt{3})^2$$
$$= 2 - 3 = -1$$

In such expressions, each factor is called the *conjugate* of the other. Thus $\sqrt{2} - \sqrt{3}$ is the conjugate of $\sqrt{2} + \sqrt{3}$, and vice versa.

It should be apparent that *to rationalize a binomial denominator, we must multiply the numerator and denominator by the conjugate of the denominator.* For example,

$$\frac{\sqrt{2}}{\sqrt{2} + \sqrt{3}} = \frac{\sqrt{2}(\sqrt{2} - \sqrt{3})}{(\sqrt{2} + \sqrt{3})(\sqrt{2} - \sqrt{3})} = \frac{2 - \sqrt{6}}{-1} = \sqrt{6} - 2$$

Note that we *cannot* rationalize such a denominator by squaring it. Try it and see!

Example 12.57. $\dfrac{\sqrt{12}}{\sqrt{3}} = \sqrt{\dfrac{12}{3}} = \sqrt{4} = 2$

Example 12.58. $\dfrac{\sqrt[4]{27}}{\sqrt[4]{2/3}} = \sqrt[4]{\dfrac{27(3)}{2}} = \sqrt[4]{\dfrac{3^4}{2}\left(\dfrac{2^3}{2^3}\right)} = \dfrac{3}{2}\sqrt[4]{8}$

Example 12.59. $\dfrac{\sqrt[5]{16}}{\sqrt[3]{4}} = \dfrac{(2^4)^{1/5}}{(2^2)^{1/3}} = \dfrac{(2^4)^{3/15}}{(2^2)^{5/15}} = \left(\dfrac{2^{12}}{2^{10}}\right)^{1/15} = \sqrt[15]{4}$

Example 12.60. $\dfrac{\sqrt[3]{x^2/a}}{\sqrt[6]{x^3/a^2}} = \left(\dfrac{x^2}{a}\right)^{2/6}\left(\dfrac{a^2}{x^3}\right)^{1/6} = \sqrt[6]{\dfrac{x^4}{\not{a}^2}\dfrac{\not{a}^2}{x^3}} = \sqrt[6]{x}$

Example 12.61. $\dfrac{2\sqrt{x}+\sqrt{5}}{2\sqrt{x}-\sqrt{5}} = \dfrac{(2\sqrt{x}+\sqrt{5})(2\sqrt{x}+\sqrt{5})}{(2\sqrt{x}-\sqrt{5})(2\sqrt{x}+\sqrt{5})} = \dfrac{4x+4\sqrt{5x}+5}{4x-5}$

EXERCISE 12.6

In each problem, perform the indicated division by any of the methods shown in Sec. 12.7.

1. $\dfrac{\sqrt{15}}{\sqrt{3}}$
2. $\dfrac{\sqrt{3}}{\sqrt{15}}$
3. $\dfrac{\sqrt{8}}{\sqrt{5}}$
4. $\dfrac{\sqrt[3]{16}}{\sqrt[3]{2}}$
5. $\dfrac{\sqrt[3]{25}}{\sqrt[3]{9}}$
6. $\dfrac{\sqrt[3]{9}}{\sqrt{3}}$
7. $\dfrac{\sqrt[4]{36}}{\sqrt[3]{12}}$
8. $\dfrac{\sqrt{6}}{\sqrt{1/2}}$
9. $\dfrac{\sqrt[3]{25/9}}{\sqrt[3]{3/5}}$
10. $\dfrac{\sqrt[6]{8x}}{\sqrt[3]{4x^2}}$
11. $\dfrac{\sqrt[6]{3a}}{\sqrt[3]{a^2/3}}$
12. $\dfrac{\sqrt{2h}}{\sqrt[5]{162h^4}}$
13. $\dfrac{4-\sqrt{2}}{\sqrt{2}}$
14. $\dfrac{\sqrt{6}+\sqrt{3}}{\sqrt{3}}$
15. $\dfrac{\sqrt{6+3}}{\sqrt{3}}$
16. $\dfrac{\sqrt{20}-\sqrt{3}}{\sqrt{5}}$
17. $\dfrac{\sqrt{5}}{\sqrt{15}-\sqrt{5}}$
18. $\dfrac{1}{1+\sqrt{2}}$
19. $\dfrac{\sqrt{2}+3}{\sqrt{2}-3}$
20. $\dfrac{\sqrt{a}-\sqrt{b}}{\sqrt{a}+\sqrt{b}}$
21. $\dfrac{\sqrt{a-b}}{\sqrt{a+b}}$, $a > b > 0$
22. $\dfrac{\sqrt{3x}-\sqrt{3}}{2\sqrt{x}-\sqrt{3}}$
23. $\dfrac{\sqrt{a-1}}{\sqrt{a}-\sqrt{a-1}}$
24. $\dfrac{\sqrt[3]{125x^6}}{\sqrt[4]{64x^2}+\sqrt{3x}}$
25. $\dfrac{1}{\sqrt[3]{x}-1}$ [*Hint:* Find the "conjugate" of $\sqrt[3]{x}-1$ by using the factors of the difference between two cubes: $a^3-1=(a-1)(a^2+a+1)$.]
26. $\dfrac{1}{\sqrt[3]{a^2}-\sqrt[3]{a}+1}$ See the hint for Prob. 25.

TRIGONOMETRY

UNIT 3

CHAPTER 13

TRIGONOMETRIC FUNCTIONS OF ANY ANGLE; SOLUTIONS OF TRIANGLES

In Chap. 5 we used the trigonometric functions to solve right triangles. In Unit 3 (Chaps. 13 through 16) we shall examine these functions in more detail and see their more general applications.

13.1 TRIGONOMETRIC FUNCTIONS OF A GENERAL ANGLE

For reference, we repeat here the general definitions of the six trigonometric functions. Recall that these definitions, as illustrated in Fig. 13.1, apply to the coordinates of any point on the terminal side of an angle in standard position. For a more detailed review of the trigonometric functions, see Sec. 5.1.

$$\sin\theta = \frac{\text{ordinate}}{\text{radius}} = \frac{y}{r} \qquad (13.1)$$

$$\cos\theta = \frac{\text{abscissa}}{\text{radius}} = \frac{x}{r} \qquad (13.2)$$

$$\tan\theta = \frac{\text{ordinate}}{\text{abscissa}} = \frac{y}{x} \qquad (13.3)$$

$$\sec\theta = \frac{\text{radius}}{\text{abscissa}} = \frac{r}{x} \qquad (13.4)$$

$$\csc\theta = \frac{\text{radius}}{\text{ordinate}} = \frac{r}{y} \qquad (13.5)$$

$$\cot\theta = \frac{\text{abscissa}}{\text{ordinate}} = \frac{x}{y} \qquad (13.6)$$

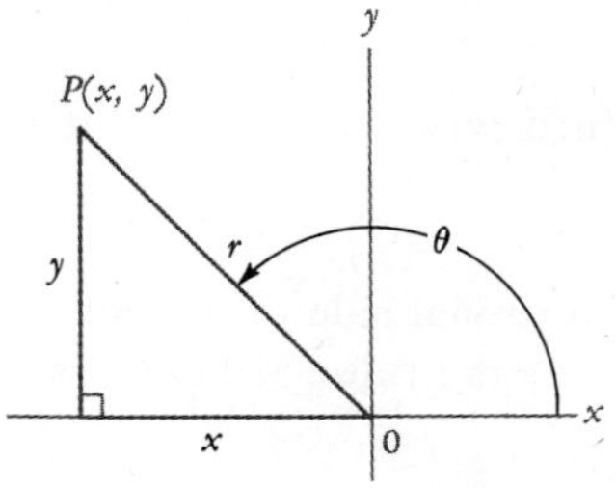

Fig. 13.1 Coordinates of a point on an angle in standard position.

Although Fig. 13.1 shows it in the second quadrant, angle θ could terminate in any quadrant; the definitions would still apply.

A. THE SIGNS OF THE FUNCTIONS

The algebraic sign of any trigonometric function is governed by the signs of the defining coordinates. For example, since $\sin\theta = y/r$ and since r is always positive, $\sin\theta$ takes the sign of y: *If y is positive, then* $\sin\theta$ *is positive; if y is negative, then* $\sin\theta$ *is negative; if y is zero, then* $\sin\theta$ *is zero.*

We could develop some general rules concerning the signs of the various functions in the various quadrants. But instead, we shall rely upon the definitions as expressed in Eqs. (13.1) to (13.6) and the common-sense approach outlined above.

The following examples illustrate the values of the trigonometric functions.

Example 13.1. **Determine the values of $\sin\theta$, $\cos\theta$, and $\tan\theta$, if the terminal side of θ passes through the point $(x,y) = (-8,6)$.**

First, we use the Pythagorean formula to evaluate r. A sketch such as Fig. 13.1 helps us to visualize what is given and what we need to do.

$$r = \sqrt{x^2 + y^2} = \sqrt{100} = 10$$

Then we employ Eqs. (13.1) to (13.3) directly, to evaluate each function, as follows:

$$\sin\theta = \frac{y}{r} = \frac{6}{10} = 0.60$$

$$\cos\theta = \frac{x}{r} = \frac{-8}{10} = -0.80$$

$$\tan\theta = \frac{y}{x} = \frac{6}{-8} = -0.75$$

Example 13.2. **If $\cos\theta = -2/5$ and θ terminates in the third quadrant, determine the values of the other five functions of θ.**

As shown in Fig. 13.2, one point on the terminal side of θ can be located by assigning to x the value -2 and to r the value 5. Then $\cos\theta = -2/5$, as specified. Then

$$y = -\sqrt{r^2 - x^2} = -\sqrt{21}$$

and the other five functions are as follows:

$$\sin\theta = \frac{y}{r} = -\frac{\sqrt{21}}{5} \qquad \csc\theta = \frac{r}{y} = -\frac{5}{\sqrt{21}} = -\frac{5\sqrt{21}}{21}$$

$$\tan\theta = \frac{y}{x} = \frac{-\sqrt{21}}{-2} = \frac{\sqrt{21}}{2} \qquad \cot\theta = \frac{x}{y} = \frac{-2}{-\sqrt{21}} = \frac{2\sqrt{21}}{21}$$

$$\sec\theta = \frac{r}{x} = -\frac{5}{2}$$

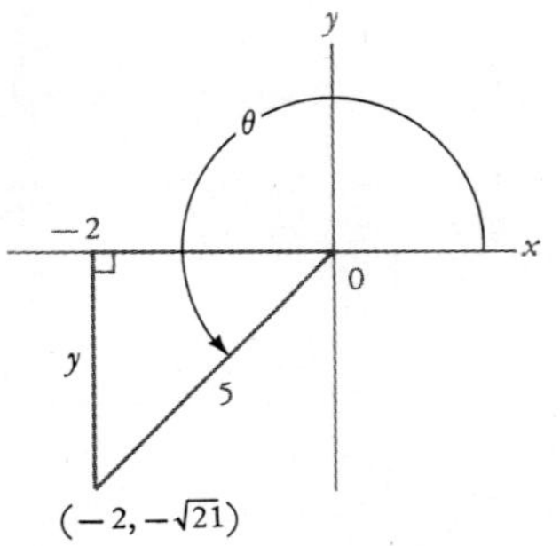

Fig. 13.2 Example 13.2.

EXERCISE 13.1

In Probs. 1 to 10, the given point is on the terminal side of an angle in standard position. Determine the values of the six trigonometric functions of each angle.

1. (3,4)
2. (−3,4)
3. (−3,−4)
4. (3,−4)
5. (−2,−2)
6. (−10,−10)
7. (−25,−25)
8. (−5,5)
9. $(-1, \sqrt{3}/3)$
10. $(5, -5\sqrt{3})$

11. What is the value of each angle in Probs. 5 to 10?

In Probs. 12 to 19, determine the values of the remaining five functions of the angle θ. Use the method shown in Examples 13.1 and 13.2.

12. $\sin\theta = 2/3$
    $90° < \theta < 180°$
13. $\tan\theta = 1$
    $180° < \theta < 270°$
14. $\tan\theta = -\sqrt{5}/4$
    $90° < \theta < 180°$
15. $\cot\theta = -5\sqrt{3}/3$
    $270° < \theta < 360°$
16. $\cos\theta = 0.8$
    $270° < \theta < 360°$
17. $\csc\theta = 1.25$
    $90° < \theta < 180°$
18. $\sec\theta = -2.0$
    $180° < \theta < 270°$
19. $\sin\theta = -0.5$
    $270° < \theta < 360°$

B. TABLES OF TRIGONOMETRIC FUNCTIONS

In Chap. 5 we learned how to use the tables of trigonometric functions of angles between 0 and 90°. Now we shall use those tables to find values of the functions of any general angle.

This system of using the tables is based on a simple fact: the functions of any general angle can always be related to the corresponding functions of some first-quadrant (0 to 90°) angle. The preceding problems 1 to 4 illustrate this fact. For example, let's compare functions of the point $P_1(3,4)$ on a first-quadrant angle θ_1 to those of the point $P_2(-3,4)$ on a second-quadrant angle θ_2 (see Fig. 13.3). We see that

$$\sin\theta_1 = |\sin\theta_2| = 4/5$$

$$\cos\theta_1 = |\cos\theta_2| = 3/5$$

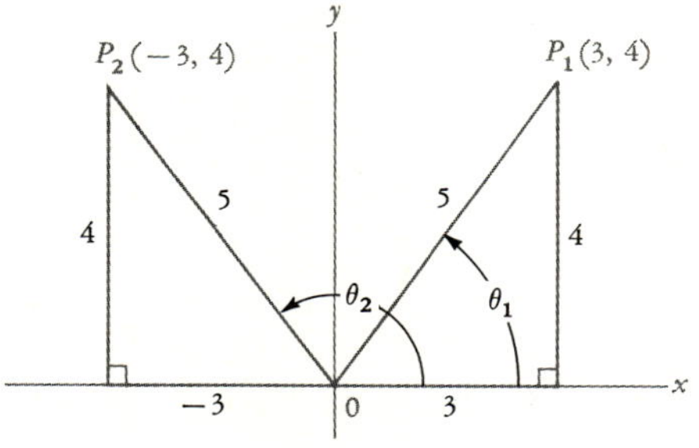

Fig. 13.3 The trigonometric functions of θ_2 are related to those of θ_1.

We use absolute values because some functions of second-quadrant angles are negative; otherwise, the values are identical. The other four functions could be similarly related. And in general, *for any angle, there exists a first-quadrant angle such that the absolute values of their corresponding trigonometric functions are equal.*

We can express this relationship with the following formula, in which the symbol TF represents any one of the six trigonometric functions, θ any general angle, and θ_1 the corresponding first-quadrant angle. We shall refer to θ_1 as the *reference angle.*

$$|\text{TF } \theta| = \text{TF } \theta_1 \qquad (13.7)$$

Relationship between any trigonometric function of any general angle θ and the corresponding function of the first-quadrant reference angle θ_1

Our only problem now is to determine the reference angle θ_1 for any given angle θ, and vice versa. If we can do this, then we can find the TF of any angle by using the table of functions for angles between 0 and 90°. The next three paragraphs develop three very simple conversion formulas for θ and θ_1, depending on whether θ is a second-, third-, or fourth-quadrant angle.

Second-quadrant Angles: $90° < \theta < 180°$

In Fig. 13.4, we define θ_1 and θ_2 to be first- and second-quadrant angles, respectively, so that their corresponding functions are related according to Eq. (13.7). Then we can always find two points P_1 and P_2 such that $|y_2| = y_1$ and $|x_2| = x_1$, making $\triangle OP_1Q_1 \cong \triangle OP_2Q_2$. From the figure, we see that θ_2' must equal θ_1 (because they are corresponding angles of congruent triangles) and also that $\theta_2' + \theta_2 = 180°$. Therefore the following formulas relate θ_1 and θ_2:

$$\theta_1 = 180° - \theta_2 \qquad \theta_2 = 180° - \theta_1 \tag{13.8}$$

Conversion formulas for second-quadrant angles and the first-quadrant reference angles

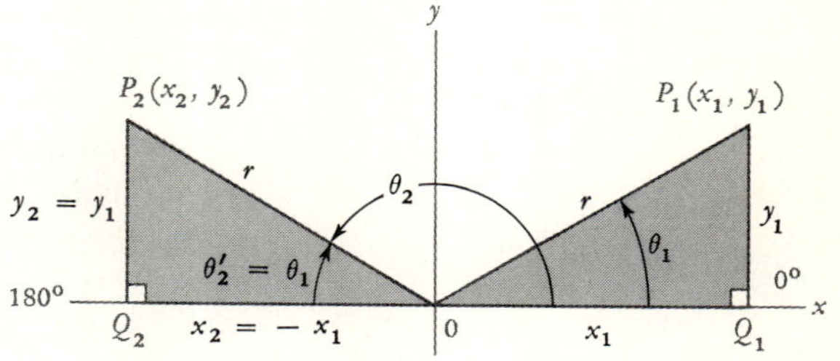

Fig. 13.4 The relationship between a second-quadrant angle and its first-quadrant reference angle.

Example 13.3. **Find the sine, cosine, and tangent of 123°.**

First we use Eq. (13.8) to find the reference angle:

$$\theta_1 = 180° - 123° = 57°$$

Then we use Table A5.4 in the Appendix to evaluate the functions as follows:

$$\sin 123° = \sin 57° = 0.8387$$

$$\cos 123° = -\cos 57° = -0.5446$$

$$\tan 123° = -\tan 57° = -1.540$$

Note that *we* must supply the appropriate sign for each function. Neither the formula nor the table will supply the sign for us.

Example 13.4. Determine the second-quadrant angle whose cotangent is -2.081.

First, we find the reference angle whose cotangent is 2.081. From Table A5.4 (in the Appendix) we see that

$$\cot 25°40' = 2.081$$

Therefore $\theta_1 = 25°40'$, and by Eq. (13.8)

$$\theta = 180° - 25°40' = 154°20'$$

Third-quadrant Angles: $180° < \theta < 270°$

In Fig. 13.5, we use a construction similar to that for the second-quadrant angles. Now we have $\triangle OP_3Q_3 \cong \triangle OP_1Q_1$, $\theta_3' = \theta_1$, and $\theta_3 - 180° = \theta_3'$. Therefore the following formulas relate θ_1 and θ_3:

$$\theta_1 = \theta_3 - 180° \qquad \theta_3 = 180° + \theta_1 \tag{13.9}$$

Conversion formulas for third-quadrant angles and the first-quadrant reference angles

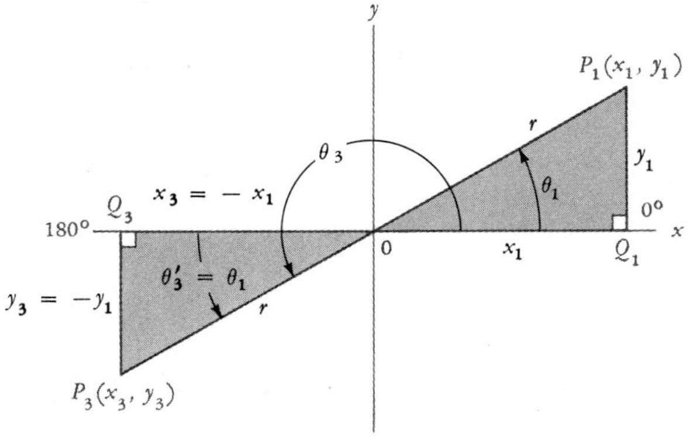

Fig. 13.5 The relationship between a third-quadrant angle and its first-quadrant reference angle.

Example 13.5. Find the sine, cotangent, and secant of 233°15′.

First, we find the reference angle:

$$\theta_1 = 233°15' - 180° = 53°15'$$

Then the values of the desired functions are as follows:

$$\sin 233°15' = -\sin 53°15' = -0.8012$$

$$\cot 233°15' = \cot 53°15' = 0.7468$$

$$\sec 233°15' = -\sec 53°15' = -1.672$$

Example 13.6. Determine the third-quadrant angle whose tangent is 0.2156.

First we find the reference angle:

$$\tan 12°10' = 0.2156$$

Therefore $\theta_1 = 12°10'$, and the required angle is

$$\theta = 180° + 12°10' = 192°10'$$

Fourth-quadrant Angles: $270° < \theta < 360°$

In Fig. 13.6, $\triangle OP_4Q \cong \triangle OP_1Q$, $\theta_4' = \theta_1$, and $\theta_4 = 360° - \theta_4'$. Therefore, the following formulas relate θ_4 and θ_1:

$$\theta_1 = 360° - \theta_4 \qquad \theta_4 = 360° - \theta_1 \tag{13.10}$$

Conversion formulas for fourth-quadrant angles and the first-quadrant reference angles

Example 13.7. Find the cosine, tangent, and cosecant of 351°.

The reference angle is

$$\theta_1 = 360° - 351° = 9°$$

and the required functions are

$$\cos 351° = \cos 9° = 0.9877$$

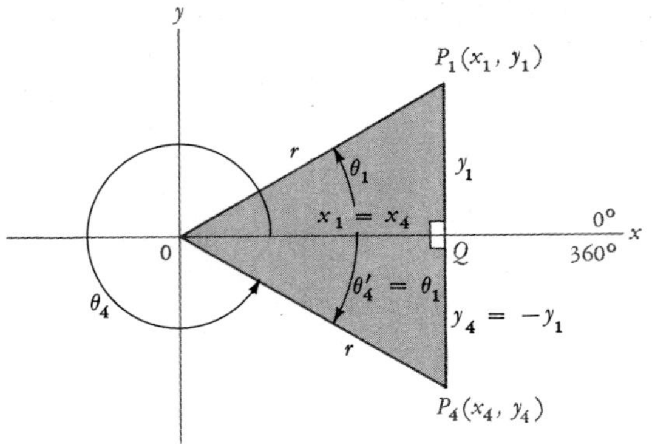

Fig. 13.6 The relationship between a fourth-quadrant angle and its first-quadrant reference angle.

$$\tan 351° = -\tan 9° = -0.1584$$

$$\csc 351° = -\csc 9° = -6.392$$

Example 13.8. Determine the fourth-quadrant angle whose sine is −0.9528.

$$\sin 72°20' = 0.9528$$

Therefore

$$\theta_1 = 72°20'$$

and

$$\theta = 360° - 72°20' = 287°40'$$

Example 13.9. Determine all angles between 0° and 360° whose cosines are −0.8030.

$$\cos 36°35' = 0.8030$$

Thus the reference angle is 36°35′. Since the cosine is negative, the angles are in the second and third quadrants:

$$\theta_2 = 180° - 36°35' = 143°25'$$

$$\theta_3 = 180° + 36°35' = 216°35'$$

C. NEGATIVE ANGLES AND ANGLES GREATER THAN 360°

The TF of either a negative angle or an angle greater than 360° is equal to the corresponding TF of the coterminal angle between 0° and 360°. For example, since 390° = 30° + 360°, we say that 390° is coterminal with 30°, and each TF of 390° equals the corresponding TF of 30°, as follows:

$$\sin 390° = \sin (390° - 360°) = \sin 30° = 0.500$$

$$\cos 390° = \cos (390° - 360°) = \cos 30° = 0.866$$

etc.

In general, we can express this relationship between corresponding functions of coterminal angles as follows:

$$\boxed{\begin{array}{l}\text{TF } \theta = \text{TF } (\theta \pm k \cdot 360°) \\ \text{where } k \text{ is any integer}\end{array}} \tag{13.11}$$

To find a TF of any angle, if $\theta > 360°$ or if $\theta < 0°$: (1) use Eq. (13.11) to determine the coterminal angle between 0° and 360°; (2) use the appropriate conversion formula, Eqs. (13.8) to (13.10), to write the function in terms of the corresponding function of the reference angle.

Example 13.10. Determine tan 868°.

$$\tan 868° = \tan (868° - 2 \cdot 360°) = \tan 148°$$

$$\tan 868° = -\tan (180° - 148°) = -\tan 32° = -0.6249$$

Example 13.11. Determine sin (−1,203°).

$$\begin{aligned}\sin (-1{,}203°) &= \sin (-1{,}203° + 4 \cdot 360°) = \sin (-1{,}203° + 1{,}440°) \\ &= \sin 237° = -\sin (237° - 180°) = -\sin 57° \\ \sin (-1{,}203°) &= -0.8387\end{aligned}$$

EXERCISE 13.2

Determine the value of each of the six trigonometric functions for the angles in Probs. 1 to 15.

1. 150°	2. 210°	3. 330°	4. 300°
5. 225°	6. 252°10′	7. 96°40′	8. 359°50′
9. 213°12′	10. 139°25′	11. 1,380°	12. 1,500°
13. −552°	14. −495°	15. −312°	

Determine all possible values of θ, $0° \leq \theta < 360°$, for each function of θ in Probs. 16 to 24.

16. $\cot \theta = 2.066$
17. $\sin \theta = 0.9250$
18. $\sec \theta = 1.181$
19. $\tan \theta = -0.1292$
20. $\csc \theta = -4.363$
21. $\cos \theta = -0.6456$
22. $\sin \theta = -0.3136$
23. $\cot \theta = -4.113$
24. $\cos \theta = 0.6456$

13.2 THE LAW OF SINES

In this section we shall develop a formula for solving certain oblique triangles (non-right triangles). But before we develop the general algebraic solutions, we can gain some helpful insights by examining graphic solutions.

A. BASIC IDEAS ABOUT SOLVING TRIANGLES

In solving triangles, we are primarily interested in six parts: the three sides and the three angles. To draw a unique triangle, we must know at least one side and any other two parts. There are just four different ways to specify these three necessary parts. The four ways are traditionally named Cases I, II, III, and IV, as shown in Fig. 13.7.

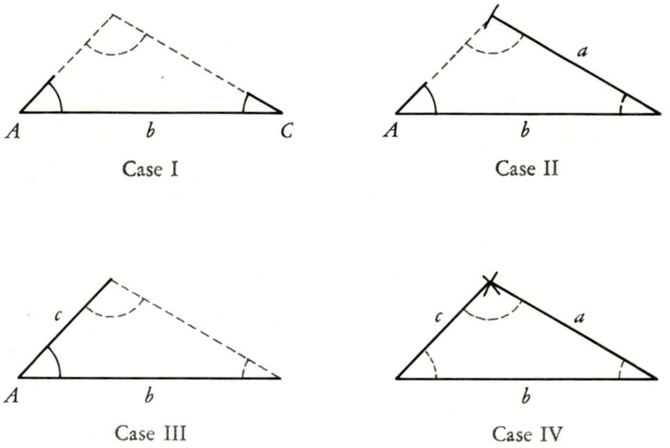

Fig. 13.7 The four unique ways to specify and draw a triangle. Case I: two angles and one side are known; Case II: two sides and the angle opposite one of them are known; Case III: two sides and the included angle are known; Case IV: all three sides are known.

In each part of Fig. 13.7, the dashed lines indicate how we can complete the triangle graphically or measure the unknown parts. For example, in Case I, Fig. 13.7, we draw the known side b, construct the known angles A and C at each end, and extend them until they intersect, as shown by the dashed lines. Then we measure the unknown sides and angle. See if you can describe how to draw the triangles and measure the unknown parts for Cases II, III, and IV.

The information provided in Cases I, III, and IV always leads to a unique solution. That is, with the given parts we can draw one and only one triangle. We assume, of course, that the sum of any two given angles is less than 180°, that the sum of any two given sides always exceeds the third given side, and that the longest side is opposite the largest angle and the shortest side is opposite the smallest angle. (Why?)

Case II has some special problems. There can be: (1) no way to draw a triangle, (2) one way to draw a triangle, or (3) two ways to draw a triangle. Figure 13.8 outlines these possibilities and the reasons why they occur.

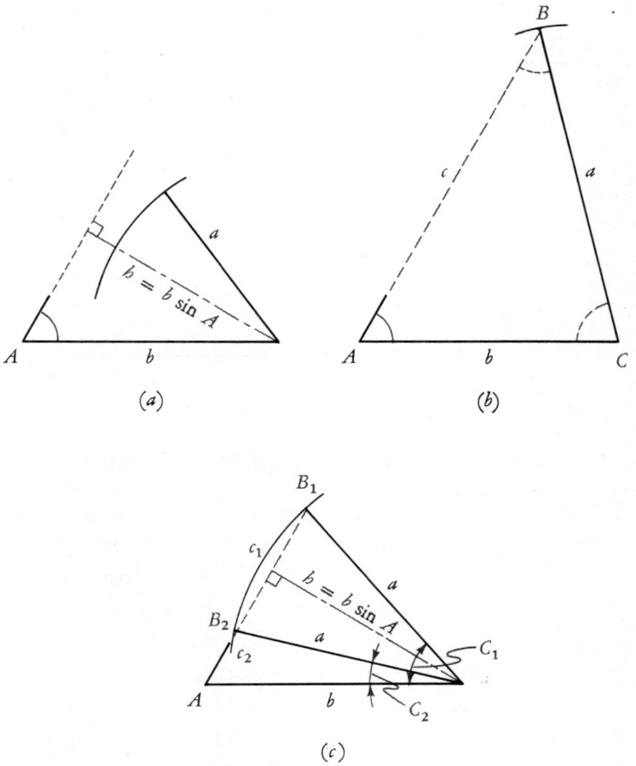

Fig. 13.8 Case II, the ambiguous case. There are three possible modes of solution for a Case II triangle: (a) If $a < h$, then a cannot reach to the third side, c. No triangle is possible. (b) If $a > b$, then a will reach to exactly one point on the third side. One triangle is possible. (c) If $h < a < b$, then a will reach to two points on the third side. Two triangles are possible.

EXERCISE 13.3

In each problem, the letters a, b, c, A, B, and C correspond to those of Fig. 13.9. Use a compass, scale, and protractor to construct the triangles with the given parts and to measure the missing parts. If no unique triangle can be drawn, explain why.

1. $A = 30°$, $C = 70°$, $b = 2$ in.
2. $A = 45°$, $B = 100°$, $b = 2$ in.
3. $A = 60°$, $a = 3$ in., $b = 1$ in.
4. $a = 1.5$ in., $b = 2.0$ in., $c = 2.5$ in.
5. $B = 50°$, $C = 70°$, $c = 4$ in.
6. $A = 60°$, $b = 1$ in., $c = 1$ in.
7. $C = 53°$, $a = 4$ in., $c = 2$ in.
8. $B = 37°$, $b = 2\frac{1}{2}$ in., $c = 3$ in.
9. $a = 5$ in., $b = 1\frac{1}{2}$ in., $c = 3$ in.
10. $A = 110°$, $B = 35°$, $C = 35°$
11. $C = 80°$, $a = 1\frac{1}{4}$ in., $b = 2\frac{1}{2}$ in.
12. $B = 120°$, $a = 1$ in., $b = \frac{3}{4}$ in.
13. $A = 150°$, $a = 10$ in., $c = 15$ in.

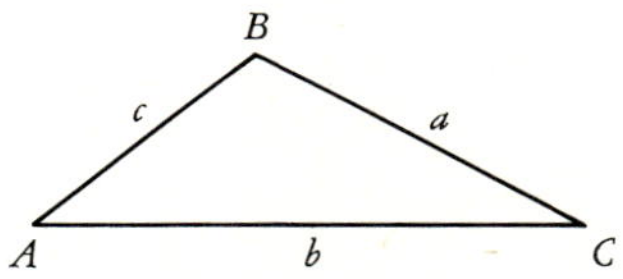

Fig. 13.9 Probs. 1 to 13.

B. A GENERAL SOLUTION: THE LAW OF SINES

General algebraic solutions for the triangles of Fig. 13.7 can be derived. As shown in Fig. 13.10, we can always form two right triangles (ACD and BCD) from any oblique triangle (ABC) by drawing an altitude (h). By solving both right triangles, we can find general formulas that will solve the oblique triangle directly. We shall develop here the formula for Cases I and II and in the next section the formula for Cases III and IV.

At this point we should recall the special definition of the sine function for an angle in a right triangle:

$$\sin A = \frac{\text{side opposite } A}{\text{hypotenuse}}$$

In $\triangle ACD$ of Fig. 13.10, h is opposite A, and b is the hypotenuse.

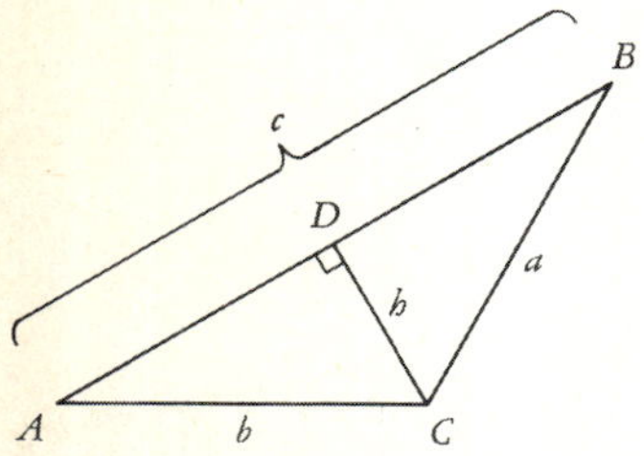

Fig. 13.10 Two right triangles, ACD and BCD, can be formed by drawing an altitude (such as h) of the oblique triangle ABC.

Thus we have $\sin A = h/b$, or

$$h = b \sin A$$

But from $\triangle BCD$ we have

$$h = a \sin B$$

Thus

$$a \sin B = b \sin A$$

or

$$\frac{a}{\sin A} = \frac{b}{\sin B}$$

By using a different altitude we could also show that

$$\frac{b}{\sin B} = \frac{c}{\sin C}$$

By combining these last two equations we have the following general formula, called the law of sines. It will solve Case I and Case II triangles.

LAW OF SINES $$\frac{a}{\sin A} = \frac{b}{\sin B} = \frac{c}{\sin C} \qquad (13.12)$$

The message of this formula is as follows: *The ratio of any side to the sine of the opposite angle is the same for all three pairs of sides and angles.* Notice that the ratio is $a/(\sin A)$, not a/A.

Example 13.12. **Solve triangle RST of Fig. 13.11, if $x = 12$ in., $R = 75°$, and $S = 40°$.**

Notice that to use the law of sines, we must know one of the sides and the opposite angle. In this example no such combination is given, but in Case I, we can always find the third angle. Thus our first step in Case I is always as follows:

$$T = 180° - (R + S) = 180° - 115° = 65°$$

Then because T is the angle opposite x, S is opposite y, and R is opposite z, we can apply Eq. (13.12) as follows:

$$\frac{y}{\sin S} = \frac{x}{\sin T} \longrightarrow y = \frac{x \sin S}{\sin T} = \frac{12 \sin 40°}{\sin 65°} = 8.52 \text{ in.}$$

$$\frac{z}{\sin R} = \frac{x}{\sin T} \longrightarrow z = \frac{x \sin R}{\sin T} = \frac{12 \sin 75°}{\sin 65°} = 12.8 \text{ in.}$$

Answers:

$T = 65°$
$y = 8.52$ in.
$z = 12.8$ in.

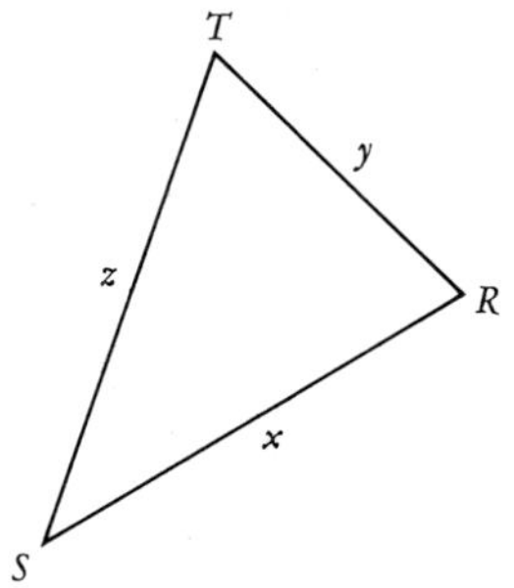

Fig. 13.11 Examples 13.12 to 13.14.

Analysis of Case II Triangles

Because of the ambiguous nature of Case II triangles (see Fig. 13.8, repeated on p. 295), we should analyze each one carefully before attempting to solve it. We would look very foolish trying to solve a triangle where no triangle exists, or claiming to have *the* solution where another solution exists.

Let's review the possible solutions as outlined by Fig. 13.8. First, the critical factor is the length of the side *opposite the known angle.* To determine the number of solutions, we compare this side a (in the figure) to the other known side b and to the altitude h, drawn perpendicular to the *unknown side.*

Since the two sides a and b are given, it is expedient to compare them first. If (as in Fig. 13.8b) the side opposite the known angle is greater than or equal to the side adjacent (if $a \geq b$), then one solution exists, and we proceed to use the law of sines.

$$\text{If} \quad a \geq b \ . \ . \ . \text{ then one solution exists} \qquad (13.13)$$

But if, as in Fig. 13.8a and c, the side opposite the known angle is less than the other side (if $a < b$), then either no triangle exists or two triangles exist.

$$\text{If} \quad a < b \ . \ . \ . \text{ then no solution or two solutions exist}$$

To find out, we must determine the altitude h on the unknown side c and compare it to the side a opposite the known angle. As shown,

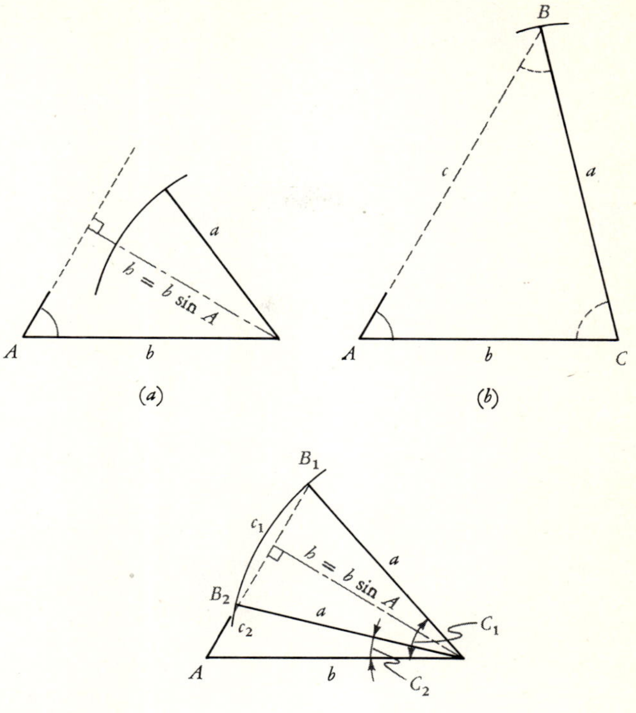

Fig. 13.8 (Repeated) (c)

we can always compute $h = b \sin A$ in the right triangle formed by the altitude. Then the remaining possibilities are

$$\text{If} \quad a < h \ldots \text{ then no solution exists} \qquad (13.14)$$

$$\text{If} \quad h < a < b \ldots \text{ then two solutions exist} \qquad (13.15)$$

where $h = b \sin A$.

What if $a = h$?

Figure 13.12 is a flowchart that traces the procedures for analyzing and solving the Case II triangles.

Comment: *Be sure to* ***understand the significance*** *of statements (13.13), (13.14), and (13.15) and the questions in Fig. 13.12. Do not just memorize the formulas, but learn the geometry of the three possible situations. Compare the sides of the triangles in terms of their geometric roles, such as "the side opposite the known angle." To make these comparisons, it is essential that you make a neat sketch. If you merely memorize the sides in terms of the letters a, b, and h used in this presentation, you will be lost when different letters are used or when the triangle is in a different position.*

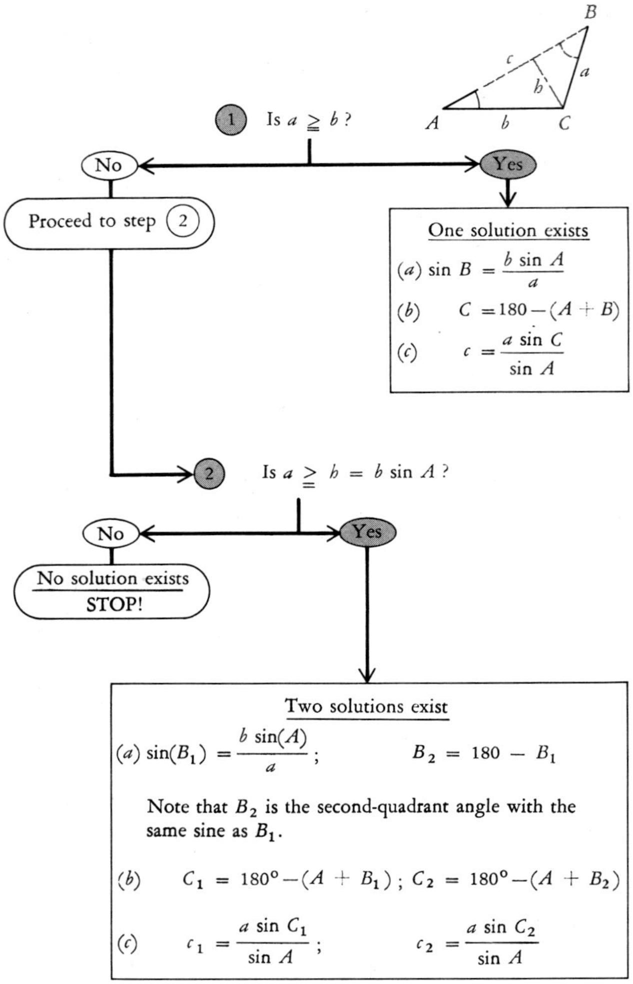

Fig. 13.12 A flow diagram for solving law of sines, Case II triangles.

The following examples illustrate how to solve Case II triangles. Steps 1, 2, *a*, *b*, and *c* correspond to those of Fig. 13.12.

Example 13.13. **Solve triangle *RST* of Fig. 13.11 if $R = 50°$, $y = 10$, and $z = 7.0$.**

1. **Compare the side opposite the known angle to the other known side.**

$$z = 7.0 < 10 = y$$

Therefore either no solution exists or two solutions exist.

2. Make a sketch to determine the appropriate altitude and compare it to side z. See Fig. 13.13.

$$h = y \sin 50° = 7.7 > 7.0 = z$$

No solution

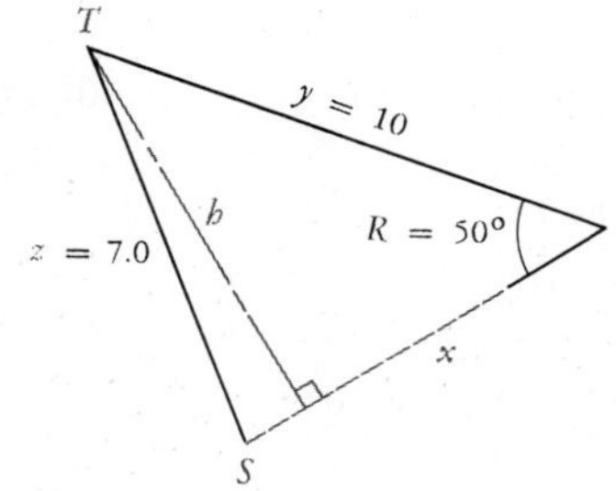

Fig. 13.13 Example 13.13.

Example 13.14. Solve triangle RST of Fig. 13.11 if $T = 50°$, $x = 9.30$, and $y = 10.0$

1. Compare x to y:

$$x = 9.3 < 10 = y$$

Therefore either no solution exists or two solutions exist.

2. Make a sketch to determine the appropriate altitude to compare to side x. See Fig. 13.14.

$$h = y \sin 50° = 7.66 < 9.3 = x$$

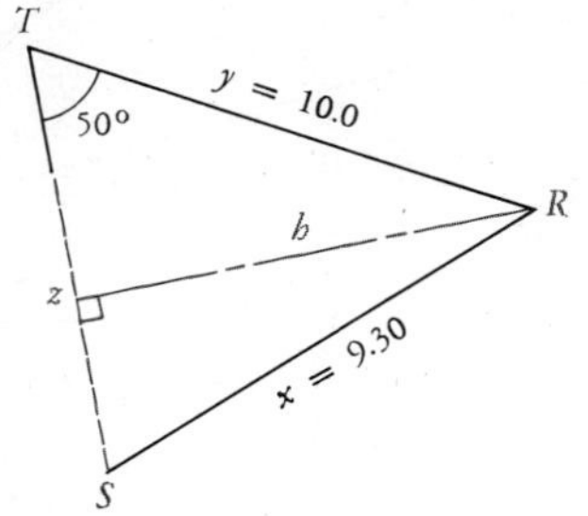

Fig. 13.14 Example 13.14.

Therefore two solutions exist.

$$a. \quad \sin S_1 = \frac{y \sin 50°}{x} = \frac{7.66}{9.30} = 0.824$$

$$S_1 = 55°30' \qquad S_2 = 180° - S_1 = 124°30'$$

b. $R_1 = 180° - (T + S_1) = 74°30'$
$R_2 = 180° - (T + S_2) = 5°30'$

c. $$z_1 = \frac{x \sin R_1}{\sin T} = \frac{9.30 \sin 74°30'}{\sin 50°} = 11.7$$

$$z_2 = \frac{x \sin R_2}{\sin T} = \frac{9.30 \sin 5°30'}{\sin 50°} = 1.16$$

Example 13.15. A pilot wishes to fly in the direction 60° west of north. The indicated airspeed is 520 mph, and the wind is 55 mph from due southwest. Find the heading and the groundspeed of the plane.

The absolute velocity $\mathbf{V}_p$ (the groundspeed) is the *vector* sum of the airspeed $\mathbf{V}_{p/a}$ and the velocity of the wind $\mathbf{V}_w$.

$$\mathbf{V}_p = \mathbf{V}_w + \mathbf{V}_{p/a} \qquad \text{(vector equation)}$$

Figure 13.15 shows how these vectors must be arranged in order that the true heading be 60° west of north. The figure indicates that a Case II triangle must be solved. Since $V_{p/a} = 520 > 55 = V_w$, one solution exists.

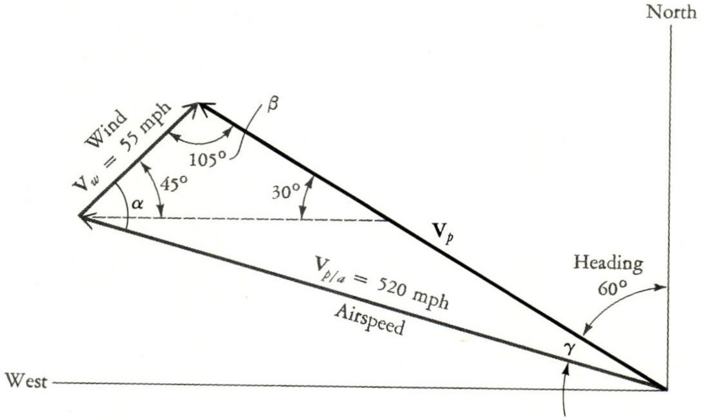

Fig. 13.15 Example 13.15.

a. $$\sin \gamma = \frac{V_w \sin \beta}{V_{p/w}} = \frac{55 \sin 105°}{520} = \frac{11}{104} \sin 75°$$

$\sin \gamma = 0.102 \longrightarrow \gamma = 5°50'$

Heading: $\gamma + 60° = 65°50'$

b. $\alpha = 180° - (\beta + \gamma) = 69°10'$

c. $V_p = \dfrac{V_w \sin \alpha}{\sin \gamma} = \dfrac{55 \sin 69°10'}{0.102} = 504 \text{ mph}$

Answer: $\mathbf{V}_p = 504$ mph @ 65°50′ west of north

EXERCISE 13.4

In Probs. 1 to 12, the letters correspond to those of Fig. 13.16. Solve each triangle for the missing parts.

1. $\theta = 30°$, $\phi = 70°$, $n = 15$
2. $\phi = 110°$, $\mu = 50°$, $e = 10$
3. $\mu = 60°$, $\theta = 20°$, $t = 30$
4. $\phi = 30°$, $e = 10$, $n = 5$
5. $\theta = 20°$, $e = 25$, $n = 100$
6. $\mu = 30°$, $\theta = 120°$, $n = 200$
7. $\mu = 15°$, $t = 300$, $n = 500$
8. $\theta = 105°$, $n = 1$, $e = 2$
9. $\phi = 35°$, $\theta = 55°$, $n = 62$
10. $\phi = 100°$, $n = 75$, $e = 80$
11. $\mu = 20°$, $\phi = 20°$, $t = 7.5$
12. $\theta = 75°$, $t = 22.0$, $e = 21.5$

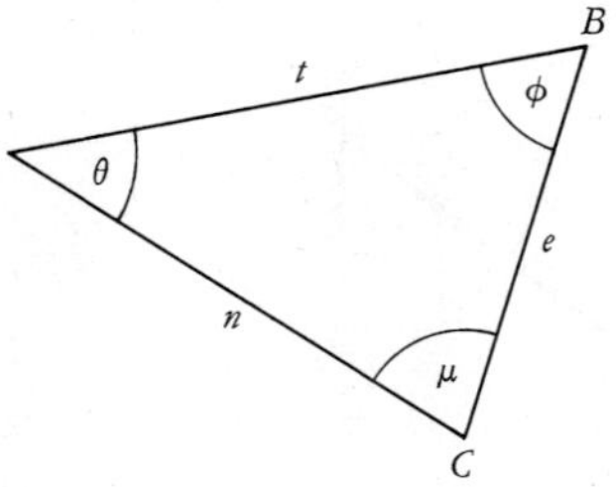

Fig. 13.16 Probs. 1 to 12.

13. Use Fig. 13.10 to show that the second part of Eq. (13.12) is true. (*Hint:* Draw an altitude from *A* to line *BC* extended.)
14. To determine the distance from *A* to *C* across a lake (Fig. 13.17), a surveyor at point *A* measures ∡*CAB*; then he moves to point *B* and measures ∡*ABC*. What is the distance *AC*?

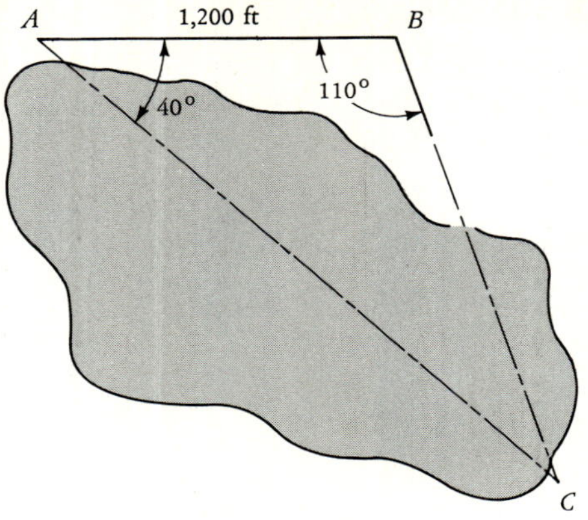

Fig. 13.17 Prob. 14.

15. For vector **A** of Fig. 13.18, find the components along the *r* and *s* axes.

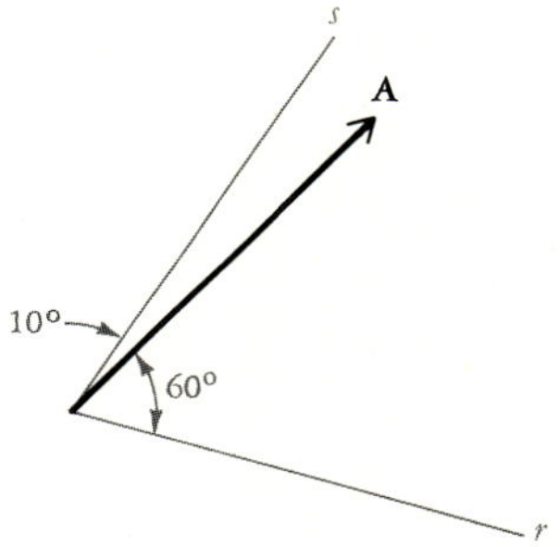

Fig. 13.18 Prob. 15.

16. If an object is in equilibrium, the resultant of all forces on it is zero, and thus the force vectors must form a closed polygon. Figure 13.19 shows the vector polygon for the forces on a

Fig. 13.19 Prob. 16. (a) A sign supported by a brace and a chain. (b) Vector polygon for the forces acting on the sign.

200-lb sign supported by a brace and a chain. Find the force in the brace and the force in the chain.

17. Two cables support a 500-lb weight as shown in Fig. 13.20. What is the tension in each cable? (*Hint:* Draw a vector triangle as described in Prob. 16.)

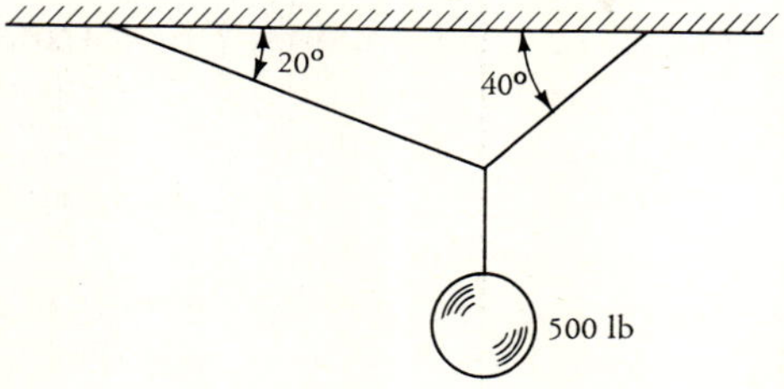

Fig. 13.20 Prob. 17. Two cables support the 500-lb weight.

18. An asymmetric derrick boom supports a 250-lb load, as shown in Fig. 13.21. Find the force along the axis of each member of the boom.

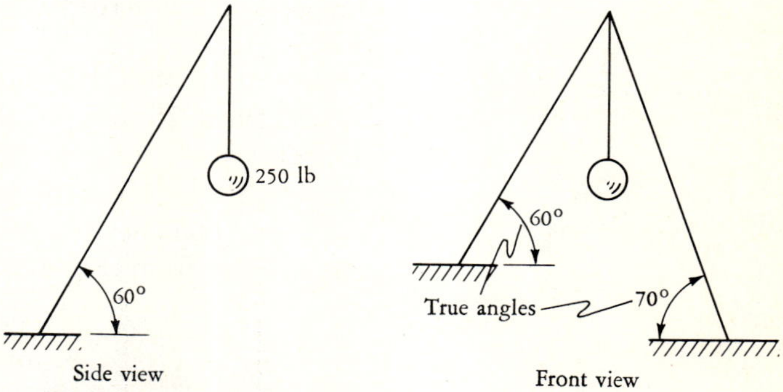

Fig. 13.21 Prob. 18. An asymmetric derrick.

19. A ship travels the course from A to B of Fig. 13.22. Then it travels from B to C. How far is it from A?

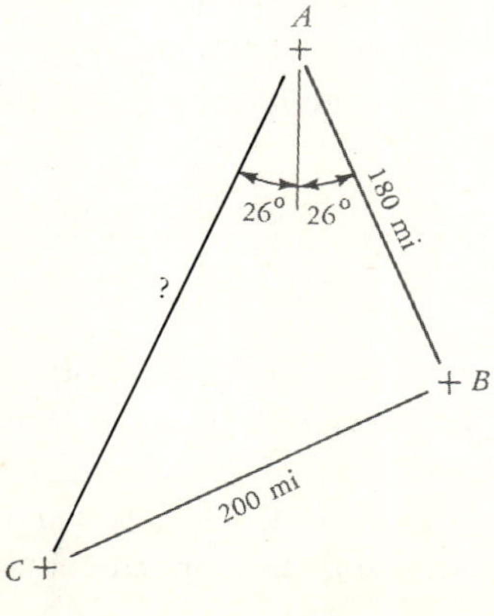

Fig. 13.22 Prob. 19.

20. Repeat Prob. 19 with the distances AB and BC interchanged.
21. A surveyor reports the following measurements of a triangular plot: One side is 230 ft, another side is 140 ft, and the angle opposite the longer side is 35°. Can this information be correct? Explain.
22. Repeat Prob. 21 if the angle opposite the shorter side is 35°.
23. On a trip from New Orleans to Atlanta, a certain airplane maintains an indicated speed (airspeed) of 300 mph. If the bearing line between these cities is 52° east of north and the wind is 75 mph due east, what heading should the airplane take? What is its groundspeed?
24. A boat that travels 8 mph in still water heads due east in a current known to be 5 mph. The boat is observed to move in the direction 30° north of east. What is the true speed of the boat, and in what direction does the current move?
25. A certain airplane is supposed to fly in the direction 37° east of north. The wind is due north at 100 mph, and the plane's airspeed indicator reads 250 mph. In what direction should the plane head?
26. Repeat Prob. 25 if the wind is due south at 100 mph.
27. Repeat Prob. 25 if the wind is 15° west of south.
28. A boat is attempting to cross from A to B in the river shown in Fig. 13.23. The speed of the boat is 7 mph relative to the water, and heavy rains have caused the current to be 8 mph. How can the boat get from A to B?

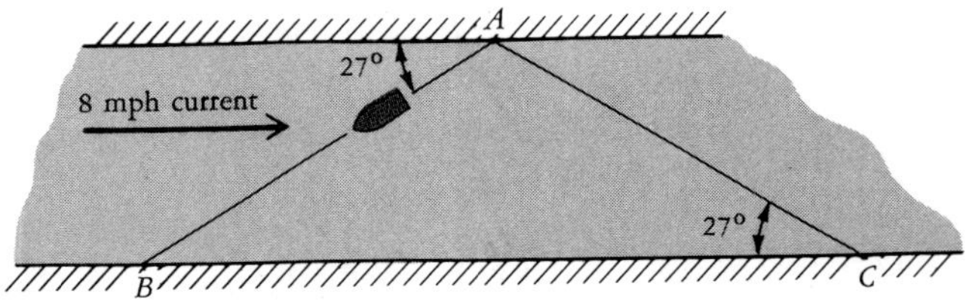

Fig. 13.23 Prob. 28. A boat in a river.

29. How can the boat of Prob. 28 reach point C? How fast will it go?

13.3 THE LAW OF COSINES

The astute observer may have noticed that the law of sines will not solve Case III or Case IV triangles. To use the law of sines, Eq. (13.12), we must know one of the ratios $a/(\sin A)$, $b/(\sin B)$, or $c/(\sin C)$. But in Cases III and IV, Fig. 13.7c and d, we are not given any combination of a side and the opposite angle.

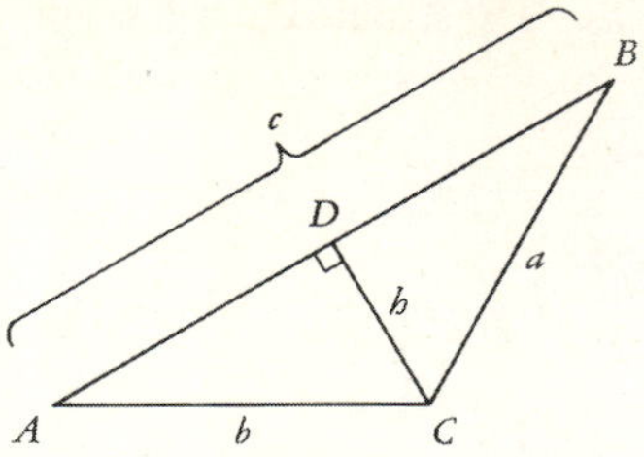

Fig. 13.10 (Repeated)

To find solutions for Cases III and IV, we draw an altitude of the oblique triangle to form two right triangles, as before. See Fig. 13.10, repeated below.

We apply the Pythagorean formula to both right triangles ACD and BCD as follows:

$$\text{Rt } \triangle ACD: \quad b^2 = h^2 + \overline{AD}^2 \longrightarrow h^2 = b^2 - \overline{AD}^2$$

$$\text{Rt } \triangle BCD: \quad a^2 = h^2 + \overline{BD}^2 \longrightarrow h^2 = a^2 - \overline{BD}^2$$

By equating the two expressions for h^2, we have

$$a^2 - \overline{BD}^2 = b^2 - \overline{AD}^2$$

$$a^2 = b^2 + \overline{BD}^2 - \overline{AD}^2$$

But $\overline{BD} = c - \overline{AD}$. Therefore

$$\begin{aligned} a^2 &= b^2 + (c - \overline{AD})^2 - \overline{AD}^2 \\ &= b^2 + c^2 - 2c\overline{AD} + \cancel{\overline{AD}^2} - \cancel{\overline{AD}^2} \end{aligned}$$

Note in Rt $\triangle ACD$ that

$$AD = b \cos A \qquad \text{(Why?)}$$

Thus we have the following result:

LAW OF COSINES	$a^2 = b^2 + c^2 - 2bc \cos A$	(13.16)

The meaning of this formula is as follows: *The square of one side of a triangle equals the sum of the squares of the other two sides minus twice their product times the cosine of the included angle.*

If A, b, and c of Fig. 13.10 were known, then we could use Eq. (13.16) to solve for a. Notice that the equation uses precisely the known parts of a Case III triangle. Notice also that the *unknown* side a is the one *opposite* the known angle A.

We could use the same formula to solve for b, if B, a, and c were known.

$$b^2 = a^2 + c^2 - 2ac \cos B$$

Now see if you can write a similar solution for c of Fig. 13.10, assuming that C, a, and b are known. See Example 13.16.

If we knew all three sides, the Case IV situation, we could solve Eq. (13.16) for cos A as follows:

ALTERNATE FORM OF LAW OF COSINES	$\cos A = \dfrac{b^2 + c^2 - a^2}{2bc}$	(13.17)

To help to remember this form, notice that the two *positive* terms come from the sides *adjacent* to the angle (b and c are adjacent to A in Fig. 13.10), and the *negative* term comes from the side *opposite* the angle (a is opposite A).

The corresponding solution for angle B would be

$$\cos B = \frac{a^2 + c^2 - b^2}{2ac}$$

Now see if you can write the solution for angle C. See Example 13.17.

The following examples illustrate how to solve oblique triangles by the law of cosines.

Example 13.16. **Solve $\triangle ABC$ of Fig. 13.10, if $C = 120°$, $a = 5.00$, and $b = 6.00$.**

According to the rule expressed by Eq. (13.16), we have

$$\begin{aligned} c^2 = a^2 + b^2 - 2ab \cos C &= 25 + 36 - 60 \cos 120° \\ &= 61 - 60(-\cos 60°) \\ &= 61 - 60(-0.500) \end{aligned}$$

Look closely here! Why is cos 120° = −cos 60°?

$$c^2 = 91.0$$

$$c \approx 9.54^*$$

Having found c, we can now use either the law of sines or Eq. (13.17) to solve for angles A and B. The law of sines is usually easier to use.

$$\frac{\sin A}{a} = \frac{\sin C}{c}$$

$$\sin A = \frac{a \sin C}{c} = \frac{5 \sin 120°}{9.54} = \frac{5(0.866)}{9.54}$$

* "≈" means "equals approximately."

$$\sin A = 0.454 \longrightarrow A = 27°$$

$$B = 180° - (A + C) = 33°$$

Answers:

$c = 9.54$
$A = 27°$
$B = 33°$

Comment: *When we used the law of sines above, we solved a Case II triangle either with none, one, or with two solutions. It was not necessary to investigate the number of solutions for the following two reasons: (1) We started with a Case III situation, which assures us of having a unique solution (why?). (2) Angle A had to be the first-quadrant* solution because C was 120°—we could not have another second-quadrant* angle. What if C had been less than 90°? (See Example 13.17.)*

Example 13.17. Solve $\triangle ABC$ of Fig. 13.10, if $a = 12.0$, $b = 8.00$, and $c = 5.00$.

According to the rule expressed by Eq. (13.17), we have

$$\cos C = \frac{a^2 + b^2 - c^2}{2ab} = \frac{144 + 64 - 25}{2(12)(8)} = \frac{183}{192}$$

$$\cos C = 0.9531 \longrightarrow C = 17°37'$$

Having found C, we can now use the law of sines (or we could use the law of cosines again) to find either A or B. If we proceed *without thinking,* we shall get the following *incorrect* answers:

$$\frac{\sin A}{a} = \frac{\sin C}{c}$$

$$\sin A = \frac{a \sin C}{c} = \frac{12.0(0.3027)}{5} = 0.7265$$

$$A = 46°36'$$

$$B = 180° - (A + C) = 115°47'$$

* It is customary to use the terms *first-quadrant, second-quadrant,* and so on, even though the angles may not be placed in standard position on a coordinate system. What we really mean here by *first-quadrant* and *second-quadrant* is "between 0° and 90°" and "between 90° and 180°," respectively.

Now think a minute! Can these angles possibly be correct? Is the largest angle opposite the longest side, as it must be? What has gone wrong? We used a formula blindly, without *thinking*. The law-of-sines solution is for the ambiguous Case II situation, and this time the angle has to be the second-quadrant angle. We found the first-quadrant reference angle. The correct answers are

$$A = 180° - 46°36' = 133°24'$$

$$B = 180° - (A + C) = 28°59'$$

Answers:

$A = 133°24'$
$B = 28°59'$
$C = 17°37'$

Comment: *There are two ways to avoid the error just illustrated.* (1) *When using the law of sines at this point in the solution,* ***solve first for the angle opposite the shorter side.*** *That angle has to be a first-quadrant angle.* (*Why?*) (2) *Use just the numerator of Eq.* (13.17) *to determine the algebraic sign of the cosine of the next angle. If it is negative, the angle has to be a second-quadrant angle.* (*Why?*) *Above all, think!*

Example 13.18. An airplane heads 36°50′ south of west, and the indicated speed (airspeed) is 500 mph. A 100-mph wind blows due east. What are the true speed (groundspeed), the track (true direction of travel), and the drift angle of the airplane?

The first step is to *make a sketch*. From the sketch we determine how to set up and solve the problem.

Figure 13.24 is the sketch. It is drawn according to the relative-velocity principle: The absolute velocity $\mathbf{V}_g$ is the vector sum of the plane's velocity relative to the air $\mathbf{V}_{p/a}$ and the wind velocity $\mathbf{V}_w$.

$$\mathbf{V}_g = \mathbf{V}_w + \mathbf{V}_{p/a} \qquad \text{(Vector sum)}$$

The figure shows us the track angle θ and the drift angle ϕ, and it shows that we have a Case III triangle.

We first solve for V_g, using the rule expressed by Eq. (13.16) as follows:

$$\begin{aligned} V_g^2 &= V_{p/a}^2 + V_w^2 - 2V_{p/a}V_w \cos 36°50' \\ &= 25 \cdot 10^4 + 1 \cdot 10^4 - 10 \cdot 10^4(0.800) \\ &= 26 \cdot 10^4 - 8 \cdot 10^4 \end{aligned}$$

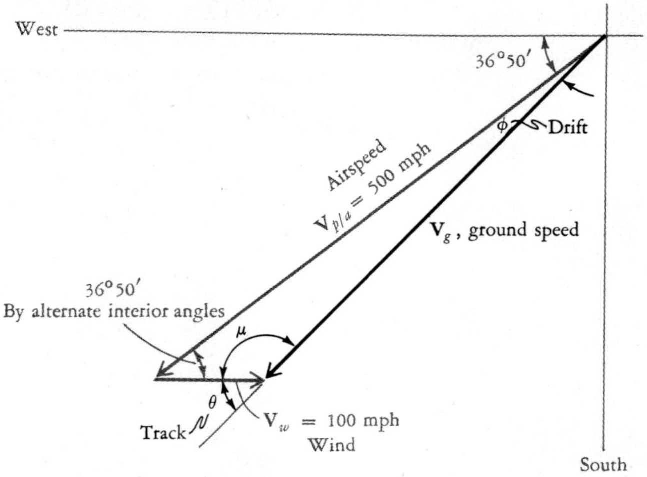

Fig. 13.24 Example 13.18.

$$V_g^2 = 18 \cdot 10^4$$

$$V_g = 4.24 \cdot 10^2 = 424 \text{ mph}$$

There are several ways to complete the solution, but to make a point, let us next use Eq. (13.17) to solve for angle μ.

$$\cos \mu = \frac{V_g^2 + V_w^2 - V_{p/a}^2}{2V_gV_w} = \frac{18 \cdot 10^4 + 1 \cdot 10^4 - 25 \cdot 10^4}{8.48 \cdot 10^4}$$
$$= \frac{-6 \cdot 10^4}{8.48 \cdot 10^4} = -0.707$$

$$\mu = 180° - 45° = 135°$$

The point is this: *The negative cosine tells us that μ has to be a second-quadrant angle.*

The track angle and the drift angle are as follows.

Track: $\theta = 180° - \mu = 45°$
Drift: $\phi = 180° - (\mu + 36°50') = 8°10'$

Answers:

True speed:	$V_g = 424$ mph
Track:	$\theta = 45°$ south of west
Drift angle:	$\phi = 8°10'$ toward the east

EXERCISE 13.5

For Probs. 1 to 12, use the triangle of Fig. 13.25 and solve for the missing parts.

1. $\beta = 60°$, $n = 5$, $s = 5$
2. $\alpha = 50°$, $p = 12$, $s = 10$
3. $\alpha = 130°$, $p = 12$, $s = 10$
4. $n = 30$, $p = 20$, $s = 30$
5. $s = 5$, $n = 6$, $p = 4$
6. $\gamma = 10°$, $n = 1.00$, $p = 0.684$
7. $p = 9$, $s = 2$, $n = 8$
8. $\beta = 120°$, $s = 200$, $n = 500$
9. $\beta = 62°$, $n = 12$, $p = 11$
10. $s = 20$, $n = 15$, $p = 25$
11. $\gamma = 75°$, $p = 1{,}000$, $n = 950$
12. $n = 137$, $p = 65$, $s = 57$

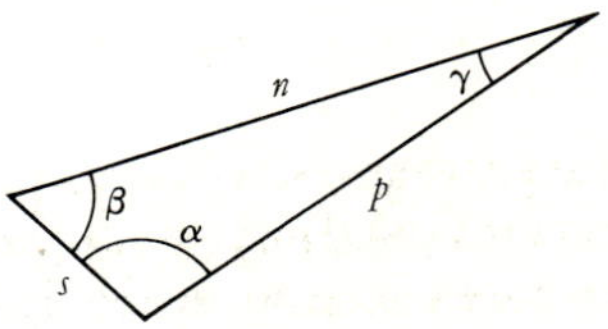

Fig. 13.25 Probs. 1 to 12.

13. In Fig. 13.26, **A** is a 35-lb force, **B** is a 43-lb force, and ϕ is 110°. What is the resultant force **R** (magnitude and direction)?

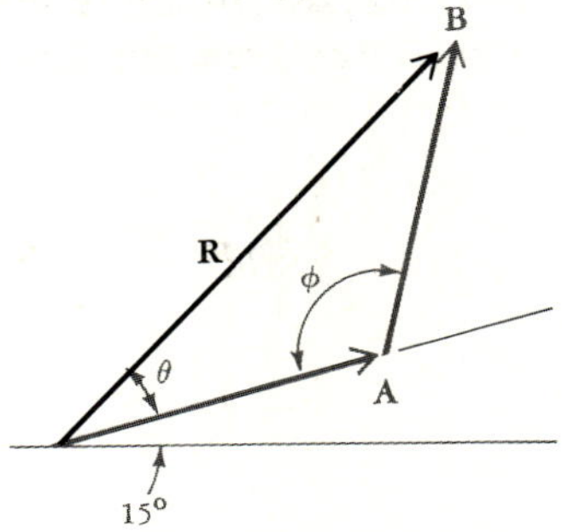

Fig. 13.26 Probs. 13 to 15.

14. If **R** of Fig. 13.26 is 80 lb, **A** is 60 lb, and θ is 60°, what is **B** (magnitude and direction)?
15. If **A** of Fig. 13.26 is 100 lb, **B** is 120 lb, and **R** is 80 lb, what are θ and ϕ?
16. If **A** and **B** of Fig. 13.27 represent forces, find the resultant force and its directional angle from the positive x axis.

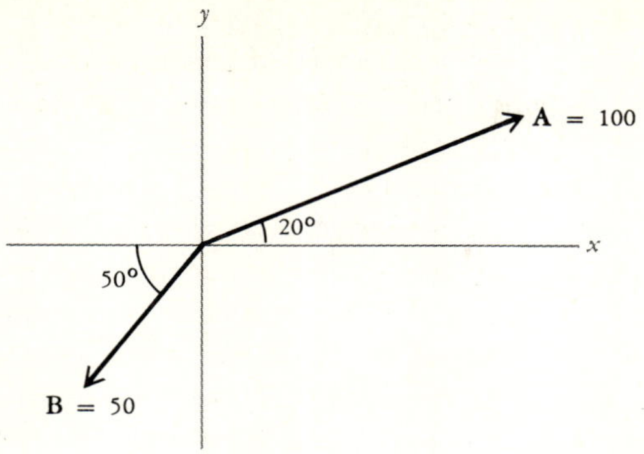

Fig. 13.27 Prob. 16.

17. Find the length of each diagonal of the parallelogram of Fig. 13.28, if $a = 8.00$ in., $b = 6.00$ in., and $\theta = 25°$.
18. If the longer diagonal of Fig. 13.28 is 21 in. and the two sides are 10 and 12 in., how long is the other diagonal?

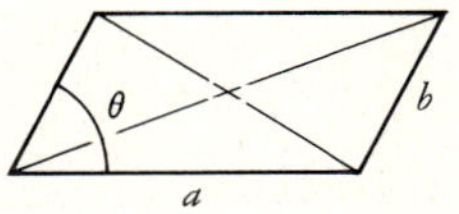

Fig. 13.28 Probs. 17 and 18.

19. Nine 1/4-in.-diameter holes are to be spaced at equal angles on a circular plate. The center of the first hole is 1/2 in. from the center of the plate, and each succeeding hole is 1/8 in. farther from the center. What is the thickness of material between the two closest holes? between the two most widely separated holes?
20. A man walks 12 miles in the direction 37° north of west and then turns and walks 3.6 miles in the direction 72° north of west. How far and in what direction is he from the starting point?
21. The equal legs of a certain A-frame are each 13 ft long, and the distance between them at the bottom is 7 ft. What is the angle between them?
22. In Fig. 13.29, the tensions in cables A and B are 600 and 500 lb, respectively. What are the angles θ_A and θ_B?

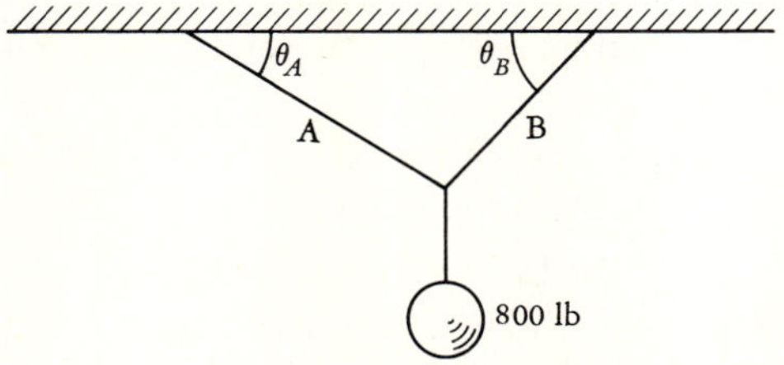

Fig. 13.29 Prob. 22. Cables A and B support the 800-lb load.

23. To determine the height of a certain tree, a surveyor measures the angle of elevation to the top (the angle up from a horizontal

line) as $26°10'$. Then he moves 100 ft farther away and measures the angle of elevation as $20°40'$. How tall is the tree?

24. A certain river flows 6 mph due east, and a boat is moving at 10 mph relative to the water. What are the true speed and direction of the boat if it heads
    a. downstream at an angle of 30° north of east?
    b. upstream at an angle of 30° north of west?
25. Repeat Prob. 24 if the river flows 15° south of east.
26. An astronaut on the moon drives his lunar rover 16 miles in the direction 60° north of east from the base. How many degrees must he turn so that he can drive 12 miles more and then turn again and return to the base on a north-south line? How far must he travel on the last leg of the trip? Can he make it back if the maximum range of the rover is 40 miles?
27. Repeat Prob. 26 if the astronaut drives 18 miles on the second leg of the trip.
28. The bearing from city A to city B is north 15° east, and the distance is 600 miles. An airplane pilot wishes to fly from A to B, but mixes up the directions and erroneously flies 15° north of east. After flying 150 miles, he discovers the error. How far and in what direction must he now fly to reach city B?
29. To determine the distance $\overline{MN}$ across a pond, a surveyor measures 600 ft from M to a third point K. Then he turns 60° and measures 800 ft from K to N. What is the distance $\overline{MN}$?
30. The directional vector between two cities R and S is 1,700 miles at 23° west of south. An airplane heads along this vector and maintains an airspeed of 450 mph. But after 1 hr, the pilot discovers that the wind from due south has been causing a steady drift of 3° toward the west. At this point, how far and in what direction is the airplane from city S?
31. What direction should the airplane of Prob. 30 take in order to reach city S? How late will it be in arriving? (*Hint:* Find the wind vector and add it to the airspeed vector to determine the groundspeed vector through S.)
32. An airplane heads $53°10'$ south of west at an indicated speed (airspeed) of 520 mph. If the wind blows due south at 100 mph, find the true speed and direction of the plane.
33. A power-boat squadron wants to lay out a triangular race course with consecutive legs of 3/8, 3/4, and 1 mile. What is the angle at each vertex of the course?

TRIGONOMETRIC IDENTITIES

CHAPTER 14

The equation

$$x^2 + 4x + 4 = (x + 2)^2$$

is true for all values of x. Similarly,

$$\frac{x^2 + x - 2}{x - 1} = x + 2$$

is true for all values of x except $x = 1$. These equations are examples of *identities.* An *identity* is an equation that is true for all values of the variable for which the variable can be defined.

There are many identities among the trigonometric functions. For example, we shall show that $(\cos \theta)^2 = 1 - (\sin \theta)^2$ is always true. A major purpose of trigonometric identities is to simplify expressions involving the functions. The applications are found in advanced engineering subjects and in calculus, and this phase of trigonometry is a prerequisite for those subjects.

Since you are not familiar with engineering subjects and calculus, we cannot demonstrate the applications of trigonometric identities in detail, for it would be absurd to expect you to understand these advanced subjects. Therefore this chapter and the next two will, of necessity, be more abstract than usual.

To combat the abstractness and to give you at least some appreciation for this trigonometry, we shall occasionally quote applications from engineering and calculus. We shall place in quotation marks the terms you are not expected to understand. You will not have to understand them in order to do the trigonometry. But perhaps you will see the usefulness of the trigonometry for future applications and thereby accept it and study it with more interest.

14.1 TRIGONOMETRIC FUNCTIONS OF ANGLES IN RADIANS

The majority of applications for trigonometric identities (and later for trigonometric graphs and equations) use radian measure for angles. For your convenience, this section is a review of radian measure.

Recall that, by definition, the number of radians in 1 rev (revolution) is 2π. Therefore, 2π radians is equivalent to 360°, π radians is equivalent to 180°, and so on. The general relationship between the number of radians θ_R in an angle and the number of degrees θ_D in the same angle is

$$\frac{\theta_R}{\pi} = \frac{\theta_D}{180°} \tag{14.1}$$

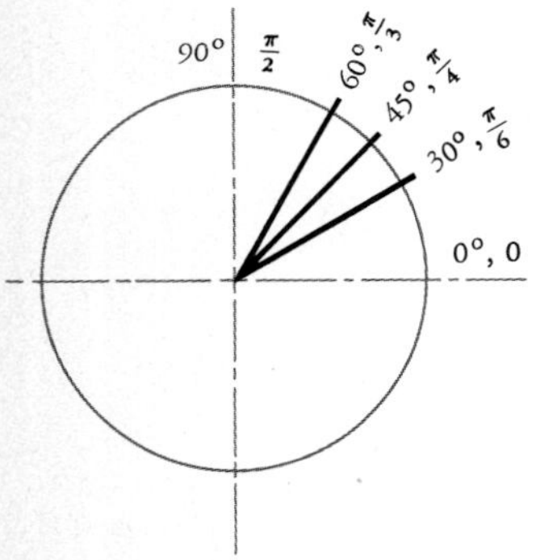

Fig. 14.1 The correspondence between radian measure and degree measure for certain commonly used angles.

Figure 14.1 gives a graphic sense of the correspondence between radian measure and degree measure for some of the commonly used angles. You should be able to show in the second, third, and fourth quadrants of Fig. 14.1 angles for which the given ones are reference angles. See Prob. 1 of Exercise 14.1.

We shall be particularly interested in trigonometric functions of angles in radians. For example

$$\sin\frac{\pi}{6} = \sin 30° = \frac{1}{2}$$

How do we know this is true? In the first place, Eq. (14.1) or Fig. 14.1 shows that $\pi/6$ radians corresponds to (but is not *equal* to) 30°. Secondly, since $\pi/6$ and 30° represent the same angle ($1/12$ of a revolution), their sine functions have to be identical. Table 14.1 lists the commonly used angles and their trigonometric functions for both degree measure and radian measure. You should be able to fill in the missing values. See Prob. 2 of Exercise 14.1.

After completing Table 14.1, you should learn the entries in it. In fact, you should learn the values of the trigonometric functions

TABLE 14.1

		FIRST QUADRANT				SECOND QUADRANT		
θ_D, DEGREES θ_R, RADIANS	0° 0	30° $\pi/6$	45° $\pi/4$	60° $\pi/3$	90° $\pi/2$	120°	 $3\pi/4$	150°
$\sin\theta$	0	$1/2$	$\sqrt{2}/2$	$\sqrt{3}/2$	1	$\sqrt{3}/2$		
$\cos\theta$	1	$\sqrt{3}/2$	$\sqrt{2}/2$	$1/2$	0	$-1/2$	$-\sqrt{2}/2$	
$\tan\theta$	0	$\sqrt{3}/3$	1	$\sqrt{3}$	—	$-\sqrt{3}$		
$\cot\theta$	—	$\sqrt{3}$	1	$\sqrt{3}/3$	0		-1	$-\sqrt{3}$
$\sec\theta$	1	$2\sqrt{3}/3$	$\sqrt{2}$	2	—		$-\sqrt{2}$	$-2\sqrt{3}/3$
$\csc\theta$	—	2	$\sqrt{2}$	$2\sqrt{3}/3$	1			2

(Continued opposite)

of these angles in radians directly, without first converting to degrees. In addition to learning the functions of 0 and $\pi/2$, you need learn only those of the first-quadrant angles; they are the reference angles for the others. For example, $\cos(7\pi/6) = -\cos(\pi/6) = -\sqrt{3}/2$. Note that, except for algebraic signs, corresponding trigonometric functions (TF) of prime multiples of $\pi/6$ are equal. That is, $\text{TF}(\pi/6) = |\text{TF}(5\pi/6)| = |\text{TF}(7\pi/6)| = |\text{TF}(11\pi/6)|$. The same is true for prime multiples of $\pi/4$ and $\pi/3$.

The following conversion formulas relate any angle in radians to the corresponding first-quadrant reference angle. Note the similarity between these formulas and Eqs. (13.8), (13.9), and (13.10), respectively.

$$\boxed{\begin{aligned}\theta_1 &= \pi - \theta_2\\ \theta_2 &= \pi - \theta_1\end{aligned}} \qquad \boxed{\begin{aligned}\theta_1 &= \theta_3 - \pi\\ \theta_3 &= \pi + \theta_1\end{aligned}} \qquad \boxed{\begin{aligned}\theta_1 &= 2\pi - \theta_4\\ \theta_4 &= 2\pi - \theta_1\end{aligned}} \tag{14.2}$$

Second quadrant	Third quadrant	Fourth quadrant
(a)	(b)	(c)

The following examples illustrate trigonometric functions of angles in radians.

Example 14.1. Find $\sin 11\pi/6$:

$$\sin\frac{11\pi}{6} = -\sin\frac{\pi}{6} = -\frac{1}{2}$$

Example 14.2. Find $\tan 8\pi/3$.

This angle exceeds 1 rev (2π radians, or 360°), and so we first find a coterminal angle by subtracting 2π.

$$\tan\frac{8\pi}{3} = \tan\left(\frac{8\pi}{3} - 2\pi\right) = \tan\frac{2\pi}{3} = -\tan\frac{\pi}{3} = -\sqrt{3}$$

TABLE 14.1 (*Continued*)

		THIRD QUADRANT				FOURTH QUADRANT			
θ_D, DEGREES θ_R, RADIANS	180° π	 $7\pi/6$	225°	 $4\pi/3$	270°	300°	 $7\pi/4$	330°	360° 2π
$\sin\theta$	0		$-\sqrt{2}/2$			$-\sqrt{3}/2$			
$\cos\theta$		$-\sqrt{3}/2$					$\sqrt{2}/2$		
$\tan\theta$	0		1		—	$-\sqrt{3}$		$-\sqrt{3}/3$	
$\cot\theta$		$\sqrt{3}$		$\sqrt{3}/3$		$-\sqrt{3}/3$	-1		
$\sec\theta$	-1		$-\sqrt{2}$	-2	—			$2\sqrt{3}/3$	
$\csc\theta$		-2		$-2\sqrt{3}/3$	-1		$-\sqrt{2}$		

Example 14.3. Find the sine, cosine, and tangent of 3.4994 radians.

First note that 3.4994 is a third-quadrant angle: $3.1416\ldots = \pi < 3.4994 < 3\pi/2 = 4.7124$. The reference angle is, according to Eq. (14.2*b*),

$$\theta_1 = 3.4994 - \pi = 3.4994 - 3.1416 = 0.3578$$

From Table A5.4 in the Appendix, reading from the radians column, we have

$$\sin 3.4994 = -\sin 0.3578 = -0.3502$$

$$\cos 3.4994 = -\cos 0.3578 = -0.9367$$

$$\tan 3.4994 = \tan 0.3578 = 0.3739$$

Example 14.4. Given that $\sin x = -(\sqrt{a-1})/2$, $\pi < x < 3\pi/2$, and $1 < a < 5$, find similar expressions for the other five functions of x.

Figure 14.2 is a diagram showing x in the third quadrant (x is the angle in this example) and showing a point on x with $-\sqrt{a-1}$ the ordinate and 2 the radius. We first compute the abscissa u.

$$u = -[2^2 - (-\sqrt{a-1})^2]^{1/2} = -(4 - a + 1)^{1/2} = -\sqrt{5-a}$$

Then by applying the definitions of the functions to the diagram, we have

$$\cos x = \frac{u}{r} = -\frac{\sqrt{5-a}}{2}$$

$$\tan x = \frac{v}{u} = \sqrt{\frac{a-1}{5-a}}$$

$$\cot x = \frac{u}{v} = \sqrt{\frac{5-a}{a-1}}$$

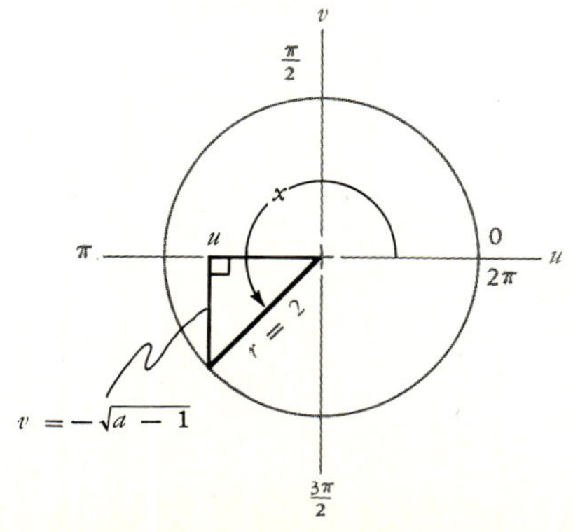

Fig. 14.2 Example 14.4.

$$\sec x = \frac{r}{u} = -\frac{2}{\sqrt{5-a}} = -\frac{2\sqrt{5-a}}{5-a}$$

$$\csc x = \frac{r}{v} = -\frac{2}{\sqrt{a-1}} = -\frac{2\sqrt{a-1}}{a-1}$$

Example 14.5. In calculus, to "integrate" the function $1/(y^2\sqrt{y^2+9})$, we let $\tan u = y/3$, $0 < u < \pi/2$. After "integrating," we have to evaluate $-(1/9) \csc u$.

Here again, a diagram is very useful. From Fig. 14.3 we see that

$$\csc u = \frac{\text{radius}}{\text{ordinate}} = \frac{\sqrt{y^2+9}}{y}$$

$$-\frac{1}{9}\csc u = -\frac{\sqrt{y^2+9}}{9y}$$

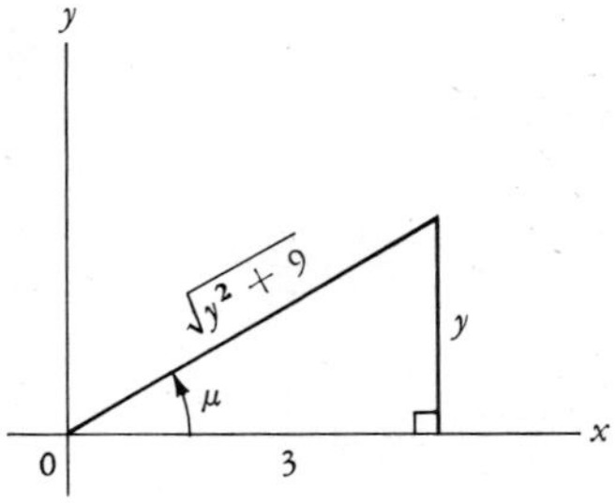

Fig. 14.3 Example 14.5.

EXERCISE 14.1

1. Make a diagram similar to Fig. 14.1, showing the angles 120°, 135°, 150°, 210°, 225°, 240°, 300°, 315°, and 330° and the radian equivalent of each of these angles.
2. Complete Table 14.1.

In Probs. 3 to 14, evaluate each function (1) in terms of the same function of one of the reference angles 0, $\pi/6$, $\pi/4$, $\pi/3$, or $\pi/2$, and (2) in terms of the numbers 1, 2, $\sqrt{2}$, and $\sqrt{3}$. For example, $\sin 4\pi/3 = -\sin \pi/3 = -\sqrt{3}/2$.

3. $\cos 5\pi/6$	4. $\tan 300°$	5. $\sin 2\pi/3$
6. $\csc 7\pi/6$	7. $\sec 4\pi/3$	8. $\cot 3\pi/4$
9. $\sin 120°$	10. $\sec 7\pi/4$	11. $\cos 225°$
12. $\tan(-11\pi/6)$	13. $\cos 20\pi/3$	14. $\sin 10\pi$

In Probs. 15 to 20, evaluate each function by using Table A5.4 in the Appendix, as in Example 14.3. Interpolate where necessary.

15. sin 0.6574	16. cos 1.1781	17. tan 1.1141
18. cos 2.2696	19. sin 4.3357	20. cot 5.6132

In Probs. 21 to 26, evaluate the remaining five functions using the diagram method as in Example 14.4.

21. $\tan x = 5/\sqrt{11},\ \pi < x < 3\pi/2$
22. $\sin x = \sqrt{5}/3,\ \pi/2 < x < \pi$
23. $\csc x = -2\sqrt{2},\ 3\pi/2 < x < 2\pi$
24. $\cos x = -1/a,\ \pi/2 < x < \pi,\ a > 1$
25. $\cot x = \dfrac{-3\sqrt{a-1}}{a-1},\ \dfrac{\pi}{2} < x < \pi,\ a > 1$
26. $\sin x = -\dfrac{1}{\sqrt{S^2+1}},\ \pi < x < \dfrac{3\pi}{2}$
27. In calculus, to "integrate" $1/(1-x^2)$, we let $x = \sin\theta$; then we have to find $\sec\theta + \tan\theta$. Do this by the diagram method, if $0 < x < \pi/2$.
28. The "phase angle" in certain kinds of ac circuits is given by $\tan\alpha = (X_L - X_C)/R$. Show that the "power factor," $\cos\alpha$, is

$$\cos\alpha = \frac{R}{\sqrt{R^2 + (X_L - X_C)^2}}$$

14.2 FUNDAMENTAL IDENTITIES

Several trigonometric identities occur as a direct result of the definitions of the functions. In this section we shall derive four sets of such identities. These sets are not necessarily complete, since many variations can be formed by combining identities.

A. THE NEGATIVE-ANGLE IDENTITIES

Figure 14.4 shows angles θ and $-\theta$ in standard position. From the congruent triangles of the figure we see that

$$y' = -y \qquad \text{and} \qquad x' = x$$

Therefore

$$\sin(-\theta) = \frac{y'}{r} = \frac{-y}{r} = -\sin\theta$$

and

$$\cos(-\theta) = \frac{x'}{r} = \frac{x}{r} = \cos\theta$$

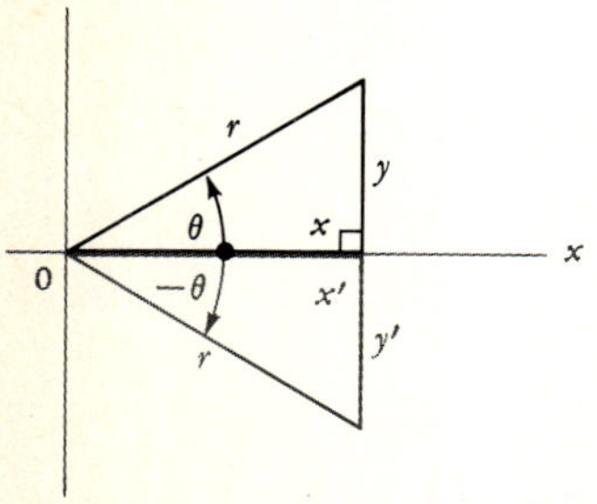

Fig. 14.4 Equal but opposite angles in standard position.

The following list summarizes the negative-angle identities.

$$\boxed{\begin{array}{ll} \sin(-\theta) = -\sin\theta & \cot(-\theta) = -\cot\theta \\ \cos(-\theta) = \cos\theta & \sec(-\theta) = \sec\theta \\ \tan(-\theta) = -\tan\theta & \csc(-\theta) = -\csc\theta \end{array}} \qquad (14.3)$$

Comment: *Four of these six equations give the false impression that we can factor −1 from the function. Notice that we never factored; instead,* ***we based the identities on the definitions of the functions.***

B. THE RECIPROCAL IDENTITIES

We repeat here the definitions of the six trigonometric functions, in terms of the abscissa x the ordinate y and the radius r of a point on the terminal side of an angle θ in standard position. See Fig. 14.5.

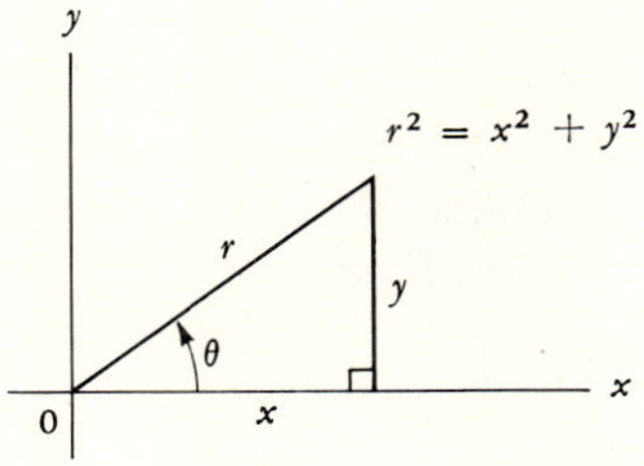

Fig. 14.5

$$\begin{bmatrix} \sin\theta = \dfrac{y}{r} \\ \\ \csc\theta = \dfrac{r}{y} \end{bmatrix} \quad \begin{bmatrix} \cos\theta = \dfrac{x}{r} \\ \\ \sec\theta = \dfrac{r}{x} \end{bmatrix} \quad \begin{bmatrix} \tan\theta = \dfrac{y}{x} \\ \\ \cot\theta = \dfrac{x}{y} \end{bmatrix}$$

The functions above are purposely grouped into pairs that use the same coordinates. From these groupings we can see the following reciprocal relationships:

$$\csc\theta = \frac{1}{\sin\theta} \qquad \sin\theta = \frac{1}{\csc\theta} \qquad \sin\theta\csc\theta = 1 \qquad (14.4)$$

$$\sec\theta = \frac{1}{\cos\theta} \qquad \cos\theta = \frac{1}{\sec\theta} \qquad \cos\theta\sec\theta = 1 \qquad (14.5)$$

$$\cot\theta = \frac{1}{\tan\theta} \qquad \tan\theta = \frac{1}{\cot\theta} \qquad \tan\theta\cot\theta = 1 \qquad (14.6)$$

We can derive any of these identities by substituting the letters x, y, or r. For example,

$$\frac{1}{\sin\theta} = \frac{1}{y/r} = 1 \cdot \frac{r}{y} = \frac{r}{y} = \csc\theta$$

Notice that each identity has several forms. Multiple forms often occur, and we must learn to recognize each identity in any of its forms.

C. THE QUOTIENT IDENTITIES

Several identities can be developed by writing the quotient of certain trigonometric functions. For example,

$$\frac{\sin\theta}{\cos\theta} = \frac{y/r}{x/r} = \frac{y}{r}\cdot\frac{r}{x} = \frac{y}{x} = \tan\theta$$

The others listed below can be derived in a similar fashion.

$$\tan\theta = \frac{\sin\theta}{\cos\theta} \qquad \cos\theta = \frac{\sin\theta}{\tan\theta} \qquad \sin\theta = \cos\theta\tan\theta \qquad (14.7)$$

$$\cot\theta = \frac{\cos\theta}{\sin\theta} \qquad \sin\theta = \frac{\cos\theta}{\cot\theta} \qquad \cos\theta = \sin\theta\cot\theta \qquad (14.8)$$

Each equation shows the identity and two other forms of it. Still other identities can be formed by combining these with Eqs. (14.4) and (14.5). See Probs. 15, 16, 18, and 19 in Exercise 14.2.

D. THE PYTHAGOREAN IDENTITIES

As shown in Fig. 14.5, the coordinates of any point define a right triangle. Consequently the coordinates can be related by the Pythagorean formula, as follows:

$$y^2 + x^2 = r^2$$

By dividing both sides by r^2, we have

$$\left(\frac{y}{r}\right)^2 + \left(\frac{x}{r}\right)^2 = 1 \qquad \text{or} \qquad \sin^2\theta + \cos^2\theta = 1^*$$

By dividing both sides of $y^2 + x^2 = r^2$ by x^2, we have

$$\left(\frac{y}{x}\right)^2 + 1 = \left(\frac{r}{x}\right)^2 \qquad \text{or} \qquad \tan^2\theta + 1 = \sec^2\theta$$

And by dividing both sides of $y^2 + x^2 = r^2$ by y^2, we have

$$1 + \left(\frac{x}{y}\right)^2 = \left(\frac{r}{y}\right)^2 \qquad \text{or} \qquad 1 + \cot^2\theta = \csc^2\theta$$

* The symbols $\sin^2\theta$ and $\cos^2\theta$ mean $(\sin\theta)^2$ and $(\cos\theta)^2$, respectively.

The following list summarizes these identities. Below each identity are two of its commonplace variations.

$$\sin^2\theta + \cos^2\theta = 1 \tag{14.9}$$

$$\sin^2\theta = 1 - \cos^2\theta \qquad \cos^2\theta = 1 - \sin^2\theta$$

$$\tan^2\theta + 1 = \sec^2\theta \tag{14.10}$$

$$\tan^2\theta = \sec^2\theta - 1 \qquad \sec^2\theta - \tan^2\theta = 1$$

$$1 + \cot^2\theta = \csc^2\theta \tag{14.11}$$

$$\cot^2\theta = \csc^2\theta - 1 \qquad \csc^2\theta - \cot^2\theta = 1$$

The following examples show how we can use the fundamental identities (1) to evaluate one function when another function is known and (2) to simplify trigonometric expressions.

Generally speaking, there are no well-defined procedures for transforming one trigonometric expression into another by using identities. Experience, insight, and foresight usually suggest the best identities to use. These qualities are acquired (1) by becoming *very* familiar with the fundamental identities and (2) by doing *many* problems. Even experts occasionally run down blind alleys in proving identities, so be patient and be willing to try several different approaches. Try to anticipate the result of each step. The more steps you can visualize in advance, the better. Above all, keep referring to the fundamental identities until you know them instinctively.

Example 14.6. **Evaluate cot x, given that tan $x = 2/3$, $\pi < x < 3\pi/2$.**

$$\cot x = \frac{1}{\tan x} = \frac{1}{2/3} = \frac{3}{2}$$

Example 14.7. **Evaluate sin 5Z, given that cos 5Z = 1/5, $3\pi/2 < 5Z < 2\pi$.**

$$\sin 5Z = -\sqrt{1 - \cos^2 5Z} = -\sqrt{1 - 1/25} = -\frac{2\sqrt{6}}{5}$$

Example 14.8. **Show that (sec u)/(csc u) = tan u.**

One method that frequently works is to change all functions to sines or cosines. Thus

$$\frac{\sec u}{\csc u} = \frac{1/(\cos u)}{1/(\sin u)} = \frac{1}{\cos u} \cdot \frac{\sin u}{1} = \frac{\sin u}{\cos u} = \tan u$$

Example 14.9. Show that

$$\left(\frac{\sec^2 x \cos^2 x}{\sin^2 x} - 1\right) \tan^2 x = 1$$

Note that [from Eq. (14.5)] $\sec^2 x \cos^2 x = 1$, and multiply through by $\tan^2 x$ (the equations used to go from one step to the next are listed between lines, on the far right).

$$\left(\frac{\sec^2 x \cos^2 x}{\sin^2 x} - 1\right) \tan^2 x$$

Eq. (14.5)

$$= \frac{\tan^2 x}{\sin^2 x} - \tan^2 x$$

Eq. (14.7)

$$= \frac{\cancel{\sin^2 x}}{\cos^2 x \cancel{\sin^2 x}} - \tan^2 x$$

Eqs. (14.5), (14.10)

$$= \sec^2 x - \tan^2 x = 1$$

In proving identities, we can never assume that the equation to be proved is true. We must prove it independently. One way to do this is to work with only one side, usually the more complicated one, as in Example 14.9 above. Example 14.10 also transforms the left side into the right side.

Example 14.10. Show that

$$\frac{\sec x - 2 \sec x \cos^2 x}{\tan x - 1} = \sin x + \cos x$$

Sometimes algebraic operations help. Note that we can factor $\sec x$.

$$\frac{\sec x - 2 \sec x \cos^2 x}{\tan x - 1}$$

$$= \frac{\sec x\,(1 - 2\cos^2 x)}{\tan x - 1}$$

Eqs. (14.5), (14.7)

$$= \frac{1}{\cos x} \cdot \frac{1 - \cos^2 x - \cos^2 x}{(\sin x/\cos x) - 1}$$

Eq. (14.9)

$$= \frac{\sin^2 x - \cos^2 x}{\sin x - \cos x}$$

Factor the difference between squares

$$= \frac{(\sin x + \cos x)\cancel{(\sin x - \cos x)}}{\cancel{(\sin x - \cos x)}}$$

$$= \sin x + \cos x$$

EXERCISE 14.2

In Probs. 1 to 9, evaluate each function by using one or more of the fundamental identities.

1. $\sin(-45°)$ 2. $\tan(-5\pi/6)$ 3. $\cos(-240°)$
4. $\sec x$, given $\cos x = 2/17$
5. $\sin y$, given $\csc y = -5$
6. $\cos M$, given $\sin M = 2/3$ and $\frac{\pi}{2} < M < \pi$
7. $\tan m$, given $\sin m = 2/3$, $\frac{\pi}{2} < m < \pi$ (*Hint:* Use Prob. 6.)
8. $\cot(3a + 5)$, given $\csc(3a + 5) = -4/3$ and $\cos(3a + 5) = -\sqrt{7}/4$
9. $\sin(-A)$, given $\cot A = 3$ and $\cos A = -3\sqrt{10}/10$
10. Derive Eqs. (14.5) and (14.6).
11. Derive Eqs. (14.7) and (14.8).
12. Verify Eqs. (14.4) to (14.6) by using $\theta = \pi/4$ and Table 14.1.
13. Verify Eqs. (14.7) and (14.8) by using $\theta = 5\pi/6$ and Table 14.1.
14. Show that

$$\tan x = \pm \frac{\sin x}{\sqrt{1 - \sin^2 x}}$$

15. Use Prob. 14 to show that

$$\sin x = \pm \frac{\tan x}{\sqrt{1 + \tan^2 x}}$$

16. Show that

$$\tan x = \pm \frac{\sqrt{1 - \cos^2 x}}{\cos x}$$

17. Use Prob. 16 to show that

$$\cos x = \pm \frac{1}{\sqrt{1 + \tan^2 x}}$$

In Probs. 18 to 32, prove each identity by transforming the left-hand expression into the right-hand expression.

18. $\cot\theta = \csc\theta/\sec\theta$

19. $\dfrac{\tan\theta}{\sin\theta} = \sec\theta$

20. $\pm\sqrt{1+\cot^2\theta} = 1/(\sin\theta)$

21. $\cos\theta\tan\theta = \sin\theta$

22. $\cot\theta\sin\theta = \cos\theta$

23. $1-\sec^2\theta = -(\sec^2\theta)/(\csc^2\theta)$

24. $\sin^2 z - \cos^2 z = 1 - 2\cos^2 z$

25. $\dfrac{\sin y}{\tan y} \div \dfrac{\cot y}{\cos y} = \sin\theta\cos\theta$

26. $\sin^4 t - \cos^4 t = \sin^2 t - \cos^2 t$

27. $(\sin^3\theta\cos\theta + \cos^3\theta\sin\theta)\tan\theta = \sin^2\theta$

28. $\sin^4 x\sec^2 x + \sin^2 x = \tan^2 x$

29. $\dfrac{1}{\sin^2 5x} - 1 = \dfrac{\cos 5x\csc 5x}{\tan 5x}$

30. $\sqrt{\dfrac{1-\sin\omega t}{1+\sin\omega t}} = \dfrac{\cos\omega t}{1+\sin\omega t}$

31. $\sqrt{\dfrac{\sec\beta+1}{\sec\beta-1}} = \dfrac{\sin\beta}{1-\cos\beta}$

32. $\dfrac{1}{2}\cos 3x\left(\dfrac{\cos 3x}{1-\sin 3x} + \dfrac{1-\sin 3x}{\cos 3x}\right) = 1$

33. In strength of materials, a formula for "maximum deflection of a buckling column" is

$$\delta = e\left[\frac{\sin^2(KL/2)}{\cos(KL/2)} + \cos\frac{KL}{2}\right]$$

Show that $\delta = e\sec(KL/2)$. This equation is called the *secant formula* for columns.

34. The "coefficient of static friction," f_s, required to hold a block on a plane inclined at an angle α is given by $mg\ sin\ \alpha = f_s mg\cos\alpha$. Show that $f_s = \tan\alpha$.

35. The "average value of current in an ac circuit" is given by $i_{\text{av}} = i_{\max}\sin(2\pi f)t$. In evaluating i_{av}, the equation $\sin^2(2\pi f)t = \cos^2(2\pi f)t$ arises. Show from this equation that $\sin^2(2\pi f)t = 1/2$ and that $i_{\text{av}} = 0.707 i_{\max}$.

14.3 SUM AND DIFFERENCE IDENTITIES

A typical reaction to expressions such as $\cos(\alpha-\beta)$ is "this is obviously $\cos\alpha - \cos\beta$." *Nothing could be more incorrect.* Remember that $\cos(\alpha-\beta)$ is a functional notation, not a multiplication. Once again, we must not jump to conclusions. Instead we return to the definitions and derive our identities on the basis of established facts, principles, and logic.

Since it is relatively easy to derive, we shall first establish the identity for cos $(\alpha - \beta)$. Figure 14.6*a* shows angles α and β with respective points P_1 and P_2 at the same radius r. The angle between α and β is $\alpha - \beta$. The chord distance $\overline{P_1P_2}$ corresponding to $\alpha - \beta$ is d. Notice that the lengths of the legs of right triangle P_1P_2Q are $y_1 - y_2$ and $x_2 - x_1$. For the right triangle we have

$$\begin{aligned} d^2 &= (x_2 - x_1)^2 + (y_1 - y_2)^2 \\ &= x_2{}^2 - 2x_1x_2 + x_1{}^2 + y_1{}^2 - 2y_1y_2 + y_2{}^2 \\ &= (x_1{}^2 + y_1{}^2) + (x_2{}^2 + y_2{}^2) - 2x_1x_2 - 2y_1y_2 \end{aligned}$$

(Note that $x^2 + y^2 = r^2$.)

$$= r^2 + r^2 - 2x_1x_2 - 2y_1y_2$$

$$d^2 = 2r^2 - 2x_1x_2 - 2y_1y_2 \tag{14.12}$$

Now let us turn our attention to Fig. 14.6*b*. This figure comes from revolving $\triangle OP_1P_2$ of Fig. 14.6*a* clockwise, until $\overline{OP}_2$—labeled $\overline{OP}_2'$ in Fig. 14.6*b*—coincides with the x axis. Note that angle $\alpha - \beta$ is now in standard position; also $\overline{P_1'P_2'} = \overline{P_1P_2} = d$, because chords of equal angles (in the same circle) are equal. For right triangle $P_1'P_2'Q'$ we have

$$\begin{aligned} d^2 &= (r - x_1')^2 + (y_1')^2 \\ &= r^2 - 2rx_1' + (x_1')^2 + (y_1')^2 \\ &= r^2 + [(x_1')^2 + (y_1')^2] - 2rx_1' \end{aligned}$$

$$d^2 = 2r^2 - 2rx_1' \tag{14.13}$$

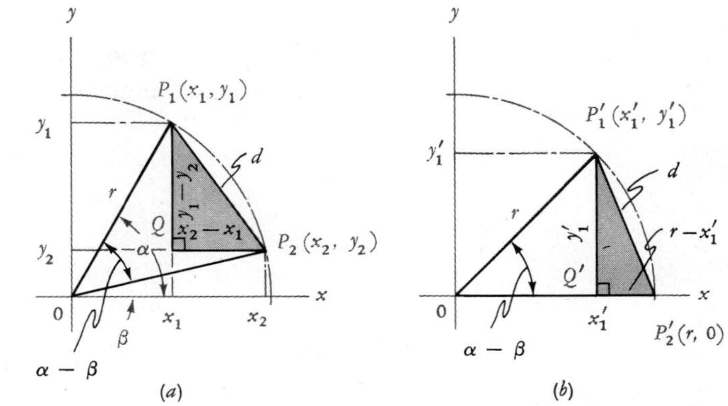

Fig. 14.6 (a); (b) angle $(\alpha - \beta)$ of part (a) is revolved until the initial side coincides with the $+x$ axis.

Now we have two different expressions, Eqs. (14.12) and (14.13), for d^2; we equate them as follows:

$$2r^2 - 2rx_1' = 2r^2 - 2x_1x_2 - 2y_1y_2$$

$$-2rx_1' = -2x_1x_2 - 2y_1y_2$$

$$rx_1' = x_1x_2 + y_1y_2$$

And we divide both sides by $r \cdot r$ as follows:

$$\frac{r}{r}\frac{x_1'}{r} = \frac{x_1}{r} \cdot \frac{x_2}{r} + \frac{y_1}{r} \cdot \frac{y_2}{r} \qquad (14.14)$$

Notice that each of these five ratios corresponds to one of the trigonometric functions of the angles in Fig. 14.6, as follows:

$$\frac{x_1'}{r} = \cos(\alpha - \beta) \qquad \frac{x_1}{r} = \cos\alpha \qquad \frac{x_2}{r} = \cos\beta$$

$$\frac{y_1}{r} = \sin\alpha \qquad \text{and} \qquad \frac{y_2}{r} = \sin\beta$$

By substituting these expressions into Eq. (14.14) we have the following identity for $\cos(\alpha - \beta)$:

$$\cos(\alpha - \beta) = \cos\alpha\cos\beta + \sin\alpha\sin\beta$$

In a similar fashion we can derive the following identity for $\sin(\alpha - \beta)$. See Prob. 38 of Exercise 14.3.

$$\sin(\alpha - \beta) = \sin\alpha\cos\beta - \cos\alpha\sin\beta$$

To find an identity for $\sin(\alpha + \beta)$, we replace β by $-\beta$ in $\sin(\alpha - \beta)$ and use the negative-angle identities as follows:

$$\begin{aligned} \sin(\alpha + \beta) &= \sin[\alpha - (-\beta)] \\ &= \sin\alpha\cos(-\beta) - \cos\alpha\sin(-\beta) \end{aligned}$$

$$\sin(\alpha + \beta) = \sin\alpha\cos\beta + \cos\alpha\sin\beta$$

The following identity for $\cos(\alpha + \beta)$ can be derived in a similar fashion. See Prob. 39 of Exercise 14.3.

$$\cos(\alpha + \beta) = \cos\alpha\cos\beta - \sin\alpha\sin\beta$$

The following equations summarize our results.

$$\sin(\alpha+\beta) = \sin\alpha\cos\beta + \cos\alpha\sin\beta \tag{14.15a}$$

$$\sin(\alpha-\beta) = \sin\alpha\cos\beta - \cos\alpha\sin\beta \tag{14.15b}$$

$$\cos(\alpha+\beta) = \cos\alpha\cos\beta - \sin\alpha\sin\beta \tag{14.16a}$$

$$\cos(\alpha-\beta) = \cos\alpha\cos\beta + \sin\alpha\sin\beta \tag{14.16b}$$

Example 14.11. Verify the identity for $\cos(\alpha+\beta)$ by using $\alpha = 60°$, $\beta = 90°$, and Table 14.1 to find $\cos 150°$.

$$\begin{aligned}\cos 150° = \cos(60°+90°) &= \cos 60° \cos 90° - \sin 60° \sin 90° \\ &= \frac{1}{2}(0) - \frac{\sqrt{3}}{2}(1)\end{aligned}$$

$$\cos 150° = -\frac{\sqrt{3}}{2}$$

Example 14.12. Find a value for $\cos 75°$ by using Table 14.1 and $75° = 120° - 45°$.

$$\begin{aligned}\cos 75° &= \cos(120° - 45°) \\ &= \cos 120° \cos 45° + \sin 120° \sin 45° \\ &= \left(-\frac{1}{2}\right)\left(\frac{\sqrt{2}}{2}\right) + \left(\frac{\sqrt{3}}{2}\right)\left(\frac{\sqrt{2}}{2}\right)\end{aligned}$$

$$\cos 75° = \frac{\sqrt{2}}{4}(\sqrt{3}-1)$$

Example 14.13. Show that $\sin(\pi/6 + x) = (\cos x + \sqrt{3}\sin x)/2$.

$$\begin{aligned}\sin\left(\frac{\pi}{6}+x\right) &= \sin\frac{\pi}{6}\cos x + \cos\frac{\pi}{6}\sin x \\ &= \frac{1}{2}\cos x + \frac{\sqrt{3}}{2}\sin x \\ &= \frac{\cos x + \sqrt{3}\sin x}{2}\end{aligned}$$

Example 14.14. Change $\cos 3x \sin x - \sin 3x \cos x$ to a single term. Note that by commuting the terms we have

$$\begin{aligned}\cos 3x \sin x - \sin 3x \cos x &= -\sin 3x \cos x + \cos 3x \sin x \\ &= -(\sin 3x \cos x - \cos 3x \sin x)\end{aligned}$$

$$= -\sin(3x - x) \qquad \text{by Eq. (14.15}b\text{)}$$
$$= -\sin 2x$$

Example 14.15. Show that

$$\sec x \csc y \sin(x - y) = \tan x \cot y - 1$$

$$\begin{aligned}\sec x \csc y \sin(x - y) &= \frac{1}{\cos x}\frac{1}{\sin y}(\sin x \cos y - \cos x \sin y)\\ &= \frac{\sin x \cos y}{\cos x \sin y} - \frac{\cos x \sin y}{\cos x \sin y}\\ &= \tan x \cot y - 1\end{aligned}$$

Example 14.16. Derive an identity for $\tan(\alpha + \beta)$ in terms of $\tan \alpha$ and $\tan \beta$.

$$\tan(\alpha + \beta) = \frac{\sin(\alpha + \beta)}{\cos(\alpha + \beta)} = \frac{\sin \alpha \cos \beta + \cos \alpha \sin \beta}{\cos \alpha \cos \beta - \sin \alpha \sin \beta}$$

Divide the numerator and the denominator by $\cos \alpha \cos \beta$.

$$\tan(\alpha + \beta) = \frac{\dfrac{\sin \alpha \cos \beta}{\cos \alpha \cos \beta} + \dfrac{\cos \alpha \sin \beta}{\cos \alpha \cos \beta}}{\dfrac{\cos \alpha \cos \beta}{\cos \alpha \cos \beta} - \dfrac{\sin \alpha \sin \beta}{\cos \alpha \cos \beta}}$$

$$\boxed{\tan(\alpha + \beta) = \frac{\tan \alpha + \tan \beta}{1 - \tan \alpha \tan \beta}}$$

As a result of Example 14.16 and other similar derivations (Probs. 40 and 41 of Exercise 14.3), we have the following identities:

$$\tan(\alpha \pm \beta) = \frac{\tan \alpha \pm \tan \beta}{1 \mp \tan \alpha \tan \beta} \qquad (14.17)$$

$$\cot(\alpha \pm \beta) = \frac{\cot \alpha \cot \beta \mp 1}{\cot \beta \pm \cot \alpha} \qquad (14.18)$$

EXERCISE 14.3

In Probs. 1 to 12, use Table 14.1 and one of Eqs. (14.15) to (14.18). Evaluate each given function by using the given sum or difference of angles.

1.	sin 150°,	60° + 90°	2.	cos 240°,	60° + 180°
3.	sin 210°,	240° − 30°	4.	cot 315°,	90° + 225°
5.	$\tan \frac{5\pi}{4}$,	$\pi + \frac{\pi}{4}$	6.	$\sin \frac{2\pi}{3}$,	$\frac{\pi}{6} + \frac{\pi}{2}$
7.	$\cos \frac{5\pi}{6}$,	$\frac{7\pi}{6} - \frac{\pi}{3}$	8.	$\tan \frac{11\pi}{6}$,	$\frac{7\pi}{6} + \frac{2\pi}{3}$
9.	sin 105°,	60° + 45°	10.	sin 105°,	150° − 45°
11.	$\cos \frac{\pi}{12}$,	$\frac{\pi}{3} - \frac{\pi}{4}$	12.	$\sin \frac{13\pi}{12}$,	$\frac{13\pi}{3} - \frac{13\pi}{4}$

In Probs. 13 to 20, prove each identity by transforming the left side into the right side.

13. $\cos\left(\frac{\pi}{6} + x\right) = \frac{\sqrt{3}\cos x - \sin x}{2}$

14. $\sin\left(x - \frac{\pi}{4}\right) = \frac{\sin x - \cos x}{\sqrt{2}}$

15. $\sin(x + \pi/2) = \cos x$

16. $\cos(x - 3\pi/2) = -\sin x$

17. $\cot(\theta + 90°) = -\tan\theta$

18. $\sin(\theta - 90°) = -\cos\theta$

19. $\cos(x - \pi) = -\cos x$

20. $\tan(\theta + 180°) = \tan\theta$

21. In calculus, to find the "derivative" of sin u, we have to change $\cos(u + \theta)(\sin\theta)/\theta$ to $(\cos u \cos\theta - \sin u \sin\theta)(\sin\theta)/\theta$. Verify that change.

22. Figure 14.7 shows a pendulum being used as an accelerometer in a car that is braking on a downhill slope. If the slope angle is θ and the pendulum angle is β, then the deceleration due to braking is

$$a = g\,\frac{\sin(\beta - \theta)}{\cos\beta}$$

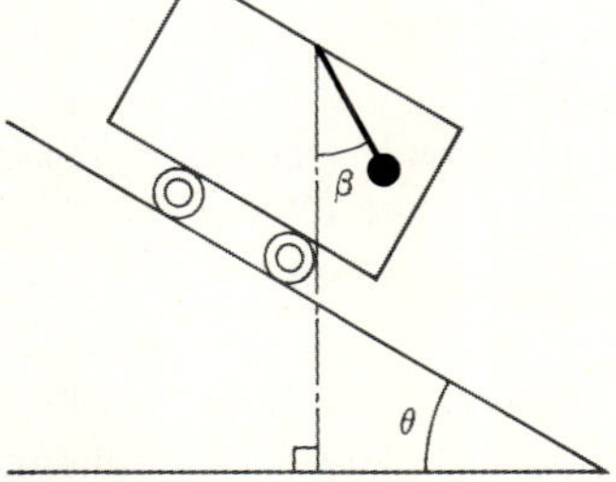

Fig. 14.7 Prob. 22.

Show that $a = g\tan\beta\cos\theta - g\sin\theta$.

23. The relationship between the bank angle, ϕ, the "coefficient of friction," tan μ, and the "centrifugal acceleration," A, of a race

car on a banked track is

$$A = g\,\frac{\tan\phi + \tan\mu}{1 - \tan\phi\,\tan\mu}$$

Show that $A = g\tan(\phi + \mu)$.

24. For the car of Prob. 23, the speed v at which sliding begins is given by

$$\frac{v^2}{gr} = \tan(\phi + \mu)$$

If $g = 32.2$ ft/sec², $r = 1{,}000$ ft, and $\tan\mu = 0.4$, what is the minimum bank angle ϕ that will prevent a 120-mph (176-ft/sec) car from sliding?

25. A weight W suspended by a spring is pulled a distance y_0 and released. If friction could be neglected, the position of the weight as a function of time would be

$$y = A\cos\left(\frac{kg}{W}t + \phi\right) + B\sin\left(\frac{kg}{W}t + \phi\right)$$

where the initial displacement is $y_0 = \sqrt{A^2 + B^2}$, $\tan\phi = A/B$, and k is the "spring rate." Show that the displacement function can be written as $y = y_0\sin(2\phi + kgt/W)$. (*Hint:* With a diagram show that $A = y_0\sin\phi$ and $B = y_0\cos\phi$.)

In Probs. 26 to 31, change each expression to a single term. See Example 14.14.

26. $\cos 3x\cos x - \sin 3x\sin x$

27. $\sin x\cos 2x + \cos x\sin 2x$

28. $\dfrac{\tan 5x - \tan 2x}{1 + \tan 5x\tan 2x}$

29. $\cos^2 x - \sin^2 x$ (*Hint:* $\sin^2 x = \sin x\sin x$ and $\cos^2 x = \cos x\cos x$.)

30. $\sin(x + y)\cos y - \cos(x + y)\sin y$

31. $\sin^2 x\cos^2 y - \cos^2 x\sin^2 y$ (*Hint:* Factor the difference between two squares.)

32. To "integrate" $\sin\alpha\sin\beta$, we change it to $\frac{1}{2}[\cos(\alpha - \beta) - \cos(\alpha + \beta)]$. Verify this change by subtracting Eq. (14.16*a*) from Eq. (14.16*b*).

33. To "integrate" $\sin 5x\cos 3x$, we can change it to $\frac{1}{2}(\sin 8x + \sin 2x)$. Use Eqs. (14.15*a*) and (14.15*b*) to verify this.

34. Problem 32 illustrates a product identity. Derive three other

product identities by combining Eqs. (14.15*a*) and (14.15*b*) or Eqs. (14.16*a*) and (14.16*b*).

35. Show that $\sin u + \sin v = 2 \sin [\frac{1}{2}(u + v)] \cos [\frac{1}{2}(u - v)]$ by adding Eqs. (14.15*a*) and (14.15*b*). (*Hint:* Let $\alpha + \beta = u$ and $\alpha - \beta = v$, and solve for α and β.)
36. Problem 35 illustrates a sum identity. Derive the other sum identity and the two difference identities below by combining Eqs. (14.15*a*) and (14.15*b*) or Eqs. (14.16*a*) and (14.16*b*).

$$\cos u + \cos v = 2 \cos \tfrac{1}{2}(u + v) \cos \tfrac{1}{2}(u - v)$$

$$\sin u - \sin v = 2 \cos \tfrac{1}{2}(u + v) \sin \tfrac{1}{2}(u - v)$$

$$\cos u - \cos v = -2 \sin \tfrac{1}{2}(u + v) \sin \tfrac{1}{2}(u - v)$$

37. Show that $\sec x \csc y \cos (x + y) = \cot y - \tan x$.
38. Derive the identity for $\sin (\alpha - \beta)$, Eq. (14.15*b*), by using Fig. 14.8. Use a procedure similar to the one in this section for $\cos (\alpha - \beta)$. (*Hint:* From Rt $\triangle OP_1R$ notice that $\sin \beta = -x_1/r$ and $\cos \beta = y_1/r$.)

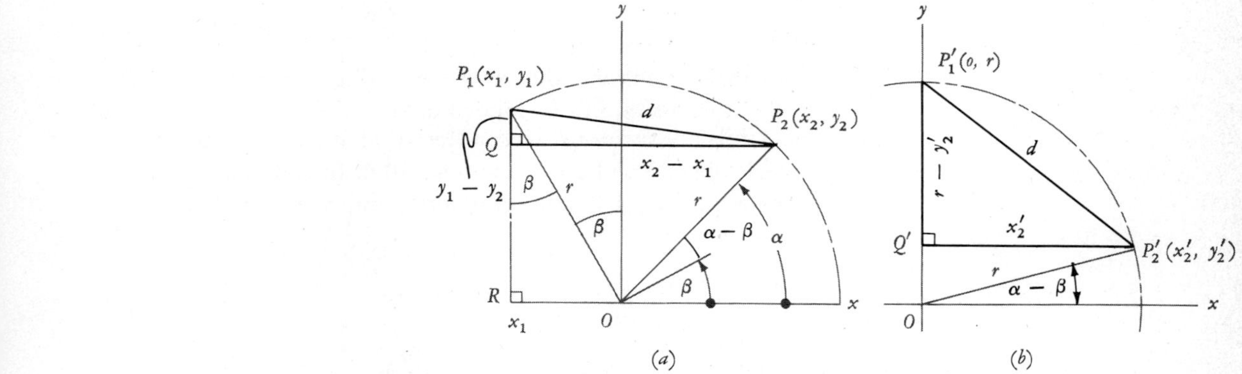

Fig. 14.8 Prob. 38. (a); (b) $\triangle OP_1P_2$ of part (a) is revolved so that $(\alpha - \beta)$ is in standard position, and $\overline{OP}_1$, now $\overline{OP}_1'$, is on the +y axis.

39. Derive the identity for $\cos (\alpha + \beta)$ by using the negative-angle identities and the identity for $\cos (\alpha - \beta)$.
40. Derive the identity for $\tan (\alpha - \beta)$ by using the result of Example 14.16 and the negative-angle identities.
41. Derive Eq. (14.18) by using procedures similar to those in Example 14.16 and Prob. 40.
42. Observe Probs. 15 to 18. See if you can discover a general reduction formula for any TF $(k\pi/2 \pm \theta)$ or TF $(k \cdot 90° \pm \theta)$ when k is an odd integer.

43. Observe Probs. 19 and 20. See if you can discover a general reduction formula for any TF $(k\pi/2 \pm \theta)$ or TF $(k \cdot 90° \pm \theta)$ when k is an even integer.

14.4 DOUBLE-ANGLE IDENTITIES

It is a very easy step to derive identities for double-angle functions from the sum functions. For example, we can think of 2α as $\alpha + \alpha$; then we apply the $\sin(\alpha + \beta)$ identity, Eq. (14.15*a*), to $\sin(\alpha + \alpha)$ as follows:

$$\begin{aligned}\sin 2\alpha = \sin(\alpha + \alpha) &= \sin\alpha\cos\alpha + \cos\alpha\sin\alpha \\ &= \sin\alpha\cos\alpha + \sin\alpha\cos\alpha \\ &= 2\sin\alpha\cos\alpha\end{aligned}$$

By a similar application of Eq. (14.16*a*) we have

$$\begin{aligned}\cos 2\alpha = \cos(\alpha + \alpha) &= \cos\alpha\cos\alpha - \sin\alpha\sin\alpha \\ &= \cos^2\alpha - \sin^2\alpha \\ &= 2\cos^2\alpha - 1 \\ &= 1 - 2\sin^2\alpha\end{aligned}$$

The last two equations are the results of using $\sin^2\alpha = 1 - \cos^2\alpha$ and $\cos^2\alpha = 1 - \sin^2\alpha$. [Remember Eq. (14.9)?]

By similar derivations we can derive identities for $\tan 2\alpha$ and $\cot 2\alpha$. See Probs. 23 and 24 of Exercise 14.4. In summary, we have the following identities for trigonometric functions of 2 times an angle.

$$\sin 2\alpha = 2\sin\alpha\cos\alpha \qquad (14.19)$$

$$\begin{aligned}\cos 2\alpha &= \cos^2\alpha - \sin^2\alpha && (14.20a)\\ &= 2\cos^2\alpha - 1 && (14.20b)\\ &= 1 - 2\sin^2\alpha && (14.20c)\end{aligned}$$

$$\tan 2\alpha = \frac{2\tan\alpha}{1 - \tan^2\alpha} \qquad (14.21)$$

$$\cot 2\alpha = \frac{\cot^2\alpha - 1}{2\cot\alpha} \qquad (14.22)$$

Comment: *After all this derivation, please don't turn around and write 2 sin α the next time you encounter sin 2α. Equation (14.19) shows very clearly that we cannot simply remove the 2 from inside the function sin 2α and place it outside as 2 sin α.*

If it were that simple, why would we bother with all the definitions and derivations that lead to Eq. (14.19)? Be equally circumspect about the other identities.

Example 14.17. Find sin 60° by using $\alpha = 30°$ in Eq. (14.19).

$$\sin 60° = \sin (2 \cdot 30°) = 2 \sin 30° \cos 30°$$
$$= 2\left(\frac{1}{2}\right)\left(\frac{\sqrt{3}}{2}\right)$$

$$\sin 60° = \frac{\sqrt{3}}{2}$$

Example 14.18. Find $\cos (4\pi/3)$ by using $\alpha = 2\pi/3$ in Eq. (14.20*a*).

$$\cos \frac{4\pi}{3} = \cos\left(2 \cdot \frac{2\pi}{3}\right) = \cos^2 \frac{2\pi}{3} - \sin^2 \frac{2\pi}{3}$$
$$= \left(-\frac{1}{2}\right)^2 - \left(\frac{\sqrt{3}}{2}\right)^2$$
$$= \frac{1}{4} - \frac{3}{4}$$

$$\cos \frac{4\pi}{3} = -\frac{1}{2}$$

Example 14.19. Change $\cos^2 3\theta - \sin^2 3\theta$ to a single function.

By comparing this expression to Eq. (14.20*a*), we see that 3θ takes the place of α. Therefore we have

$$\cos^2 3\theta - \sin^2 3\theta = \cos (2 \cdot 3\theta) = \cos 6\theta$$

Example 14.20. In strength of materials, engineers have found a relationship between "shear stress," S_s, and "axial stress," S_x, to be

$$S_s = S_x \sin \theta \cos \theta$$

The engineers have also found that the form $S_s = (S_x/2) \sin 2\theta$ is more useful. They derive this form by recognizing that

$$\sin \theta \cos \theta = \tfrac{1}{2}(2 \sin \theta \cos \theta) = \tfrac{1}{2} \sin 2\theta$$

Example 14.21. In calculus, to "integrate" $(1 - x^2)/(1 + x^2)$, we can let $x = \tan \theta$, and change $(1 - x^2)/(1 + x^2)$ to $\cos 2\theta$ as follows:

$$\frac{1 - x^2}{1 + x^2} = \frac{1 - \tan^2 \theta}{1 + \tan^2 \theta} = \frac{1 - \sec^2 \theta + 1}{\sec^2 \theta} = \frac{2}{\sec^2 \theta} - \frac{\sec^2 \theta}{\sec^2 \theta}$$

$$\frac{1-x^2}{1+x^2} = 2\cos^2\theta - 1 = \cos 2\theta$$

Example 14.22. **Find $\cos 2\theta$ if $\sin\theta = 2/3$.**

$$\cos 2\theta = 1 - 2\sin^2\theta = 1 - 2(2/3)^2 = 1 - 8/9 = 1/9$$

EXERCISE 14.4

In Probs. 1 to 6, evaluate each given function by using the given angle and one of the double-angle identities.

1. $\cos 60°$, $30°$
2. $\tan(2\pi/3)$, $\pi/3$
3. $\sin(\pi/2)$, $\pi/4$
4. $\sin 180°$, $90°$
5. $\cot 90°$, $45°$
6. $\cos(3\pi/2)$, $3\pi/4$

In Probs. 7 to 12, change each given expression to an expression containing a single term or factor. See Example 14.19.

7. $1 - 2\sin^2 3\theta$
8. $3 - 6\sin^2\theta$
9. $-10\sin(-\alpha)\cos(-\alpha)$
10. $\dfrac{4\tan 2\beta}{2 - 2\tan^2 2\beta}$
11. $\dfrac{1-\cot^2 4x}{2\cot 4x}$
12. $3\sin^2 2x - 3\cos^2 2x$

13. Find $\sin 2\theta$ if $\sin\theta = 3/5$, $\pi/2 < \theta < \pi$.
14. Find $\tan 2y$ if $\cot y = 1/2$.
15. Find $\cos 2\alpha$ if $\cos\alpha = 1/5$.
16. Find $\sin 4\beta$ if $\cos\beta = 5/13$, $3\pi/2 < \beta < 2\pi$. [*Hint:* $\sin 4\beta = \sin(2\cdot 2\beta)$.]
17. Show that

$$\frac{\cot 2\theta}{\tan\theta} = \frac{\cot^2\theta - 1}{2}$$

18. Engineers have found that a relationship between "normal stress," S_n, and "axial stress," S_x, is $S_n = S_x\cos^2\theta$. Show how they can change this to

$$S_n = \frac{S_x}{2} + \frac{S_x}{2}\cos 2\theta$$

19. To "integrate" $(x^2-1)/(x^2+1)$, we can let $x = \cot\theta$, and change $(x^2-1)/(x^2+1)$ to $\cos 2\theta$. Show how this is done. See Example 14.21.

20. To "integrate" $\sin^2\theta$ in calculus, we can change it to

$$\tfrac{1}{2}(1 - \cos 2\theta)$$

Verify this change.

21. If friction is neglected, the maximum range of a bullet fired with a muzzle velocity of v and an elevation angle θ (from the horizontal) is

$$R = \frac{2v^2 \sin\theta \cos\theta}{g}$$

Show that

$$R = \frac{v^2 \sin 2\theta}{g}$$

22. In analytic geometry, a formula used to "rotate the axes" is

$$2(C - A) \sin\alpha \cos\alpha + B(\cos^2\alpha - \sin^2\alpha) = 0$$

Show that

$$\tan 2\alpha = \frac{B}{A - C}$$

23. Use Eq. (14.17) and $2\alpha = \alpha + \alpha$ to derive Eq. (14.21).
24. Use Eq. (14.18) to derive Eq. (14.22).
25. Derive an identity for $\sin 3\theta$ in terms of $\sin\theta$ and $\cos\theta$. [*Hint:* $\sin 3\theta = \sin(\theta + 2\theta)$.]

14.5 HALF-ANGLE IDENTITIES

Equation (14.20*b*), repeated below, gives a relationship between the cosine of twice an angle and the cosine of the angle. We called it a double-angle identity.

$$\cos 2\alpha = 2\cos^2\alpha - 1 \qquad (14.20b)$$

We can also think of this equation as being a half-angle identity, because α is one-half of 2α. To make it look more like a half-angle relationship, we replace α by $\alpha/2$. Then 2α is replaced by α, and the equation becomes

$$\cos\alpha = 2\cos^2\frac{\alpha}{2} - 1$$

We solve for $\cos(\alpha/2)$ as follows:

$$2\cos^2\frac{\alpha}{2} - 1 = \cos\alpha$$

$$2\cos^2\frac{\alpha}{2} = 1 + \cos\alpha$$

$$\cos^2\frac{\alpha}{2} = \frac{1+\cos\alpha}{2}$$

$$\cos\frac{\alpha}{2} = \pm\sqrt{\frac{1+\cos\alpha}{2}}$$

The algebraic sign is chosen according to the quadrant for $\alpha/2$.

Example 14.23. $\cos 30° = \cos\frac{60°}{2} = +\sqrt{\frac{1+\cos 60°}{2}} = \sqrt{\frac{1+1/2}{2}} = \sqrt{\frac{2+1}{2(2)}}$

$$\cos 30° = \frac{\sqrt{3}}{2}$$

Example 14.24. $\cos\frac{2\pi}{3} = \cos\left[\frac{1}{2}\left(\frac{4\pi}{3}\right)\right] = -\sqrt{\frac{1+\cos(4\pi/3)}{2}} = -\sqrt{\frac{1-1/2}{2}}$

$$\cos\frac{2\pi}{3} = -\sqrt{\frac{1/2}{2}} = -\frac{1}{2}$$

By rearranging Eq. (14.20*c*) in a similar way, we can derive the half-angle sine identity below.

$$\sin\frac{\alpha}{2} = \pm\sqrt{\frac{1-\cos\alpha}{2}}$$

These two equations are the half-angle identities.

$$\cos\frac{\alpha}{2} = \pm\sqrt{\frac{1+\cos\alpha}{2}} \tag{14.23}$$

$$\sin\frac{\alpha}{2} = \pm\sqrt{\frac{1-\cos\alpha}{2}} \tag{14.24}$$

Example 14.25. $\sin\frac{7\pi}{12} = \sin\left[\frac{1}{2}\left(\frac{7\pi}{6}\right)\right] = +\sqrt{\frac{1-\cos(7\pi/6)}{2}} = \sqrt{\frac{1-(-\sqrt{3}/2)}{2}}$

$$\sin\frac{7\pi}{12} = \sqrt{\frac{2+\sqrt{3}}{2(2)}} = \frac{1}{2}\sqrt{2+\sqrt{3}}$$

Example 14.26. Simplify $\sqrt{1-\cos 8\theta}$.

$$\sqrt{1-\cos 8\theta} = \sqrt{\frac{2(1-\cos 8\theta)}{2}} = \sqrt{2}\sqrt{\frac{1-\cos 8\theta}{2}} = \sqrt{2}\sin\left[\frac{1}{2}(8\theta)\right]$$

$$\sqrt{1-\cos 8\theta} = \sqrt{2}\sin 4\theta$$

Example 14.27. In analytic geometry, to "rotate the axes" of a coordinate system when $\tan 2\phi = -4/3$, we have to find $\sin\phi$ and $\cos\phi$, $\pi/2 < 2\phi < 3\pi/4$.

First we find $\cos 2\phi$ by the diagram method. In Fig. 14.9, we let $x = -3$ and $y = 4$ so that $\tan 2\phi = -4/3$. Then $r = 5$, and $\cos 2\phi = -3/5$. From Eqs. (14.23) and (14.24) we have

$$\cos\phi = \sqrt{\frac{1+\cos 2\phi}{2}} = \sqrt{\frac{1-3/5}{2}} = \frac{\sqrt{5}}{5}$$

$$\sin\phi = \sqrt{\frac{1-\cos 2\phi}{2}} = \sqrt{\frac{1-(-3/5)}{2}} = \frac{2\sqrt{5}}{5}$$

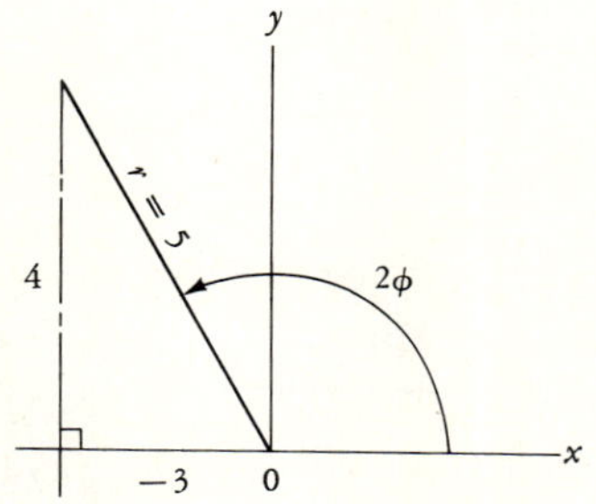

Fig. 14.9 Example 14.27.

EXERCISE 14.5

In Probs. 1 to 4, use the half-angle identities to evaluate each function.

1. $\sin 15°$ 2. $\cos(\pi/2)$ 3. $\tan(\pi/8)$ 4. $\sin(13\pi/12)$
5. Derive Eq. (14.24).

In Probs. 6 to 9, evaluate each function. See Example 14.27.

6. $\sin\theta$ if $\cos 2\theta = 1/4$, $0 < \theta < \pi/2$
7. $\cos(\beta/2)$ if $\sin\beta = -4/5$, $\pi < \beta < 3\pi/2$

8. $\tan \frac{\phi}{2}$ if $\cos \phi = -\frac{\sqrt{2}}{2}$, $\pi < \phi < 3\pi/2$

9. $\sin \alpha$ if $\cot 2\alpha = -4/3$, $0 < \alpha < \pi/2$

In Probs. 10 to 15, simplify each expression. See Example 14.26.

10. $\sqrt{\frac{1 + \cos 8\theta}{2}}$

11. $\sqrt{\frac{3 - 3 \cos 2\alpha}{6}}$

12. $-\sqrt{\frac{9 + 9 \cos 4\theta}{2}}$

13. $\sqrt{2 + 2 \cos 6\alpha}$

14. $\sqrt{18 - 18 \cos 2\beta}$

15. $\sqrt{\frac{1 - \cos 2\phi}{1 + \cos 2\phi}}$

16. Figure 14.10 shows an oscillating semicylinder. The angular speed as the cylinder crosses the "equilibrium position" is given by

$$\omega = 4 \sqrt{\frac{g}{r} \frac{1 - \cos \alpha}{9\pi - 16}}$$

Show that this can be written as

$$\omega = 4 \sin \frac{\alpha}{2} \sqrt{\frac{2g}{r(9\pi - 16)}}$$

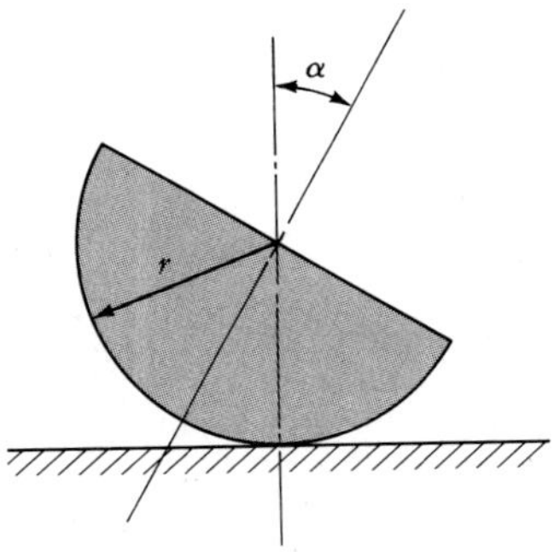

Fig. 14.10 Prob. 16.

17. In fluid mechanics, it is shown that the force required to hold a "deflection vane" in a stream is

$$F = 2\rho A v^2 \sin^2 \frac{\phi}{2}$$

Show that $F = \rho A v^2 (1 - \cos \phi)$.

18. In calculus, to "integrate" $(1 - x^2)/(1 + x^2)$, we can let $x = \tan (\theta/2)$. Show that $(1 - x^2)/(1 + x^2)$ becomes $\cos \theta$.

19. In electronics, to find the relation between "root-mean-square" voltage and "peak" voltage, we have to "integrate" $\sin^2 \omega t$. To "integrate" $\sin^2 \omega t$ we change it to $1/2 - 1/2 \cos 2\omega t$. Show how to make this change.

GRAPHS OF TRIGONOMETRIC FUNCTIONS

CHAPTER 15

Graphs provide one of the clearest means for understanding properties of the trigonometric functions. With graphs, we can literally see how the functions behave, we can demonstrate identities, and we can visualize certain problems.

15.1 BASIC IDEAS ABOUT GRAPHING ANGLES

On a cartesian graph of a trigonometric function, values of the angle are placed on the abscissa (x axis), and values of the function are placed on the ordinate (y axis). For example, to graph $\sin \theta$, values of θ are placed on the "x axis," and values of $\sin \theta$ on the "y axis."

Figure 15.1 helps us visualize how angles can be scaled on a straight-line axis. The figure depicts a taut cord pinned to a circle at point O and wrapped around the circle. Each point where the cord comes tangent to the circle can be associated with the angle θ and with the distance x through which the cord has been wrapped. We can label each point with the value of its angle θ. For example, $\pi/2$ (radians) is labeled at the point one-quarter of the way around, and θ is labeled at the arbitrary point shown. Suppose we unwrap the cord, place point O at the origin of a coordinate system, and stretch the cord along the x axis. Then each point of the cord will mark its corresponding angle on the x axis, at the distance x from the origin.

Notice that the full circumference of the circle represents an angle of 2π (radians) or 1 "cycle" of the cord around the circle. We can generate positive angles 4π, 6π, . . . without limits by continuing to wrap the cord around and around, and we can generate negative angles -2π, -4π, -6π, . . . without limits, by wrapping the cord in the opposite (clockwise) direction.

If we use θ for angles and place values of θ on the abscissa, we should label it the "θ axis." But since the abscissa is commonly labeled

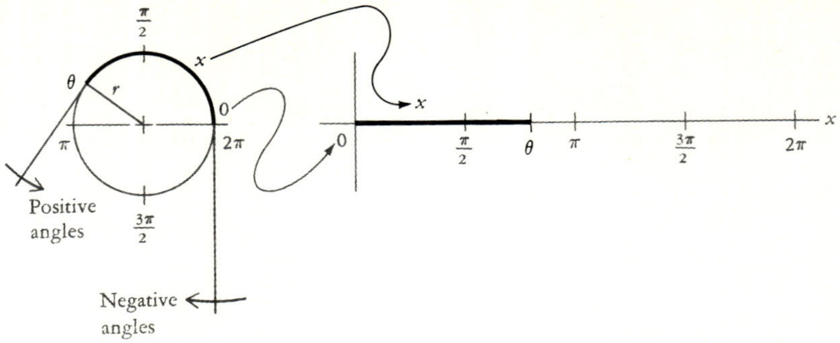

Fig. 15.1 Angle θ in standard position can be represented by distance x on the x axis.

"x axis," we shall sometimes use x for angles instead of for arcs (as we did in Fig. 15.1).* There are no stringent rules about labeling of angles and axes, and so we shall use rather loose notation and not limit ourselves to any one system.

Radian measure will be used exclusively, since nearly all applications of graphs of trigonometric functions deal with angles in radians.

15.2 GRAPHS OF A SIN θ AND A COS θ

Figure 15.2 shows how we can generate a graph of $A \sin \theta$. We imagine the radius A revolving around the circle. We locate each angle, such as $\pi/3$, on the abscissa of the graph (as we did in Fig. 15.1), and we project the corresponding ordinate, such as $y = A \sin(\pi/3) = A\sqrt{3}/2$, from the circle to the graph. As the radius completes 1 rev (revolu-

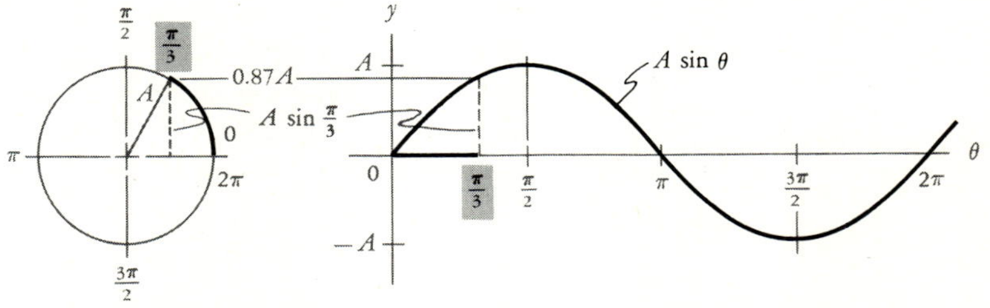

Fig. 15.2 One cycle of $A \sin \theta$. The graph of $A \sin \theta$ can be generated by projecting the ordinate from the circle to the graph.

* There is a whole system of trigonometry in which the trigonometric functions are based upon the arclengths x (for arcs on a unit circle, radius = 1), instead of angles. Such functions are sometimes called *circular functions.* Notice that for a *unit* circle, the arclength and the corresponding angle in radians have the same value. For example, the circumference is 2π and the corresponding angle is also 2π.

tion), it generates the graph of $A \sin \theta$ for 1 cycle. Because of the shape of the curve, such graphs are called *sine waves* or *sinusoidal waves.*

We could generate a similar cosine wave by taking the horizontal projection $A \cos \theta$, turning it to a vertical position, and placing it on the graph. Figure 15.3 shows a graph of $A \cos \theta$ for 1 cycle.

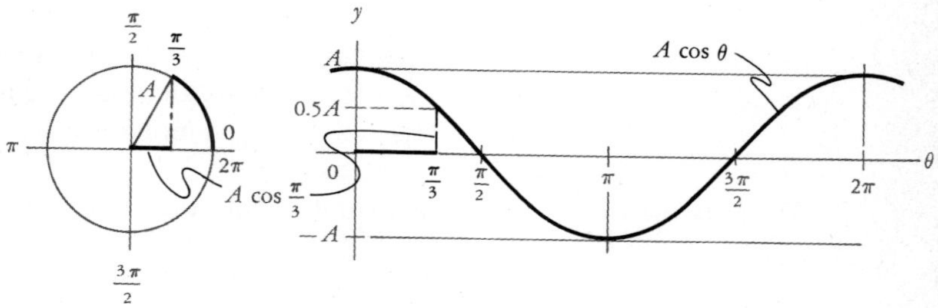

Fig. 15.3 One cycle of A cos θ.

The term *cycle* is used above because each succeeding revolution of the radius generates another wave identical to the first one. Figure 15.4 shows 5 cycles both of $\sin \theta$ and of $\cos \theta$. Each of the identical, repetitive segments or waves is called 1 *cycle* of the graph. Notice that each such wave covers a distance of 2π units on the abscissa. Although we usually think of the cycle beginning at $\theta = 0$ (and ending at $\theta = 2\pi$), it can begin anywhere. For example, we could have a cycle beginning at $\theta = -3\pi/4$ and ending at $\theta = 5\pi/4$, as shown in Fig. 15.4.

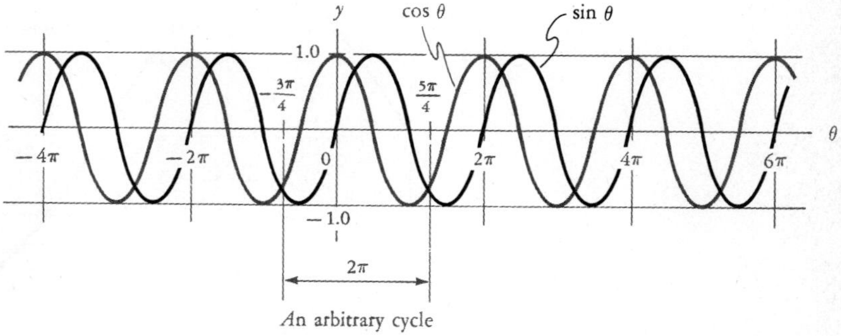

Fig. 15.4 Five cycles of sin θ and of cos θ. Each cycle is 2π units long on the θ axis. Cycles can begin at any point.

The absolute value of A, in $A \sin \theta$ or $A \cos \theta$, is called the *amplitude* of the function. For simple functions such as these, the amplitude is the maximum value of the function, or the maximum distance of the wave from the abscissa. Figures 15.2 to 15.4 illustrate this feature. Later, for more complicated functions we shall need a more rigorous definition. In any case, the sole effect of the amplitude is to stretch or compress the graph in the "vertical" direction. This effect is especially prominent in Fig. 15.6.

In actual practice, we do not construct a graph of $A \sin \theta$ by drawing the circle and projecting points from it—that is only a means for understanding the graph. Instead, we plot points from a table. Table 14.1 is an extremely convenient one because just a few values need be memorized, and as shown in Example 15.1, these values give symmetric sets of points all along the graph.

The following examples are graphs of $A \sin \theta$ and $A \cos \theta$.

Example 15.1. **To graph $5 \sin \theta$, we use the table below.**

θ	0	$\frac{\pi}{6}$	$\frac{\pi}{4}$	$\frac{\pi}{3}$	$\frac{\pi}{2}$
$\sin \theta$	0	0.50	$\sqrt{2}/2 \approx 0.71$	$\sqrt{3}/2 \approx 0.87$	1
$5 \sin \theta$	0	2.50	$5\sqrt{2}/2 \approx 3.55$	$5\sqrt{3}/2 \approx 4.35$	5

We need use only first-quadrant angles, because they are reference angles for the corresponding angles in other quadrants. Also, as shown in Fig. 15.5, the corresponding angles are symmetric about

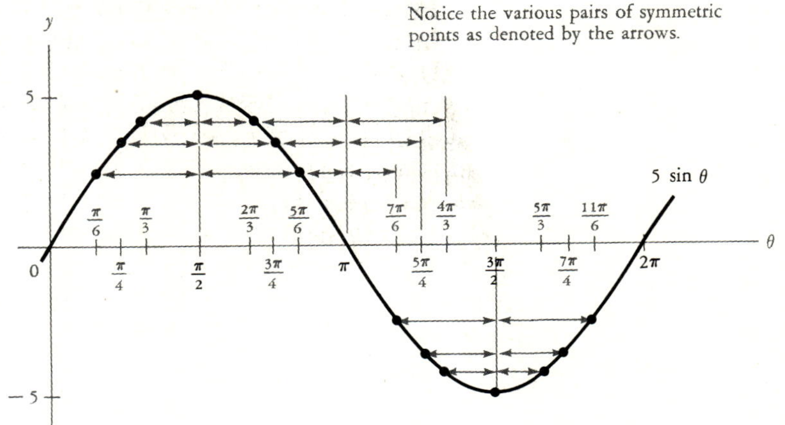

Fig. 15.5 Example 15.1. 5 sin θ.

$\pi/2$, π, and $3\pi/2$. For example, $\pi/3$ and $2\pi/3$ are the same distance ($\pi/6$ units) on either side of $\pi/2$, $2\pi/3$ and $4\pi/3$ are symmetric about π, and $4\pi/3$ and $5\pi/3$ are symmetric about $3\pi/2$. The "$\pi/6$ set" of angles and the "$\pi/4$ set" of angles are likewise symmetric. Once we recognize this symmetry, we simply "step our way" through the second, third, and

fourth quadrants by reflecting the symmetric points of the preceding quadrant.

Example 15.2. To graph $-4 \cos x$, we use the table below. And we use the same technique for plotting symmetric points as in Example 15.1. In Fig. 15.6, we have not even labeled all such points. You should be able to identify them. The amplitude is $|-4| = 4$. Notice that the negative coefficient reflects the curve about the x axis. The graph of $-\cos x$ is included for comparison.

x	0	$\frac{\pi}{6}$	$\frac{\pi}{4}$	$\frac{\pi}{3}$	$\frac{\pi}{2}$
$\cos x$	1	0.87	0.71	0.5	0
$-4 \cos x$	-4	-3.48	-2.84	-2.0	0

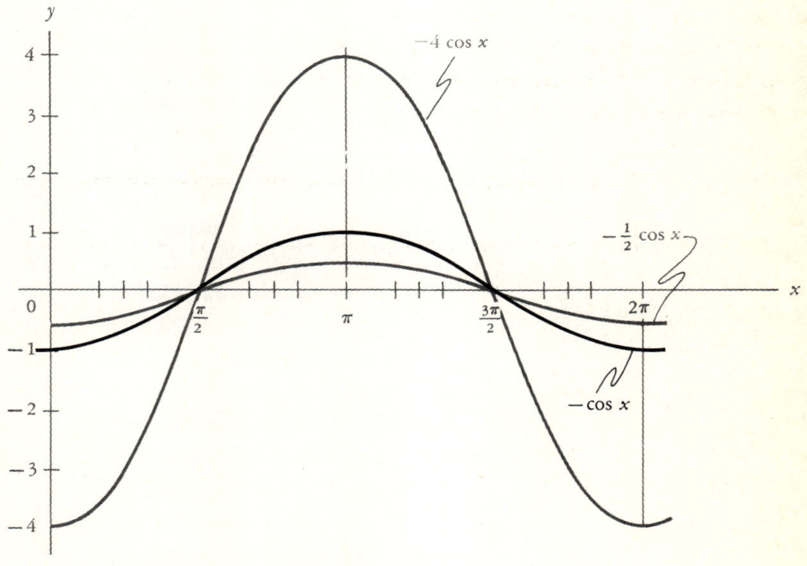

Fig. 15.6 Examples 15.2 and 15.3.

Example 15.3. To graph $-(1/2) \cos x$, we can halve the ordinates for the graph of $-\cos x$ in Fig. 15.6.

EXERCISE 15.1

Graph each function for the number of cycles indicated. Begin at the specified angle and graph in the positive x direction.

1. 1 cycle of 2 sin θ beginning at $\theta = 0$
2. 1 cycle of 2 cos θ beginning at $\theta = 0$
3. 2 cycles of (1/2) cos θ beginning at $\theta = -2\pi$
4. 3 cycles of (1/4) sin θ beginning at $\theta = -2\pi$
5. 2 cycles of 5 sin $(-\theta)$ beginning at $\theta = -\pi$ [*Hint:* Use the identity sin $(-\theta) = -$sin θ.]
6. 1 cycle of -0.60 cos x beginning at $x = 0$
7. 3 cycles of 2.50 cos t beginning at $t = -2\pi$
8. 2 cycles of 120 sin ω beginning at $\omega = 0$

15.3 GRAPHS OF *A* SIN *kx* AND *A* COS *kx*

We discovered in Sec. 15.2 that graphs of A sin θ and A cos θ are cyclic—that is, they consist of a basic wave, or cycle, that is repeated over and over. Furthermore, we found that each cycle covered a distance of 2π units on the "x axis." This distance, the minimum value of θ for which 1 cycle occurs, is called the *period* of the function. And functions that repeat, or are cyclic, are called *periodic functions.* Thus sin θ and cos θ are periodic functions of period 2π.

Sines and cosines are very important because they are used to describe any phenomenon that is periodic. Examples are voltages and currents in ac (alternating current) electric circuits, radar waves, light waves, sound waves, pressure pulses in hydraulic mechanisms (e.g., power steering), stress and strain in structures and mechanisms (e.g., the load on a railroad track due to a passing train), and, in general, characteristics of any kind of vibrating or oscillating mechanism.

In A sin kx and A cos kx, the factor k is called the *period factor.* It determines the wavelength or minimum value of x for which 1 cycle of the function occurs. This value of x is the period. As we shall see, the period is not always 2π. (Remember, we *cannot* factor the k from sin kx to obtain k sin x.

To investigate the effect of the period factor, let us compare the graphs of sin $2x$ and sin x. To graph sin $2x$, we make the table shown below. In order to use the convenient values of Table 14.1, we begin the table by assigning values to $2x$ instead of x. Then we work up to find x and down to find sin $2x$.

x	0	$\frac{\pi}{12}$	$\frac{\pi}{8}$	$\frac{\pi}{6}$	$\frac{\pi}{4}$	$\frac{\pi}{3}$	$\frac{3\pi}{8}$	$\frac{5\pi}{12}$	$\frac{\pi}{2}$	$\frac{7\pi}{12}$	$\frac{5\pi}{8}$	$\frac{2\pi}{3}$	$\frac{3\pi}{4}$	$\frac{5\pi}{6}$	$\frac{7\pi}{8}$	$\frac{11\pi}{12}$	π
$2x$	0	$\frac{\pi}{6}$	$\frac{\pi}{4}$	$\frac{\pi}{3}$	$\frac{\pi}{2}$	$\frac{2\pi}{3}$	$\frac{3\pi}{4}$	$\frac{5\pi}{6}$	π	$\frac{7\pi}{6}$	$\frac{5\pi}{4}$	$\frac{4\pi}{3}$	$\frac{3\pi}{2}$	$\frac{5\pi}{3}$	$\frac{7\pi}{4}$	$\frac{11\pi}{6}$	2π
sin $2x$	0	0.50	0.71	0.87	1.0	0.87	0.71	0.50	0	−0.50	−0.71	−0.87	−1.0	−0.87	−0.71	−0.50	0

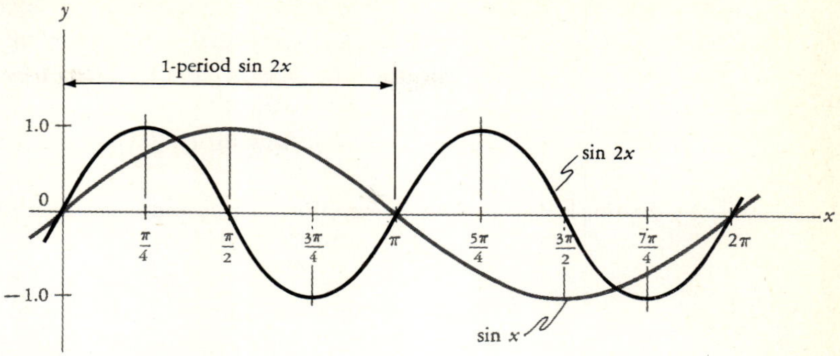

Fig. 15.7 sin 2x and sin x compared.

From the table and the corresponding graph, Fig. 15.7, we see that sin $2x$ completes 1 cycle as x varies from 0 to π. The effect of $k = 2$ is *to halve the period* of sin $2x$ as compared to that of sin x, or to double the number of cycles in a distance of 2π units on the x axis. The graph of sin $2x$ oscillates "twice as fast" as that of sin x.

We might suspect that the period of sin $3x$ is $1/3$ that of sin x, the period of sin $[(1/2)x]$ is twice that of sin x, and in general, the period of sin kx is $1/|k|$ that of sin x. To verify these statements, recall that the period of sin θ is 2π. That is, sin θ completes 1 cycle as θ varies from 0 to 2π (recall Figure 15.2). Now replace θ by kx and repeat the last sentence: sin kx completes 1 cycle as kx varies from 0 to 2π. When kx varies from 0 to 2π, x must vary from 0 to $2\pi/k$. Therefore *the period of sin kx is $2\pi/|k|$*. For example, the period of sin $2x$ is $2\pi/2 = \pi$, as we have seen. All this discussion applies as well to cos kx.

With the preceding ideas in mind, we can make the following definitions concerning the period of a function:

PERIOD OF A FUNCTION

1. The period of a function is the smallest absolute value of the variable for which the function completes a cycle.
2. The period of A sin kx (or of A cos kx) is $2\pi/|k|$, the smallest absolute value of x for which A sin kx completes a cycle.
3. The period of a function is the wavelength, the distance along the abscissa, between corresponding points on adjacent cycles of the graph of the function.

We have seen that in $A \sin kx$ or in $A \cos kx$ the amplitude A stretches or compresses the curve in the vertical direction. Now we find that the period factor k stretches or compresses the curve in the horizontal direction.

The following examples are graphs of $A \sin kx$ and $A \cos kx$.

Example 15.4. **Determine the periods and compare the graphs of $3 \cos [(1/2)x]$, $3 \cos x$, and $3 \cos 2x$.**

According to definition 2 above, the respective periods are 4π, 2π, and π. The graphs in Fig. 15.8 were made from tables similar to the one in this section.

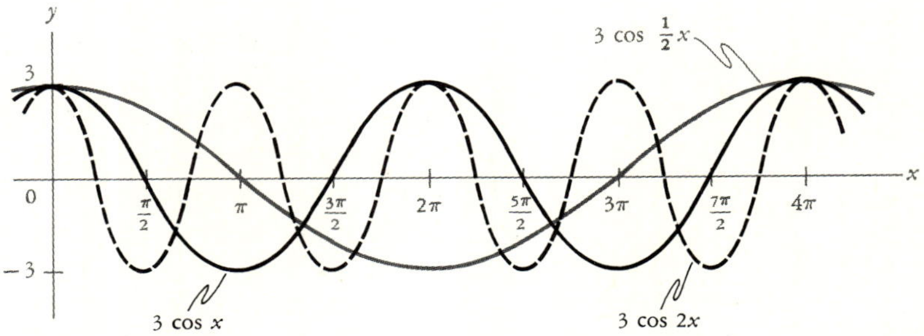

Fig. 15.8 Example 15.4.

Example 15.5. **Determine the period and graph 1 cycle of $-5 \sin (-3x/2)$.**

The period is:

$$\frac{2\pi}{|-3/2|} = \frac{4\pi}{3}$$

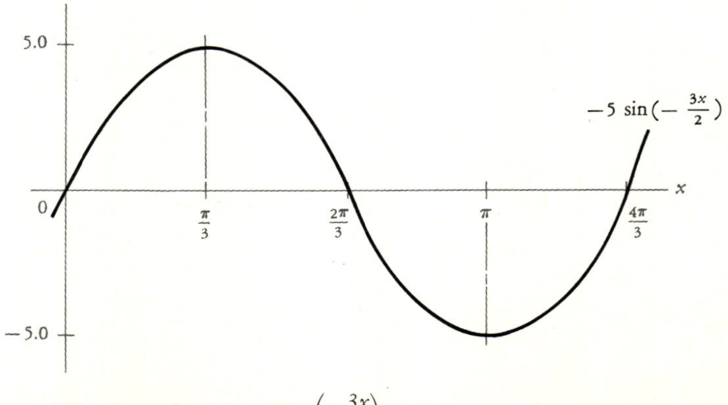

Fig. 15.9 Example 15.5. $-5 \sin \left(-\frac{3x}{2}\right)$.

By using the negative-angle identity, we have $-5 \sin(-3x/2) = 5 \sin(3x/2)$.

Let us try a different technique for constructing the graph. Instead of making a table, we can use any previous graph of $A \sin x$ and simply multiply each value of x by $2/3$. The period of 2π then becomes $4\pi/3$. Figure 15.9 is Fig. 15.5 modified in this manner.

EXERCISE 15.2

In Probs. 1 to 6, determine the period and graph 1 cycle of each function by making a table.

1. $\cos 4x$	2. $\sin 4x$	3. $\sin(-2x)$
4. $-(1/3)\cos(-3x)$	5. $2 \sin[(1/2)x]$	6. $5 \sin 3x$

7. Graph $\cos 5x$ by using any previous graph of $\cos x$ and changing the values of x as in Example 15.5.
8. Graph $\sin(2x/5)$ by using any previous graph of $\sin x$ as in Example 15.5.

15.4 SKETCHING SINE AND COSINE FUNCTIONS

We have constructed accurate graphs of $A \sin kx$ and $A \cos kx$ by making tables. If only approximate results are desired, however, this process is too tedious and time-consuming.

Figures 15.2 through 15.9 reveal some characteristics of the sine and cosine functions that enable us to sketch approximate graphs very quickly. For example, considering that a cycle begins at $x = 0$ where A and k are positive, the maximum value of $A \sin kx$ occurs at the quarter-cycle point, $x = \pi/(2k)$. Table 15.1 lists the zero, maxi-

TABLE 15.1 CHARACTERISTICS OF THE SINE AND COSINE FUNCTIONS

	VALUE OF THE FUNCTION		
FUNCTION	ZERO	MAXIMUM VALUE A	MINIMUM VALUE $-A$
$A \sin kx$	1. beginning of cycle 2. one-half cycle 3. end of cycle	1. one-quarter of cycle	1. three-quarters of cycle
$A \cos kx$	1. one-quarter of cycle 2. three-quarters of cycle	1. beginning of cycle 2. end of cycle	1. one-half cycle

$A > 0, k > 0$

mum, and minimum values of the sine and cosine functions in terms of the fraction of a cycle at which they occur.

Note that both functions continually alternate in quarter-cycle steps, from zero to maximum to zero to minimum, etc. To quickly make a fairly accurate graph, we need only sketch the rounded waves through these quarter-cycle steps. For even more accuracy, we can plot the one-eighth-cycle points halfway between the quarter-cycle points. The y value of these points is always $\pm A\sqrt{2}/2 \approx \pm 0.71A$.

The following examples show how to sketch $A \sin kx$ and $A \cos kx$.

Example 15.6. To sketch two cycles of $6 \sin 4x$, we note that the amplitude is 6 and the period is $2\pi/4 = \pi/2$. As shown in Fig. 15.10, we mark 6 and -6 on the y axis. On the x axis, we mark $\pi/2$ for the first cycle and π for the second cycle. Then we mark the quarter-cycle divisions: $\pi/8$, $\pi/4$, and $3\pi/8$ (first cycle) and $5\pi/8$, $3\pi/4$, and $7\pi/8$. Figure 15.10 shows the neat, smooth, rounded waves that alternate through the zeros, the peaks, and the valleys, at quarter-cycle intervals.

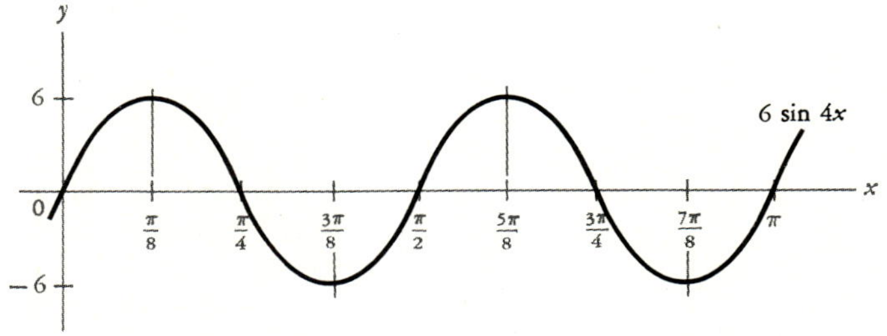

Fig. 15.10 Example 15.6. A sketch of 6 sin 4x made by using only the quarter-cycle points.

Example 15.7. To sketch $1\frac{1}{2}$ cycles of $-\cos 3x$ in both directions, we determine the period $2\pi/3$ and the quarter-cycle step $(1/4)(2\pi/3) = \pi/6$. Then we begin at $y = -1$ (because of the negative factor) and sketch the waves to $\pm\pi$. See Fig. 15.11.

EXERCISE 15.3

For Probs. 1 to 8, use Probs. 1 to 8 of Exercise 15.2, but sketch each graph instead of using a table.

For Probs. 9 to 12, sketch the graph of each function as indicated.

9. 2 cycles of $10 \sin (3x/2)$, beginning at $x = -\pi$

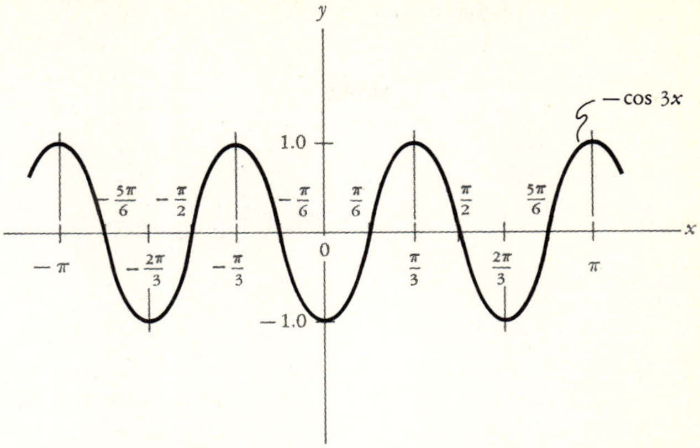

Fig. 15.11 Example 15.7. A sketch of − cos 3x made by using only the quarter-cycle points.

10. 2 cycles of 120 cos $120\pi t$, beginning at $t = -1/120$
11. 2 cycles of cos $\pi\theta$, beginning at $\theta = 0$
12. 3 cycles of sin $(8x/3)$, beginning at $x = -3\pi/8$

15.5 PHASE ANGLES AND DISPLACEMENTS; GRAPHS OF A SIN $(kx + \alpha)$ AND A COS $(kx + \alpha)$

In the expression $A \sin (x + \alpha)$, α is a constant angle and is called the *phase angle.* Figure 15.12 explains the phase angle and shows its role in the graph of $A \sin (x + \alpha)$. In Fig. 15.12, the revolving radii that generate $A \sin (x + \alpha)$ and $A \sin x$ are always separated by the angle α. As the two radii revolve, they generate identical sine waves, but one wave always leads the other by α units. Therefore we say that the two sine waves are α units *out of phase,* and we call α the phase angle.

If α is positive as shown, then the graph of $A \sin (x + \alpha)$ is displaced $-\alpha$ units relative to the graph of $A \sin x$. This term is called

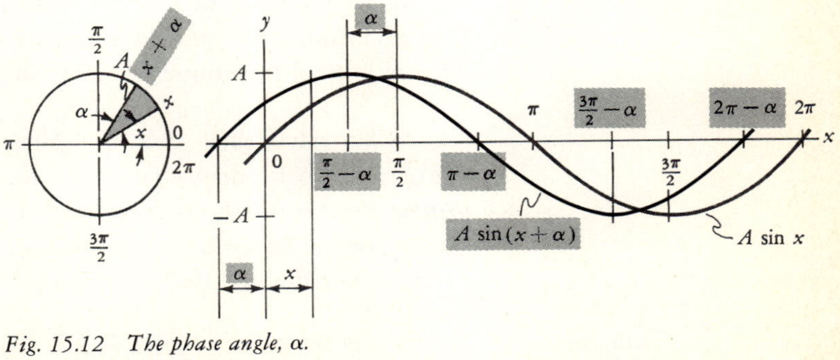

Fig. 15.12 The phase angle, α.

Table for Example 15.8

x	$-\frac{\pi}{4}$	$-\frac{\pi}{12}$	0	$\frac{\pi}{12}$	$\frac{\pi}{4}$	$\frac{5\pi}{12}$	$\frac{\pi}{2}$	$\frac{7\pi}{12}$	$\frac{3\pi}{4}$
$x+\frac{\pi}{4}$	0	$\frac{\pi}{6}$	$\frac{\pi}{4}$	$\frac{\pi}{3}$	$\frac{\pi}{2}$	$\frac{2\pi}{3}$	$\frac{3\pi}{4}$	$\frac{5\pi}{6}$	π
$\sin\left(x+\frac{\pi}{4}\right)$	0	0.50	0.71	0.87	1.0	0.87	0.71	0.50	0

x	$\frac{11\pi}{12}$	π	$\frac{13\pi}{12}$	$\frac{5\pi}{4}$	$\frac{17\pi}{12}$	$\frac{3\pi}{2}$	$\frac{19\pi}{12}$	$\frac{7\pi}{4}$	$\frac{23\pi}{12}$	2π
$x+\frac{\pi}{4}$	$\frac{7\pi}{6}$	$\frac{5\pi}{4}$	$\frac{4\pi}{3}$	$\frac{3\pi}{2}$	$\frac{5\pi}{3}$	$\frac{7\pi}{4}$	$\frac{11\pi}{6}$	2π	$\frac{13\pi}{6}$	$\frac{9\pi}{4}$
$\sin\left(x+\frac{\pi}{4}\right)$	−0.50	−0.71	−0.87	−1.0	−0.87	−0.71	−0.50	0	0.50	0.71

the *displacement* of the function $A \sin(x+\alpha)$. To better understand why the displacement is $-\alpha$, consider the point at sin 0 to be the "starting point" for any sine wave. For $\sin x$ to be sin 0, x has to be zero, but for $\sin(x+\alpha)$ to be sin 0, x has to be $-\alpha$. Therefore the "starting point" (and all other points) for $A\sin(x+\alpha)$ is displaced α units to the left.

A phase angle has the same effect on cosine waves as on sine waves.

Example 15.8. A graph of $\sin(x+\pi/4)$

To verify the preceding analysis, let us make a detailed graph of $\sin(x+\pi/4)$. We construct the table above by assigning the convenient values 0, $\pi/6$, $\pi/4$, $\pi/3$, and so on, to $x+\pi/4$ (instead of to x). We work up to find x and down to find $\sin(x+\pi/4)$.

Figure 15.13 shows the graph of $\sin(x+\pi/4)$, and for comparison, the graphs of $\sin x$ and $\sin(x-\pi/4)$. As predicted, $\sin(x+\pi/4)$ is displaced $\pi/4$ radians to the left, and $\sin(x-\pi/4)$ is displaced $\pi/4$ radians to the right of $\sin x$. We could have graphed $\sin(x+\pi/4)$ by first graphing $\sin x$ and then moving each point $\pi/4$ units to the left. This is a good technique for sketching sine waves with phase angles.

We have seen that the phase angle α causes the graph of $\sin(x+\alpha)$ to be displaced $-\alpha$ units. However, *for* $\sin(kx+\alpha)$, *the displacement is* $-\alpha/k$. To see why, let us consider the starting point again. For $\sin(kx+\alpha)$ to be sin 0, x has to be $-\alpha/k$. The following example verifies the displacement.

Example 15.9. A graph of $3\sin(2x+\pi/4)$

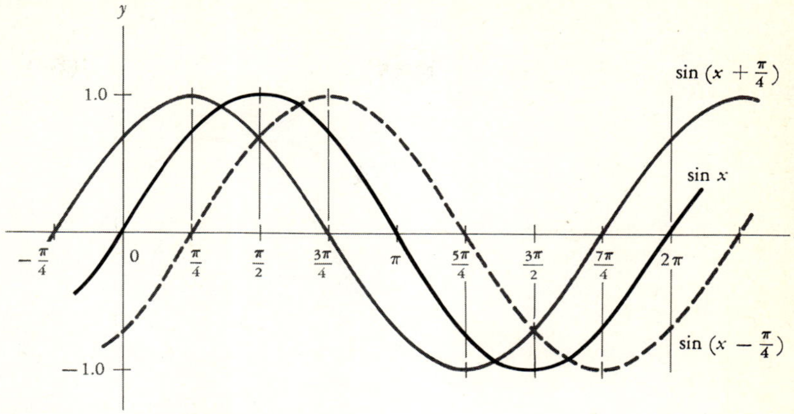

Fig. 15.13 Example 15.8.

The table for this graph is given below. It was constructed by assigning the values 0, $\pi/6$, $\pi/4$, $\pi/3$, and so on, to $2x + \pi/4$. Then x and $3 \sin (2x + \pi/4)$ were found. Note the similarity to the table in Example 15.8. The graph is shown in Fig. 15.14.

Table for Example 15.9

x	$-\frac{\pi}{8}$	$-\frac{\pi}{24}$	0	$\frac{\pi}{24}$	$\frac{\pi}{8}$	$\frac{5\pi}{24}$	$\frac{\pi}{4}$	$\frac{7\pi}{24}$	$\frac{3\pi}{8}$
$2x$	$-\frac{\pi}{4}$	$-\frac{\pi}{12}$	0	$\frac{\pi}{12}$	$\frac{\pi}{4}$	$\frac{5\pi}{12}$	$\frac{\pi}{2}$	$\frac{7\pi}{12}$	$\frac{3\pi}{4}$
$2x + \frac{\pi}{4}$	0	$\frac{\pi}{6}$	$\frac{\pi}{4}$	$\frac{\pi}{3}$	$\frac{\pi}{2}$	$\frac{2\pi}{3}$	$\frac{3\pi}{4}$	$\frac{5\pi}{6}$	π
$\sin\left(2x + \frac{\pi}{4}\right)$	0	0.50	0.71	0.87	1.0	0.87	0.71	0.50	0
$3\sin\left(2x + \frac{\pi}{4}\right)$	0	1.50	2.13	2.61	3.0	2.61	2.13	1.50	0

x	$\frac{11\pi}{24}$	$\frac{\pi}{2}$	$\frac{13\pi}{24}$	$\frac{5\pi}{8}$	$\frac{17\pi}{24}$	$\frac{3\pi}{4}$	$\frac{19\pi}{24}$	$\frac{7\pi}{8}$	$\frac{23\pi}{24}$	π
$2x$	$\frac{11\pi}{12}$	π	$\frac{13\pi}{12}$	$\frac{5\pi}{4}$	$\frac{17\pi}{12}$	$\frac{3\pi}{2}$	$\frac{19\pi}{12}$	$\frac{7\pi}{4}$	$\frac{23\pi}{12}$	2π
$2x + \frac{\pi}{4}$	$\frac{7\pi}{6}$	$\frac{5\pi}{4}$	$\frac{4\pi}{3}$	$\frac{3\pi}{2}$	$\frac{5\pi}{3}$	$\frac{7\pi}{4}$	$\frac{11\pi}{6}$	2π	$\frac{13\pi}{6}$	$\frac{9\pi}{4}$
$\sin\left(2x + \frac{\pi}{4}\right)$	−0.50	−0.71	−0.87	−1.0	−0.87	−0.71	−0.50	0	0.50	0.71
$3\sin\left(2x + \frac{\pi}{4}\right)$	−1.50	−2.13	−2.61	−3.0	−2.61	−2.13	−1.50	0	1.50	2.13

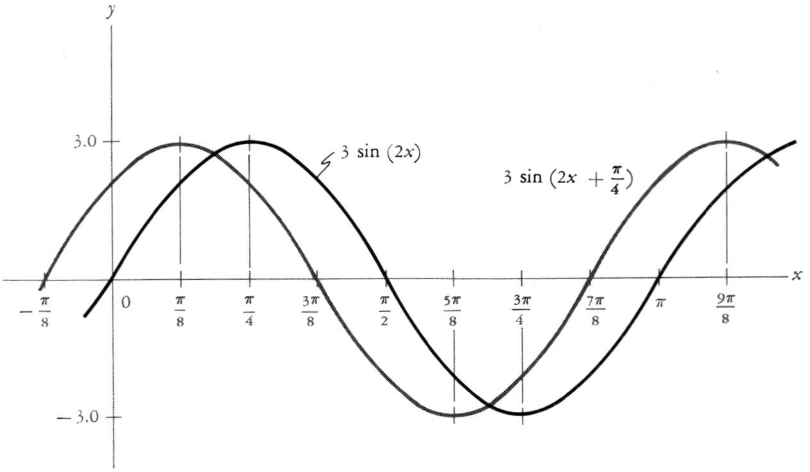

Fig. 15.14 Example 15.9. $3 \sin\left(2x + \frac{\pi}{4}\right)$

From the preceding analysis, we expect the displacement to be: $-\alpha \div k = -\pi/4 \div 2 = -\pi/8$; and Fig. 15.14 confirms this.

Although the displacement of $\sin(kx + \alpha)$ and $\cos(kx + \alpha)$ is different for each different period factor k, there is an area of consistency. The ratio *displacement/period* is the same for all such functions with the *same phase angle.* For example, in $\sin(x + \pi/4)$, this ratio is $-\pi/4 \div 2\pi = -1/8$, and in $\sin(2x + \pi/4)$ it is $-\pi/8 \div \pi = -1/8$. Therefore for sinusoidal functions, a phase angle of $\pm\pi/4$ radians always represents a displacement of $\mp$ one-eighth period, and in general a phase angle $\pm\alpha$ represents a displacement of $\mp\alpha/(2\pi)$ period.

The following list summarizes the characteristics of the functions $A \sin(kx + \alpha)$ and $A \cos(kx + \alpha)$.

1. Shape of graph : repetitive waveform; each wave curves up and down from zero to peak, from peak to zero, from zero to valley, and from valley to zero, in four equal-spaced, symmetric steps.
2. Amplitude, $|A|$: maximum value of function and maximum height of wave. $-|A|$ is minimum value.
3. Period, $|2\pi/k|$: length of each wave, or distance between corresponding points on adjacent waves.
4. Displacement, $-\alpha/k$: amount the graph is shifted relative to the graph of $A \sin kx$ or $A \cos kx$.

Example 15.10. **A sketch of $10 \cos(3x/2 - 3\pi/4)$**

To make a quick sketch, first determine the amplitude, the period, and the displacement.

Amplitude	*Period*	*Displacement*
$A = 10$	$\left\lvert\frac{2\pi}{k}\right\rvert = 2\pi \cdot \frac{2}{3} = \frac{4\pi}{3}$	$-\frac{\alpha}{k} = -\left(-\frac{3\pi}{4} \cdot \frac{2}{3}\right) = \frac{\pi}{2}$

Next, as shown in Fig. 15.15, label ± 10 on the y axis, label the period $4\pi/3$ on the x axis, and locate the three other quarter-cycle points, $\pi/3$, $2\pi/3$, and π.

Finally, mark all zero, maximum, and minimum points for the characteristic wave of $10 \cos (3x/2)$, and then shift each point to the *right* $\pi/2$ units. Sketch the wave through the displaced points, and backtrack it to the origin.

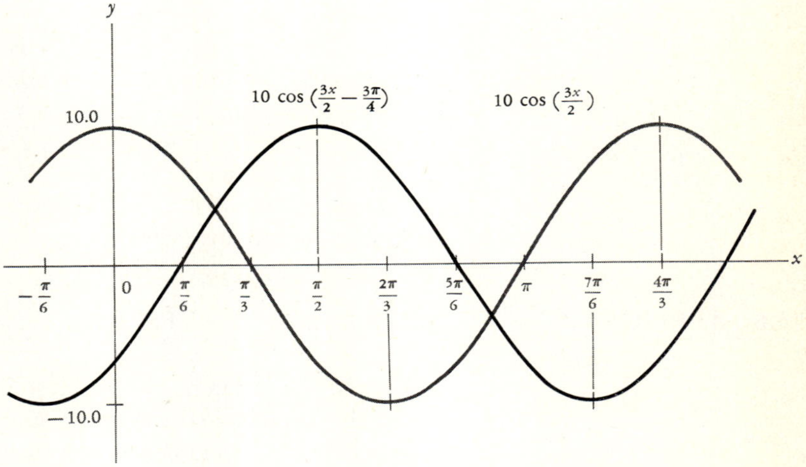

Fig. 15.15 Example 15.10. $10 \cos \left(\frac{3x}{2} - \frac{3\pi}{4}\right)$

EXERCISE 15.4

In Probs. 1 to 6, determine the amplitude, the period, and the displacement. Then verify these values by making an accurate graph from a table, as in Examples 15.8 and 15.9.

1. $\cos (x + \pi/4)$
2. $\cos (x - \pi/4)$
3. $5 \sin (2x - \pi/2)$
4. $5 \sin (2x + \pi/2)$
5. $2 \cos (x/2 + \pi/2)$
6. $-6 \cos (4x - \pi)$

In Probs. 7 to 15, use the amplitude, the period, and the displacement to *sketch* a graph of each function, as in Example 15.10.

7. $-\cos(3x + \pi)$
8. $\sin(4x - \pi/4)$
9. $\sin(x/4 - 3\pi)$
10. $\frac{3}{2}\cos\left(\frac{4x}{3} - \frac{2\pi}{3}\right)$
11. $-\frac{1}{5}\sin\left(\frac{8x}{3} + \frac{\pi}{8}\right)$
12. $\frac{1}{4}\cos\left(3x + \frac{\pi}{6}\right)$
13. $120\sin(120\pi t - 40\pi)$
14. $-10\sin\left(2\pi\omega + \frac{\pi}{2}\right)$
15. $60\cos\left(\pi t + \frac{3\pi}{4}\right)$

16. Sketch 4 cycles of $(1/2)\sin(3x - \pi/2)$, beginning at $x = -\pi$.
17. Sketch 3 cycles of $-2\cos(x/5 + \pi)$, beginning at $x = -6\pi$.
18. Demonstrate the identity $\cos(90° - \theta) = \sin\theta$ by sketching graphs of both functions.
19. Sketch a graph of $\sin(\theta + \pi/2)\cos(\theta + \pi/2)$. (*Hint:* Use the identity $\sin 2x = 2\sin x\cos x$.)
20. Sketch a graph of $\pm\sqrt{[1 + \cos(x - \pi/3)]/8}$. (*Hint:* Use an identity to simplify the expression.)
21. Sketch a graph of $(\sqrt{2}/2)\cos x - (\sqrt{2}/2)\sin x$. [*Hint:* Note that $\sin(\pi/4) = \cos(\pi/4) = \sqrt{2}/2$, and use an identity.]
22. Sketch a graph of $(1/2)\sin\theta - (\sqrt{3}/2)\cos\theta$. See hint for Prob. 21.
23. Demonstrate that 2π is the period for $\cos x$ by sketching graphs of $\cos x$ and $\cos(x + 2\pi)$.

15.6 APPLICATIONS OF SINUSOIDAL WAVES

Sinusoidal waves can be used to describe a variety of physical actions. Examples of sinusoidal waves are mechanical vibrations and electric currents. In their simplest form, such phenomena are based upon *simple harmonic motion.* This type of motion has to do with a point that travels with *constant speed* around a circle. The motion of the *projection of the point* onto a diameter of the circle is simple harmonic motion.

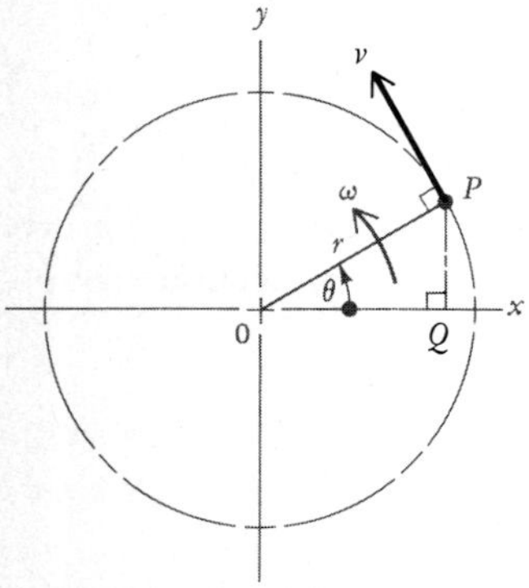

Fig. 15.16 Simple harmonic motion. As P revolves around the circle, Q moves with simple harmonic motion.

In Fig. 15.16, point P moves with constant linear speed v around the circle, and the radius therefore moves with constant angular speed ω so that $v = r\omega$ [recall Eq. (4.11)]. Point Q is the projection of p onto the x axis. The motion of Q is simple harmonic motion.

From Fig. 15.16, we see that the position of Q is given by $x = r\cos\theta$. For a constant angular speed, we have $\theta = \omega t$, where t stands for time. In other words, angular "distance" equals angular speed multiplied by time. Therefore, in this case we can write the position of Q as a function of time, as follows:

$$x = r\cos\theta = r\cos\omega t \tag{15.1}$$

This equation assumes that P and Q start at $x = r$ when $t = 0$. If we want Q to start at any other point, then we have to insert a phase angle into the equation. For example, if P were to start at 30° or $\pi/6$ rad, the coordinates of P would be $(r\sqrt{3}/2, r/2)$, and the position of Q would be

$$x = r \cos\left(\omega t + \frac{\pi}{6}\right)$$

Note that ω, the angular speed in radians per unit time, is the period factor. The *period,* or time for 1 cycle, is $T = 2\pi/\omega$, and the *frequency,* or number of cycles per unit time, is $f = 1/T = \omega/(2\pi)$.

Example 15.11. Motion of a mass on a spring

Figure 15.17 shows an object suspended on a spring. Suppose the object is pulled down a distance Y_0 from the "at rest" position and released. It is a well-established fact that if no friction acts to retard it, the object oscillates up and down in simple harmonic motion. The amplitude of the motion is Y_0, and the period factor is $\sqrt{kg/W}$, where W is the weight of the object, k is the "spring constant" (e.g., the number of pounds required to stretch or compress the spring 1 in.), and g is the acceleration due to gravity. Therefore, we can describe the position of the object as a function of time, as follows:

$$y = Y_0 \sin\left(\sqrt{\frac{kg}{W}}\, t - \frac{\pi}{2}\right) \tag{15.2}$$

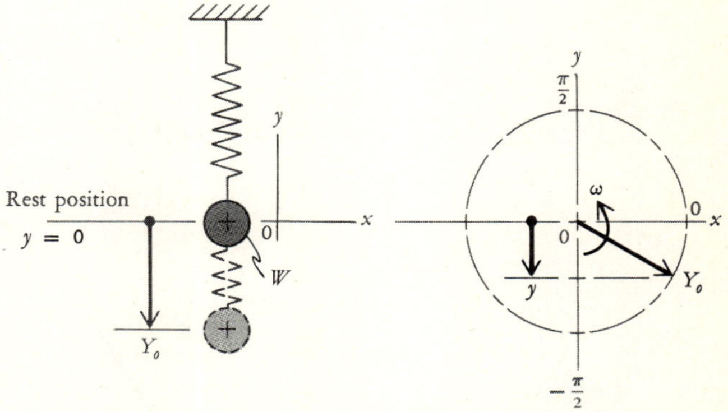

Fig. 15.17 Example 15.11. The motion of a mass on a spring.

Suppose that the object weighs 2 lb, the spring constant is 50 lb/in., and the initial displacement is 0.50 in.

a. How long does the object take to complete 1 cycle? This amount of

time is the period of the function

$$T = \frac{2\pi}{\sqrt{kg/W}} = 2\pi \sqrt{\frac{W}{kg}}$$

$$T = 2\pi \sqrt{\frac{2 \text{ lb}}{50 \text{ lb/in.} \cdot 386 \text{ in./sec}^2}} = 0.064 \text{ sec/cycle}$$

b. How fast does the object oscillate? This quantity is the frequency, the number of cycles completed in 1 unit of time [e.g., Hz (formerly cycles/sec)]. It is the reciprocal of the period.

$$f = \frac{1}{2\pi} \sqrt{\frac{kg}{W}} = \frac{1}{T} = 15.6 \text{ Hz}$$

c. What is the position of the object 0.30 sec after it is released?

$$y = 0.50 \sin \left[\sqrt{\frac{50 \text{ lb/in.} \cdot 386 \text{ in./sec}^2}{2 \text{ lb}}} \cdot (0.30 \text{ sec}) - \frac{\pi}{2}\right]$$
$$= 0.50 \sin (27.90 \text{ rad})$$
$$= 0.50 \sin (0.378 \text{ rad})$$
$$= 0.50(0.369)$$

$$y = 0.18 \text{ in.}$$

d. Display the motion of the object on a graph. Figure 15.18 is a sketch of the graph for one period. Note that the displacement is

$$\frac{\pi}{2} \div \sqrt{\frac{kg}{W}} = 0.016 \text{ sec}$$

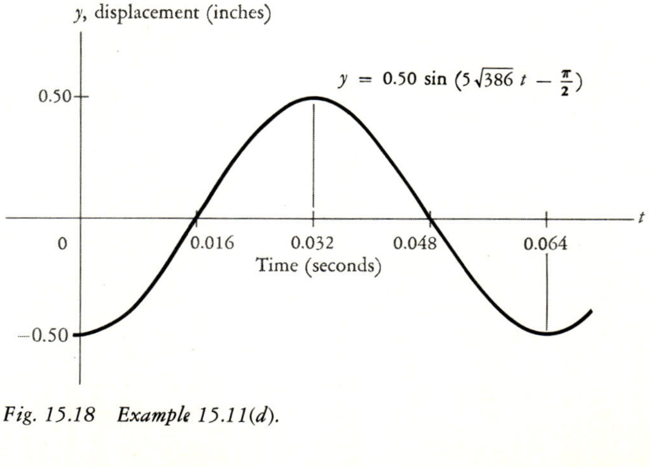

Fig. 15.18 Example 15.11(d).

or exactly one-quarter period, as it should be for a phase angle of $\pi/2$ rad.

Example 15.12. Electric generators

If a wire rotates within a magnetic field, an electric current and voltage are produced in the wire. This fact is the basis upon which electric generators operate.

Figure 15.19 depicts a wire that rotates with constant speed v between the poles of two magnets. The magnitudes of the current and voltage generated are proportional to the speed at which the wire crosses lines of magnetic force. As the figure shows, this rate is the x component of v, which is $v_x = -v \sin \theta = -v \sin \omega t$. Thus in general the instantaneous current can be expressed as

$$i = I \sin \omega t \tag{15.3}$$

where I is the maximum value of current. Likewise, the instantaneous voltage can be expressed as

$$e = E \sin \omega t \tag{15.4}$$

By comparing these equations to Eq. (15.1), we see that the magnitude of instantaneous current or voltage behaves like the displacement of a point in harmonic motion.

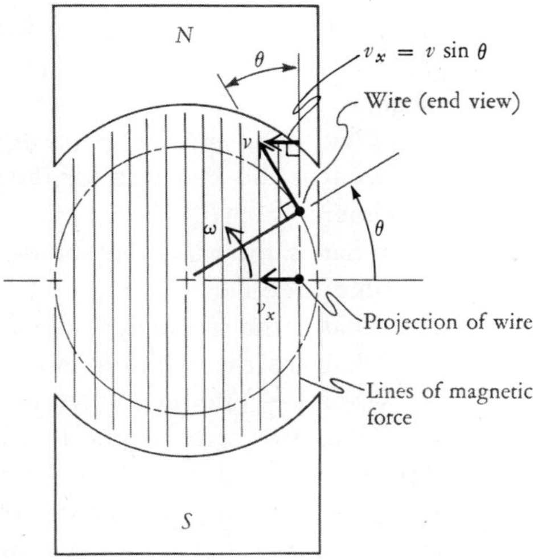

Fig. 15.19 Example 15.12. A two-pole generator.

The maximum voltage produced by a certain generator is 120 volts, and the generator turns at 120π rad/sec.

a. What is the equation for the instantaneous voltage as a function of time? From Eq. (15.4) we have

$$e = E \sin \omega t = 120 \sin 120\pi t$$

b. What is the frequency of the voltage wave?

$$f = \frac{\omega}{2\pi} = \frac{120\pi}{2\pi} = 60 \text{ Hz}$$

c. Sketch the voltage wave. Figure 15.20 shows 1 cycle of the voltage.

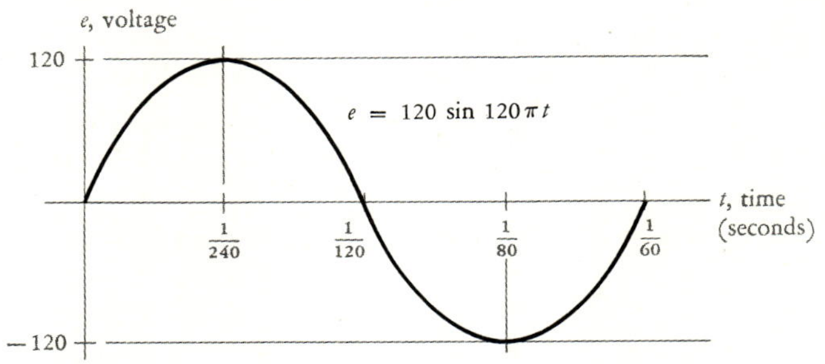

Fig. 15.20 Example 15.12(c).

EXERCISE 15.5

1. In Fig. 15.16, suppose that $r = 5$ in., $\theta = 0$ when $t = 0$, and point P completes 1 rev (revolution) in 2 sec with uniform speed.
   a. What is the angular speed of the radius to point P?
   b. What is the equation for the position of Q as a function of time?
   c. What is the equation for the speed of Q as a function of time? See Fig. 15.19.
   d. What is the frequency of the motion of Q?
   e. What are the position and speed of Q when $t = 0.50$ sec? When $t = 1.00$ sec? When $t = 1.50$ sec? When $t = 2.00$ sec?
   f. Graph two cycles of the position of Q and of the speed of Q.
2. Repeat Prob. 1 if $\theta = \pi/4$ when $t = 0$.
3. In Fig. 15.16, suppose that $r = 0.60$ cm, Q oscillates at $8\frac{1}{3}$ Hz in simple harmonic motion, and the coordinates of P are $(0.30, -0.30\sqrt{3})$ when $t = 0$.
   a. What is the angular speed of the radius to P?

b. What is the equation for the position of Q as a function of time?
c. What is the equation for the speed of Q as a function of time? See Fig. 15.19.
d. What is the period of the motion of Q?
e. What are the position and the speed of Q when $t = 0.0125$ sec? When $t = 10.080$ sec?
f. Graph 2 cycles of the position and the speed of Q.

4. Suppose that the object in Fig. 15.17 weighs 1.00 gm (gram), the spring constant is 100 gm/cm, and the initial displacement is up 0.010 cm. (In the metric system, g is 980 cm/sec^2.)
   a. What is the period of the motion of the object?
   b. What is the frequency of the motion?
   c. If timing begins when the object is released, what is the position 0.50 sec after it is released?
   d. What are the maximum speed of the object and the speed 0.50 sec after it is released?
5. Repeat Prob. 4 if the initial displacement is 0.020 cm and timing begins when the object is at $y = 0$ and is moving upward.
6. A two-pole electric generator (as in Fig. 15.19) is to produce alternating current with a maximum value of 100 amp at 60 Hz. What is the equation for instantaneous current as a function of time? At how many rpm must the armature turn? In how many seconds after the current is zero will it reach 50 percent of the maximum current?

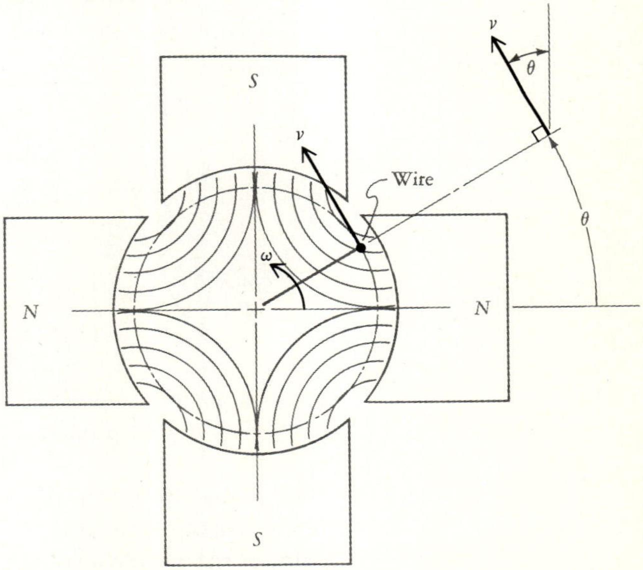

Fig. 15.21 Prob. 7. A four-pole generator.

7. Figure 15.21 depicts an electric generator with four poles. If the maximum voltage is 120 volts and the armature turns at 100 rpm, what is the equation for instantaneous voltage? How fast should the armature turn if the voltage frequency is to be 60 Hz? (*Hint:* The voltage reaches the maximum absolute value where the wire crosses the center of a pole, and it is zero where the wire is halfway between poles. Therefore each revolution of the wire produces 2 voltage cycles.)
8. Repeat Prob. 7 if the generator has six poles.
9. The motion of each piston in a reciprocating engine closely approximates simple harmonic motion. The speed of each piston is $\omega Y_0 \sin(\omega t + \pi/2)$, where ω is the angular speed of the crankshaft in radians per unit time and Y_0 is the maximum displacement of the piston from the midpoint of its path. Graph the speed vs. time, in seconds, of a piston of the Jaguar V-12 engine at 6,500 rpm. The stroke (distance between extreme positions) of each piston is 2.76 in.
10. Figure 15.22 shows a shaft supported by two ball bearings. The load on the shaft causes a sinusoidal load distribution in each outer ring as shown in the end view. This load is shared by all the balls in the *lower half* of the bearing. The load on each ball acts in the radial direction and varies from zero to a maximum of F_{max} and back to zero as the ball rotates through 180°. Suppose that $F_{max} = 30$ lb and the set of balls turns at 2,000 rpm relative to the center line of the shaft. Graph load vs. time, in seconds, for any one ball for 3 rev.

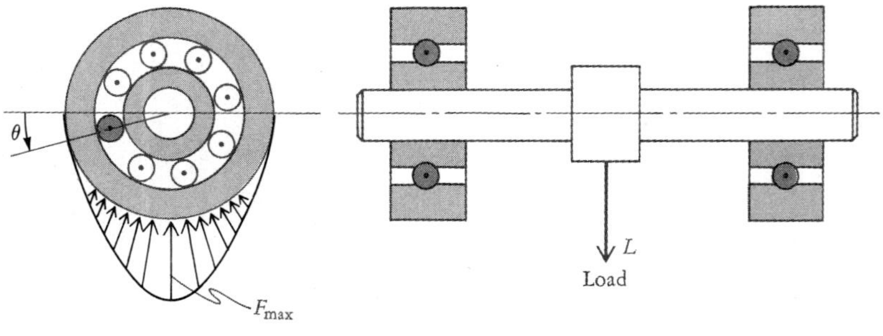

Fig. 15.22 Prob. 10. A ball bearing.

11. What total load L in the vertical direction can the bearings of Prob. 10 support? (*Hint:* The total load on each bearing is the sum of the *vertical components* of the loads on all the balls in any one position.)
12. Repeat Prob. 11 if each bearing has 12 balls.

13. Show by using trigonometric identities that the total load on each bearing of Fig. 15.22 is $2F_{max}$ no matter what the angle θ.
14. If each bearing of Fig. 15.22 had 12 balls, what total load could each bearing carry in terms of F_{max}?

15.7 COMPOSITE TRIGONOMETRIC FUNCTIONS

Sometimes a function consists of the sum of two or more other functions, such as $\sin\theta + \cos\theta$. One way to graph such a *composite function* is to sketch the individual functions and add them graphically. This method is called *composition of ordinates.*

Figure 15.23 is a graph of $\sin\theta + \cos\theta$ by composition of ordinates. To construct the curve, first $\sin\theta$ and $\cos\theta$ were sketched. Then the ordinates of these curves were added graphically at various points along the x axis. Two such additions are illustrated in the figure. On graph paper we can easily add ordinates at any point by counting squares.

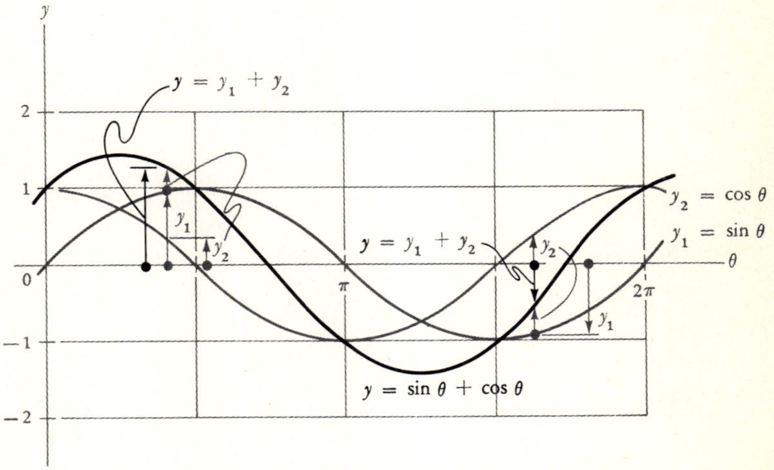

Fig. 15.23 A graph of sin θ + cos θ by composition of ordinates.

The *period of a composite function* is the *lowest common multiple* (*LCM*) of the individual periods. For $\sin\theta + \cos\theta$, both periods are 2π, and the composite period is 2π. For $\sin\theta + \cos 2\theta$, the periods are 2π and π, respectively, and the composite period is 2π. To graph 1 cycle of the composite function, 1 cycle of $\sin\theta$ and 2 cycles of $\cos 2\theta$ are required. Figure 15.24 shows this graph. Notice that $\cos 2\theta$ has to complete 2 cycles before $\sin\theta$ and $\cos 2\theta$ complete 1 cycle simultaneously.

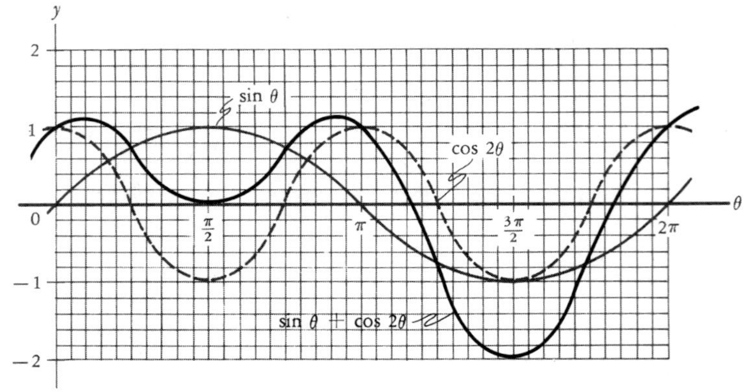

Fig. 15.24 A graph of sin θ + cos 2θ.

As a final example, let's graph 2 cos $4x$ − sin $6x$. The individual periods are $\pi/2$ and $\pi/3$, respectively. Therefore, the composite period is π (note that π is the LCM of $\pi/2$ and $\pi/3$); 2 cycles of 2 cos $4x$ and 3 cycles of sin $6x$ will be required. Figure 15.25 shows this graph.

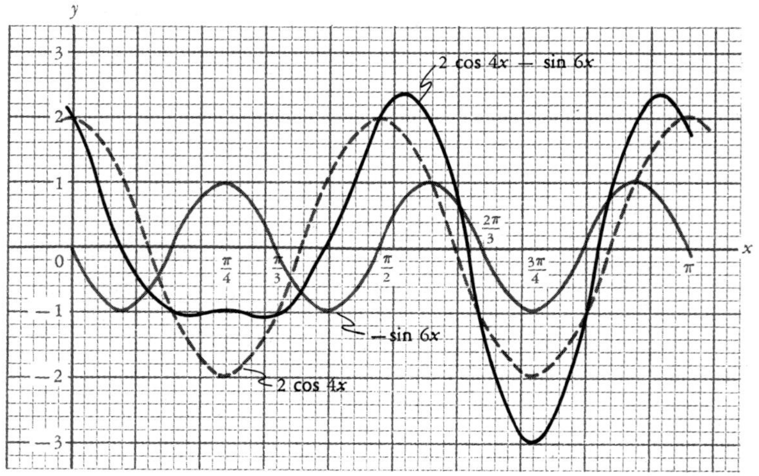

Fig. 15.25 A graph of 2 cos 4x − sin 6x.

EXERCISE 15.6

For Probs. 1 to 16, graph 1 cycle of the composite function by the method of composition of ordinates.

1. $3 \cos \theta + 2 \sin \theta$
2. $\sin 2\theta + \sin \theta$

3. $\cos \phi + \cos (\phi/2)$
4. $\cos (x/2) + \sin (x/2)$
5. $\sin x - \cos x$
6. $\cos \pi t - \sin \pi t$
7. $1 - \sin 2\beta$
8. $10 + 10 \cos (\phi/4)$
9. $5 \cos 6x + 4 \sin 4x$
10. $\sin (9x/2) + \cos 3x$
11. $1.6 \sin \dfrac{\pi}{2} t + 1.2 \sin \dfrac{2\pi}{3} t$
12. $\cos (x + \pi) + \sin x$
13. $\sin (2x - \pi/4) - \cos (x/2)$
14. $x + \sin x$
15. $1 - \sin x \cos x$ (*Hint:* Use an identity to simplify the product.)
16. $2 - \dfrac{\sqrt{2}}{2} \cos x + \dfrac{\sqrt{2}}{2} \sin x$ (*Hint:* What are $\cos (3\pi/4)$ and $\sin (3\pi/4)$? Use an identity to simplify.)
17. See if you can devise a general formula for the amplitude of a periodic function. What are the period and the amplitude of each composite function in Probs. 1 through 16?
18. The power dissipated by an electric resistor is $w = ei$, where w is the power in watts, e is the instantaneous voltage in volts, and i is the instantaneous current in amps. If $e = 120 \sin 120\pi t$ and $i = 5 \sin 120\pi t$, graph the power loss for 1 voltage cycle. (*Hint:* Use an identity to simplify $ei = 600 \sin^2 120\pi t$.)
19. The current through an electric inductor is always retarded by one-quarter cycle. Therefore, the instantaneous current in a 60-Hz line with an inductor is $i = I_0 \sin (120\pi t - \pi/2)$. The voltage is $e = E_0 \sin 120\pi t$, and the power is ei. Graph the power for 1 cycle of the voltage wave. (*Hint:* Use identities to simplify the product ei.)
20. The current through an electric capacitor is given by $i = I_0 \sin (120\pi t + \pi/2)$, and the voltage is given by $e = E_0 \sin 120\pi t$. Graph the power $w = ei$ for 1 cycle of the voltage wave. (*Hint:* Use identities to simplify the product ei.)
21. A sawtooth waveform can be closely approximated by the following sum of sine functions:

$$\sin x - \frac{\sin 3x}{3^2} + \frac{\sin 5x}{5^2} - \cdots$$

To see the sawtooth forming, graph just the first term, then the first two terms, and finally the first three terms on one set of axes.

15.8 GRAPHS OF OTHER TRIGONOMETRIC FUNCTIONS

The tangent, cotangent, secant, and cosecant functions do not have the widespread applications of the sine and cosine. We shall examine the graphs of these other functions, but we shall not deal with them extensively.

Since sec $x = 1/(\cos x)$, these functions have several characteristics in common. For example, their periods and displacements are identical. The term *amplitude* does not apply to sec x; however, the amplitude of $A \cos x$ is the *minimum* absolute value of $A \sec x$.

Example 15.13. A graph of sec x

A convenient way to sketch a graph of $y = \sec x$ is to first sketch $y = \cos x$. Then plot the reciprocal of each y value. Figure 15.26 is such a sketch.

An important fact is that the absolute value of sec x becomes indefinitely large as cos x approaches zero—or as x approaches an odd multiple of $\pi/2$. [Remember, sec $x = 1/(\cos x)$.] The vertical dashed lines in Fig. 15.26 indicate the boundaries that sec x approaches but can never touch; sec x is not defined at these values of x.

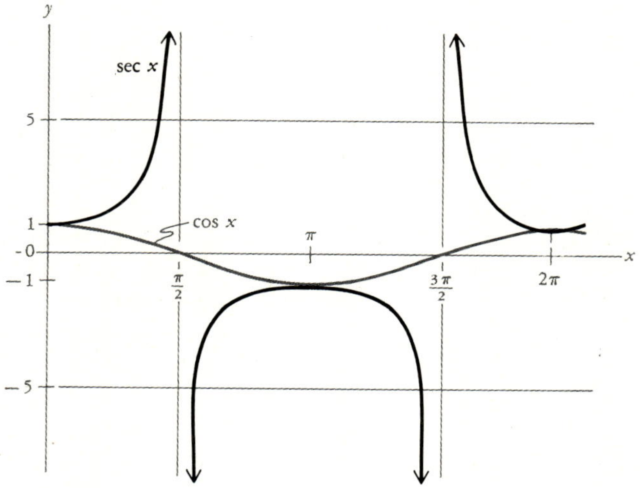

Fig. 15.26 Example 15.13. A sketch of y = sec x.

Example 15.14. A sketch of 2 csc $(x/2)$

Since $2 \csc (x/2) = 2/[\sin (x/2)]$, we can use the technique of Example 15.13. As shown in Fig. 15.27, we first sketch sin $(x/2)$, then we invert each y value to obtain csc $(x/2)$. Finally, we multiply the resulting y values by 2.

For more accurate graphs, points should be plotted accurately from the secant and cosecant tables.

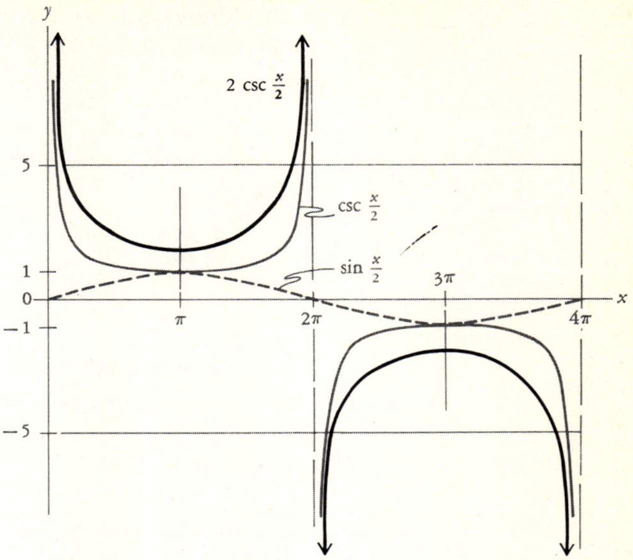

Fig. 15.27 Example 15.14. A sketch of $y = 2 \csc \frac{x}{2}$.

The functions tan x and cot x are different from the other trigonometric functions in that their period is π rather than 2π. Example 15.15 illustrates this fact.

Example 15.15. **A graph of tan x**

To make an accurate graph of tan x, we first make a table of values. The table below is based upon the commonly used angles, 0, $\pi/6$, $\pi/4$, etc. From the table, we plot the graph of tan x shown in Fig. 15.28.

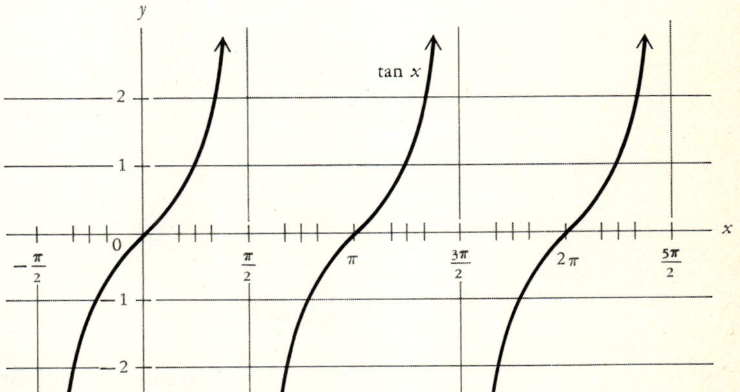

Fig. 15.28 Example 15.15. A graph of three cycles of $y = \tan x$.

Note that the graph has 2 complete cycles as x varies from 0 to 2π. Therefore the period of tan x is π.

x	0	$\frac{\pi}{6}$	$\frac{\pi}{4}$	$\frac{\pi}{3}$	$\frac{\pi}{2}$	$\frac{2\pi}{3}$	$\frac{3\pi}{4}$	$\frac{5\pi}{6}$	π	$\frac{7\pi}{6}$	$\frac{5\pi}{4}$	$\frac{4\pi}{3}$	$\frac{3\pi}{2}$	$\frac{5\pi}{3}$	$\frac{7\pi}{4}$	$\frac{11\pi}{6}$	2π
tan x	0	$\sqrt{3}/3$	1	$\sqrt{3}$	—	$-\sqrt{3}$	-1	$-\sqrt{3}/3$	0	$\sqrt{3}/3$	1	$\sqrt{3}$	—	$-\sqrt{3}$	-1	$-\sqrt{3}/3$	0
Decimal approximation	0	0.58	1.0	1.73	—	-1.73	-1.0	-0.58	0	0.58	1.0	1.73	—	-1.73	-1.0	-0.58	0

Like sec x, the absolute value of tan x becomes indefinitely large as x approaches $\pi/2$ or any odd multiple of $\pi/2$. Why?

Example 15.16. A sketch of cot x

To graph cot x accurately, we could make a table similar to the one in Example 15.15. Or, since cot $x = 1/(\tan x)$, we can sketch $y = \tan x$ and plot the reciprocal of each y value. Figure 15.29 was constructed in this manner.

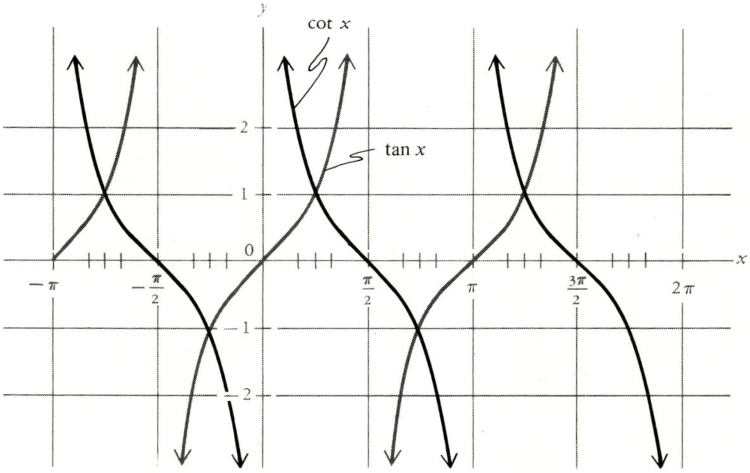

Fig. 15.29 Example 15.16. A graph of three cycles of $y = \cot x$.

EXERCISE 15.7

In Probs. 1 to 6, make an accurate graph of each function by first constructing a table.

1. sec x
2. csc x
3. cot x
4. 5 tan $2x$
5. (½) csc $4x$
6. 2 sec $(x - \pi/4)$

In Probs. 7 to 15, sketch a graph of each function. Where applicable, first sketch a graph of the reciprocal function.

7. $\csc 2x$
8. $\sec (x/2)$
9. $(1/2) \sec (2x - \pi/2)$
10. $2 \csc (x + \pi/4)$
11. $(1/4) \sec (x + \pi/2)$
12. $-\csc (3x - \pi/2)$
13. $-2 \tan x$
14. $(1/2) \cot 2x$
15. $\tan (2x + \pi/4)$

CHAPTER 16
TRIGONOMETRIC EQUATIONS AND INVERSE TRIGONOMETRIC FUNCTIONS

16.1 TRIGONOMETRIC EQUATIONS

A trigonometric equation is an equation that contains one or more trigonometric functions. An example is $2 \sin x - 1 = 0$. The usual objective in solving such an equation is to find all values of the *angle* x that satisfy it. To do so, we first solve for the value of the *function*—in this case $\sin x = 1/2$—and then from experience or by use of the trigonometric tables, we determine the values of x.

In this case, since $\sin x = 1/2$, x can be $\pi/6$, or $5\pi/6$, or any angle that differs from these by a multiple of the period of $\sin x$. Since the period of $\sin x$ is 2π rad (radians), we can indicate all possible solutions as follows.

For $$2 \sin x - 1 = 0$$

$$x = \frac{\pi}{6} + 2k\pi \qquad \text{or} \qquad x = \frac{5\pi}{6} + 2k\pi \tag{16.1}$$

where k is any integer: $k = 0, \pm 1, \pm 2, \pm 3, \ldots$

Figure 16.1 shows some of the graphical solutions of $2 \sin x - 1 = 0$. This graph helps to explain (1) why we have infinitely many solutions for a trigonometric equation and (2) why certain of the solutions differ by one period of the function. Every point where the line $y = 1/2$ crosses the graph of $y = \sin x$ is a solution. In this case each such point belongs to one of two sets, and in each set the points are corresponding points on successive cycles of the sine wave; therefore successive points of the set differ by one period.

To avoid the multiplicity of solutions, we define the *fundamental solutions* to be *all solutions in the first cycle* (beginning at $x = 0$ and extending to the right). If P is the period of the function, then

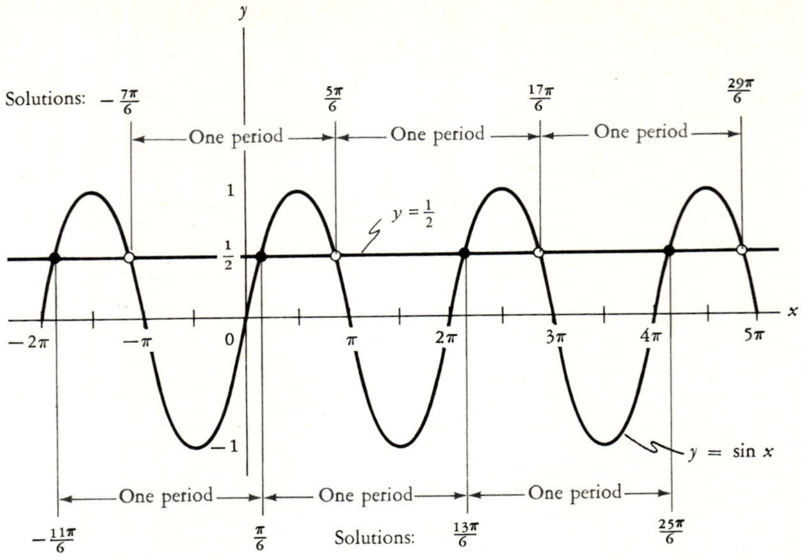

Fig. 16.1 Graphical solutions of $2 \sin x - 1 = 0$.

the fundamental solutions are those from 0 to P (excluding P). Thus the fundamental solutions for Eq. (16.1) are all solutions from 0 to 2π, which are $x = \pi/6$ and $x = 5\pi/6$. All other solutions can be found by adding $\pm P, \pm 2P, \pm 3P, \ldots$ to the fundamental solutions.

In an equation such as $\sin 4x + \sin 6x + 2 = 0$, the fundamental solutions are based upon the composite period, π.

Example 16.1. Solutions of $2 \sin^2 (\alpha/12) - 1 = 0$

To solve for α in $2 \sin^2 (\alpha/12) - 1 = 0$, we first solve for $\sin (\alpha/12)$.

$$2 \sin^2 \frac{\alpha}{12} - 1 = 0$$

$$2 \sin^2 \frac{\alpha}{12} = 1$$

$$\sin^2 \frac{\alpha}{12} = \frac{1}{2}$$

$$\sin \frac{\alpha}{12} = \pm \frac{1}{\sqrt{2}} = \pm \frac{\sqrt{2}}{2}$$

Next we reason that since $\sin (\pi/4) = \sqrt{2}/2$, then $\sin (\alpha/12) = \pm\sin (\pi/4)$, and $\alpha/12$ can be any angle having $\pi/4$ as a reference angle.

Therefore

$$\frac{\alpha}{12} = \frac{\pi}{4} + 2k\pi, \frac{3\pi}{4} + 2k\pi, \frac{5\pi}{4} + 2k\pi, \frac{7\pi}{4} + 2k\pi$$

Notice that we solved for $\alpha/12$, not α. Finally we solve for α by multiplying each set of solutions by 12 to obtain the following general solutions:

$$\alpha = 3\pi + 24k\pi, 9\pi + 24k\pi, 15\pi + 24k\pi, 21\pi + 24k\pi$$

where $k = 0, \pm1, \pm2, \pm3, \ldots$

The period of sin $(\alpha/12)$ is 24π; therefore the fundamental solutions are

$$\alpha = 3\pi, 9\pi, 15\pi, 21\pi$$

Techniques for Solving Trigonometric Equations

There is no general method or formula that will solve all trigonometric equations; each equation must be treated individually. However, the following techniques are usually applicable. With practice, the student can develop the skill to apply these techniques.

1. *Identities.* If the equation contains two or more different trigonometric functions, we may be able to simplify it with identities. We transform one or more of the functions until only one kind of function remains.

   For example, to solve

   $$2 \sin \theta - \cos^2 \theta - 2 = 0$$

   we use the identity [Eq. (14.9)] $\cos^2 \theta = 1 - \sin^2 \theta$. Then, after we rearrange terms, the original equation becomes

   $$\sin^2 \theta + 2 \sin \theta - 3 = 0$$

   which has only one kind of function, $\sin \theta$. This technique is also useful if two different angles occur. See Example 16.2 below.
2. *Algebraic methods.* We can frequently use algebraic methods such as factoring or the quadratic formula to solve trigonometric equations. To continue the example above, we observe that $\sin^2 \theta + 2 \sin \theta - 3$ has the same form as $x^2 + 2x - 3$, which factors into $(x - 1)(x + 3)$. Therefore, we solve the equation as follows.

$$\sin^2 \theta + 2 \sin \theta - 3 = 0$$

$$(\sin \theta - 1)(\sin \theta + 3) = 0$$

from which

$$\sin \theta = 1 \longrightarrow \theta = \frac{\pi}{2} + 2k\pi \qquad k = 0, \pm 1, \pm 2, \pm 3, \ldots$$

and

$$\sin \theta = -3 \longrightarrow \text{no solution, because } |\sin \theta| \leq 1$$

3. *Reference angles.* In certain equations in which the functions are the same but the angles are different, the following solutions can be used. These solutions are based upon two facts: (1) the trigonometric functions of coterminal angles are equal, and (2) certain related angles have equal trigonometric functions. For example, $\sin (5\pi/6) = \sin (\pi - 5\pi/6) = \sin (\pi/6) = 1/2$. In each equation below, α and β represent two different angles, and k is any integer, $0, \pm 1, \pm 2, \pm 3, \ldots$.
   a. If $\sin \beta = \sin \alpha$ or $\csc \beta = \csc \alpha$,

   then $\boxed{\beta = \alpha + 2k\pi}$ or $\boxed{\beta = (\pi - \alpha) + 2k\pi}$ (16.2)

   These situations are illustrated by Fig. 16.2. Example 16.1 shows how to apply this technique.
   b. If $\cos \beta = \cos \alpha$ or $\sec \beta = \sec \alpha$,

   then $\boxed{\beta = \alpha + 2k\pi}$ or $\boxed{\beta = (2\pi - \alpha) + 2k\pi}$ (16.3)

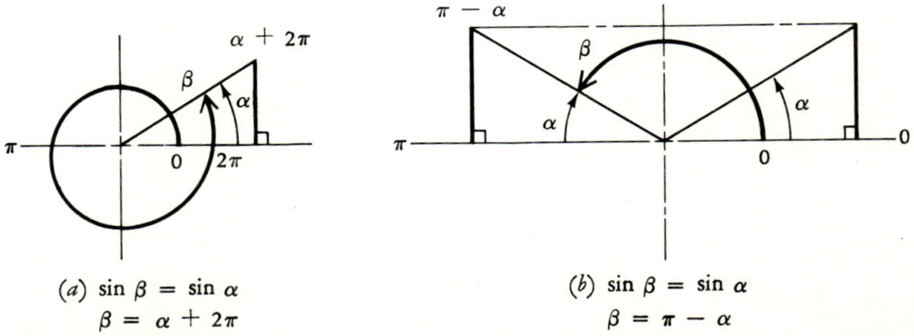

(*a*) $\sin \beta = \sin \alpha$, $\beta = \alpha + 2\pi$ (*b*) $\sin \beta = \sin \alpha$, $\beta = \pi - \alpha$

Fig. 16.2 Related angles for the sine function.

These situations are illustrated by Fig. 16.3.

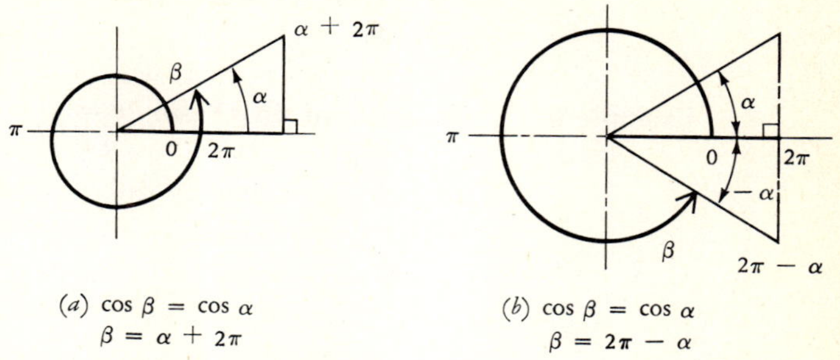

Fig. 16.3 Related angles for the cosine function.

c. If $\tan\beta = \tan\alpha$ or $\cot\beta = \cot\alpha$,

then $\beta = \alpha + 2k\pi$

or $\beta = (\alpha + \pi) + 2k\pi = \alpha + (2k + 1)\pi$

We can combine these two equations to obtain*

$$\boxed{\beta = \alpha + k\pi} \tag{16.4}$$

Figure 16.4 illustrates these situations.

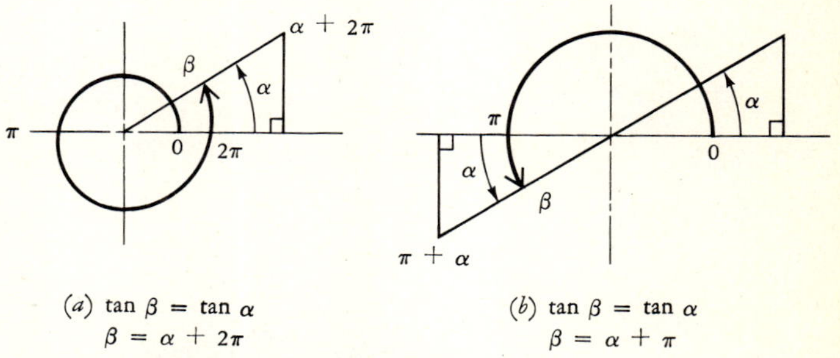

Fig. 16.4 Related angles for the tangent function.

4. *Graphs.* Almost any trigonometric equation can be solved by graphing. Sometimes, however, the construction is tedious, and results are limited by the accuracy of the graph. Since approximate graphs can be quickly sketched, this method is useful for predicting solutions and for checking solutions obtained by other methods. Figure 16.1 is an example of solutions by graphing. Example 16.4 uses the method of composition of ordinates.

* Since $2k\pi$ is an *even* multiple of π and $(2k + 1)\pi$ is an *odd* multiple of π, both expressions together represent any multiple of π, which is $k\pi$.

In the following three examples we shall solve the same equation three different ways, to illustrate the different techniques just outlined.

Example 16.2. Solutions of $\sin\theta - \sin 2\theta = 0$ by identities and algebra

By using the identity $\sin 2\theta = 2\sin\theta\cos\theta$, we can form an equation with only one angle, θ, instead of two angles, θ and 2θ.

$$\sin\theta - \sin 2\theta = 0$$

$$\sin\theta - 2\sin\theta\cos\theta = 0$$

At this point we might be tempted to use an identity to transform $\cos\theta$ into $\sqrt{1-\sin^2\theta}$; for some equations this would indeed be the best course to follow. (See Example 16.5.) However, in this case, we can separate the two different functions by factoring $\sin\theta$. (This kind of insight or skill is almost a necessity for success in mathematics and many professional endeavors both technical and nontechnical.) Therefore, we have

$$\sin\theta\,(1 - 2\cos\theta) = 0$$

$$\sin\theta = 0 \longrightarrow \theta = 0 + 2k\pi$$

$$\theta = \pi + 2k\pi$$

$$\cos\theta = \frac{1}{2} \longrightarrow \theta = \frac{\pi}{3} + 2k\pi$$

$$\theta = \frac{5\pi}{3} + 2k\pi$$

The *fundamental solutions* are

$$\boxed{\theta = 0, \frac{\pi}{3}, \pi, \frac{5\pi}{3}}$$

Example 16.3. Solutions of $\sin\theta - \sin 2\theta = 0$ by reference angles

$$\sin\theta - \sin 2\theta = 0$$

$$\sin\theta = \sin 2\theta$$

From Eq. (16.2) we have

$$2\theta = \theta + 2k\pi$$

$$2\theta - \theta = 2k\pi$$

$$\theta = 2k\pi \tag{16.5}$$

or

$$2\theta = \pi - \theta + 2k\pi$$

$$2\theta + \theta = \pi + 2k\pi$$

$$3\theta = \pi + 2k\pi$$

$$\theta = \frac{\pi}{3} + \frac{2k\pi}{3} \tag{16.6}$$

The *fundamental solutions* are those from 0 to 2π, and are obtained by letting $k = 0$ in Eq. (16.5) and $k = 0$, 1, and 2 in Eq. (16.6).

$$\boxed{\theta = 0, \frac{\pi}{3}, \pi, \frac{5\pi}{3}}$$

Example 16.4. Solutions of $\sin \theta - \sin 2\theta = 0$ by graphing

The solutions of this equation are the values of θ where the function $y = \sin \theta - \sin 2\theta$ crosses the θ axis. In Fig. 16.5 this composite func-

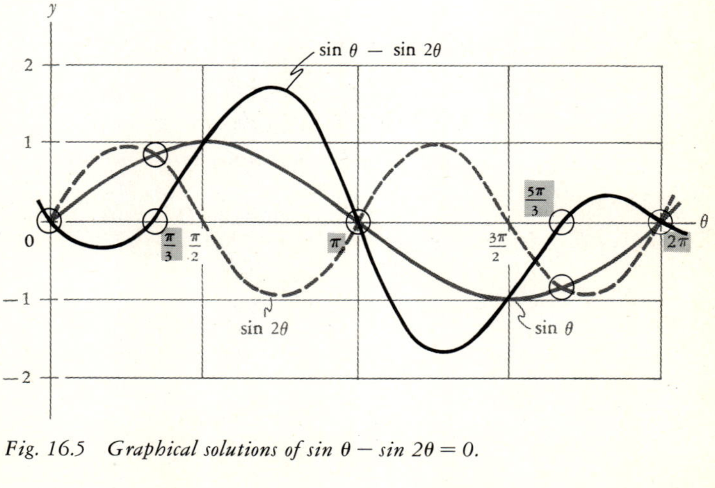

Fig. 16.5 Graphical solutions of sin θ − sin 2θ = 0.

tion has been graphed by composition of ordinates. As shown in the figure, the *fundamental solutions* are

$$\boxed{\theta = 0, \frac{\pi}{3}, \pi, \frac{5\pi}{3}}$$

An alternate technique is to consider the equation in the form $\sin\theta = \sin 2\theta$ and read the values of θ where the graphs of $\sin\theta$ and $\sin 2\theta$ intersect.

Example 16.5. Solutions of $\sin\theta - \cos\theta = 0$

To solve this equation, we change $\cos\theta$ to $\pm\sqrt{1-\sin^2\theta}$ and then square both sides, as follows:

$$\sin\theta - \cos\theta = 0$$

$$\sin\theta = \cos\theta = \pm\sqrt{1-\sin^2\theta}$$

$$\sin^2\theta = 1 - \sin^2\theta$$

$$2\sin^2\theta = 1$$

$$\sin\theta = \pm\frac{1}{\sqrt{2}} = \pm\frac{\sqrt{2}}{2}$$

The fundamental solutions *appear* to be

$$\theta = \frac{\pi}{4}, \frac{3\pi}{4}, \frac{5\pi}{4}, \frac{7\pi}{4}$$

Note: It is very important to check all fundamental solutions, because *extraneous* (or false) *solutions* occur frequently. In this case, $\theta = 3\pi/4$ and $\theta = 7\pi/4$ will *not* work in the original equation. Therefore the *fundamental solutions* are

$$\boxed{\theta = \frac{\pi}{4}, \frac{5\pi}{4}}$$

Example 16.6. Solutions of $\tan 3z - \cot 3z + 2 = 0$

We use the identity $\cot 3z = 1/(\tan 3z)$ and then multiply both sides by $\tan 3z$ to obtain a quadratic equation, as follows:

$$\tan 3z - \cot 3z + 2 = 0$$

$$\tan 3z - \frac{1}{\tan 3z} + 2 = 0$$

$$\tan^2 3z + 2 \tan 3z - 1 = 0$$

Since the left side is not directly factorable, we use the quadratic formula to solve for tan $3z$.

$$\tan 3z = \frac{-2 \pm \sqrt{4+4}}{2}$$
$$= -1 \pm \sqrt{2}$$

$$\tan 3z = 0.414, -2.414$$

With the aid of the trigonometric tables we find that the solutions for $\tan 3z = 0.414$ are

$$3z = 0.3927 + k\pi$$

and for $\tan 3z = -2.414$, the solutions are

$$3z = \pi - 1.1781 + k\pi = 1.9635 + k\pi$$

Therefore

$$z = 0.1309 + \frac{k\pi}{3}, \ 0.6545 + \frac{k\pi}{3}$$

The *fundamental solutions* are those from 0 to $\pi/3$, which are

$$\boxed{z = 0.1309 \text{ rad}, 0.6545 \text{ rad}}$$

Example 16.7. Solutions of $\sin 8x + \sin 6x = 0$

As a final example, we shall demonstrate an important application of the identity

$$\sin u + \sin v = 2 \sin [\tfrac{1}{2}(u + v)] \cos [\tfrac{1}{2}(u - v)]$$

(See Prob. 35 of Exercise 14.3.)

Replacing u by $8x$ and v by $6x$ in this identity, we have

$$\sin 8x + \sin 6x = 0$$

$$2 \sin [\tfrac{1}{2}(8x + 6x)] \cos [\tfrac{1}{2}(8x - 6x)] = 0$$

$$2 \sin 7x \cos x = 0$$

From $\sin 7x = 0$:

$$7x = 0 + 2k\pi \quad \text{or} \quad 7x = \pi + 2k\pi$$

$$x = 0 + \frac{2k\pi}{7} \quad \text{or} \quad x = \frac{\pi}{7} + \frac{2k\pi}{7}$$

From $\cos x = 0$:

$$x = \frac{\pi}{2} + 2k\pi$$

$$x = \frac{3\pi}{2} + 2k\pi$$

Since π is the period of the composite function $\sin 8x + \sin 6x$, the *fundamental solutions* are those from 0 to π, as follows:

$$\boxed{x = 0, \frac{\pi}{7}, \frac{2\pi}{7}, \frac{3\pi}{7}, \frac{\pi}{2}, \frac{4\pi}{7}, \frac{5\pi}{7}, \frac{6\pi}{7}}$$

EXERCISE 16.1

In Probs. 1 to 10, find the fundamental solutions by graphing as in Fig. 16.1 or Example 16.4.

1. $2 \sin x + 1 = 0$
2. $2 \cos x - 1 = 0$
3. $4 \cos^2 2\beta = 1$
4. $4 \sin^2 (\alpha/12) - 1 = 0$
5. $\sin \theta - \cos \theta = 0$
6. $\cos \theta + \sin \theta = 0$
7. $\cos \theta + \sin 2\theta + 1 = 0$
8. $2 \sin \theta + \cos 2\theta - 1 = 0$
9. $\sin (4x - \pi/2) + \cos (2x - \pi/4) = 0$
10. $\cos (4x/3 + \pi/3) - \cos 2x = 0$

For Probs. 11 through 20, use Probs. 1 through 10, respectively. Find the fundamental solutions by any methods except graphing. Problem 17 (using Prob. 7) cannot be solved directly. But work it as far as you can and note the value of the graphical method.

In Probs. 21 to 32, find the general solutions and the fundamental solutions. Use any method.

21. $\tan 2x - 1 = 0$
22. $2 \sec \alpha + 4 = 0$
23. $\sin 3\theta + \csc 3\theta + 2 = 0$
24. $4 \cot^2 3\beta = 12$
25. $\sec 2\alpha - \sec 3\alpha = 0$
26. $2 \cos 4\theta - \sec 4\theta + 1 = 0$
27. $\cot 3x = \cot 4x$
28. $\tan 2x = \tan 3x$
29. $5 \sin^2 t + 2 \sin t - 1 = 0$
30. $\cos 2\theta + \sin 4\theta + 1 = 0$
31. $10 \sin \beta = \beta$
32. $\sin (\alpha - \pi/4) = \sin 2\alpha$

16.2 INVERSE TRIGONOMETRIC FUNCTIONS

A. BASIC IDEAS ABOUT INVERSE FUNCTIONS

The function $y = 2x$ expresses a certain relationship between the two variables x and y. With the relationship in this form, we can find y if we know x. However, if we know y and wish to find x, then we have to write the relationship in the *inverted* form, $x = y/2$. We call this an inverted form or *inverse function* because the roles of the independent and dependent variables are inverted (reversed or interchanged) *relative to those of the original function.* Whereas x was the independent variable before, now it is the dependent variable.

In general, if two functions represent exactly the same relationship, but the roles of the independent and dependent variables are inverted, we say that each function is the inverse of the other.

We can find the inverse of a function by solving for the "other variable." For instance, to find the inverse function of $y = 5x - 3$, we solve for x in terms of y. The resulting inverse function is $x = (y + 3)/5$. The two functions $y = 5x - 3$ and $x = (y + 3)/5$ are inverses of each other.

B. INVERSES OF TRIGONOMETRIC FUNCTIONS

The statements (1) "$1/2$ is the sine of $\pi/6$" and (2) "$\pi/6$ is an angle whose sine is $1/2$" are inverses; they are two different ways of expressing exactly the same relationship between an angle, $\pi/6$ rad, and the value of its sine, $1/2$. We have a symbolic notation for the first statement: $1/2 = \sin (\pi/6)$.

The second statement expresses the idea of an *inverse sine function.* But so far, we have no symbolic means for writing it. There are problems in which we need to write an expression for the *angle* in a sine function. To do so, we need an inverse-sine symbol. Therefore, we define the following notations.

$\sin^{-1} y$ and $\arcsin y$
Inverse-sine notations: an angle whose sine is y

Each of these expressions means "an angle whose sine is y." The first one is read "inverse sine of y," and the second one "arc sine of y." Both forms are in common usage, and we shall use them interchangeably.*

By using either of these inverse-sine symbols, we can now write the inverse-sine statement "$\pi/6$ is an angle whose sine is ½," as follows:

$$\frac{\pi}{6} = \sin^{-1}\frac{1}{2} \qquad \text{or} \qquad \frac{\pi}{6} = \arcsin\frac{1}{2}$$

And in general, we can say that

$$\text{if} \quad y = \sin\theta \qquad \text{then} \quad \theta = \sin^{-1} y \quad \text{or} \quad \theta = \arcsin y$$

Notice that the expressions $y = \sin\theta$ and $\theta = \sin^{-1} y$ represent exactly the same relationship between the variables θ and y. Having both notations, we can now write expressions for (1) the value of the sine of an angle in terms of the angle, and (2) the value of the *angle* itself, in terms of the value of the sine of the angle.

Example 16.8. **Find all values of $\arcsin(\sqrt{3}/2)$ in radians.**

Since $\arcsin(\sqrt{3}/2)$ means "an *angle* whose sine is $\sqrt{3}/2$," we can reword the problem as follows: Find all angles θ such that $\sin\theta = \sqrt{3}/2$. Therefore, since $\sin(\pi/3) = \sqrt{3}/2$, $\arcsin(\sqrt{3}/2)$ is any first- or second-quadrant angle for which $\pi/3$ is the reference angle. The set of all possible angles is

$$\theta = \arcsin\frac{\sqrt{3}}{2} = \frac{\pi}{3} + 2k\pi, \frac{2\pi}{3} + 2k\pi$$

where $k = 0, \pm1, \pm2, \pm3, \ldots$

* There is a technical difference in the meanings of $\sin^{-1} y$ and $\arcsin y$, but that difference is not important enough to consider here. Many textbooks do not make any distinction between the two symbols. Note in $\sin^{-1} y$ that the -1 is not an exponent. This is the only exception to the exponent notation for trigonometric functions. The -1 power of $\sin y$ is written $(\sin y)^{-1}$; $\sin^{-1} y$ always means the inverse sine of y, and not $1/\sin y$.

We have similar symbols for the inverses of the other five trigonometric functions. For example, $\cos^{-1} y$ or $\arccos y$ means "an angle whose cosine is y," and $\tan^{-1} y$ or $\arctan y$ means "an angle whose tangent is y."

Example 16.9. Find all the values of $\cos^{-1}(-\sqrt{2}/2)$ in degrees.

In terms of the cosine function, this statement means "find all angles θ, such that $\cos\theta = -\sqrt{2}/2$." Since $\cos 45° = \sqrt{2}/2$, θ can be any second- or third-quadrant angle for which 45° is the reference angle. Therefore, the solutions are

$$\theta = \arccos\left(-\frac{\sqrt{2}}{2}\right) = 135° + k \cdot 360°,\ 225° + k \cdot 360°$$

where $k = 0, \pm 1, \pm 2, \pm 3, \ldots$

Example 16.10. Determine the angle θ with the smallest absolute value, such that $\theta = \tan^{-1}(-1.428)$.

Since $\tan\theta = -1.428$, we first find the reference angle, 0.9599 rad (or 55°), from the trigonometric tables. Ordinarily, we would subtract from π (or 180°) to find the second-quadrant angle. However, since we are to find the angle with the smallest absolute value, we must use the negative angle in the fourth quadrant. Thus,

$$\boxed{\theta = -0.9599 \text{ rad} \quad \text{or} \quad \theta = -55°}$$

Example 16.11. Solve for x: $2y = \sin(2x - \pi/4)$

We make use of the inverse-sine notation to solve for the *angle*, $2x - \pi/4$, as follows:

$$2x - \frac{\pi}{4} = \arcsin 2y$$

$$2x = \frac{\pi}{4} + \arcsin 2y$$

$$\boxed{x = \frac{\pi}{8} + \frac{1}{2}\arcsin 2y}$$

EXERCISE 16.2

In Probs. 1 to 6, determine all possible values of the given expressions in the units indicated.

1. $\cos^{-1}(\sqrt{3}/2)$, degrees
2. $\arcsin(1/2)$, radians
3. $\text{arccot}(-1)$, radians
4. $\tan^{-1}(-\sqrt{3})$, degrees
5. $\sin^{-1}(-\sqrt{2}/2)$, degrees
6. $\text{arcsec}\ \sqrt{2}$, radians

In Probs. 7 to 12, determine in radians and in degrees the value of the angle with the smallest absolute value.

7. $\sin^{-1} 1$
8. $\arccos(-1/2)$
9. $\arctan 1$
10. $\text{arccsc}(-2)$
11. $\cos^{-1} 0.8090$
12. $\sin^{-1}(-0.7648)$

In Probs. 13 to 18, solve for x.

13. $-\sqrt{3}/3 = \tan x$
14. $2u = \sin 4x$
15. $\pi/2 = \cos^{-1} 3x$
16. $\cot^{-1}(5x - 4) = -\pi/4$
17. $30° + \sin^{-1} 2x = 0$
18. $y/2 = 1 - \sec(x + 2)$

16.3 THE PRINCIPAL INVERSE TRIGONOMETRIC FUNCTIONS

Examples 16.8 and 16.9 point out the fact that the inverse of a trigonometric function has infinitely many values. This situation duplicates that involving the solutions of trigonometric equations. For example, the inverse statement $\theta = \sin^{-1}(\sqrt{3}/2)$ corresponds exactly to the trigonometric equation $\sin\theta = \sqrt{3}/2$, and the general solutions (Example 16.8) are identical.

To avoid the multiplicity of solutions, we defined in Chap. 15 the fundamental solutions of a trigonometric equation. Now we shall make similar definitions for the values of the inverse trigonometric functions. However, the applications of inverse trigonometric functions differ from those of trigonometric equations (the applications are too complicated to discuss here); therefore the definitions are different. To avoid confusion, we shall use the term *principal inverse functions,* whereas we previously used *fundamental solutions.*

The definitions of the principal inverse trigonometric functions are as follows. Notice that we always *capitalize the first letter to distinguish the principal inverse function* from the general inverse of the function.

$$\theta = \mathrm{Sin}^{-1}\, y \quad \text{or} \quad \theta = \mathrm{Arcsin}\, y$$
$$\text{where } y = \sin\theta \quad \text{and} \quad -\frac{\pi}{2} \leqslant \theta \leqslant \frac{\pi}{2} \tag{16.7}$$

$$\theta = \mathrm{Cos}^{-1}\, y \quad \text{or} \quad \theta = \mathrm{Arccos}\, y$$
$$\text{where } y = \cos\theta \quad \text{and} \quad 0 \leqslant \theta \leqslant \pi \tag{16.8}$$

$$\theta = \mathrm{Tan}^{-1}\, y \quad \text{or} \quad \theta = \mathrm{Arctan}\, y$$
$$\text{where } y = \tan\theta \quad \text{and} \quad -\frac{\pi}{2} < \theta < \frac{\pi}{2} \tag{16.9}$$

$$\theta = \mathrm{Cot}^{-1}\, y \quad \text{or} \quad \theta = \mathrm{Arccot}\, y$$
$$\text{where } y = \cot\theta, \quad -\frac{\pi}{2} \leqslant \theta \leqslant \frac{\pi}{2}, \quad \theta \neq 0 \tag{16.10}$$

$$\theta = \mathrm{Sec}^{-1}\, y \quad \text{or} \quad \theta = \mathrm{Arcsec}\, y$$
$$\text{where } y = \sec\theta, \quad 0 \leqslant \theta \leqslant \pi, \quad \theta \neq \frac{\pi}{2} \tag{16.11}$$

$$\theta = \mathrm{Csc}^{-1}\, y \quad \text{or} \quad \theta = \mathrm{Arccsc}\, y$$
$$\text{where } y = \csc\theta, \quad -\frac{\pi}{2} \leqslant \theta \leqslant \frac{\pi}{2}, \quad \theta \neq 0 \tag{16.12}$$

The Principal Inverse Trigonometric Functions

The objectives in these definitions are (1) to provide exactly one value for any given inverse trigonometric function and (2) to limit the range of values to the smallest possible continuous set of angles.

Example 16.12. **Solve for x in radians and in degrees, given that $2x = \mathrm{Tan}^{-1}\, 1$.**

First, we reason that $\tan(\pi/4) = 1$. Therefore $\mathrm{Tan}^{-1}\, 1 = \pi/4$. Next we solve for x.

$$2x = \mathrm{Tan}^{-1}\, 1 = \frac{\pi}{4}$$

$$\boxed{x = \frac{\pi}{8} \qquad \text{or} \qquad x = 22\tfrac{1}{2}^\circ}$$

Note that although $\tan(5\pi/4) = 1$ also, $5\pi/4$ is not allowed as a value for $\mathrm{Tan}^{-1}\, 1$ because it is outside the specified range as defined in Eq. (16.9).

Example 16.13. Find the value of Arcsin $(-\tfrac{1}{2})$.

Since $\sin(\pi/6) = \tfrac{1}{2}$, $\pi/6$ is a reference angle for Arcsin $(-\tfrac{1}{2})$. Therefore

$$\boxed{\mathrm{Arcsin}\left(-\frac{1}{2}\right) = -\frac{\pi}{6}}$$

Note that we cannot use $7\pi/6$ or $11\pi/6$ as solutions because of the range of values dictated by the definition, Eq. (16.7).

Example 16.14. Solve for y in degrees, if $y/3 + 45^\circ = \mathrm{Cos}^{-1}(-0.7808)$.

First, from the trigonometric tables we find that $\cos 38^\circ 40' = 0.7808$. Therefore $\mathrm{Cos}^{-1}(-0.7808)$ is the second-quadrant angle: $180^\circ - 38^\circ 40' = 141^\circ 20'$. Thus

$$\frac{y}{3} + 45^\circ = \mathrm{Cos}^{-1}(-0.7808) = 141^\circ 20'$$

$$\frac{y}{3} = 141^\circ 20' - 45^\circ = 96^\circ 20'$$

$$\boxed{y = 289^\circ}$$

Example 16.15. Evaluate Arccos $[\sin(11\pi/6)]$.

$$\mathrm{Arccos}\left(\sin\frac{11\pi}{6}\right) = \mathrm{Arccos}\left(-\frac{1}{2}\right) = \frac{2\pi}{3}$$

Example 16.16. Evaluate $\cos(\pi + \mathrm{Sin}^{-1}\, x)$.

A good procedure in evaluating such expressions is to substitute a letter for the inverse function. Let $\text{Sin}^{-1} x = \theta$. Then use an identity to evaluate the function.

$$\cos(\pi + \theta) = \cancelto{-1}{\cos \pi} \cos \theta - \cancelto{0}{\sin \pi} \sin \theta = -\cos \theta$$

Finally, we must evaluate $\cos \theta$. Since $\theta = \text{Sin}^{-1} x$, then $x = \sin \theta$ and

$$\cos \theta = \sqrt{1 - \sin^2 \theta} = \sqrt{1 - x^2}$$

Therefore the required expression is

$$\boxed{\cos(\pi + \text{Sin}^{-1} x) = -\sqrt{1 - x^2}}$$

Note that we use only the positive root in the expression for $\cos \theta$ above, because $\text{Sin}^{-1} x$ is between $\pi/2$ and $-\pi/2$, and the cosine is positive in this range.

Example 16.17. What is $\sec(\text{Arcsec}\, x)$?

This expression means "the secant of the angle whose secant is x." The question is much like, What is the name of the man whose name is Jones? The answer should be obvious. However, to proceed in a more analytical vein, we make the substitution $\theta = \text{Arcsec}\, x$. Then we try to find $\sec \theta$. But by definition, $\theta = \text{Arcsec}\, x$ means that $\sec \theta = x$. Therefore

$$\boxed{\sec(\text{Arcsec}\, x) = x}$$

This example illustrates the point that *the inverse function of a function of a number is the number itself.* In other words, the function undoes to the number what the inverse function does, and vice versa.

Example 16.18. Evaluate $\sin[\text{Cos}^{-1}(1/4) - \text{Sin}^{-1}(1/5)]$.

Again, we make substitutions for the inverse functions. Let

$$\alpha = \text{Cos}^{-1} \frac{1}{4} \qquad \text{and} \qquad \beta = \text{Sin}^{-1} \frac{1}{5}$$

Then $\qquad \cos \alpha = \dfrac{1}{4} \qquad\qquad \sin \beta = \dfrac{1}{5}$

and

$$\sin \alpha = \frac{1}{4}\sqrt{15} \qquad \cos \beta = \frac{2}{5}\sqrt{6}$$

Now we have

$$\begin{aligned}\sin(\alpha - \beta) &= \sin \alpha \cos \beta - \cos \alpha \sin \beta \\ &= \frac{1}{4}\sqrt{15} \cdot \frac{2}{5}\sqrt{6} - \frac{1}{4} \cdot \frac{1}{5} = \frac{6\sqrt{10}}{20} - \frac{1}{20}\end{aligned}$$

$$\boxed{\sin\left(\text{Cos}^{-1}\frac{1}{4} - \text{Sin}^{-1}\frac{1}{5}\right) = \frac{1}{20}(6\sqrt{10} - 1)}$$

EXERCISE 16.3

In Probs. 1 through 18, evaluate each expression in radians and in degrees.

1. $\text{Sin}^{-1}(1/2)$
2. $\text{Cos}^{-1}(\sqrt{3}/2)$
3. $\text{Arccos}(-\sqrt{3}/2)$
4. Arccot 0
5. $\text{Tan}^{-1}\ 2$
6. $\text{Csc}^{-1}\ 2$
7. $\text{Sec}^{-1}\sqrt{2}$
8. Arctan (-1)
9. $\text{Arcsin}(-\sqrt{2}/2)$
10. $\text{Sin}^{-1}\ 0.5225$
11. $\text{Cos}^{-1}(-0.9948)$
12. $\text{Tan}^{-1}(-1.393)$
13. $\text{Tan}^{-1}[\cos(3\pi/4)]$
14. $\text{Arcsin}[\sin(2\pi/3)]$
15. $\text{Cos}^{-1}[\sin(5\pi/4)]$
16. $\text{Sec}^{-1}[\tan(-60°)]$
17. $\text{Csc}^{-1}(\csc 36°)$
18. $\text{Arccot}[\sin(\pi/6)]$

In Probs. 19 to 27, evaluate each expression.

19. $\sin[\text{Tan}^{-1}(-1)]$
20. $\tan[\text{Cos}^{-1}(1/2)]$
21. $\cot[\text{Arccot}(-10)]$
22. $\cos\left[\frac{\pi}{2} - \text{Sin}^{-1}\left(-\frac{1}{2}\right)\right]$
23. $\sin(\text{Cos}^{-1} u)$
24. $\cos(2\ \text{Sin}^{-1} x)$
25. $\tan(\pi/4 + \text{Cot}^{-1} 2z)$
26. $\sin[2\ \text{Cos}^{-1}(3/5)]$
27. $\cos(\text{Sin}^{-1} x + \text{Cos}^{-1} x)$

ADVANCED TOPICS IN ALGEBRA AND PRECALCULUS MATHEMATICS

UNIT 4

COMPLEX NUMBERS

CHAPTER 17

In this chapter we shall examine a system of numbers called *complex numbers.* But first, we shall try to dispel certain unfortunate connotations in the language of complex numbers. The name "complex number" itself, for example, conjures up a notion of something too complicated for any mortal student to comprehend. But such is not the case. Complex numbers are not complex in the sense of being extraordinarily complicated.

A brief look at the development of numbers in general will help to put complex numbers in proper perspective. First, we should understand that all types of numbers are invented; they are products of man's imagination. In this sense, all numbers are artificial and imaginary. All are man-made and made to perform some service, such as counting. Indeed, the very first numbers were used for counting. Man invented abstract symbols (which we call numerals) to represent numbers of concrete objects. The counting numbers freed man from tasks such as collecting pebbles in order to keep count of his belongings and business transactions.

As specific needs developed, man invented new kinds of numbers such as fractions and negative numbers. For example, the need to break standard units of measure down into parts gave rise to concepts and symbols for fractions.

Some types of numbers evolved as a result of an invented system of numbers, no practical application being apparent initially. For example, it is probable that an early-day mathematician invented negative numbers when he needed answers for the problems of "taking away" one number from a smaller number (as in $4 - 5 = -1$).

Such inventions required many centuries to become universally accepted. Negative numbers are known to date back to the time of Christ; yet as late as the 1500s they were called *fictitious* numbers. We can well imagine that people of that era were accustomed to using numbers for counting and measuring tangible objects. The idea of a negative object must have seemed preposterous and "fictitious," an invention of crackpot mathematicians and certainly of no practical use. Yet today we readily accept negative numbers. We realize that there are no negative objects, but we make good use of negative numbers as answers to problems and as directional indicators in coordinate systems.

Complex numbers have the form $a + bj$, where j stands for $\sqrt{-1}$. The history of misunderstanding and rejection of complex numbers is very similar to that of negative numbers. Part of the misunderstanding centers on the so-called "imaginary unit" $\sqrt{-1}$. The name *imaginary* is a misnomer; its use in mathematics is a disaster which has not yet been overcome. Except inasmuch as all numbers are products of man's imagination, $\sqrt{-1}$ is no more imaginary than -1 is fictitious. Both numbers were invented to provide answers to certain types of problems. With -1 we can represent answers to problems such as $x = 4 - 5$, and with $\sqrt{-1}$ we have a convenient means of representing answers to problems such as $x^2 = -1$. Furthermore, both types of numbers have practical applications. Although complex numbers cannot be used for counting or measuring objects, they are, like negative numbers, used as directional indicators in coordinate systems. They are also used to describe very real concepts in electricity and in vibratory motion.

17.1 BASIC IDEAS ABOUT IMAGINARY NUMBERS

The entire universe of numbers can be subdivided into two types: *real* and *imaginary.* The real numbers are the integers, the rational numbers, and the irrational numbers, as outlined in Chap. 2. Every real number can be written in a form that does *not* contain a root of a negative number. Imaginary numbers are defined as follows:

IMAGINARY NUMBER	An imaginary number is any number that can be written in the form $b\sqrt{-1}$, where b is a real number.

For example, $\sqrt{-25} = \sqrt{(25)(-1)} = 5\sqrt{-1}$ is an imaginary number.

To simplify the writing of imaginary numbers, we define the letter j to mean $\sqrt{-1}$, and we also define j^2 to be -1. All other letters in this book will stand for real numbers. By definition, then, we have

$$j = \sqrt{-1} \tag{17.1}$$

$$j^2 = -1 \tag{17.2}$$

Therefore $2j$, for example, means $2\sqrt{-1}$, and

$$(5j)^2 = 25j^2 = 25(-1) = -25$$

Traditionally, mathematicians have used i for $\sqrt{-1}$, and most mathematics books do so, but books on electricity use i for current and j for $\sqrt{-1}$. Therefore as a concession to the predominant application for imaginary numbers, we shall use the symbol j for $\sqrt{-1}$.

Powers of j

A very simple and useful operation is to evaluate any positive, integral power of j. All such powers are based upon the definitions of Eqs. (17.1) and (17.2) and the law of exponents $a^m \cdot a^n = a^{m+n}$. The first eight powers are listed below.

$$j^1 = j \qquad j^5 = j \cdot j^4 = j \cdot 1 = j$$

$$j^2 = (\sqrt{-1})^2 = -1 \qquad j^6 = j^2 \cdot j^4 = (-1)(1) = -1$$

$$j^3 = j \cdot j^2 = j(-1) = -j \qquad j^7 = j^3 \cdot j^4 = (-j)(1) = -j$$

$$j^4 = j^2 \cdot j^2 = (-1)(-1) = 1 \qquad j^8 = j^4 \cdot j^4 = (1)(1) = 1$$

This list shows that the powers of j go through repetitive cycles of j, -1, $-j$, and 1. And for exponents that differ by 4 (or a multiple of 4), the powers of j are equal. For example, $j^3 = j^7 = j^{11} = j^{15}$, and so on. The following rule enables us to evaluate such powers quickly:

> Any positive integral power j^n equals j^r, where r is the remainder after dividing n by 4; if the remainder is zero, then $j^n = j^r = 1$.

Example 17.1. Powers of j

a. Since $7 \div 4 = 1 +$ remainder of 3, $j^7 = j^3 = -j$.
b. Since $38 \div 4 = 9 +$ remainder of 2, $j^{38} = j^2 = -1$.
c. Since $52 \div 4 = 13 +$ remainder of 0, $j^{52} = j^4 = 1$.

Operations with Imaginary Numbers

The following laws of exponents (or radicals) are *not always valid* if imaginary numbers are concerned:

$$(ab)^{1/n} = a^{1/n}b^{1/n} \quad \text{or} \quad \sqrt[n]{ab} = \sqrt[n]{a}\,\sqrt[n]{b}$$

$$(b^{1/n})^k = (b^k)^{1/n} \quad \text{or} \quad (\sqrt[n]{b})^k = \sqrt[n]{b^k}$$

Difficulty arises for certain combinations of even-numbered indices n and for negative values of a and b. However, by using the imaginary unit j and the following guidelines, we can always avoid errors.

1. Any number of the form $\sqrt{-a}$, for $a > 0$, can always be separated into a real factor $\sqrt{a}$ and the imaginary unit j. In other words, $\sqrt{-a} = \sqrt{a} \cdot \sqrt{-1} = \sqrt{a}j$ is always true.
2. If imaginary numbers are to be operated on, always write them in the form bj; then perform the indicated operations.

   Example:

   $$\sqrt{-2}\sqrt{-8} = (\sqrt{2}j)(\sqrt{8}j) = \sqrt{2 \cdot 8}j^2 = \sqrt{16}(-1) = -4$$

   Note that $\sqrt{-2}\sqrt{-8}$ *cannot* be written as $\sqrt{(-2)(-8)} = \sqrt{16} = 4$.
3. If operations on *negative* numbers are indicated under a radical sign, *perform the operations first;* then evaluate the radical or change to bj form.

   Example: $\sqrt{(-2)(-8)} = \sqrt{+16} = 4$.

   Note that $\sqrt{(-2)(-8)}$ *cannot* be written as

   $$\sqrt{-2}\sqrt{-8} = (\sqrt{2}j)(\sqrt{8}j) = 16j^2 = -4.$$

Example 17.2. Evaluate $(\sqrt{-9})^2$.

$$(\sqrt{-9})^2 = (\sqrt{9}j)^2 = (\sqrt{9})^2j^2 = 9(-1) = -9$$

Notice that we *cannot* write $(\sqrt{-9})^2$ as $\sqrt{(-9)^2} = \sqrt{81} = 9$; 9 is not equal to -9.

Example 17.3. *a.* $\sqrt{-6} \cdot \sqrt{2} = \sqrt{6}j \cdot \sqrt{2} = \sqrt{12}j = 2\sqrt{3}j$

b. $\sqrt{-4} \cdot \sqrt{-9} = \sqrt{4}j \cdot \sqrt{9}j = 2 \cdot 3j^2 = -6$

Notice that we *cannot* write $\sqrt{-4} \cdot \sqrt{-9}$ as $\sqrt{(-4)(-9)} = \sqrt{36} = 6$.

EXERCISE 17.1

In Probs. 1 through 12, evaluate each power of j as either j, -1, $-j$, or 1.

1. j^9	2. j^{10}	3. j^{20}	4. j^{15}
5. j^{69}	6. j^{-16}	7. $-j^{58}$	8. j^{244}
9. j^{-1}	10. $(-j)^{13}$	11. $-j^{350}$	12. j^{-7}

In Probs. 13 to 20, simplify each expression. If the result is an imaginary number, write it in terms of j.

13. $\sqrt{-4}$	14. $\sqrt{2(-6)}$	15. $\sqrt{(-2)(-6)}$	16. $\sqrt{-4/3}$
17. $(\sqrt{-2})^2$	18. $\sqrt{-0.16}$	19. $\sqrt{27/(-5)}$	20. $(\sqrt{-3})^3$

17.2 BASIC IDEAS ABOUT COMPLEX NUMBERS

A complex number is a combination of a real number and an imaginary number. There are various forms in which a complex number can be written, but they all stem from the same concept. We shall first describe a complex number in terms of the coordinates of a point in a special rectangular coordinate plane.

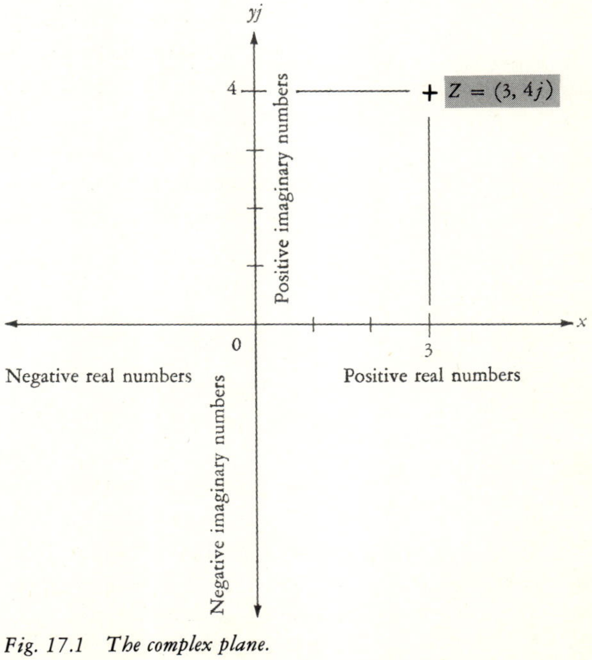

Fig. 17.1 The complex plane.

We can place all real numbers on the abscissa of a coordinate plane and all imaginary numbers on the ordinate, as shown in Fig. 17.1. Such a coordinate plane is called a *complex plane,* and each point in the plane represents a complex number. Thus a *complex number* is an ordered pair of numbers (x, yj) in which x is the *real part* and yj is the *imaginary part.* For example, the complex number Z of Fig. 17.1 can be described as $Z = (3,4j)$. From this description we see that a complex "number" is not a single number in the usual sense, but is instead a combination of a real number and an imaginary number.

The complex numbers include the pure-real numbers and the pure-imaginary numbers. Thus every point on the real or x axis is a pure-real number and has coordinates $(x,0)$, and every point on the imaginary or yj axis is a pure-imaginary number and has coordinates $(0, yj)$.

Algebraic Form

A frequently used notation for complex numbers is of the form $x + yj$. This notation is called *algebraic form* or *rectangular form.* The coordinate notation (x, yj) is equivalent to $x + yj$. Either form can be used to locate a point in the complex plane, and the two forms are interchangeable.

The addition sign in $x + yj$ is somewhat misleading because the real part and the imaginary part cannot be added as scalar numbers. However, we can perform algebraic operations on complex numbers in algebraic form just as we operate on pure-real numbers such as $ax + by$. We shall consider operations with complex numbers in the next section.

Algebraic form is convenient for writing solutions to certain quadratic equations. In Sec. 11.3 we found that the solutions to the general quadratic equation $ax^2 + bx + c = 0$ are given by the quadratic formula

$$x = \frac{-b \pm \sqrt{b^2 - 4ac}}{2a}$$

The radicand $b^2 - 4ac$ is called the *discriminant.* The solutions can be pure-real numbers or more general complex numbers, depending on whether the discriminant is positive (or zero) or negative. If the discriminant is positive or zero, then x is a pure-real number; if the discriminant is negative, then x is a more general complex number of the form $A \pm Bj$.

Example 17.4. Solve $x^2 - 3x + 5 = 0$.

From the quadratic formula we have

$$x = \frac{-(-3) \pm \sqrt{(-3)^2 - 4(1)(5)}}{2(1)} = \frac{3}{2} \pm \frac{\sqrt{-11}}{2}$$

$$x = \frac{3}{2} \pm \frac{1}{2}\sqrt{11}j$$

Vector Interpretation

As shown in Fig. 17.2, we can interpret the sum $x + yj$ as *vector* addition. If $\mathbf{Z} = \mathbf{x} + \mathbf{yj}$, then $\mathbf{x}$ and $\mathbf{yj}$ are the rectangular components of the complex vector $\mathbf{Z}$ from the origin to the point (x, yj). We find the *magnitude* or *absolute value* of $\mathbf{Z}$ as follows, by using the Pythagorean formula:

$$|\mathbf{Z}| = \sqrt{x^2 + y^2}$$

And we find the directional angle from the relationship $\tan\theta = y/x$.

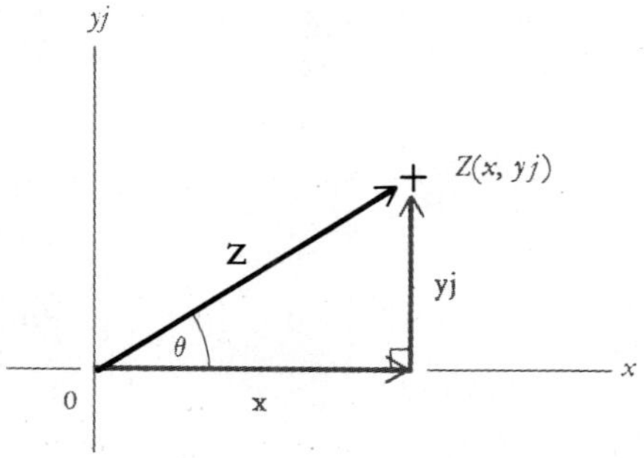

Fig. 17.2 A complex number $x + yj$ can be used to represent a vector: $\mathbf{Z} = x + yj$.

Example 17.5. Find the absolute value and direction of the vector $\mathbf{Z} = 3 + 4j$.

$$|\mathbf{Z}| = \sqrt{3^2 + 4^2} = 5 \qquad \text{and} \qquad \theta = \mathrm{Tan}^{-1}\frac{4}{3} = 53°10'$$

Equality

Two complex numbers are defined to be equal if they correspond to the same point in the complex plane. Thus the real parts must be equal, and the imaginary parts must be equal. In general,

EQUALITY OF COMPLEX NUMBERS	If	$a + bj = c + dj$
	then	$a = c$ and $b = d$

Example 17.6. Solve for x and for y: $4 + yj = 2x - 3j$.

By equating the real parts we have

$$4 = 2x \qquad \text{or} \qquad x = 2$$

and from the imaginary parts we have $y = -3$.

EXERCISE 17.2

In Probs. 1 to 6, solve each quadratic equation. If the solutions are not pure-real numbers, write them in the form $A + Bj$.

1. $x^2 + x + 1 = 0$
2. $x^2 - x + 1 = 0$
3. $x^2 + 2 = 0$
4. $3x^2 - 4x + 2 = 0$
5. $5x^2 - 13x + 6 = 0$
6. $x^2 + 16 = 0$

In Probs. 7 to 14, find the magnitude and the directional angle of the vector corresponding to each given complex number. Sketch each vector in the complex plane.

7. $1 + j$
8. $\sqrt{2} + \sqrt{2}j$
9. $(2, \sqrt{5}j)$
10. $(3, j)$
11. $3 + 4j$
12. $-3 + 4j$
13. $-3 - 4j$
14. $3 - 4j$

In Probs. 15 to 20, solve each equation for x and y.

15. $x + yj = 5 + 3j$
16. $3x - yj = 9 + 2j$
17. $15 + 4yj = -[(3/2)x + 6j]$
18. $(x, yj) = (-3, 7j)$
19. $(-5, 2yj) = (10x, -j)$
20. $3a - 4byj = 2a^2x + 8b^2j$

17.3 OPERATIONS WITH COMPLEX NUMBERS

Algebraic operations with complex numbers are much the same as those with pure-real numbers. But in order to perform them correctly, we must handle the imaginary unit j properly.

Addition and Subtraction

To add two complex numbers, combine the real parts and the imaginary parts separately, as follows:

$$(a + bj) + (c + dj) = (a + c) + (b + d)j \tag{17.3}$$

Example 17.7. *a.* $(3+5j)+(6+7j)=(3+6)+(5+7)j=9+12j$

b. $(2-j)-(3-3j)=(2-3)+(-1+3)j=-1+2j$

c. $-(5+4j)+(-17+j)=(-5-17)+(-4+1)j=-22-3j$

Multiplication

Multiply two complex numbers just as you would any other binomial expressions. Then evaluate the powers of j according to Sec. 17.1, and combine the real and imaginary parts as follows:

$$\begin{aligned}(a+bj)(c+dj) &= ac+bdj^2+(ad+bc)j\\ &= ac+bd(-1)+(ad+bc)j\\ (a+bj)(c+dj) &= ac-bd+(ad+bc)j \qquad (17.4)\end{aligned}$$

Since Eq. (17.4) provides the general expression for the product, you can use it directly without having to work through each step.

Example 17.8. *a.* $(3+5j)(6+7j)=(3\cdot 6-5\cdot 7)+(3\cdot 7+5\cdot 6)j$
$=-17+51j$

b. $(2-j)(3+7j)=[2\cdot 3-(-1)7]+[2\cdot 7+(-1)3]j$
$=13+11j$

c. $(-1+2j)(12-13j)=[(-1)12-2(-13)]+[(-1)(-13)+2(12)]j$
$=14+37j$

Division

If the divisor in a division problem contains an imaginary number, then you should simplify the quotient by rationalizing the denominator. The process is like that for rationalizing the denominator of a fraction. (See Sec. 12.4.) The idea is to multiply the numerator and the denominator by the *conjugate* of the denominator. If $c+dj$ is one complex number, then its *complex conjugate* is $c-dj$. The general procedure for division is as follows:

$$\begin{aligned}\frac{a+bj}{c+dj} = \frac{(a+bj)(c-dj)}{(c+dj)(c-dj)} &= \frac{(ac-bdj^2)+(bc-ad)j}{c^2-cdj+cdj-d^2j^2}\\ &= \frac{ac+bd}{c^2+d^2}+\frac{bc-ad}{c^2+d^2}\,j \qquad (17.5)\end{aligned}$$

Notice why this procedure works: *The product of a complex number and its conjugate is always the sum of two perfect-square, real numbers.* That is, $(c+dj)(c-dj)=c^2+d^2$.

Example 17.9. *a.*

$$\frac{2+3j}{5+4j}=\frac{(2+3j)(5-4j)}{(5+4j)(5-4j)}=\frac{(2\cdot 5-3\cdot 4j^2)+(3\cdot 5-2\cdot 4)j}{5^2+4^2}$$

$$=\frac{22}{41}+\frac{7}{41}j$$

b. In this example we shall use Eq. (17.5) directly, without going through the intermediate steps.

$$\frac{-12+5j}{4-3j}=\frac{(-12)(4)+5(-3)}{4^2+3^2}+\frac{5(4)-(-12)(-3)}{4^2+3^2}j=-\frac{63}{25}-\frac{16}{25}j$$

EXERCISE 17.3

Perform the indicated operations and write each answer in the form $a+bj$.

1. $(2+3j)+(1+2j)$
2. $(5-3j)-(3+j)$
3. $(4-j)-(4+2j)$
4. $(4j)(6j)$
5. $(5+\sqrt{-7})\sqrt{-4}$
6. $j(16-27j)$
7. $(1+j)(1-j)$
8. $(1-j)^2$
9. $(2+3j)^3$
10. $(-6+3j)(5-4j)$
11. $6j/(3j)$
12. $(2+3j)/j$
13. $(1-\sqrt{-1})/\sqrt{-2}$
14. $\sqrt{-15}/(3-\sqrt{-3})$
15. $(1+j)/(1-j)$
16. $\dfrac{1-j}{1+j}$
17. $\dfrac{3+2j}{5-4j}$
18. $-5+9j+\dfrac{1+j}{2+j}$
19. $\dfrac{(7-3j)(3+5j)}{2+2j}$
20. $\dfrac{5-2j}{(1+j)(2-j)}$
21. $\dfrac{(2-3j)(1+4j)}{(7+j)(5-6j)}$

17.4 COMPLEX VECTORS

We have seen in Sec. 17.2 that a complex number can be interpreted as components of a vector. Now we shall discuss the meaning of complex vectors and see how to do certain operations with them.

In Chap. 2 we discussed the significance of negative numbers. We saw that no physical quantity can have a negative magnitude, just as no yardstick can be -3 ft long. The negative sign indicates that we have measured 3 ft in a certain *direction*. Likewise, no physical quantity can have an imaginary magnitude; no yardstick can be $3j$ ft long.

But we can use the j factor to indicate a certain *direction* for measuring 3 ft. Furthermore, we can use a complex number $x + yj$ to indicate the magnitudes and directions of *two* quantities at *right angles.* In this sense we use complex numbers and the complex plane as a coordinate system, and in this sense we talk about complex vectors.

The only difference between complex vectors and real vectors is the notation used to describe them. We can use the $x + yj$ notation (some people just use the more simple coordinate notation $[x, yj]$) to represent any kind of real vector quantity such as velocity or force. However, the preeminent application is the treatment of alternating voltages, currents, and impedances as vectors. The important point is this: *We use the complex number system to indicate directions of vector components; no vector component has an imaginary magnitude.* A complex number is merely a convenient and compact means to describe a vector in terms of its rectangular components.

Given two complex numbers $Z_1 = x_1 + y_1 j$ and $Z_2 = x_2 + y_2 j$, we find their sum algebraically by adding the real parts and adding the imaginary parts. Thus $Z_1 + Z_2 = (x_1 + x_2) + (y_1 + y_2)j$. Figure 17.3 shows that the graphical counterpart of this addition is a vector addition. By adding the rectangular components as scalars, we find the rectangular components of the vector sum. The vector difference would be similar. This process is no different from that for adding real vectors as in Sec. 6.3. The graphical method is no different from that for real vectors as in Sec. 6.2.

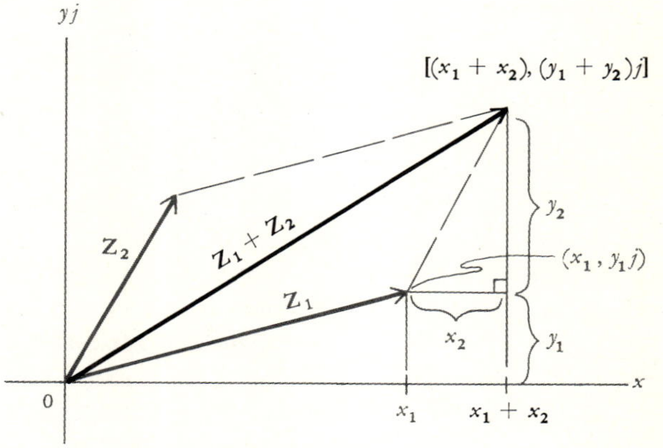

Fig. 17.3 Addition of complex vectors.

Example 17.10. Determine algebraically and graphically the sum of the two vectors $\mathbf{Z}_1 = 6 + 8j$ and $\mathbf{Z}_2 = -12 - 5j$.

The algebraic solution is as follows:

$$\mathbf{Z} = \mathbf{Z}_1 + \mathbf{Z}_2 = (6 + 8j) - (12 + 5j) = -6 + 3j$$

The magnitude and direction are

$$|\mathbf{Z}| = \sqrt{(-6)^2 + 3^2} = 3\sqrt{5}$$

$$\tan \theta_z = \frac{3}{-6} = -0.500$$

$$\theta_z = 180° - 26°30' = 153°30'$$

Note that θ_z has to be a second-quadrant angle because of the signs of the components of $\mathbf{Z}$.

The graphical solution is shown in Fig. 17.4.

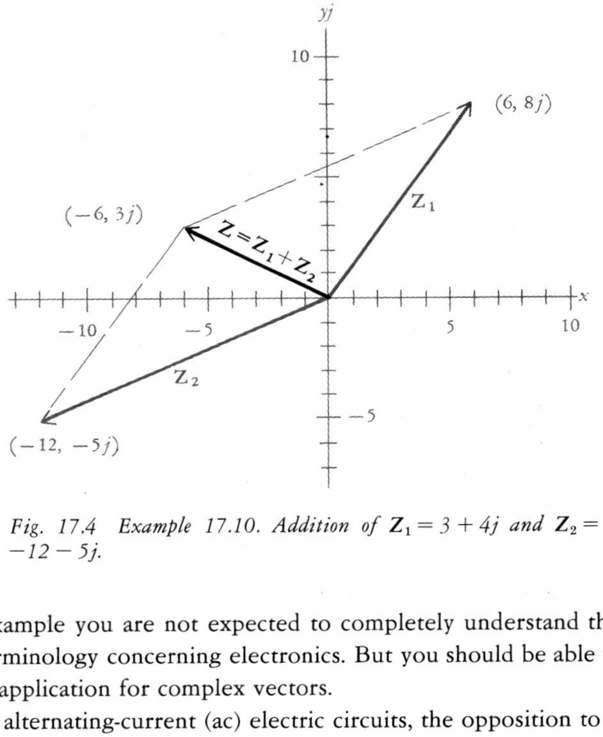

Fig. 17.4 Example 17.10. Addition of $\mathbf{Z}_1 = 3 + 4j$ *and* $\mathbf{Z}_2 = -12 - 5j$.

Example 17.11. In this example you are not expected to completely understand the theory or the terminology concerning electronics. But you should be able to appreciate the application for complex vectors.

In alternating-current (ac) electric circuits, the opposition to the flow of electrons is called *impedance* (not simply *resistance*). Impedance consists of three components; resistance R, capacitive reactance X_C, and inductive reactance X_L. It is a well-established fact that these components act out of phase timewise. X_L is always one-quarter period ahead of R, and X_C is always one-quarter period behind R. In other words, X_L leads R by 90°, and X_C lags R by 90°. The impedance can be treated as the *vector* sum of R, X_C, and X_L.

In the complex plane, the angle between a pure-imaginary vector and a pure-real vector is 90°. Therefore, complex vectors are ideally suited for describing impedances and their components. Resistance becomes the "real" component of impedance, and the reactances X_C and X_L become "imaginary" components. The resulting impedance can be described mathematically as follows:

$$\boldsymbol{Z} = \boldsymbol{R} + \boldsymbol{X}_1 j + \boldsymbol{X}_C(-j) = \boldsymbol{R} + (\boldsymbol{X}_L - \boldsymbol{X}_C)j$$

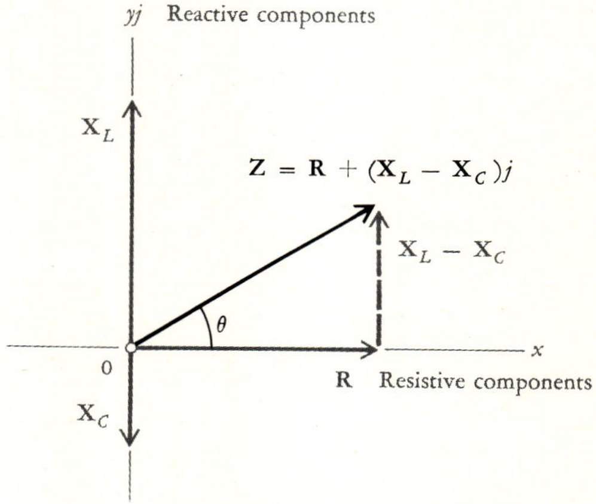

Fig. 17.5 Vector representation of impedance, **Z**, *and its components,* **R**, $\boldsymbol{X}_C$, *and* $\boldsymbol{X}_L$.

Figure 17.5 gives the graphical interpretation.

If $\boldsymbol{R} = 6$ ohms, $\boldsymbol{X}_C = 2(-j)$, and $\boldsymbol{X}_L = 5j$, then $\boldsymbol{Z} = 6 + (5 - 2)j = 6 + 3j$ ohms. The magnitude of $\boldsymbol{Z}$ is

$$|\boldsymbol{Z}| = \sqrt{6^2 + 3^2} = 3\sqrt{5} \text{ ohms}$$

and the phase angle θ is

$$\theta = \text{Tan}^{-1} (3/6) = 26°30'$$

EXERCISE 17.4

In Probs. 1 to 16, find the resultant of each set of complex vectors. Express each resultant in $x + yj$ form and determine its magnitude and directional angle.

1. $(3 + 0j) + (0 + 4j)$
2. $(5 + 2j) + (5 - 2j)$
3. $(6 - 7j) - (6 + 3j)$
4. $(0 - 8j) - (6 + 0j)$
5. $(-5 + 2j) + (3 - 2j)$
6. $(-1 + 12j) + (1 - 3j)$

7. $(2+j)-(5-5j)$
8. $(1+j)+(1+j)$
9. $(2+j)+(5-5j)$
10. $(1.2-0.5j)+(0.3-1.0j)$
11. $-(2.6+3.5j)-(1.4-0.8j)$
12. $(-5.6+1.0j)+(3.0+2.1j)$
13. $(1+0j)+(0-3j)+(0+5j)$
14. $(2+2j)-(3+0j)+(0-5j)$
15. $(5+4j)+(3-6j)-(4+j)$
16. $(8-3j)+(2-5j)+(-3+2j)$

In Probs. 17 to 32, use the vectors of Probs. 1 to 16, respectively, and find each resultant graphically.

In Probs. 33 to 36, find each impedance vector and its magnitude and phase angle (as in Example 17.11) for the given values of **R**, $\mathbf{X}_C$, and $\mathbf{X}_L$.

33. $\mathbf{R}=5$, $\mathbf{X}_C=-3j$, $\mathbf{X}_L=4j$
34. $\mathbf{R}=10$, $\mathbf{X}_C=-5j$, $\mathbf{X}_L=5j$
35. $\mathbf{R}=1.5$, $\mathbf{X}_C=-2.3j$, $\mathbf{X}_L=0.9j$
36. $\mathbf{R}=0.4$, $\mathbf{X}_C=-0.4j$, $\mathbf{X}_L=0.6j$

37. If in Example 17.11, $|\mathbf{Z}|=10$, $\theta=30°$, $\mathbf{R}=8.7$, and $\mathbf{X}_L=7j$, what is $\mathbf{X}_C$? (*Hint:* $\mathbf{X}_L-\mathbf{X}_C=|\mathbf{Z}|\sin\theta$.)
38. If in Example 17.11, $|\mathbf{Z}|=10$, $\theta=37.7°$, $\mathbf{R}=8$, and $\mathbf{X}_C=-3j$, what is $\mathbf{X}_L$?

17.5 POLAR FORM OF COMPLEX NUMBERS

Adding complex numbers (or vectors) in their algebraic, $x+yj$, form is a simple procedure. However, other notations called *polar form* or *trigonometric form* are more suitable when we multiply or divide complex numbers or find powers and roots of them.

From the right triangle of Fig. 17.6 we see that x, the real part of the number Z, is $r\cos\theta$ and that y, the imaginary part, is $jr\sin\theta$. Therefore we can specify the complex number (or point) Z as follows:

$$Z=x+yj=r\cos\theta+jr\sin\theta=r(\cos\theta+j\sin\theta)$$

POLAR FORM OF A COMPLEX NUMBER	$Z=r(\cos\theta+j\sin\theta)$	(17.6)

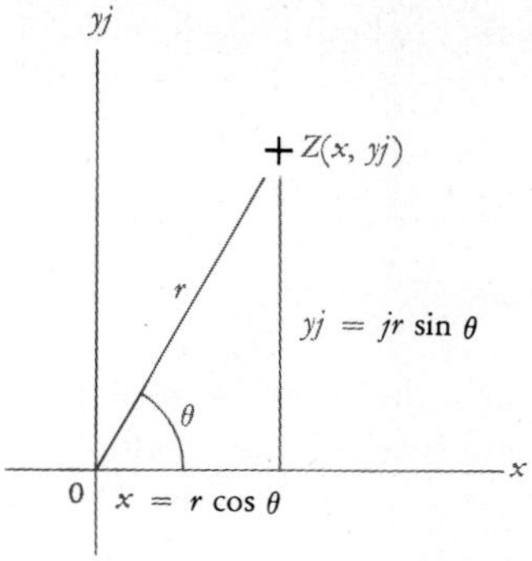

Fig. 17.6 Algebraic and polar coordinates.

Other frequently used notations for polar form are the following:

$$Z = r \operatorname{cis} \theta \tag{17.6a}$$

$$Z = r \angle \theta \tag{17.6b}$$

No matter which notation is used, the purpose of polar form is to specify a number in terms of a *radius* and an *angle* rather than rectangular coordinates. The radius r is equivalent to $|Z|$ and is called the *absolute value* or *modulus* of the complex number Z. The angle θ is called the *angle* or *argument.*

The symbol "cis θ" comes from the i notation for $\sqrt{-1}$ (see Sec. 17.1) and is a short form for $\cos \theta + i \sin \theta$ (or in our case, $\cos \theta + j \sin \theta$).

Multiplication

We derive the product of two complex numbers in polar form as follows.

Let $Z_1 = r_1(\cos \theta_1 + j \sin \theta_1)$ and $Z_2 = r_2(\cos \theta_2 + j \sin \theta_2)$ be the two numbers. Then

$$\begin{aligned} Z_1Z_2 &= r_1r_2(\cos \theta_1 + j \sin \theta_1)(\cos \theta_2 + j \sin \theta_2) \\ &= r_1r_2[(\cos \theta_1 \cos \theta_2 - \sin \theta_1 \sin \theta_2) \\ &\quad + j(\sin \theta_1 \cos \theta_2 + \cos \theta_1 \sin \theta_2)] \end{aligned}$$

Now recall the sum identities [Eqs. (14.15*a*) and (14.16*a*)] for the sine and cosine functions: $\cos \theta_1 \cos \theta_2 - \sin \theta_1 \sin \theta_2 = \cos (\theta_1 + \theta_2)$ and $\sin \theta_1 \cos \theta_2 + \cos \theta_1 \sin \theta_2 = \sin (\theta_1 + \theta_2)$. By using these identities in the equation above, we obtain the following result.

$$Z_1Z_2 = r_1r_2[\cos (\theta_1 + \theta_2) + j \sin (\theta_1 + \theta_2)]$$

or

$$Z_1Z_2 = r_1r_2 \operatorname{cis} (\theta_1 + \theta_2) \tag{17.7}$$

In other words, to multiply two complex numbers, we need only multiply the absolute values and *add* the angles.

Example 17.12. Multiply 5 cis 30° by 3 cis 45° and give the answer in both polar form and in algebraic form.

$$(5 \text{ cis } 30°)(3 \text{ cis } 45°) = 5(3) \text{ cis } (30° + 45°) = 15 \text{ cis } 75°$$

$$15 \text{ cis } 75° = 15 \cos 75° + 15j \sin 75° = 3.882 + 14.49j$$

Example 17.13. Multiply $Z_1 = 3 + 3j$ by $Z_2 = 1 + \sqrt{3}j$ and give the result in polar form. Check by changing both numbers to polar form first and then multiplying.

The product in algebraic form is

$$\begin{aligned} Z_1 Z_2 = (3 + 3j)(1 + \sqrt{3}j) &= (3 \cdot 1 - 3\sqrt{3}) + (3 + 3\sqrt{3})j \\ &= 3(1 - \sqrt{3}) + 3(1 + \sqrt{3})j \end{aligned}$$

The absolute value is

$$r = \sqrt{[3(1 - \sqrt{3})]^2 + [3(1 + \sqrt{3})]^2} = 6\sqrt{2}$$

and the angle is found as follows:

$$\tan \theta = \frac{3(1 + \sqrt{3})}{3(1 - \sqrt{3})} = \frac{(1 + \sqrt{3})(1 + \sqrt{3})}{(1 - \sqrt{3})(1 + \sqrt{3})} = \frac{4 + 2\sqrt{3}}{-2} = -(2 + \sqrt{3})$$

$$\tan \theta = -3.732 \qquad \theta = 180° - 75° = 105°$$

Thus $$Z_1 Z_2 = 6\sqrt{2} \text{ cis } 105°$$

To check, we change Z_1 and Z_2 to polar form.

$$r_1 = \sqrt{3^2 + 3^2} = 3\sqrt{2} \qquad \tan \theta_1 = \frac{3}{3} = 1 \qquad \theta_1 = 45°$$

$$r_2 = \sqrt{1^2 + (\sqrt{3})^2} = 2 \qquad \tan \theta_2 = \frac{\sqrt{3}}{1} = \sqrt{3} \qquad \theta_2 = 60°$$

$$Z_1 Z_2 = r_1 r_2 \text{ cis } (\theta_1 + \theta_2) = 3\sqrt{2}(2) \text{ cis } (45° + 60°)$$

$$Z_1 Z_2 = 6\sqrt{2} \text{ cis } 105°$$

Perhaps you have noticed that multiplication is much easier in polar form than in algebraic form.

Division

We derive the quotient of two complex numbers by rationalizing the divisor, as explained in Sec. 17.3.

$$\begin{aligned}\frac{Z_1}{Z_2} &= \frac{r_1(\cos\theta_1 + j\sin\theta_1)}{r_2(\cos\theta_2 + j\sin\theta_2)} \\ &= \frac{r_1}{r_2}\frac{(\cos\theta_1 + j\sin\theta_1)(\cos\theta_2 - j\sin\theta_2)}{(\cos\theta_2 + j\sin\theta_2)(\cos\theta_2 - j\sin\theta_2)} \\ &= \frac{r_1}{r_2}\frac{(\cos\theta_1\cos\theta_2 + \sin\theta_1\sin\theta_2) + (\sin\theta_1\cos\theta_2 - \sin\theta_1\sin\theta_2)j}{\cos^2\theta_2 + \sin^2\theta_2}\end{aligned}$$

At this point we use three trigonometric identities: $\cos^2\theta_2 + \sin^2\theta_2 = 1$, $\cos\theta_1\cos\theta_2 + \sin\theta_1\sin\theta_2 = \cos(\theta_1 - \theta_2)$, and $\sin\theta_1\cos\theta_2 - \cos\theta_1 \sin\theta_2 = \sin(\theta_1 - \theta_2)$. The quotient Z_1/Z_2 becomes

$$\frac{Z_1}{Z_2} = \frac{r_1}{r_2}[\cos(\theta_1 - \theta_2) + j\sin(\theta_1 - \theta_2)]$$

or (17.8)

$$\frac{Z_1}{Z_2} = \frac{r_1}{r_2}\operatorname{cis}(\theta_1 - \theta_2)$$

In other words, to divide one complex number by another, we need only divide the absolute values and *subtract* the angles.

Example 17.14. Divide 6 cis 40° by 4 cis 60° and give the answer in both polar form and algebraic form.

$$\frac{6\operatorname{cis}40^\circ}{4\operatorname{cis}60^\circ} = \frac{6}{4}\operatorname{cis}(40^\circ - 60^\circ) = \frac{3}{2}\operatorname{cis}(-20^\circ)$$

or

$$\frac{6\operatorname{cis}40^\circ}{4\operatorname{cis}60^\circ} = \frac{3}{2}\operatorname{cis}340^\circ$$

$$\frac{3}{2}\operatorname{cis}340^\circ = \frac{3}{2}\cos 340^\circ + \frac{3}{2}j\sin 340^\circ = 1.410 - 0.5130j$$

Example 17.15. Divide $Z_1 = 10 - 5j$ by $Z_2 = 3 + 2j$ and give the result in polar form. Check by changing both numbers to polar form first and then dividing.

$$\frac{Z_1}{Z_2} = \frac{10 - 5j}{3 + 2j} = \frac{(10 - 5j)(3 - 2j)}{(3 + 2j)(3 - 2j)} = \frac{20}{13} - \frac{35}{13}j$$

The absolute value is:

$$r = \sqrt{\left(\frac{20}{13}\right)^2 + \left(-\frac{35}{13}\right)^2} = \frac{5}{13}\sqrt{65}$$

The angle is found as follows:

$$\tan\theta = \frac{-35/13}{20/13} = -\frac{7}{4} = -1.750$$

$$\theta = -60°15' \quad \text{or} \quad 299°45'$$

Thus

$$\frac{Z_1}{Z_2} = \frac{5}{13}\sqrt{65}\ \text{cis}\ (-60°15')$$

To check, we change Z_1 and Z_2 to polar form.

$$r_1 = 5\sqrt{2^2 + (-1)^2} = 5\sqrt{5} \qquad \tan\theta_1 = -1/2 \qquad \theta_1 = -26°34'$$

$$r_2 = \sqrt{3^2 + 2^2} = \sqrt{13} \qquad \tan\theta_2 = 2/3 \qquad \theta_2 = 33°41'$$

$$\frac{Z_1}{Z_2} = \frac{r_1\ \text{cis}\ \theta_1}{r_2\ \text{cis}\ \theta_2} = \frac{r_1}{r_2}\ \text{cis}\ (\theta_1 - \theta_2) = \frac{5\sqrt{5}}{\sqrt{13}}\ \text{cis}\ (-26°34' - 33°41')$$

$$\frac{Z_1}{Z_2} = \frac{5}{13}\sqrt{65}\ \text{cis}\ (-60°15')$$

Example 17.16. In this example you do not have to understand the electronics in order to appreciate the application of complex numbers.

The ac current across an electric inductor is the quotient of the voltage and impedance, as follows:

$$\boldsymbol{I} = \frac{\boldsymbol{E}}{\boldsymbol{Z}}\ \text{amp}$$

where $\boldsymbol{E} = \boldsymbol{E}_0 \angle \theta_E$

$\boldsymbol{Z} = \boldsymbol{R} + j\boldsymbol{X}_L$

(See Example 17.11.) If $\boldsymbol{E} = 8 \angle 20°$ volts, $\boldsymbol{R} = 6$ ohms, and $\boldsymbol{X}_L = 3$ ohms, find $\boldsymbol{I}$.

$$I = \frac{E}{Z} = \frac{8\ \text{cis}\ 20°}{6 + 2j} = \frac{8\ \text{cis}\ 20°}{2\sqrt{10}\ \text{cis}\ 18°26'}$$

$$I = \frac{4\sqrt{10}}{10}\ \text{cis}\ (20° - 18°26') = \frac{2\sqrt{10}}{5}\ \text{cis}\ 1°34'$$

or

$$I = \frac{2\sqrt{10}}{5} \angle 1°34'$$

EXERCISE 17.5

In Probs. 1 to 12, $Z_1 = 2$ cis 23°, $Z_2 = 5$ cis 37°, and $Z_3 = 10$ cis 45°. Perform the indicated operations and give each answer in both polar form and algebraic form.

1. Z_1Z_2	2. Z_1Z_3	3. Z_2Z_3	4. $Z_1Z_2Z_3$
5. $(2Z_1)Z_3$	6. $Z_1(-Z_2)$	7. Z_2/Z_1	8. Z_1/Z_2
9. Z_1Z_2/Z_3	10. Z_3/Z_1	11. Z_1/Z_3	12. $Z_2/(Z_1Z_3)$

In Probs. 13 to 18, perform the operations with the numbers in algebraic form and give each answer in polar form. Check each answer by first changing the numbers to polar form and then performing the operations, as in Example 17.15.

13. $(1+j)(1-j)$ 14. $(1+j)^2$ 15. $(5-7j)(-3+4j)$
16. $(1-j)/(1+j)$ 17. $(8-4j)/(5+10j)$
18. $(2+2j)(3-4j)/(4+3j)$

19. Find **I** of Example 17.16 if $\mathbf{E} = 120 \angle 30°$, $\mathbf{R} = 16$, and $\mathbf{X}_L = 8$.
20. The current caused by a voltage **E** applied to a capacitor is $\mathbf{I} = \mathbf{E}/\mathbf{Z}$. Determine the current if $\mathbf{E} = 50 \angle 0°$ and $\mathbf{Z} = 4 - 3j$.
21. Repeat Prob. 20 if $\mathbf{E} = 40 \angle 30°$ and $\mathbf{Z} = 5 \angle -60°$.
22. Solve for **E** of Prob. 20 if $\mathbf{I} = 3$ cis 10° and $\mathbf{Z} = 3 - 3j$.
23. Solve for **Z** of Prob. 20 if $\mathbf{E} = 10$ cis 30° and $\mathbf{I} = 5$ cis 45°.
24. The current in a circuit with voltage **E**, resistance **R**, capacitive reactance $\mathbf{X}_C$, and inductive reactance $\mathbf{X}_L$ is

$$\mathbf{I} = \mathbf{E}/[\mathbf{R} + (\mathbf{X}_L - \mathbf{X}_C)j]$$

Determine **I** if $\mathbf{E} = 12$ cis 25°, $\mathbf{R} = 3$, $\mathbf{X}_L = 4$, and $\mathbf{X}_C = 6$.

17.6 POWERS AND ROOTS OF COMPLEX NUMBERS

A. POWERS

We can use Eq. (17.7) to find an expression for the *n*th power or the *n*th root of any number. For example, we can find the square of any complex number $Z = r(\cos\theta + j\sin\theta)$ by using Eq. (17.7) as follows:

$$Z^2 = Z \cdot Z = r \cdot r\,[\cos(\theta+\theta) + j\sin(\theta+\theta)]$$

$$Z^2 = r^2(\cos 2\theta + j\sin 2\theta)$$

To find third, fourth, and other powers, we could continue multiplying by Z. The general result, which is presented here without proof, is as follows:

$$Z^n = r^n(\cos n\theta + j \sin n\theta)$$

or

$$Z^n = r^n \text{ cis } n\theta \tag{17.9}$$

where n is any rational number

This result is known as *De Moivre's theorem,* in honor of the English mathematician Abraham De Moivre (1667–1754) who first derived it. Our experience with multiplication indicates that the use of polar form will be ideal in finding powers of complex numbers.

Example 17.17. Find $(5 \text{ cis } 26°)^3$.

From Eq. (17.9) we have

$$(5 \text{ cis } 26°)^3 = 5^3 \text{ cis } (3 \cdot 26°) = 125 \text{ cis } 78°$$

Example 17.18. Find Z^5 if $Z = 1 - 2j$ and give the answer in algebraic form.

It is far easier to convert Z to polar form first and convert Z^5 back to algebraic form than it is to work the problem entirely in algebraic form. Therefore we convert Z to polar form as follows:

$$r = \sqrt{1 + (-2)^2} = \sqrt{5} \qquad \tan\theta = \frac{-2}{1} = -2 \qquad \theta = -63°26'$$

$$Z = \sqrt{5} \text{ cis } (-63°26')$$

$$Z^5 = (\sqrt{5})^5 \text{ cis } [5(-63°26')] = 25\sqrt{5} \text{ cis } (-315°130')$$

$$Z^5 = 25\sqrt{5} \text{ cis } (-317°10')$$

$$Z^5 = 25\sqrt{5} \text{ cis } 42°50'$$

Note that $42°50'$ is $360° - 317°10'$, the smallest positive angle coterminal with $-317°10'$. This value is preferred.

Finally we change back to algebraic form:

$$Z^5 = 25\sqrt{5}\cos 42°50' + j25\sqrt{5}\sin 42°50'$$

$$\boxed{Z^5 = 41.0 + 38.0j \qquad \text{to three digits}}$$

EXERCISE 17.6

In Probs. 1 to 8, use De Moivre's theorem to find each indicated power. Give each answer in polar form.

1. $(1 + j)^3$
2. $(3 \text{ cis } 30°)^3$
3. $(0 + 2j)^4$
4. $(5 + 0j)^3$
5. $[\sqrt{2} \text{ cis } (-45°)]^2$
6. $(-1 - j)^3$
7. $(4 \text{ cis } 60°)^3$
8. $(-\sqrt{3} + \sqrt{3}j)^2$

For Probs. 9 to 16, use Probs. 1 to 8, respectively. Work each problem in algebraic form; then check each answer by changing it to polar form.

In Probs. 17 to 24, use De Moivre's theorem to find each indicated power in polar form.

17. $(3 + 4j)^5$
18. $(5 - 5\sqrt{3}j)^4$
19. $(\sqrt{13} \text{ cis } 40°)^4$
20. $(-1 + j)^8$
21. $\left(\dfrac{\sqrt{3}}{2} + \dfrac{1}{2}j\right)^7$
22. $[5 \text{ cis } (-20°)]^3$
23. $(1 \text{ cis } 45°)^{11}$
24. $(2 \text{ cis } 30°)^9$

B. ROOTS

Recall that a fractional exponent in the form $1/n$ represents an nth root. In Eq. (17.9), n can be any rational number. Therefore we can obtain De Moivre's theorem for roots of complex numbers by replacing n by $1/n$ in Eq. (17.9) and restricting n to be any positive integer. At first glance, the result would *seem* to be

$$Z^{1/n} = r^{1/n}\left(\cos\frac{\theta}{n} + j\sin\frac{\theta}{n}\right)$$

Although this equation is correct, it is not complete. There are other nth roots of Z. The key to the concept of multiple roots lies in the fact that, in polar form, there are many ways to write a given complex number. To illustrate, let's examine the number 8 and some of its cube roots, keeping in mind that 8 is $8 + 0j$ or $8 \text{ cis } 0°$.

We can write the number 8 in any of the following ways: 8 cis 0°, 8 cis 360°, 8 cis (2 · 360°), and 8 cis (3 · 360°). We could continue this list indefinitely by continuing to use multiples of 360°. In general we can say that $8 + 0j = 8 \text{ cis } (0° + k \cdot 360°)$ where k is any integer. This fact is so because of the periodic properties of cos θ and sin θ. The period of these functions is 360°. Therefore cis $(\theta) = \text{cis } (\theta + k \cdot 360°)$. The important point is this: *All numbers* r *cis* $(\theta + \text{k} \cdot 360)$, *where* $\text{k} = 0, \pm 1, \pm 2, \ldots$, *are equal, are completely interchangeable, and represent the same point in the complex plane.*

To determine all the nth roots of a number, we must consider all possible forms of the number. For example, we can find three different cube roots of 8 as follows, by using three different forms of 8 in the equation for $Z^{1/n}$ above.

1. $8^{1/3} = (8 \text{ cis } 0°)^{1/3} = 8^{1/3} \text{ cis } \dfrac{0°}{3} = 2 \text{ cis } 0° = 2 + 0j$

2. $8^{1/3} = (8 \text{ cis } 360°)^{1/3} = 8^{1/3} \text{ cis } \dfrac{360°}{3} = 2 \text{ cis } 120° = -1 + \sqrt{3}j$

3. $8^{1/3} = (8 \text{ cis } 720°)^{1/3} = 8^{1/3} \text{ cis } \dfrac{720°}{3} = 2 \text{ cis } 240° = -1 - \sqrt{3}j$

We could verify that each of these numbers is a cube root of 8 by cubing them. Try it and see. For example,

$$\begin{aligned}(-1 + \sqrt{3}j)^3 &= (-1 + \sqrt{3}j)^2(-1 + \sqrt{3}j) \\ &= (1 + 3j^2 - 2\sqrt{3}j)(-1 + \sqrt{3}j) \\ &= (-2 - 2\sqrt{3}j)(-1 + \sqrt{3}j) \\ &= 2 - 2(3)j^2 - 2\sqrt{3}j + 2\sqrt{3}j \\ &= 2 + 6 + 0j \\ &= 8 + 0j\end{aligned}$$

These three cube roots correspond to the values $k = 0$, 1, and 2 in $8 = 8 \cos (0° + k \cdot 360°)$. We could look for other cube roots of 8 by using other values of k. However, we would find that after the first three values of k the corresponding cube roots would duplicate the first three. For example, with $k = 3$ (the fourth value in the series 0, 1, 2, 3, . . .) we would have

$$\begin{aligned}8^{1/3} = [8 \text{ cis } (0° + 3 \cdot 360°)]^{1/3} &= 8^{1/3} \text{ cis } \left(\frac{0°}{3} + \frac{3 \cdot 360°}{3}\right) \\ &= 2 \text{ cis } 360°\end{aligned}$$

But 2 cis 360° = 2 cis 0°, which duplicates the first cube root above.

We conclude that there are exactly three *distinct* cube roots of 8 (or of any number). In general we would find that there are exactly n distinct nth roots of any number.

To find the n distinct nth roots of any number r cis $(\theta + k \cdot 360°)$, we need use only the first n values of k, $k = 0, 1, 2, 3, \ldots (n - 1)$. Based upon this consideration, the general equation for the n nth roots of a number $Z = r$ cis θ is as follows:

$$Z^{1/n} = r^{1/n}\left(\cos\frac{\theta + k \cdot 360°}{n} + j \sin\frac{\theta + k \cdot 360°}{n}\right) \qquad k = 0, 1, 2, 3, \ldots (n - 1) \tag{17.10}$$

Example 17.19. Find the five fifth roots of 1.

Taking 1 as 1 cis 0° and using successive values 0, 1, 2, 3, and 4 for k in Eq. (17.10), we have

$$k = 0: \quad 1^{1/5} = 1^{1/5} \text{ cis } \frac{0°}{5} = 1 \text{ cis } 0° = 1 + 0j$$

$$k = 1: \quad 1^{1/5} = 1^{1/5} \text{ cis } \frac{360°}{5} = 1 \text{ cis } 72° = 0.3090 + 0.9511j$$

$$k = 2: \quad 1^{1/5} = 1^{1/5} \text{ cis } \frac{720°}{5} = 1 \text{ cis } 144° = -0.8090 + 0.5878j$$

$$k = 3: \quad 1^{1/5} = 1^{1/5} \text{ cis } \frac{1{,}080°}{5} = 1 \text{ cis } 216° = -0.8090 - 0.5878j$$

$$k = 4: \quad 1^{1/5} = 1^{1/5} \text{ cis } \frac{1{,}440°}{5} = 1 \text{ cis } 288° = 0.3090 - 0.9511j$$

Example 17.20. Find the three cube roots of $3 - 4j$.

In polar form $3 - 4j$ is 5 cis $[\text{Tan}^{-1}(-4/3)] = 5$ cis $(-53°08')$.

$$(3 - 4j)^{1/3} = 5^{1/3} \text{ cis } \left(\frac{-53°08'}{3} + \frac{k \cdot 360°}{3}\right)$$
$$= 1.710 \text{ cis } (-17°43' + k \cdot 120°) \qquad \text{where } k = 0, 1, 2$$

$$k = 0: \quad (3 - 4j)^{1/3} = 1.710 \text{ cis } (-17°43') = 1.630 - 0.5204j$$

$k = 1$: $(3 - 4j)^{1/3} = 1.710 \text{ cis } 102°17' = -0.3639 + 1.671j$

$k = 2$: $(3 - 4j)^{1/3} = 1.710 \text{ cis } 222°17' = -1.265 - 1.150j$

EXERCISE 17.7

1. Use Eq. (17.10) to find the two square roots of each of the following numbers.

   a. 1 b. -1 c. $4j$ d. $-9j$ e. 8

2. Use Eq. (17.10) to find the three cube roots of each of the following numbers.

   a. 1 b. -1 c. -8 d. $-27j$ e. j

3. Find the five fifth roots of -32.
4. Find the four fourth roots of 16.

In Probs. 5 to 10, find all possible distinct values for each given expression.

5. $\sqrt[6]{1 \text{ cis } 180°}$
6. $\sqrt{8 \text{ cis } 90°}$
7. $(1 + \sqrt{3}j)^{1/2}$
8. $(-\sqrt{2} + \sqrt{2}j)^{1/3}$
9. $(1 + \sqrt{3}j)^{1/3}$
10. $(16 \text{ cis } 240°)^{1/4}$

EXPONENTIAL AND LOGARITHMIC FUNCTIONS

CHAPTER 18

We have previously encountered the concept of an inverse function. For example, if $y = 2x$ is a given function, then the inverse function is $x = y/2$ (found by solving $y = 2x$ for x in terms of y). In the inverse function the roles of the independent and dependent variables are inverted relative to those of the given function.

Sometimes we have had to invent special inverse symbols because there was no way to find a certain inverse function algebraically (as above). An example is the radical sign $\sqrt{\ }$. When we first encountered power functions such as $y = x^3$, we had no means for expressing x in terms of y. So we invented the radical sign, wrote $x = \sqrt[3]{y}$, and called x the *cube root of* y. This simple notation is far more efficient than the written statement "x is the number which, when cubed, produces y" (which is what $x = \sqrt[3]{y}$ means). In this chapter we shall examine a similar situation concerning functions of the form $y = b^x$ and their inverses.

18.1 BASIC IDEAS ABOUT LOGARITHMS

Equations of the form $y = b^x$ are called *exponential equations.* The independent variable x is the exponent, and the base b is a constant. Such equations should not be confused with power equations, $y = x^b$, in which the independent variable is the base.

There are various applications for exponential functions, and among them are cases in which y and b are known and x is to be found. Until now, we have not had a symbol or notation that would enable us to write the exponent x as a function of the base b and the number y. Mathematicians encountered this problem long ago, and they used a notation, called the *logarithm,* to handle it. The notation is used as follows:

If	$y = b^x$	exponential equation	(18.1)
then	$x = \log_b y$	logarithmic equation	(18.2)

where x is the *logarithm*
b is the *base*, $b > 0$ and $b \neq 1$
y is the *number*

In the logarithmic equation, x is called the logarithm to the base b of the number y. In the exponential equation, the same number x is called the exponent. Each function is the inverse of the other.

If you can keep the following ideas in mind, you should not have much difficulty understanding logarithms.

1. The logarithm of any number to a given base is the power (exponent) to which the base must be raised to produce that number. For example, $\log_{10} 100 = 2$ because $10^2 = 100$.
2. *A logarithm is an exponent;* logarithms and exponents are exactly the same thing. The form of an equation is the basis for using the word "logarithm" instead of "exponent."
3. $y = b^x$ means $x = \log_b y$.
4. $x = \log_b y$ means $y = b^x$.

The following examples show how we can use these ideas to evaluate logarithms, to change logarithmic equations into exponential equations (and vice versa), and to solve logarithmic equations.

Example 18.1. Evaluate $\log_2 8$.

We can use the definition as expressed by Statement 1 above, and ask the question, To what power must 2 be raised to produce 8? Since $2^3 = 8$, we have by definition, $\log_2 8 = 3$.

Another approach is to let $x = \log_2 8$, as in Statement 4 above. Then change the logarithmic equation to an exponential equation as follows:

$$x = \log_2 8 \qquad \text{means} \qquad 8 = 2^x$$

Since $8 = 2^3$, we have $2^3 = 2^x$, or $x = 3$.

Example 18.2. Solve for b: $2/3 = \log_b 25$

First, change the logarithmic equation to an exponential equation as follows: If $2/3 = \log_b 25$, then $25 = b^{2/3}$. Next, solve for b by raising both sides to the power $3/2$.

$$b^{2/3} = 25$$

$$b = 25^{3/2} = (\sqrt{25})^3 = 5^3 = 125$$

$$b = 125$$

Example 18.3. Solve for y: $1/4 = \log_{16} 10y$

From $$1/4 = \log_{16} 10y$$

we have $$10y = 16^{1/4} = \sqrt[4]{16} = 2$$

$$y = 2/10 = 1/5$$

Example 18.4. Change to a logarithmic equation: $64 = 16^{3/2}$

$3/2$ is the exponent or *logarithm,* 16 is the *base,* and 64 is the *number.* Therefore, according to Eqs. (18.1) and (18.2) we have

$$\frac{3}{2} = \log_{16} 64$$

As a consequence of the definition of a logarithm, particularly as expressed in Statement 1 above, the following elementary but important property occurs.

BASIC PROPERTY OF LOGARITHMS	$\log_b b^n = n$	(18.3)

This equation is true because $\log_b b^n$ means "the power to which b (the base b) must be raised to produce the number b^n." Since b must be raised to the nth power to produce b^n, $\log_b b^n$ is n. The simplicity of the question, What is $\log_b b^n$?, is like that of the question, What is the name of the man whose name is Jones?

Example 18.5. Evaluate $\log_2 32$ by using Eq. (18.3).

In order to use Eq. (18.3) we must rewrite the number 32 as a power of the base 2. In this case we have

$$32 = 2^5$$

$$\log_2 32 = \log_2 2^5 = 5$$

EXERCISE 18.1

In Probs. 1 to 12, change each exponential equation to its corresponding logarithmic equation.

1. $16 = 2^4$	2. $27 = 3^3$	3. $7 = 49^{1/2}$
4. $1/25 = 5^{-2}$	5. $1{,}000 = 10^3$	6. $2 = 10^{0.3010}$
7. $1/16 = 32^{-4/5}$	8. $36 = 216^{2/3}$	9. $3^{-4} = 1/81$
10. $1/100 = 10^{-2}$	11. $N = b^L$	12. $2Z = 3^{ay}$

In Probs. 13 to 18, change each logarithmic equation to its corresponding exponential equation.

13. $2 = \log_5 25$	14. $3 = \log_4 64$	15. $1/5 = \log_{32} 2$
16. $4 = \log_{10} 10^4$	17. $0.4771 = \log_{10} 3$	18. $-3/4 = \log_{16} (1/8)$

In Probs. 19 to 30, evaluate each logarithm by using Eq. (18.3).

19. $\log_3 9$	20. $\log_4 64$	21. $\log_{1/2} (1/32)$
22. $\log_{64} 8$	23. $\log_5 (1/25)$	24. $\log_8 (1/2)$
25. $\log_{10} 1$	26. $\log_{10} 10$	27. $\log_{10} 100$
28. $\log_{10} 0$	29. $\log_{10} 0.1$	30. $\log_{10} 0.01$

In Probs. 31 to 39, solve for the letter b, x, or y.

31. $x = \log_{10} 100{,}000$	32. $7 = \log_2 y$
33. $3 = \log_b 8$	34. $\log_{1/3} (1/81) = x$
35. $\log_5 (1/125) = x$	36. $\log_{36} 2y = 3/2$
37. $\log_{10} (y/5) = -3$	38. $\log_b \sqrt{2} = 1/2$
39. $\log_b 32 = -5$	

18.2 GRAPHS OF EXPONENTIAL AND LOGARITHMIC FUNCTIONS

The general form of an exponential function is as follows:

$$y = ab^{cx} \tag{18.4}$$

General exponential function

where a, b, and c are non-zero constants and $b > 0$.

To graph such a function, we make a table; we select convenient values for x and solve for y.

Example 18.6. **Graph $y = 2(4^x)$, $-3 \leqslant x \leqslant 3$.**

By choosing $x = 3, 2, 1, \ldots, -3$ the following table was constructed. Figure 18.1 is the corresponding graph.

x	4^x	y
3	64	128
2	16	32
1	4	8
0	1	2
−1	1/4	1/2
−2	1/16	1/8
−3	1/64	1/32

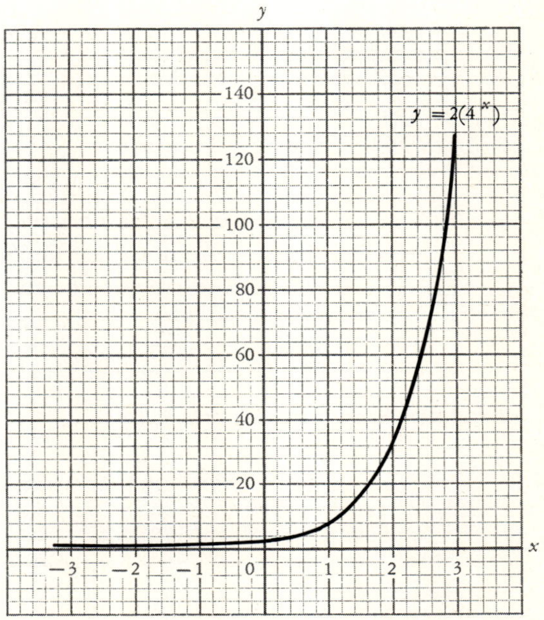

Fig. 18.1 A graph of the exponential function $y = 2(4^x)$.

Figure 18.1 illustrates the following characteristic features of the general exponential function ab^{cx}, assuming that a, b, and c are all *positive.*

1. As x increases, ab^{cx} will eventually begin to increase rapidly due to small increases of x. In Fig. 18.1 this feature can be observed where $x > 2$.
2. As x decreases, ab^{cx} approaches, but will never equal, zero. If a and b are positive, ab^{cx} can never be negative. (What if a is negative? See Prob. 4 of Exercise 18.2.) In Fig. 18.1 this feature can be observed where $x < 0$.

The general form of a logarithmic function is as follows:

$$y = m \log_b kx \tag{18.5}$$

General logarithmic function

A convenient way to graph such functions is to first determine the inverse function—the corresponding exponential function. To do so, solve for $\log_b kx$ and use Eqs. (18.1) and (18.2), as follows:

$$y = m \log_b kx$$

$$\frac{y}{m} = \log_b kx \longrightarrow kx = b^{y/m}$$

$$x = \frac{1}{k} b^{y/m} \tag{18.6}$$

Although x is the *independent* variable in $y = m \log_b kx$, to graph this function we must use Eq. (18.6) and treat x as the *dependent* variable. We choose convenient values for y and solve for x. Then we graph y vs. x, as shown in Example 18.7, so that x is restored to the role of independent variable.

Example 18.7. **Graph $y = \log_4 (x/2)$.**

First, determine the corresponding exponential equation, as follows:

$$y = \log_4 \frac{x}{2} \longrightarrow \frac{x}{2} = 4^y$$

$$x = 2(4^y)$$

Notice that this equation is the same as that of Example 18.6, except that x and y are interchanged. By choosing $y = 3, 2, 1, \ldots, -3$, the following table was constructed. For the graph, Fig. 18.2, values of x are placed on the abscissa, so that x is the independent variable as implied by the given equation.

y	4^y	x
3	64	128
2	16	32
1	4	8
0	1	2
-1	$1/4$	$1/2$
-2	$1/16$	$1/8$
-3	$1/64$	$1/32$

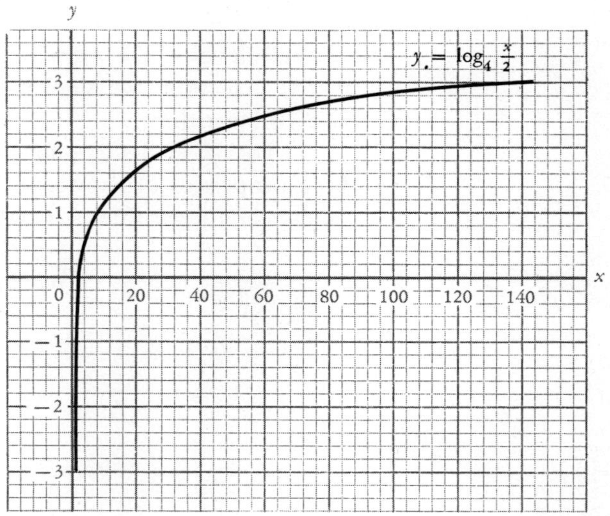

Fig. 18.2 A graph of the logarithmic function $y = \log_4 \left(\frac{x}{2}\right)$.

The graphs of Figs. 18.1 and 18.2 are inverses of each other. Figure 18.2 can be made to duplicate Fig. 18.1 by (1) turning the entire book 90° clockwise and (2) turning the page so that Fig. 18.2 is viewed from the back. Notice in Fig. 18.2, that *x can never be zero or negative;* it is impossible to have a logarithm of a negative number if the base is positive.

EXERCISE 18.2

In Probs. 1 to 6, graph each exponential function over the range $-3 \leqslant x \leqslant 3$.

1. $y = 2^x$
2. $y = 2(3^x)$
3. $y = 3(4^{-x/2})$
4. $y = -8^{2x/3}$
5. $y = (1/2)^x$
6. $y = (3/2)^{-x}$

In Probs. 7 to 12, graph each logarithmic function over the range $-3 \leqslant y \leqslant 3$, and compare to Probs. 1 to 6, respectively.

7. $y = \log_2 x$
8. $y = \log_3 (x/2)$
9. $y = -2 \log_4 (x/3)$
10. $y = (3/2) \log_8 x$
11. $y = \log_{1/2} x$
12. $y = -\log_{3/2} x$

18.3 PROPERTIES OF LOGARITHMS

Since logarithms are exponents, they have properties that correspond to those of exponents (for example, $b^n \cdot b^k = b^{n+k}$). These properties will be valuable aids when we perform computations with logarithms and when we solve exponential equations. In this section we shall develop the properties of logarithms in general, and in the next few sections we shall use them to perform computations with base 10 logarithms.

To aid in developing properties of logarithms, we list below three properties of exponents and also the relationship between exponents and logarithms for two numbers x and y.

Properties of exponents

$$b^n \cdot b^k = b^{n+k} \tag{18.7}$$

$$\frac{b^n}{b^k} = b^{n-k} \tag{18.8}$$

$$(b^n)^p = b^{n \cdot p} = b^{p \cdot n} \tag{18.9}$$

Definition of a logarithm

$$\text{if } x = b^n \qquad \text{then} \qquad n = \log_b x \tag{18.10}$$

$$\text{if } y = b^k \qquad \text{then} \qquad k = \log_b y \tag{18.11}$$

In order to develop the properties of logarithms, we let $x = b^n$ and $y = b^k$, as in (18.10) and (18.11) above. Then we apply the properties of exponents and the definition of a logarithm as follows:

Let $\qquad x = b^n \quad \text{and} \quad y = b^k$

Then $\qquad n = \log_b x \quad \text{and} \quad k = \log_b y$

From (18.7), (18.8), and (18.9) above we have, respectively,

$$x \cdot y = b^n \cdot b^k = b^{n+k} \tag{18.7a}$$

$$\frac{x}{y} = \frac{b^n}{b^k} = b^{n-k} \tag{18.8a}$$

and

$$x^p = (b^n)^p = b^{p \cdot n} \tag{18.9a}$$

Now, by applying the definition of a logarithm to each of these equations, we have

$$(x \cdot y) = b^{n+k} \longrightarrow \log_b (x \cdot y) = n + k \tag{18.7b}$$

$$\frac{x}{y} = b^{n-k} \longrightarrow \log_b \frac{x}{y} = n - k \tag{18.8b}$$

$$x^p = b^{pn} \longrightarrow \log_b x^p = pn \tag{18.9b}$$

But according to (18.10) and (18.11), $n = \log_b x$ and $k = \log_b y$. Therefore we can write these logarithms in terms of logarithms, as follows:

PROPERTIES OF LOGARITHMS

$$\log_b (x \cdot y) = \log_b x + \log_b y \tag{18.12}$$

$$\log_b \frac{x}{y} = \log_b x - \log_b y \tag{18.13}$$

$$\log_b x^p = p \log_b x \tag{18.14}$$

Equations (18.12), (18.13), and (18.14) are the respective logarithmic counterparts of Eqs. (18.7), (18.8), and (18.9) above. For example, (18.7) means "the exponent of a product (of two numbers with the same base) is the sum of the exponents," and (18.12) means "the logarithm of a product of two numbers is the sum of the logarithms of the numbers." You should be able to interpret Eqs. (18.13) and (18.14) in a similar fashion.

Comment: *In deriving Eqs. (18.12), (18.13), and (18.14), we applied the definition of a logarithm to Eqs. (18.7a), (18.8a), and (18.9a). For example, we said that, by definition,* $xy = b^{n+k}$ *means that* $\log_b xy = n + k$. *Another way to achieve this result is to consider a logarithm as an* **operation.** *In other words, we can "take the logarithm" of both sides of the equation* $xy = b^{n+k}$. *Then we can use the basic property, Eq. (18.3), as follows:*

given

$$xy = b^{n+k}$$

take $\log_b$ *of both sides*

$$\log_b xy = \log_b (b^{n+k})$$

apply Eq. (18.3)

$$\log_b xy = n + k$$

Example 18.8. **Use the properties of logarithms to write $\log_3 16$ as (a) a sum, (b) a difference, and (c) a multiple of logarithms.**

a. **Sum:** $\log_3 16 = \log_3 (2 \times 8) = \log_3 2 + \log_3 8$

b. **Difference:** $\log_3 16 = \log_3 (32/2) = \log_3 32 - \log_3 2$

c. **Multiple:** $\log_3 16 = \log_3 4^2 = 2 \log_3 4$

Example 18.9. **Use the properties of logarithms to write $4 + \log_2 6 - \log_2 12$ as a single logarithm, and evaluate it.**

$$4 = 4 \log_2 2 = \log_2 2^4 = \log_2 16$$

Therefore

$$\begin{aligned} 4 + \log_2 6 - \log_2 12 &= \log_2 16 + \log_2 6 - \log_2 12 \\ &= \log_2 \frac{(16)(6)}{12} \end{aligned}$$

$$4 + \log_2 6 - \log_2 12 = \log_2 8$$

$$\log_2 8 = \log_2 2^3 = 3$$

Example 18.10. **Simplify $\log_5 1{,}250$.**

$$\log_5 1{,}250 = \log_5 (625 \times 2) = \log_5 5^4 + \log_5 2$$

$$\log_5 1{,}250 = 4 + \log_5 2$$

Example 18.11. **Given that $\log_{10} 2 = 0.3010$ and $\log_{10} 3 = 0.4771$, find (a) $\log_{10} 32$, (b) $\log_{10} 600$, and (c) $\log_{10} (2/3)$.**

a. $\log_{10} 32 = \log_{10} 2^5 = 5 \log_{10} 2 = 5(0.3010) = 1.5050$

b. $\log_{10} 600 = \log_{10} (100 \times 2 \times 3)$

$= \log_{10} 10^2 + \log_{10} 2 + \log_{10} 3$

$= 2 + 0.3010 + 0.4771$

$= 2.7781$

c. $\log_{10} (2/3) = \log_{10} 2 - \log_{10} 3$

$= 0.3010 - 0.4771$

$= -0.1761$

Example 18.12. Solve for y in terms of x (see Comment below).

$$2 \log_b y = \log_b 4 + \log_b x \qquad x > 0, \quad y > 0$$

Equation (18.14) can be used on the left side and Eq. (18.12) on the right side to obtain

$$\log_b y^2 = \log_b 4x$$

Therefore

$$y^2 = 4x$$

$$y = 2\sqrt{x}$$

Comment: The basis for solving equations as in Example 18.12 is this: If $\log_b R = \log_b S$, then $R = S$. A frequent **error** *is to take an equation such as $\log_b R = \log_b S + \log_b T$ to mean that $R = S + T$. There is no basis for such an assumption; we must always rely on the established properties of mathematics. Another frequent error is to assume that $p \log_b y$ is equal to $\log_b py$; for example, is $(1/2) \log_b 16$ equal to $\log_b (1/2 \cdot 16)$ or $\log_b 8$? For the answer, study Eq. (18.14).*

EXERCISE 18.3

In Probs 1 to 12, write each logarithm as some combination of a sum, difference, or multiple of logarithms. If possible, evaluate each logarithm.

1. $\log_6 6$	2. $\log_3 9$	3. $\log_2 (2/3)$
4. $\log_5 15$	5. $\log_b 54b$	6. $\log_5 (2\frac{1}{5})$
7. $\log_4 64$	8. $\log_{10} 500$	9. $\log_2 3x^2$
10. $\log_3 (27y^2/z)^3$	11. $\log_{10} 0.2$	12. $\log_b \sqrt{x/y}$

In Probs. 13 to 18, condense each expression into a single logarithm.

13. $\log_b 5 + \log_b 3$	14. $2 \log_b 3 - \log_b 6$

15. $(1/2) \log_3 x - \log_3 2$
16. $3 \log_2 x + 3 \log_2 y - \log_2 z$
17. $2 - \log_5 10 - \log_5 (1/2)$
18. $(1/3) \log_{10} 27 + (1/4) \log_{10} 16$

In Probs. 19 to 24, solve each equation for y.

19. $\log_b y = \log_b 25$
20. $\log_2 y = 3 \log_2 2$
21. $\log_3 5y = \log_3 25 + \log_3 2$
22. $(1/2) \log_b 3y = (1/4) \log_b 6x^2$
23. $2 \log_b y/2 = \log_b 25 - \log_b x$
24. $\log_b y^2 = \log_b 2y + \log_b 4 + \log_b x$

25. Given that $\log_{10} 2 = 0.3010$, $\log_{10} 3 = 0.4771$, and $\log_{10} 5 = 0.6990$, evaluate each of the following logarithms.

a. $\log_{10} 20$ b. $\log_{10} 15$ c. $\log_{10} 7.5$
d. $\log_{10} 225$ e. $\log_{10} (2/5)$ f. $\log_{10} 0.006$

18.4 THE BASE 10 (COMMON) LOGARITHM SYSTEM

A. A FUNDAMENTAL PROPERTY OF THE BASE 10 SYSTEM

The number 10 is especially convenient for a logarithm base. To see why, recall the *scientific notation* with which we write a number such as 231 in the form 2.31×10^2. This example illustrates the following fundamental property of our decimal (base 10) system: *Any positive number can be written in the form of a number from 1 to 10, multiplied by an integral power of 10.*

Other examples are

$$20 = 2 \times 10^1$$

$$253.1 = 2.531 \times 10^2$$

$$0.00395 = 3.95 \times 10^{-3}$$

In algebraic notation we can state this property as follows:

$$N = T \times 10^k$$

where N is any positive number
T is a number from 1 to 10
k is an integer

By taking the base 10 logarithm of both sides and applying Eqs. (18.12) and (18.3), we can derive a corresponding property for logarithms, as follows:

$$N = T \times 10^k$$

$$\log_{10} N = \log_{10} (T \times 10^k) = \log_{10} T + \log_{10} 10^k = k + \log_{10} T$$

In writing base 10 logarithms it is customary to omit the base number; thus $\log N$ means $\log_{10} N$. The equation above is then written as

$$\boxed{\log N = k + \log T} \tag{18.15}$$

where N is any positive number and $N = T \times 10^k$
T is a number from 1 to 10
k is an integer

The idea expressed by Eq. (18.15) is this: *The base 10 logarithm of any positive number can be written as the sum of an integer plus the logarithm (base 10) of a number from 1 to 10.*

For example, $\log 20 = \log (2 \times 10^1) = \log 10^1 + \log 2 = 1 + \log 2$. Likewise, $\log 200 = 2 + \log 2$, $\log 2{,}000 = 3 + \log 2$, and so on. The importance of this property is this: We can find common logarithms of all numbers using the common logarithms of only the numbers from 1 to 10.

Base 10 logarithms are called *common logarithms.* The integral part k is called the *characteristic,* and $\log T$ is called the *mantissa. The mantissa is always a number from 0 to 1.* Why?

Example 18.13. **Given that $\log 2 = 0.3010$, determine (*a*) $\log 0.02$, (*b*) $\log 0.2$, (*c*) $\log 20$, (*d*) $\log 200$, and (*e*) $\log 2{,}000$.**

a. $\log 0.02 = \log (2 \times 10^{-2}) = -2 + \log 2 = -2 + 0.3010$
b. $\log 0.20 = \log (2 \times 10^{-1}) = -1 + \log 2 = -1 + 0.3010$
c. $\log 20 = \log (2 \times 10^{1}) = 1 + \log 2 = 1.3010$
d. $\log 200 = \log (2 \times 10^{2}) = 2 + \log 2 = 2.3010$
e. $\log 2{,}000 = \log (2 \times 10^{3}) = 3 + \log 2 = 3.3010$

Notice the similarity of these computations, and notice that the resulting logarithms differ only by integers.

Example 18.14. **Find $\log 800$, given that $\log 2 = 0.3010$.**

$$\log 800 = \log 100 + \log 8 = 2 + \log 2^3 = 2 + 3 \log 2$$
$$= 2 + 3(0.3010)$$

$$\log 800 = 2.9030$$

EXERCISE 18.4

In Probs. 1 to 8, use log 2 = 0.3010, log 3 = 0.4771, and log 5 = 0.6990. Evaluate each of the given logarithms. In addition to Eq. (18.15), also use Eqs. (18.12), (18.13), and (18.14) if necessary.

1. log 30	2. log 5,000	3. log 0.5	4. log 0.03
5. log 60	6. log 125	7. log ($^1/_{64}$)	8. log ($^{15}/_8$)

B. TABLES OF COMMON LOGARITHMS

Mathematicians have devised formulas from which tables of logarithms have been constructed. Table A5.2 in Appendix 5 is a typical four-place table of common logarithms of numbers from 1.000 to 10.000. With it we can find, to four-digit accuracy, the common logarithm of any positive number of four digits or less, and we can find any antilog—a number whose logarithm is given—of four digits or less.

Figure 18.3 is an excerpt from Table A5.2 showing how to read log 3.64 = 0.5611.

N	3	4	5 . . .
35	5478	5490	5502
36	5599 →	5611	5623
37	5717	5729	5740

log 3.64 = 0.5611

Fig. 18.3 An excerpt from Table A5.2.

In order to use this table we must observe the following features:

1. Table A5.2 in Appendix 5 is for logarithms of numbers from 1.000 to 10.000 *only*. Therefore the tabulated logarithms are numbers from 0 to 1.000 *only*, and decimal points are not shown, they are understood. For example, in Fig. 18.3, the number 36 means 3.6, and the logarithm 5611 means 0.5611.
2. Logarithms of numbers other than those from 1 to 10 can be found by adding the appropriate characteristic to the tabulated mantissa. *The table does not list characteristics.* For example, log 364 = 2 + log 3.64 = 2.5611. Only the 5611 is tabulated; you have to supply the 2 and the decimal point. Also see Examples 18.15 to 18.17 which follow.

3. The first two digits of the number N are in the column below **N**, and the third digit is in the row beside **N**. A fourth digit can be assumed by interpolating between third digits. For example, log 3,642 is ²⁄₁₀ of the "distance" between log 3,640 and log 3,650. See Example 18.16 below.

Example 18.15. Find log 80,500. See Table A5.2 in Appendix 5.

$$\log 80{,}500 = 4 + \log 8.05 = 4 + 0.9058 = 4.9058$$

Example 18.16. Find log 3,642.

Since four digits are involved, we will have to interpolate between log 3,640 and log 3,650. The following format is an aid. See Fig. 18.3.

$$\begin{array}{cc} & N \qquad \log N \\ \text{10 parts difference} & \left[\begin{array}{l} 2\left[\begin{array}{ll} 3640 & 5611 \\ 3642 & ? \end{array}\right] \\ \quad\ \ 3650 \quad\ 5623 \end{array}\right. \end{array} \begin{array}{l} \left.\begin{array}{l} \\ \end{array}\right] d \\ \\ \end{array} \left.\begin{array}{l} \\ \\ \\ \end{array}\right] \text{12 parts difference}$$

$$d/12 = 2/10 \longrightarrow d = 2(12)/10 = 2.4$$

We round d off to 2, since we are working to the fourth digit. Therefore

$$\log 3{,}642 = 3.5611 + 0.0002 = 3.5613$$

Example 18.17. Find log 0.0162. See Table A5.2.

$$\log 0.0162 = -2 + \log 1.62 = -2 + 0.2095 = -1.7905$$

Negative Logarithms

Notice in Example 18.17 that we lose both the characteristic -2 and the mantissa $+0.2095$ when we combine them to form -1.7905. For this reason it is often preferable to write a negative logarithm in the form $\log T + k$, where $\log T$ is the mantissa, a *positive number* from 0 to 1, and k is the characteristic (in this case, a *negative integer*). Thus instead of writing $\log 0.0162 = -1.7905$, we could write it as $0.2095 - 2$. It is customary to place a negative characteristic after the mantissa in this manner, to avoid misreading $-2 + 0.2095$ as -2.2095.

EXERCISE 18.5

In Probs. 1 to 8, use Table A5.2 of Appendix 5 to determine the common logarithm of each given number.

1. 2.54	2. 25.4	3. 254	4. 0.254
5. 96,100	6. 0.00125	7. 3.03×10^{-5}	8. 72.60×10^{6}

In Probs. 9 to 16, interpolate in Table A5.2 to determine the common logarithm of each number.

9. 52.15	10. 1.006	11. 999.3	12. 0.06521
13. 5.377×10^{-3}	14. 43,980	15. 0.0008794	16. 700.9

Antilogs

Another application of the common logarithm table is to find an unknown number when the value of its logarithm is known. The number is called the *antilog* of the known logarithm. For example, from Fig. 18.3 we see that 364 is the tabulated number corresponding to a tabulated mantissa of 5611, meaning that 3.64 is the antilog of 0.5611. In other words, given that $\log N = 0.5611$, then $N = \text{antilog}\ (0.5611) = 3.64$.

In general, to find an antilog we should proceed as follows:

1. Locate the given *mantissa* in the body of Table A5.2. (Remember that *characteristics are not given* in the table.)
2. Read the digits of the number N from the column below and the row beside the heading **N**.
3. To locate the decimal point, write N as a number from 1 to 10 and multiply it by 10^k, where k is the characteristic of the given logarithm.

Example 18.18. Find antilog 2.5611.

Referring to Fig. 18.3 and the steps above, we proceed as follows:

1. Locate 5611 in the table. Note that 5611 is the tabulated mantissa, *not* 25611. The 2 is the characteristic.
2. Read the digits 364 from the N column and row.
3. Write the number N as 3.64×10^2.

$$\text{antilog } 2.5611 = 3.64 \times 10^2 \quad \text{or} \quad 364$$

Example 18.19. Find antilog 4.7000. (In other words, find the number N for which $\log N = 4.7000$.)

First, remember that the table lists only the mantissa. Therefore immediately separate the logarithm into its characteristic and mantissa:

$4.7000 = 4 + 0.7000$. Since 7000 is not tabulated, interpolate between the two nearest values, as follows:

$$10\left[\begin{array}{l} d\left[\begin{array}{ll} 5010 & 6998 \\ ? & 7000 \end{array}\right]2 \\ \quad\; 5020 \quad 7007 \end{array}\right]9$$

with column headings N and $\log N$.

$$d/10 = 2/9 \qquad d = 2(10)/9 = 2.2 \text{ or } 2 \text{ parts}$$

$$\text{antilog } 0.7000 = 5.010 + 0.002 = 5.012$$

$$\text{antilog } 4.7000 = 5.012 \times 10^4 \quad \text{or} \quad 50{,}120$$

Example 18.20. Find P if $\log P = -2.3036$. [In other words, find antilog (-2.3036).]

Since the logarithm is in the form of a single negative number, neither the characteristic nor the mantissa is explicit. Notice that -2.3036 is *not* $-2 + 0.3036$, -2 is *not* the characteristic, and 0.3036 or -0.3036 is *not* the mantissa.

To find the characteristic and the mantissa, we must change -2.3036 to the form $\log T + k$, where $\log T$ is a positive number from 0 to 1, and k is a *negative integer*. To do this we add a "well-chosen zero" to -2.3036. In this case we add 3 and subtract 3 as follows:

$$-2.3036 = (+3 - 2.3036) - 3 = 0.6964 - 3$$

Now we can see the characteristic -3 and the mantissa $+0.6964$. We can find P with the aid of Table A5.2 as follows:

$$P = \text{antilog } (-2.3036) = \text{antilog } (0.6964 - 3)$$

$$P = 4.97 \times 10^{-3} \quad \text{or} \quad 0.00497$$

EXERCISE 18.5 (*Continued*)

In Probs. 17 to 24, use Table A5.2 to determine the antilog of each number.

17. 0.5855	18. 1.5855	19. 3.5855	20. 0.5855 − 2
21. 2.7251	22. 5.9489	23. −0.1733	24. −3.2007

In Probs. 25 to 32, interpolate in Table A5.2 to find the antilog of each number.

25. 0.1000	26. 1.4352	27. 0.6972 − 1	28. 0.8900 − 5
29. 4.0025	30. −2.6788	31. −4.0557	32. 3.9998

18.5 COMPUTATIONS WITH LOGARITHMS

By using a table of common logarithms and the properties of logarithms, Eqs. (18.12) to (18.14) repeated below, we can greatly simplify certain types of computations. Certain operations and the corresponding logarithmic methods of computation are listed below. Notice that the use of logarithms (1) reduces a multiplication to an addition, (2) reduces a division to a subtraction, and (3) reduces the raising to a power (or taking of a root) to a multiplication (or a division). The logarithmic approach would be a waste of time for computations with simple numbers. But for numbers of more than several digits, a reduction of effort and an increase of accuracy are achieved by using logarithms [five-place (and higher) tables are available].

Operation	Logarithmic computation	
1. Multiplication		
$N = xy$	$\log N = \log x + \log y$	(18.12)
2. Division		
$N = \dfrac{x}{y}$	$\log N = \log x - \log y$	(18.13)
3. Powers and roots		
$N = x^p$	$\log N = p \log x$	(18.14)

Operations involving combinations of those listed above can be performed by combining Eqs. (18.12) to (18.14). For example,

if

$$N = \frac{y\sqrt[5]{x}}{z^3}$$

then

$$\log N = \log y + \frac{1}{5}\log x - 3\log z$$

A great deal of efficiency can be gained by setting up a computational scheme *before* looking up the logarithms. One such scheme is outlined below. It consists of (1) writing the logarithmic equation, (2) setting aside a place for each logarithm and each computation, and (3) writing each characteristic in place. The boxed spaces indicate where mantissas and results of computations would be filled in later; see Example 18.21.

To compute $(32.9)(5{,}140)/1.56$

let $N = (32.9)(5{,}140)/1.56$

Then $\log N = \log 32.9 + \log 5{,}140 - \log 1.56$

$$\begin{array}{rcl}
\log 32.9 & = & 1.\boxed{} \\
+\log 5{,}140 & = & +3.\boxed{} \\
\hline
 & & \boxed{.} \\
-\log 1.56 & = & -0.\boxed{} \\
\hline
\log N & = & \boxed{.}
\end{array}$$

$$N = \text{antilog}\ \boxed{.} = \boxed{.}$$

$$\boxed{(32.9)(5{,}140)/1.56 = \boxed{.}}$$

A typical format for a logarithmic computation

The following examples are detailed illustrations of computations with logarithms.

Example 18.21. **Compute $(32.9)(5{,}140)/1.56$ to three digits.**

Let

$$N = \frac{(32.9)(5{,}140)}{1.56}$$

Then

$$\log N = \log 32.9 + \log 5{,}140 - \log 1.56$$

$$\begin{array}{rcl}
\log\ 32.9 & = & 1.5172 \\
+\log 5{,}140 & = & +3.7110 \\
\hline
 & & 5.2282 \\
-\log 1.56 & = & -0.1931 \\
\hline
\log N & = & 5.0351
\end{array}$$

$$N = \text{antilog } 5.0351 = 1.08 \times 10^5$$

$$\boxed{(32.9)(5{,}140)/1.56 = 1.08 \times 10^5}$$

Example 18.22. Compute $\sqrt[3]{0.01532}$ to four digits.

Let

$$N = \sqrt[3]{0.01532} = 0.01532^{1/3}$$

Then

$$\log N = (1/3) \log 0.01532$$

$$\begin{array}{llcl} & & +1 \quad\quad -1 & \\ \log 0.01532 = & & 0.1853 - 2 = 1.1853 - 3 & \\ \underline{\qquad\qquad \div 3} & & \div 3 \qquad\qquad \underline{\quad \div 3} & \\ \log N & & = 0.3951 - 1 & \end{array}$$

$$N = \text{antilog } (0.3951 - 1) = 2.484 \times 10^{-1} = 0.2484$$

$$\boxed{\sqrt[3]{0.01532} = 0.2484}$$

Comment: *In Example 18.22, before dividing log 0.01532 by 3, we added and subtracted 1. To maintain the* ***positive mantissa*** *and* ***negative characteristic,*** *we must divide them separately, and the resulting characteristic* ***has to be an integer.*** *In this case, to make the characteristic divisible by 3, we subtract 1 from the original characteristic of −2, and to compensate we add 1 to the mantissa. In general, we would* ***add and subtract the smallest positive integer such that the characteristic is evenly divisible.***

Another method is to combine the mantissa and characteristic into a single number (log 0.01532 = 0.1853 − 2 = −1.8147), divide by the index (−1.8147 ÷ 3 = −0.6049), and convert back to positive-mantissa, negative-characteristic form (1 − 0.6049) − 1 = 0.3951 − 1).

Example 18.23. Compute $(0.5931)^2/(200.1)^3$.

Let

$$N = \frac{(0.5931)^2}{(200.1)^3}$$

Then

$$\log N = 2 \log 0.5931 - 3 \log 200.1$$

$$\begin{array}{rcrcl} & & & & +6 \qquad -6 \\ 2 \log 0.5931 = & & 2(0.7731 - 1) = & & 1.5462 - 2 \\ \underline{-\,3 \log 200.1} & = & -3(2.3012) & = & \underline{-\,6.9036 \qquad} \\ \log N & & & = & 0.6426 - 8 \end{array}$$

$$N = \text{antilog } (0.6426 - 8) = 4.391 \times 10^{-8}$$

$$\boxed{(0.5931)^2/(200.1)^3 = 4.391 \times 10^{-8}}$$

Notice that we added and subtracted 6 in order to subtract 6.9036 from 1.5462 and obtain a positive mantissa between 0 and 1. This technique is preferable to that of obtaining a combined negative mantissa, negative characteristic.

Example 18.24. When a gas undergoes an adiabatic (constant-heat) process, the relationship between final and initial temperatures and pressures is as follows:

$$\frac{T_2}{T_1} = \left(\frac{P_1}{P_2}\right)^{(1-\gamma)/\gamma}$$

where γ is the ratio of specific heats at constant pressure and constant volume.

In a certain engine the temperature and pressure at the beginning of the compression stroke are $T_1 = 560°\text{F}$ abs (absolute) and $P_1 =$ 15 psi (pounds per square inch) abs; at the end of the stroke the pressure is $P_2 = 665$ psi abs.

Determine T_2, if $\gamma = 1.40$.

$$T_2 = 560 \left(\frac{15}{665}\right)^{(1-1.40)/1.40} = 560 \left(\frac{15}{665}\right)^{-0.286}$$

$$\begin{aligned} \log T_2 &= \log 560 - 0.286(\log 15 - \log 665) \\ &= \log 560 + 0.286(\log 665 - \log 15) \end{aligned}$$

$$\begin{array}{rr} \log 665 = & 2.8228 \\ -\log 15 = & -1.1761 \\ \hline & 1.6467 \\ & \times 0.286 \\ \hline & 0.4710 \end{array} \quad \longrightarrow \quad \begin{array}{rr} \log 560 = & 2.7482 \\ & +0.4710 \\ \hline \log T_2 = & 3.2192 \end{array}$$

$$\boxed{T_2 = 1660°\text{F abs}}$$

Example 18.25. Installment loans are usually arranged to be repayed in equal monthly payments. Part of each payment is principal and part is interest. But because the principal is reduced each month, the interest part of each succeeding equal payment is continually decreasing, and the principal part is correspondingly increasing.

The following formula gives the monthly installment N on a loan of P dollars at an annual rate of interest r percent for n months.

$$N = \frac{rP/1{,}200}{1 - (1 + r/1{,}200)^{-n}}$$

To the nearest \$0.1, what is the monthly payment on a \$15,000 loan at 6 percent annual interest for 10 years?

The fraction $r/1{,}200$ is $6/1{,}200 = 0.005$. Therefore the formula reduces to

$$N = \frac{0.005(15{,}000)}{1 - 1.005^{-120}} = \frac{75.00}{1 - 1.005^{-120}}$$

The term 1.005^{-120} is computed as follows:

$$\begin{aligned} \log 1.005 &= \quad 0.00217 \\ \times \quad (-120) &\quad \times\ (-120) \\ \hline \log 1.005^{-120} &= -0.2604 = 0.7396 - 1 \end{aligned}$$

$$1.005^{-120} = \text{antilog}\ (0.7396 - 1) = 0.549$$

Then N becomes

$$N = \frac{75.00}{1 - 0.549} = \frac{75.00}{0.451}$$

$$\begin{aligned} \log 75.00 &= \quad 1.8751 \\ -\log 0.451 &= -(0.6532 - 1) \\ \hline \log N &= \quad 1.2209 + 1 \\ &= \quad 2.2209 \end{aligned}$$

Note that when the -1 is subtracted, it becomes $+1$.

$$\boxed{N = \$166.30 \text{ per month}}$$

EXERCISE 18.6

In Probs. 1 to 18, use logarithms to perform the indicated computations.

1. $(92.3)(6.27)$
2. $(0.00316)(0.0193)$
3. $(0.002531)(81.35)$
4. $(-35.08)(253.7)$
5. $356/21.9$
6. $200/0.133$
7. $137.2/(-335.6)$
8. $0.002653/9{,}659$
9. 7.19^4
10. $(-0.0319)^5$
11. $\sqrt[3]{296.8}$
12. $0.0736^{1/5}$
13. $0.171^{2/3}$
14. $8{,}672^{-5/7}$
15. $69.31^{0.42}$
16. $\dfrac{71.6(84.2)^{0.65}}{6.33}$
17. $\dfrac{-25.21\sqrt{0.01795}}{1.637}$
18. $\dfrac{\sqrt[3]{19.07^4}}{-0.01872}$
19. Refer to Example 18.24. If $P_1 = 15$ psi abs, $P_2 = 579$ psi abs, and $T_1 = 580°$F abs, what is T_2?

20. In Example 18.24, if $P_1 = 14.9$ psi abs, $T_1 = 560$°F abs, and $T_2 = 1525$°F abs, what is P_2?
21. What is the monthly payment on a loan of \$7,000 for 15 years at 5½ percent annual interest? How much interest will be paid in the 15 years? See Example 18.25.
22. What is the monthly payment on a home mortgage of \$23,500 for 25 years at 7¾ percent annual interest? How much interest will be paid in the 25 years? See Example 18.25.
23. A formula for sound intensity is $I = (1/2)P^2\sqrt{k/\rho}$. Determine I, if $P = 265$, $k = 7.33 \times 10^{-6}$, and $\rho = 1.27 \times 10^{-3}$.
24. The buoyant force on a sphere submerged in a liquid is $F = 4\pi r^3 \rho g/3$, where r is the radius of the sphere, ρ is the mass density of the liquid, and g is the gravitational constant. Determine F if $r = 5$ ft 8 in., $\rho = 1.96$ slugs/ft^3, and $g = 32.2$ ft/sec^2.
25. The gravitational force between two masses m and M is given by the formula $F = GmM/r^2$. G is the universal gravitational constant, 3.44×10^{-8} lb-ft^2/slug^2, and r is the distance in feet between the masses.

    The mass of the earth is 4.12×10^{23} slugs, the mass of the moon is 4.94×10^{21} slugs, and the distance between them is approximately 227,000 miles. What is the gravitational force between the earth and the moon? Remember, there are 5,280 ft in 1 mile.
26. The orbital period of a "small" satellite is given by the formula $T = 2\pi\sqrt{r^3/(GM)}$, where r is the radius of the orbit, G is the universal gravitational constant, and M is the mass of the body around which the satellite orbits. Determine the period of an earth satellite at a radius of 4,119 miles. For data, see Prob. 25.
27. The discharge rate of a liquid flowing out of a pipe is $Q = \pi R^4(P_1 - P_2)/(8\eta L)$, where R is the radius of the pipe, P_1 is the inlet pressure, P_2 is the outlet pressure, η is the viscosity, and L is the length of the pipe.

    What is the discharge rate from a 3¼-in.-diameter pipe 14 ft long, if $P_1 = 250$ psi, $P_2 = 14.7$ psi, and $\eta = 1.45 \times 10^{-8}$ lb-sec/in.^{2}?

18.6 NATURAL LOGARITHMS AND EXPONENTIAL EQUATIONS

Although common logarithms are the most convenient type for most computations, they were not the first type to be used, and base 10 is not the best for describing certain physical relationships. In this section we shall examine logarithmic and exponential functions having the irrational number 2.7182818 . . . for a base. Such functions occur in surprisingly diverse areas of science, technology, and other fields.

It is customary to use e for 2.71828 . . . just as π is used for 3.14159 . . . , and it is customary to use "ln" for $\log_e$, just as log is used for $\log_{10}$. Therefore we make the following definitions:

$$e = 2.71828 \ldots$$

$$\ln N = \log_e N$$

Base e logarithms are called *natural logarithms.* According to our definition of a logarithm, we have the following relationship between an exponential function with base e and the corresponding logarithmic function:

$$\text{if} \quad y = e^x$$

$$\text{then} \quad x = \ln y$$

Because of the awkwardness of dealing with the number 2.71828 . . . , natural logarithms are not as easy to evaluate or to use as common logarithms. The two methods of handling natural logarithms are: (1) to make a change to common logarithms and (2) to use a table of natural logarithms. Because the first method is more general and the second method is easier, we shall present both.

A. CHANGES OF BASE NUMBERS

It is a simple matter to change from one logarithmic base system to another. Suppose that we know the logarithm of a number N in the base b and we wish to find the logarithm of N in the base a. In other words, we know the value of $\log_b N$ and we wish to find $\log_a N$. To find $\log_a N$, we let $N = a^x$, or $x = \log_a N$, and take $\log_b N$ as follows:

$$\log_b N = \log_b a^x = x \log_b a = (\log_a N)(\log_b a)$$

Upon solving for $\log_a N$, we have

$$\boxed{\log_a N = \frac{\log_b N}{\log_b a}} \qquad (18.16)$$

Example 18.26. **Show that $\log_8 64 = (\log_2 64)/(\log_2 8)$.**

Since $64 = 8^2$, $64 = 2^6$, and $8 = 2^3$, we have $\log_8 64 = 2$, $\log_2 64 = 6$, and $\log_2 8 = 3$.

$$\log_8 64 = 2 = \frac{6}{3} = \frac{\log_2 64}{\log_2 8}$$

Example 18.27. **Evaluate $\log_5 70$.**

$$\log_5 70 = \frac{\log 70}{\log 5} = \frac{1.8451}{0.6990} = 2.6396$$

Example 18.28. **Determine N if $\log_6 N = 2.3394$.**

$$\log_6 N = \frac{\log N}{\log 6}$$

$$\log N = (\log 6)(\log_6 N) = (0.7782)(2.3394) = 1.8205$$

$$N = 66.15$$

We shall be especially interested in using Eq. (18.16) to evaluate natural logarithms For the special case where $a = e = 2.7183$ (rounded off to the fourth decimal place) and $b = 10$, Eq. (18.16) becomes

$$\ln N = \frac{\log N}{\log e} = \frac{\log N}{0.4343} = 2.3026 \log N$$

Therefore we can convert from common logarithms to natural logarithms and vice versa with the following equations:

$$\ln N = 2.3026 \log N \tag{18.17}$$

$$\log N = 0.4343 \ln N \tag{18.18}$$

Natural log–common log conversion formulas

Example 18.29. **Determine ln 20.**

$$\begin{aligned} \ln 20 &= 2.3026 \log 20 \\ &= 2.3026(1.3010) \end{aligned}$$

$$\ln 20 = 2.9957$$

Example 18.30. **Determine N if $\ln N = -1.1180$.**

From Eq. (18.18) we have

$$\begin{aligned}\log N &= 0.4343 \ln N \\ &= 0.4343(-1.1180)\end{aligned}$$

$$\log N = -0.4855 = 0.5145 - 1$$

$$N = 0.3270$$

EXERCISE 18.7

In Probs. 1 to 6, evaluate each indicated logarithm by changing to a common logarithm.

1. $\log_2 32$
2. $\log_5 125$
3. $\ln 2$
4. $\log_8 0.0176$
5. $\log_{2.7} 100$
6. $\ln 1{,}000$

In Probs. 7 to 10, determine each number N.

7. $\log_5 N = 2.3174$
8. $\log_3 N = -1.3920$
9. $\ln N = 5.6200$
10. $\ln N = 6.0314$
11. Solve for x by taking the natural logarithm of both sides:

$$16.2 = 5.13e^{2x}$$

12. Solve the equation of Prob. 11 by taking the common logarithm of both sides.
13. Change the following logarithmic equation to an exponential equation:

$$t = \ln A - \ln B$$

B. TABLES OF NATURAL LOGARITHMS

The difficulty in presenting natural logarithm tables is that no two numbers have the same mantissa as in the base 10 system. Therefore every number has to be listed separately, and the tables must be extensive if many numbers are listed. For example, $\log 2{,}000 = 3 + \log 2$ because $2{,}000 = 2 \times 10^3$, but $\ln 2{,}000$ is *not* $3 + \ln 2$ because 2,000 is *not* $2 \times e^3$.

To overcome this difficulty we can use a limited table in conjunction with the properties of logarithms. For example, with a table of natural logarithms of numbers from 1 to 10, we could find ln 2,000 as follows: $\ln 2{,}000 = \ln (2 \times 10^3) = \ln 2 + 3 \ln 10$. In general we can find the natural logarithm of any positive number N by using a table of numbers from 1 to 10. First, we write N in the form $T \times 10^k$, where T is a number from 1 to 10 and k is an integer. Then we use the properties of logarithms as follows:

$$N = T \times 10^k$$

$$\begin{aligned} \ln N &= \ln (T \times 10^k) \\ &= \ln T + \ln 10^k \\ &= k \ln 10 + \ln T \end{aligned}$$

$$\boxed{\ln N = 2.30259k + \ln T} \qquad (18.19)$$

where N is any positive number and $N = T \times 10^k$
T is a number from 1 to 10
k is an integer
2.30259 is ln 10

Example 18.31. **Given that ln 2 = 0.6931, find (*a*) ln 0.2, and (*b*) ln 2,000.**

a. $\ln 0.2 = \ln (2 \times 10^{-1}) = (-1) \ln 10 + \ln 2$
$= (-1)(2.3026) + 0.6931$

$$\ln 0.2 = -1.6095$$

b. $\ln 2{,}000 = \ln (2 \times 10^3) = 3(2.3026) + 0.6931$
$= 6.9078 + 0.6931$

$$\ln 2{,}000 = 7.6009$$

Table A5.3 in Appendix 5 is a table of natural logarithms of numbers from 1 to 10. This table is read like a common logarithm table except that (1) *the decimal points are given in the table;* (2) *each tabulated logarithm includes the characteristic* as well as the mantissa; and (3) *it can be used for only those numbers tabulated.* It cannot be extended to numbers less than unity or greater than 10 by shifting the decimal point. Equation (18.19) can be used to extend the table.

Example 18.32. **Use Table A5.3 to find ln 6.058.**

Since interpolation is required, we use the following scheme. Note that the asterisks beside the tabulated logarithms in Table A5.3 indicate that the first two digits come from the row below.

		N	$\ln N$		
10	8	6.050	1.8001	d	16
		6.058	$\square$		
		6.060	1.8017		

$$\frac{d}{16} = \frac{8}{10}$$

$$d = \frac{8(16)}{10} = 12.8 \longrightarrow 13$$

$$\ln 6.058 = 1.8001 + 0.0013 = 1.8014$$

Example 18.33. Use Table A5.3 to find ln 103.5.

By using Eq. (18.19) and interpolating for ln 1.035 as in Example 18.32, we have

$$\begin{aligned} \ln 103.5 &= 2 \ln 10 + \ln 1.035 \\ &= 2(2.3026) + 0.0344 \end{aligned}$$

$$\ln 103.5 = 4.6396$$

Example 18.34. Use Table A5.3 to find antiln 5.0647; that is, find N given that $\ln N = 5.0647$.

First we must break 5.0647 up into the form $2.3026k + \ln T$ according to Eq. (18.19), and solve for $\ln T$.

$$\ln N = 2.3026k + \ln T$$

$$5.0647 = 2.3026k + \ln T$$

$$\ln T = 5.0647 - 2.3026k$$

Now we must choose k, an integer, such that $0 \leqslant \ln T < 2.3026$. (Why? Look at the smallest and largest logarithms tabulated in Table A5.3.) In this case we can see by inspection that $k = 2$. Therefore

$$\ln T = 5.0647 - 2.3026(2) = 0.4595$$

Next we interpolate in Table A5.3 to find T as follows:

		T	$\ln T$		
10	d	1.580	0.4574	21	63
		$\square$	0.4595		
		1.590	0.4637		

$$d = \frac{10(21)}{63} = 3.3 \longrightarrow 3$$

$$T = 1.580 + 0.003 = 1.583$$

$$N = T \times 10^k = 1.583 \times 10^2 = 158.3$$

EXERCISE 18.8

In Probs. 1 to 15, use Table A5.3 in Appendix 5 to find the natural logarithm of each number.

1. 2.53	2. 5.69	3. 8.18	4. 1.17
5. 0.633	6. 17.2	7. 185	8. 0.00345
9. 42×10^6	10. 42×10^{-6}	11. 7.032	12. 1.156
13. 27.54	14. 0.8145	15. 0.003267	

In Probs. 16 to 27, find the natural antilogarithm of each given number as in Example 18.34.

16. 1.2238	17. 0.0770	18. 2.2061	19. 1.1151
20. 1.4000	21. 2.0000	22. 3.6290	23. 4.5612
24. 4.6322	25. −1.3600	26. −3.5670	27. 6.7930

C. EXPONENTIAL EQUATIONS

We might justifiably wonder how the natural number e comes about and what it can be used for. An explanation of its existence is beyond the scope of this book. But we can state a general reason for its use. First let's look at some specific examples.

Example 18.35. **Uranium U^{238} decays according to the formula**

$$M = M_0 e^{-(1.59 \times 10^{-10})t}$$

where M_0 is the original amount, M is the new amount, and t is the time in years for M_0 to decay to M.

Example 18.36. **The intensity of light passing through a transparent material decreases according to the formula**

$$I = I_0 e^{-kt}$$

where I_0 is the original intensity, k is a constant, and t is the thickness of the material.

Example 18.37. **When an electric capacitor is discharged through a resistor, the voltage** across the capacitor decays according to the formula

$$E = E_0 e^{-t/(RC)}$$

where E_0 is the original voltage, and t is the time for E_0 to decrease to E.

There are but a few examples of a whole class of physical phenomena called *exponential decay.* Such phenomena occur whenever a quantity continually decreases at a rate proportional to the amount presently existing. Other phenomena involve an exponential growth. In general, exponential decay or growth phenomena can be described by an exponential function of the form $Y = Y_0 e^{kx}$, where Y_0 is the original amount (at $x = 0$), Y is the new amount, k is a constant, and x is a variable such as time. Natural logarithms are very useful for solving such equations (common logarithms can be used also).

Example 18.38. Atmospheric pressure decreases exponentially according to the formula $p = 760e^{-0.144h}$ where p is pressure in millimeters (mm) of mercury and h is altitude in kilometers (km). Determine the altitude at which the pressure is one-half the surface pressure of 760 mm.

$$p = 760e^{-0.144h}$$

$$380 = 760e^{-0.144h}$$

$$e^{-0.144h} = 0.500$$

$$-0.144h = \ln 0.500 \qquad \text{taking ln of both sides}$$

$$h = \frac{\ln 0.500}{-0.144} = \frac{-0.6932}{-0.144}$$

$$h = 4.81 \text{ km}$$

Example 18.39. In certain types of dc (direct-current) electric circuits, current exhibits an exponential growth according to the formula $i = (E/R)(1 - e^{-Rt/L})$. Determine i if $E = 12.0$ volts, $R = 60.0$ ohms, $t = 0.0100$ sec, and $L = 5.00$ henrys.

First we evaluate $e^{-Rt/L}$ as follows:

$$\ln e^{-Rt/L} = -Rt/L = -\frac{60.0(0.0100)}{5.00} = -0.120$$

Putting -0.120 in the form $\ln 10^k + \ln T$, we have

$$\ln e^{-Rt/L} = -2.303 + 2.183$$
$$= \ln 10^{-1} + \ln T$$

From Table A5.3 in Appendix 5, $T = 8.87$, and therefore

$$e^{-Rt/L} = 8.87 \times 10^{-1} = 0.887$$

$$i = \frac{12.0}{60.0}(1 - 0.887)$$
$$= (0.200)(0.113)$$

$$i = 0.0226 \text{ amp}$$

Example 18.40. It has been shown that the relationship between pressure P and volume V in certain combustion processes is

$$c_v \ln P + c_p \ln V = k$$

where c_v, c_p, and k are constants. Determine an exponential equation for P as a function of V.

By using the properties of logarithms we have

$$c_v \ln P + c_p \ln V = k$$

$$\ln P^{c_v} + \ln V^{c_p} = k$$

$$\ln P^{c_v} V^{c_p} = k$$

$$P^{c_v} V^{c_p} = e^k$$

$$P^{c_v} = V^{-c_p} e^k$$

$$P = V^{-c_p/c_v} e^{k/c_v}$$

EXERCISE 18.9

1. The *half-life* of a radioactive substance is the time required for a given amount to decay to one-half of that amount. What is the half-life of Uranium U^{238}? See Example 18.35.
2. Determine the time for uranium U^{238} to decay to 10 percent of its original amount. See Example 18.35.
3. How many years are required for uranium U^{238} to lose 1 percent of its mass? See Example 18.35.

4. The half-life of uranium U^{239} is 24 min. Determine the exponential function for the decay of U^{239}. See Prob. 1.
5. Determine the time for the voltage across a capacitor to decay to 0.1 percent of its original value if $R = 200{,}000$ ohms and $C = 4 \times 10^{-6}$ farad. See Example 18.37.
6. Determine the current in a series RL circuit 0.005 sec after the switch is closed, if $E = 24$ volts, $R = 100$ ohms, and $L = 20$ henrys. See Example 18.39.
7. What is the time required for the current of Prob. 6 to reach 0.216 amp?
8. What is the atmospheric pressure in mm of mercury, at an altitude of 5 miles (1 mile $\rightarrow$ 1.61 km)? See Example 18.38.
9. At what altitude in feet is the atmospheric pressure equal to 500 mm of mercury (1 km $\rightarrow$ 3,281 ft)? See Example 18.38.

18.7 GRAPHS ON LOGARITHMIC PAPER

Figure 18.1 (repeated below) illustrates the rapid increase typical of exponential functions. On ordinary rectangular graph paper we cannot graph such functions accurately over a wide range of values. But on certain special graph papers we can graph exponential functions over a wide range of values while maintaining the accuracy percentagewise, and the graphs are straight lines. In addition, we can graph power functions such as $y^k = x^n$ as straight lines, and we can greatly extend the graphs of other functions.

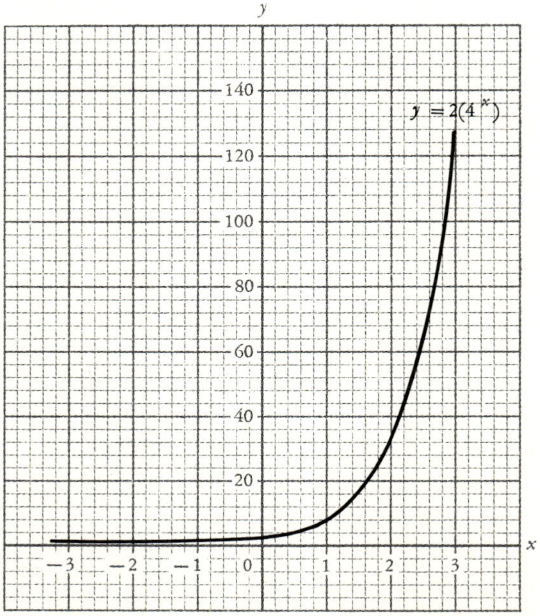

Fig. 18.1 (*Repeated*)

The special graph papers, called *logarithmic paper,* use base 10 logarithmic scales. As shown in Fig. 18.4, equally spaced divisions, called *cycles,* on a logarithmic scale represent increases in x by a *factor* of 10. We could think of such a scale as a *linear* scale in log x where values of x are labeled, instead of values of log x. If just one axis of a graph paper is a logarithmic scale, the paper is called *semilog paper,* and if both axes are logarithmic scales, the paper is called *log-log paper.*

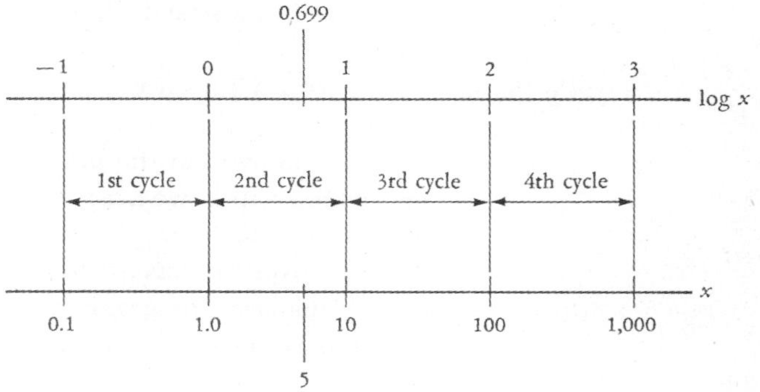

Fig. 18.4 A four-cycle logarithmic scale.

In using commercial logarithmic graph papers we must observe the following limitations:

1. Each end value of a cycle is a power of 10 such as 0.001, 0.01, 0.1, 1.0, 10, or 100.
2. The first value can be as small (absolute value) or as large as we choose, but the absolute value of each succeeding cycle must increase by a *factor* of 10.
3. The distance between two numbers on the scale is proportional to the difference between their *logarithms* (not simply the difference between the numbers themselves).

For example, on an ordinary linear scale, the number 5 is located at $4/9$ of the distance between 1 and 10: $(5 - 1) \div (10 - 1) = 4/9$. But on a logarithmic scale the number 5 is located at $\log 5 - \log 1 = 0.699 - 0 = 0.699$ times the distance between 1 and 10. If in Fig. 18.4 the distance between 1 and 10 on the logarithmic scale were 1 in., then the number 5 would be located 0.699 in. from the number 1.0.

Semilog Paper

The graph of an exponential equation $y = ab^{cx}$ will be a straight line on semilog paper. The reason becomes apparent when we take the logarithm of both sides, as follows:

given $$y = ab^{cx}$$

then $$\log y = \log a + cx \log b$$

Since a, b, and c are *constants,* we see that $\log y$ is a linear function of x. With x on a linear scale and y on a logarithmic scale, the graph of $y = ab^{cx}$ is a straight line.

Example 18.41. **Graph $y = 2(4^x)$ from $x = -2$ to $x = +4$.**

Since we know that the graphs of all such functions are straight lines on semilog paper, we need plot only two points. We need not write the logarithmic equation. A suitable choice of values on the axes ensures that we cover the specified range. The graph is shown in Fig. 18.5.

Compare this graph to that of Fig. 18.1, which shows the same function graphed on rectangular paper.

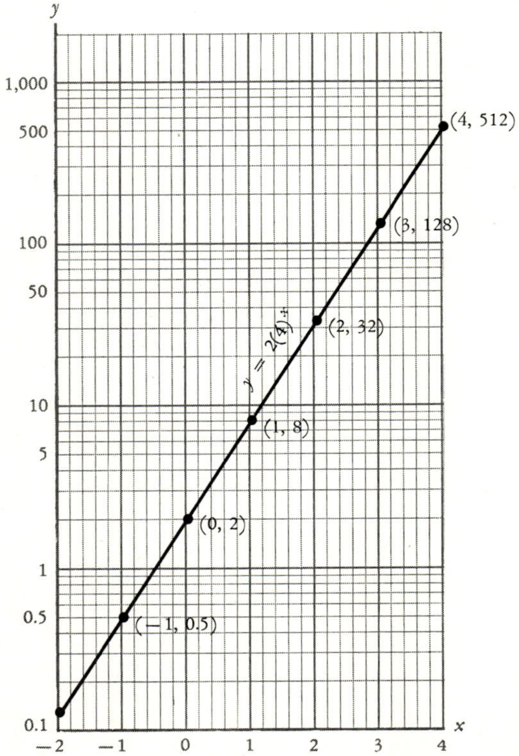

Fig. 18.5 A graph of $y = 2(4^x)$ on semilogarithmic paper.

Example 18.42. The voltage across a certain discharging capacitor is given by $E = 36e^{-0.8t}$, where t is time in seconds. Graph the voltage from $t = 0$ to $t = 10$ sec.

Since $\log E = \log 36 - 0.8t \log e$, the graph will be linear on semilog paper. We can use both the exponential equation and the logarithmic equation to compute values of E as follows:

$t = 0$: $E = 36e^0 = 36$ volts

$t = 10$:
$$\log E = \log 36 - 8 \log e$$
$$= 1.5563 - 8(0.4343)$$

$$\log E = 0.0819 - 2$$

$$E = 0.0121 \text{ volts}$$

The graph is shown in Fig. 18.6.

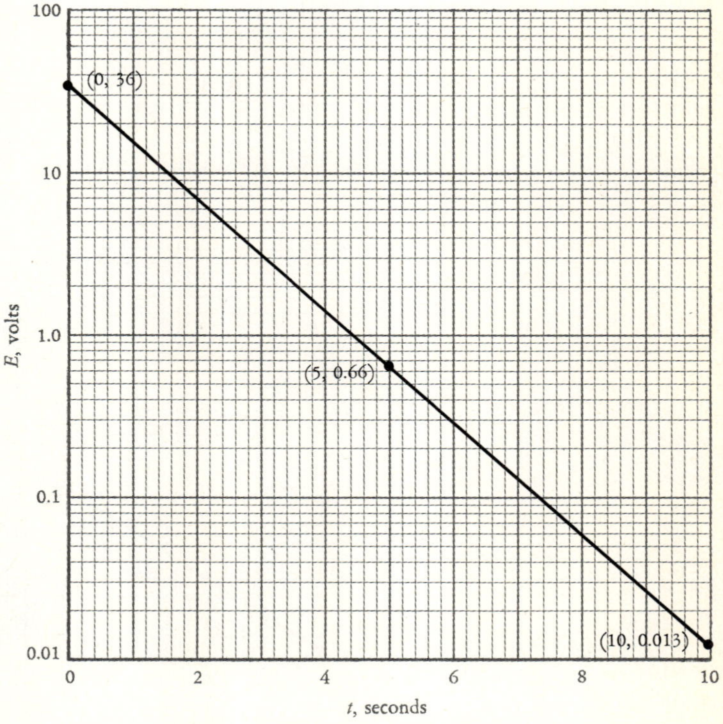

Fig. 18.6 A graph of $E = 36e^{-0.8t}$ *on semilogarithmic paper.*

Log-log Paper

The graph of a power equation $y^k = ax^n$ will be a straight line on log-log paper. To see why, take the logarithm of both sides as follows:

given $$y^k = ax^n$$

then $$k \log y = \log a + n \log x$$

Since a, k, and n are constants, $\log y$ is a linear function of $\log x$; if both x and y are placed on logarithmic scales, the graph of $y^k = ax^n$ will be a straight line.

Example 18.43. **Graph $y^5 = 8x^3$, $0.1 \leqslant x \leqslant 100$.**

From the following logarithmic equation we derive the table of values shown below.

$$\log y = \tfrac{1}{5}(\log 8 + 3 \log x) = \tfrac{1}{5}(0.9031 + 3 \log x)$$

x	0.1	1.0	10	100
$\log x$	-1	0	1	2
$\log y$	$0.5806 - 1$	0.1806	0.7806	1.3806
y	0.38	1.52	6.03	24.0

Figure 18.7 shows the graph, which is a straight line.

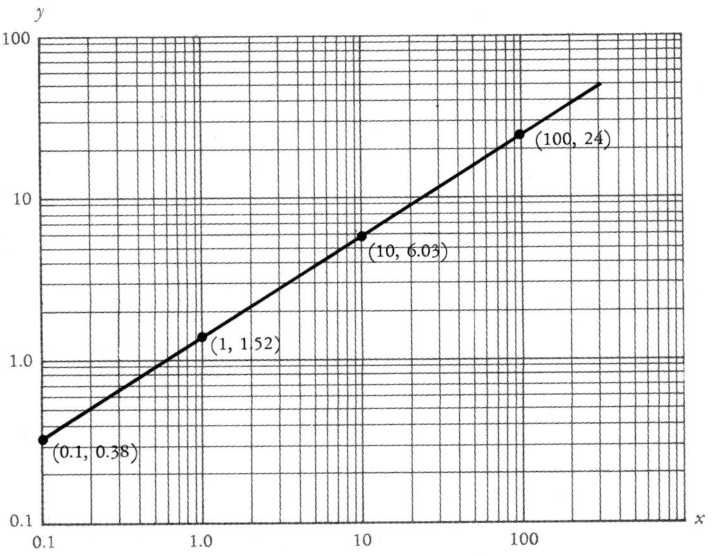

Fig. 18.7 A graph ot $y^5 = 8x^3$ on log-log paper.

Example 18.44. The costs of producing a certain machine part on an automatic screw machine are \$100 for setting up the machine, \$85 for tooling, and \$0.015 for each part. Graph cost as a function of the number of parts, from 1,000 to 1 million parts.

The equation for cost c vs. the number of parts n is $c = 185 + 0.015n$. On regular rectangular paper this equation would graph as a straight line, but due to the wide range of values, accuracy would be very poor. By graphing the equation on log-log paper we can expand the accuracy of the graph; however, the graph will not be a straight line. Figure 18.8*a* shows the linear graph, and Fig. 18.8*b* shows the log-log graph.

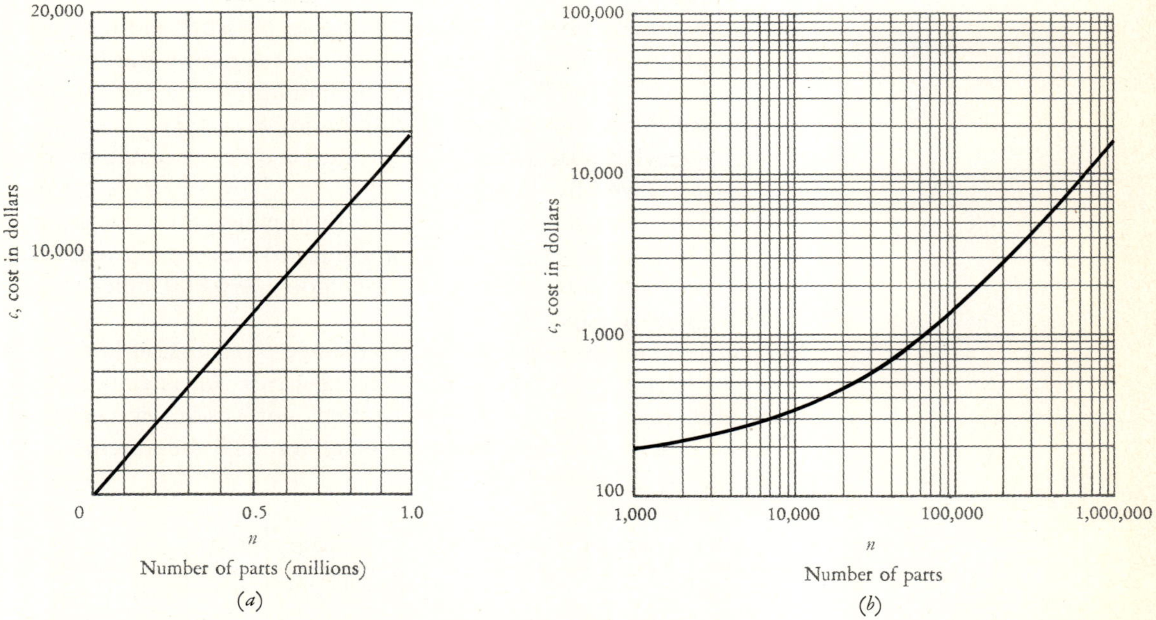

Fig. 18.8 Example 4. (a) Graph of $c = 185 + 0.015n$ on rectangular paper; (b) graph of $c = 185 + 0.015n$ on log-log paper.

EXERCISE 18.10

Graph each equation in Probs. 1 to 6 on semilog paper.

1. $y = 3^x$
2. $y = (1/2)^x$
3. $4y = 3(5^x)$
4. $y = (1/2)10^x$
5. $y = 10^{-x}$
6. $y = 5e^x$

Graph each equation in Probs. 7 to 12 on log-log paper.

7. $y^2 = x^3$
8. $y^3 = x^2$
9. $y = 0.1x^5$
10. $xy = 1$
11. $xy^2 = 64$
12. $x^3y = 64$

13. The voltage across a certain capacitor is given by the formula $E = 8e^{-2t}$. Graph E vs. t, from $t = 0$ to $t = 3$; use semilog paper.
14. The earth's atmospheric pressure is given by the formula $p = 760e^{-0.144h}$, where p is pressure in millimeters (mm) of mercury, and h is altitude in kilometers (km). Graph p vs. h up to an altitude of 10 miles. (1 mile → 1.61 km.)
15. Uranium U^{239} decays according to the formula $M = M_0e^{-0.0289t}$, where M_0 is the original amount and M is the amount remaining after t min.

    Graph M vs. t for 10 gm (grams) of U^{239}. How many minutes are required for U^{239} to decay to one-half the original amount?
16. In a certain internal combustion engine, pressure and temperature during the combustion stroke are related by the formula $T = 560(15/p)^{-0.286}$, where T is temperature in °F abs and p is pressure in psi abs. Graph T vs. p on log-log paper.
17. The gravitational acceleration produced on a spaceship by the earth is given by the formula $g = (5.08 \times 10^8)/r^2$, where r is the distance from the center of the earth in miles and g is acceleration in ft/sec². Graph g vs. r for a ship that travels between the earth (4,000-mile radius) and the moon 230,000 miles away. Use log-log paper.
18. In designing waveguides for microwave transmission lines, engineers are concerned with reflected waves. An engineer measured the "coefficients of reflection" for various waveguides and determined the percent of error for each measurement as follows:

COEFFICIENT OF REFLECTION	0.01	0.05	0.10	0.50	1.00
Percent error	54.0	11.2	5.50	1.12	0.55

    Use these data to make a graph of percent error vs. Γ. Use log-log paper. From the graph determine an equation for percent of error as a function of Γ.
19. A microwave technician measured the power losses in decibels (db) in a certain waveguide component for various "voltage–standing-wave ratios" *VSWR*. His measurements are given in the table below. Use these data to make a graph of power loss vs. *VSWR* on semilog paper.

VSWR	2	3	4	5	6	7	8	9	10
Power loss, db	0.50	1.24	1.94	2.55	3.10	3.60	4.05	4.48	4.90

20. An important aspect of heating elements is the *watt density*—the number of watts per square inch of surface to maintain a certain surface temperature. A certain manufacturer gives the following data for maintaining the element at 1000°F:

AIR TEMPERATURE, °F	10	200	400	600	800	1000
Watts/in.^{2}	25	22.5	19	14	8.5	0

On semilog paper, graph watt density as a function of air temperature for this heating element.

CHAPTER 19 INEQUALITIES AND LINEAR PROGRAMMING

In Chap. 2 we had a brief introduction to inequalities and the inequality symbols $>$, $<$, $\geqslant$, $\not<$, and so on. Now we shall examine relationships involving inequalities and problems that require us to solve inequalities.

Whereas an equation is concerned with discrete values of each variable (that is, one or at most a few distinct numbers), an inequality is concerned with entire ranges of values. For example, if we are to find the values of x that make y *equal* zero in the equation $y = \sqrt{1 - x^2}$, then we must solve the equation $\sqrt{1 - x^2} = 0$; the solutions are the discrete values $x = \pm 1$. But if we are to find the values of x for which y is any real number, then we must solve the *inequality* $1 - x^2 \geqslant 0$. Perhaps we can intuitively see that the solution is the *entire range* of numbers between and including $+1$ and -1. We express this infinite range of values with inequality symbols as follows: $-1 \leqslant x \leqslant 1$.

The methods of this chapter will help us solve such inequalities in a more analytical fashion. But first we shall review the basic concepts of inequalities and examine properties of inequalities. We shall also see how to graph inequalities. Then we shall solve inequalities and examine practical applications for them.

19.1 BASIC DEFINITIONS AND SYMBOLS

Before we begin an in-depth study of inequalities, we must have a clear understanding of the basic definitions and symbols. We define the basic inequality relationships "greater than" and "less than" in terms of a standard number line such as the one shown in Fig. 19.1. On a standard number line, positive numbers are placed to the right of zero and negative numbers to the left of zero. Then we define the basic inequality relationships and symbols as follows:

Greater than $a > b$ — If a is *greater than* b, then a is *to the right* of b on the number line. (19.1)

Examples: $5 > 2 \quad 1 > 0 \quad 3 > -10$ and $-2 > -5$

Less than $a < b$ — If a is *less than* b, then a is *to the left* of b on the number line. (19.2)

Examples: $2 < 5 \quad 0 < 1 \quad -10 < 3$ and $-5 < -2$

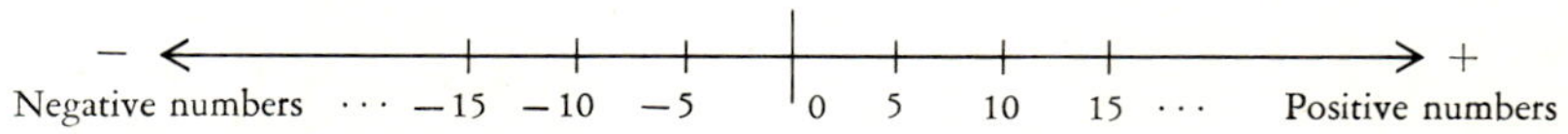

Fig. 19.1 The standard number line. Positive numbers are placed in order to the right *of zero, and negative numbers are ordered to the* left *of zero.*

As a consequence of these definitions, we have special means for indicating *positive* numbers or *negative* numbers. Since positive numbers are to the right of zero on the number line and negative numbers are to the left, we have the following special cases of definitions (19.1) and (19.2):

$a > 0$ — a is a positive number (19.3)

$a < 0$ — a is a negative number (19.4)

Notice that *zero is neither positive nor negative.*

Many other inequality statements can be formed by combining the basic inequality symbols $>$ and $<$ with the equals sign $=$ and the negation sign /. Thus $a \geqslant 0$ means "a is positive or zero" and $a \nless 0$ means "a is not negative"; these two statements are equivalent.

Example 19.1. **English statements of inequality can be translated into symbols as shown below.**

a. **x is any number greater than 5: $x > 5$.**
b. **y is any number between -3 and 7 inclusive: $-3 \leqslant y \leqslant 7$.**
c. **z is any number from -10 (inclusive) to 0 (exclusive): $-10 \leqslant z < 0$.**
d. **x is any number less than 2 *and* greater than -2: $x < 2$ and $x > -2$, which means $-2 < x < 2$, or $|x| < 2$.**

e. **All numbers N greater than 2 *or* less than -2: $N > 2$ or $N < -2$; also $|N| > 2$ and $|N| \nleq 2$.**

f. **All numbers greater than 2 *and* less than -2: No number can be both greater than 2 *and* also less than -2.**

Examples 19.1*d*, *e*, and *f* illustrate the meanings of the words "and" and "or" in mathematics. When a number N must satisfy two conditions connected by *and,* then *all* values of N must satisfy *both* conditions as in Examples 19.1*d* and *f*. But if N must satisfy two conditions connected by *or*, then each value of N must satisfy *one condition or the other* but not necessarily both conditions (see Example 19.1*e*).

Graphs of Inequalities with One Variable

We can make a graph of any inequality with one variable by drawing a heavy line over the appropriate range of values on the number line. Isolated points or endpoints that are *included* in a range of values are marked with a solid dot ●. Endpoints *excluded* from a range of values are marked with an open circle ○.

Example 19.2. **The graph of $-10 \leq x < 10$ is drawn by making a heavy line between -10 and $+10$ and by making a solid dot at -10 and an open circle at $+10$. See Fig. 19.2.**

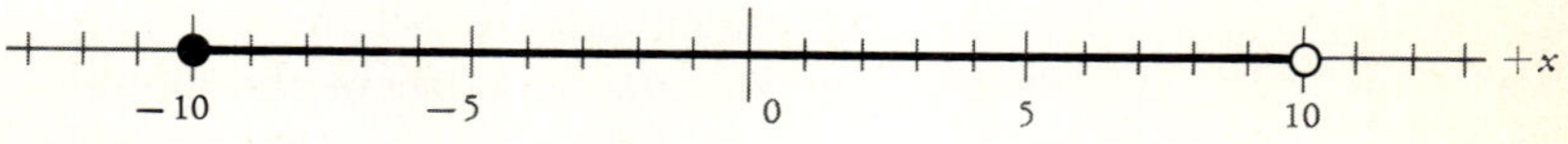

Fig. 19.2 A graph of $-10 \leq x < 10$.

Example 19.3. **The graph of $x > -5$ is made by drawing a heavy arrow from -5 toward the right. An open circle is drawn at -5. See Fig. 19.3.**

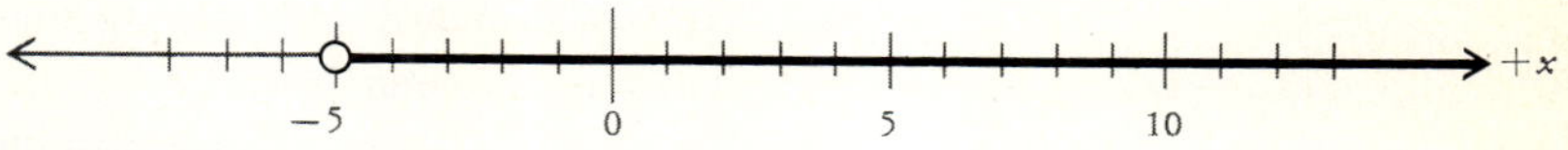

Fig. 19.3 A graph of $x > -5$.

EXERCISE 19.1

1. Rewrite each of the following statements by using inequality symbols as in Example 19.1. In (*h*) to (*k*) also find the range of values for each variable x, z, y, and t.

   a. N is any number less than 10.
   b. Y is any number greater than 25.

c. Z is any number between 12 and 30 inclusive.
d. K is any number less than 10 and greater than -10.
e. P can be a number greater than 7 or less than -7.
f. S is any number greater than 14 or less than -10.
g. B is any number less than 5 and greater than 6.
h. $x + 5$ is a positive number.
i. $z - 2$ is a negative number or zero.
j. $y - 4$ is not a positive number.
k. $t + 10$ is not a negative number.

2. Translate each of the following statements into an English sentence, as in Prob. 1.

a. $A < 12$ b. $B \geqslant -5$ c. $-10 < C \leqslant 10$
d. $-12 \leqslant D \leqslant -3$ e. $E - 5 > 0$ f. $F + 2 \leqslant 0$
g. $G + 10 \nless 0$ h. $-8 < H < 8$ i. $|I| < 8$ Compare with (*h*).
j. $J > 5$ or $J < -5$ k. $|K| > 5$

3. On a number line, graph each inequality statement of Prob. 2.

Absolute Value and Inequalities

The absolute value sign is useful for writing certain kinds of inequality statements. Recall that the absolute value of any real number is the *magnitude* or *size* of the number. It is also the distance between the number and the origin of the number line. The following geometric interpretation will help us to understand absolute values and inequalities. (For a review of absolute value, see Sec. 2.1C.)

We can interpret the difference between two numbers as meaning the *distance* between them on the number line. For example, $5 - 3 = 2$ means that the distance between 3 and 5 is 2 units. The distance between 5 and 3 is also 2 units, but if we write it as $3 - 5$, we get -2 units. Since distance is a magnitude or absolute value and is not related to direction, *the correct expression for distance is the absolute value of the difference between two numbers.* Thus $|3 - 5| = |-2| = 2$ is the correct way to arrive at the distance between 5 and 3. We can think of distance as an *absolute difference* between two numbers.

Of particular interest are statements such as $|x - a| < n$. This statement means "the distance between a and x is less than n units" or "x is within n units of a." A special case is $|x| < n$, meaning $|x - 0| < n$. For example, $|x| < 4$ means that x is within 4 units of the origin. Therefore $x < 4$ and $x > -4$, which we can combine and write as $-4 < x < 4$. Figure 19.4 is a graph of this inequality. You should check to see that any value of x in Fig. 19.4 satisfies the inequality $|x| < 4$ and also $-4 < x < 4$. Figure 19.5 is a similar graph of the inequality $|x| \geq 4$. This statement means that x is 4 units or more from zero. Therefore $x \geq 4$ or $x \leq -4$.

With these interpretations of the absolute value in mind, we make the following definitions:

$\lvert x - a \rvert < n$ means $a - n < x < a + n$.	The distance between a and x is less than n.	(19.5)
$\lvert x - a \rvert > n$ means $x > a + n$ or $x < a - n$.	The distance between a and x is more than n, where n is any positive number.	(19.6)

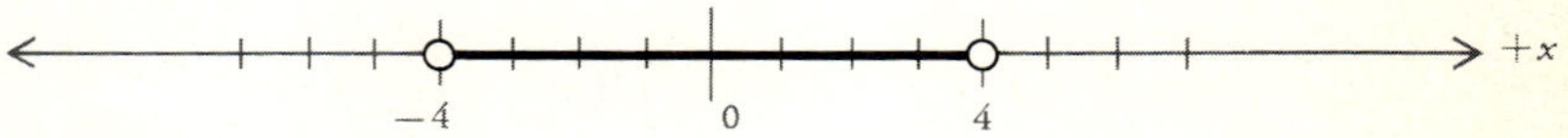

Fig. 19.4 A graph of $|x| < 4$; *also* $-4 < x < 4$.

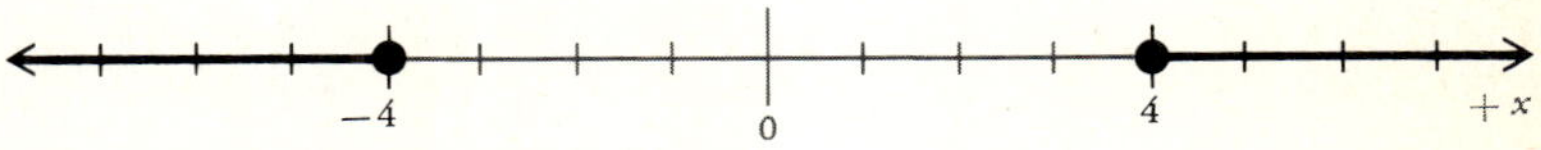

Fig. 19.5 A graph of $|x| \geq 4$; *also* $x \geq 4$ *or* $x \leq -4$.

Example 19.4. $|x - 5| \geq 2$ **means that the distance between** x **and 5 is greater than or equal to 2 units, or** x **is at least 2 units from 5. Therefore** $x \geq 5 + 2$ **or** $x \leq 5 - 2$. **The result is** $x \geq 7$ **or** $x \leq 3$ **(see Fig. 19.6).**

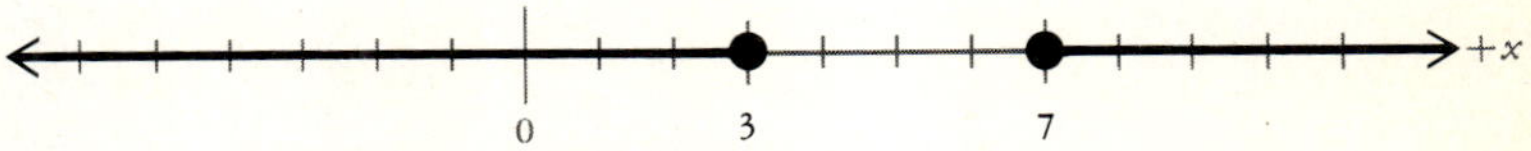

Fig. 19.6 Example 19.4. A graph of $|x - 5| \geq 2$; *also* $x \geq 7$ *or* $x \leq 3$.

Example 19.5. **Determine the largest and smallest values of** x **for which** $|x + 5| \leq 2$, **and graph all such values of** x **on the number line.**

Solution. Writing $|x + 5|$ in the form $|x - a|$, we have $|x - (-5)| \leq 2$. Therefore according to inequality (19.5), we have

$$-5 - 2 \leq x \leq -5 + 2 \qquad \text{or} \qquad -7 \leq x \leq -3$$

The largest value is -3; the smallest value is -7. The graph is shown in Fig. 19.7.

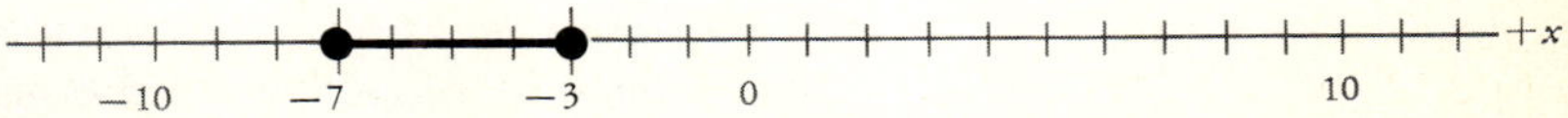

Fig. 19.7 Example 19.5. A graph of $|x + 5| \leq 2$; *also* $-7 \leq x \leq -3$.

Example 19.6. Change the statement $-5 < x < 8$ to the form $|x - a| < n$.

Solution. $|x - a| < n$ can be written as $a - n < x < a + n$ [see inequality (19.5)], and the given statement is $-5 < x < 8$. Therefore $a - n = -5$ and $a + n = 8$. We solve these equations simultaneously for a and n as follows:

$$a - n = -5$$

$$a + n = 8$$

add $\quad 2a = 3$

$$a = 1.5$$

by substitution $\quad n = 6.5$

The result is $\quad |x - 1.5| < 6.5$

EXERCISE 19.1 (*Continued*)

4. In each exercise remove the absolute-value sign by rewriting the inequality in a form such as $b < x < a$.

   a. $|x| < 9$ b. $|x| > 6.5$ c. $|x - 4| \geqslant 10$
   d. $|x + 2| \leqslant 10$ e. $|x + 1| > 8$ f. $|x - 3| \geqslant 3$

5. Rewrite each of the following inequalities as an absolute value in a form such as $|x - a| < n$. See Example 19.6.

   a. $-12 \leqslant x \leqslant 12$ b. $x > 2^1/_2$ or $x < -2^1/_2$ c. $0 < x < 4$
   d. $-2 \leqslant x \leqslant 10$ e. $x \geqslant -5$ or $x \leqslant -10$ f. $-8 \leqslant x \leqslant 0$

6. Graph each inequality of Prob. 4 on the number line.
7. Graph each inequality of Prob. 5 on the number line.

19.2 PROPERTIES OF INEQUALITIES

We can solve inequalities in much the same way that we solve equations; that is, we perform algebraic operations on both sides. But certain operations that apply to equations do *not* apply to inequalities. In this section we shall see how algebraic operations affect inequalities. The effects of these operations on inequalities are called *properties of inequalities*.

First, the properties of inequalities involve the *direction* or *sense* of the inequalities. The *sense* of an inequality is the direction of the inequality symbol. If the inequality symbols of two statements point in the *same direction,* then the inequality statements have the *same sense;* if the symbols point in *opposite directions,* then the inequality statements have *different senses.* The statements below illustrate the concept of inequality sense.

$$\left.\begin{matrix} a > b \\ c > d \end{matrix}\right] \text{inequalities of the } \textit{same sense}$$

$$\left.\begin{matrix} p < q \\ r > s \end{matrix}\right] \text{inequalities of } \textit{different senses}$$

The following four properties are *axioms;* that is, they are defined to be true statements. They cannot be derived from more basic propositions or from any force of logic. We agree to accept them because they seem intuitively correct. Having accepted these basic propositions, we can build many other useful ones. We should note that these properties apply only to real numbers.

A. THE FOUR BASIC PROPERTIES OF INEQUALITIES

Property 1. The Trichotomy Property

One and only one of the following statements regarding two numbers a and b can be *true:*

$$a > b \quad \text{or} \quad a = b \quad \text{or} \quad a < b \tag{19.7}$$

This property is the basis for statements such as if $a > b$, then $a \nless b$.

Property 2. The Transitive Property

$$\text{If } \quad a > b \text{ and } b > c, \quad \text{then } \quad a > c \tag{19.8}$$

In words, inequality (19.8) means "If one number is greater than a second number and the second number is greater than a third number, then the first number is greater than the third number."

Property 3. The Addition Property

$$\text{If } \quad a > b, \quad \text{then } \quad a + c > b + c \tag{19.9}$$

$$(\text{and } \quad a - c > b - c)$$

The sense of an inequality is *not changed* if the *same number* is added to (or subtracted from) *both* sides.

Property 4. **The Multiplication Property**

$$\text{If} \quad a > b \text{ and } c > 0, \qquad \text{then} \quad ac > bc \tag{19.10}$$

$$(\text{and} \quad a/c > b/c)$$

The sense of an inequality is *not changed* if *both* sides are multiplied (or divided) by the *same positive* number.

The statement $a > b$ can be read either "a is greater than b" or "b is less than a." Therefore the "less than" sign may be substituted for the "greater than" sign in any of the basic properties or those that follow. For example, Property 3 could also be read as follows:

$$\text{If} \quad a < b, \qquad \text{then} \quad a + c < b + c$$

Another implied condition is that any expression may be substituted for an equal expression. For example,

$$\text{If} \quad a > b \text{ and } b = c, \qquad \text{then} \quad a > c$$

Example 19.7. To find the values of x that satisfy the statement $3x - 2 < 10$, we apply Properties 3 and 4 as follows:

given: $$3x - 2 < 10$$

add 2 (Property 3): $$3x - 2 + 2 < 10 + 2$$

$$3x < 12$$

÷ 3 (Property 4): $$\frac{3x}{3} < \frac{12}{3}$$

$$\boxed{x < 4}$$

B. OTHER PROPERTIES OF INEQUALITIES

Innumerable other properties can be derived from the four basic ones. It is not our primary goal to study mathematical proofs, nor do we wish to dwell upon mathematical rigor. But in this section we shall present several properties by showing proofs of them. In so doing we shall familiarize ourselves with them, and we shall establish the validity of these properties and the following work that uses them. An important point in this lesson is that we should not take anything for granted. Even the most seemingly obvious statements must be traceable to the basic properties. In any field of endeavor if we cannot justify our work, then it may well be invalid.

Property 5. *a.* If $a > b$, then $-a < -b$.
b. If $-a < -b$, then $a > b$.

Proof of (*a*): Add $-a - b$ to both sides and combine similar terms.

Given $a > b$

add $-a - b$ $a + (-a - b) > b + (-a - b)$ (by Property 3)

and combine terms $a - a - b > b - b - a$

$-b > -a$

$-a < -b$

Example: $5 > 2$ and $-5 < -2$

Property 6. *a.* If $a > 0$, then $-a < 0$.
b. If $a < 0$, then $-a > 0$.

Proof of (*a*): Use Property 5*a* and the fact that $0 = -0$.

Given $a > 0$

then $-a < -0$ (by Property 5*a*)

and $-a < 0$ (because $-0 = 0$)

Example: $5 > 0$ and $-5 < 0$

Property 7. If $a > b$ and $c < 0$, then $ac < bc$ (19.11)

(and $a/c < b/c$)

The sense of an inequality is *reversed* if *both* sides are multiplied (or divided) by the same *negative* number.

Proof:

Given $c < 0$

then $\quad -c > 0 \quad$ (by Property 6*b*)

(In other words, $-c$ is positive.)

Given $\quad a > b$

then $\quad a(-c) > b(-c) \quad$ (by Property 4)

$$-ac > -bc$$

$ac < bc \quad$ (by Property 5*b*)

Example: $\quad 6 > 5$

$6(-3) < 5(-3) \quad$ (note the *reversed* sense)

$$-18 < -15$$

Property 8. If $a > b$ and $c > d$, then $a + c > b + d$

Proof:

Given $\quad a > b$

then $\quad a + c > b + c \quad$ (by Property 3)

Given $\quad c > d$

then $\quad b + c > b + d \quad$ (by Property 3)

and $\quad a + c > b + c > b + d$

or $\quad a + c > b + d \quad$ (by Property 2)

Example: $\quad -5 > -12 \quad$ and $\quad 4 > 3$

$$-5 + 4 > -12 + 3$$

$$-1 > -9$$

Property 9. If a, b, c, and d are *positive* numbers, $a > b$, and $c > d$, then $ac > bd$

Proof:

Given $a > b$ and $c > 0$

then $ac > bc$ (by Property 4)

Given $c > d$ and $b > 0$

then $bc > bd$ (by Property 4)

and $ac > bc > bd$

or $ac > bd$ (by Property 2)

Example: $8 > 5$ and $6 > 2$

$8(6) > 5(2)$

$48 > 10$

Property 10. If $a > 0, b > 0, n > 0$, and $a > b$, then $a^n > b^n$

Proof: The proof of this property involves *mathematical induction,* a topic that we shall not take up. However, we can demonstrate the property for specific values of n, as follows:

Proof for $n = 2$:

Given: $a > 0 \quad b > 0 \quad a > b$

$a \cdot a > b \cdot b$ (Property 9)

$a^2 > b^2$

Examples: *a.* $5 > 2$ and $5^3 = 125 > 8 = 2^3$
b. $27 < 64$ and $27^{1/3} < 64^{1/3}$

or $\sqrt[3]{27} < \sqrt[3]{64}$

$3 < 4$

EXERCISE 19.2

1. Verify each of Properties 1 to 4 by using suitable numbers.
2. Verify each of Properties 5 to 9 by using suitable numbers.

3. Prove Property 5*b*. Use the proof of Property 5*a* as a guide.
4. Prove Property 6*b*. Use the proof of Property 6*a* as a guide.
5. Show with suitable numbers that the following statement can be either true or false depending upon the numbers. If $a > b$, then $1/a < 1/b$. (*Hint:* Include some negative numbers in your selections.)
6. Find the values of x that make each of the following inequalities true. Use Property 3.

   a. $x - 3 > 5$ b. $x - 3 < 5$ c. $x + 2 < -7$
   d. $x + 6 > -1$ e. $x + 2 > 2$ and $x + 2 < 7$

7. Find the values of x that make each of the following inequalities true. Use Properties 4, 5, or 7.

   a. $2x > 12$ b. $-3x < -9$ c. $-x/5 < 1$
   d. $3x/4 > 25$

8. Show that the following statement is true: If $0 < x < a$, then $x^2 < a^2$. (*Hint:* Since $x > 0$ and $a > x$, then x is positive and a is positive. Use Property 9.)
9. Show that the following statement is true: If $0 < x < 1$, then $x^2 < x$. In other words, the square of a *positive* number less than unity is less than the number itself.

19.3 SOLVING INEQUALITIES

The general procedure for solving an inequality is like that for solving an equation: we isolate the unknown variable by performing the appropriate operations—addition, subtraction, multiplication, division, and so on—to *both sides*. However, we must take care to perform each operation in accordance with the properties of inequalities. In particular, we must note whether a given operation involves positive or negative numbers. The following examples show how to use the properties to solve inequalities.

Example 19.8. Solve for x: $2x - 5 < -6$

Given	$2x - 5 < -6$
add 5	$2x < -1$
÷ 2	$\boxed{x < -1/2}$

Example 19.9. Solve for y: $2 < 3 - \frac{y}{2} < 10$

Given $$2 < 3 - \frac{y}{2} < 10$$

subtract 3 $$-1 < -\frac{y}{2} < 7$$

multiply by -2 and *reverse* the inequalities according to Property 7. $$2 > y > -14$$

or $$\boxed{-14 < y < 2}$$

Example 19.10. Solve for z: $z + 5 > 2z - 3$

Given $$z + 5 > 2z - 3$$

subtract $2z$ and 5 $$z - 2z > -3 - 5$$

$$-z > -8$$

by Property 5*b* $$\boxed{z < 8}$$

The answer can be spot-checked by substituting values for z. Any value of z less than 8 should make the original inequality true, and any other value should make it false.

a. Let $z = 7$: $$7 + 5 \overset{?}{>} 2(7) - 3$$

$$12 \overset{?}{>} 14 - 3$$

$$12 > 11$$

b. Let $z = 8$: $$8 + 5 \overset{?}{>} 2(8) - 3$$

$$13 \overset{?}{>} 16 - 3$$

$$13 \overset{?}{>} 13$$

Since $13 \not> 13$, the check is good.

Example 19.11. Solve for x: $1/(x - 2) < 5$

If this statement were an equation, we could take the reciprocal of both sides; however such an operation is not generally valid for an inequality (see Prob. 5 of Exercise 19.2). We must use the properties of inequalities to work out a valid scheme.

First, we need to get x out of the denominator and into a numerator. We could do so by multiplying both sides by $x - 2$, except that we do not know if $x - 2$ is positive or negative. In order to get around this limitation, we invoke Property 1 and examine all possibilities as follows:

Property 1: $x - 2 < 0 \qquad x - 2 = 0 \qquad \text{or} \qquad x - 2 > 0$

$x - 2$ cannot be zero because division by zero is undefined.

<u>If $x - 2 < 0$:</u>

then use Property 7

given $\dfrac{1}{x-2} < 5$

then $1 > 5(x - 2)$

$5x < 11$

$x < 2.2$

But since $x - 2 < 0$ was a condition for using Property 7, we already have $x < 2$. This more restrictive limitation must take precedence.

<u>If $x - 2 > 0$:</u>

then use Property 4

given $\dfrac{1}{x-2} < 5$

then $1 < 5(x - 2)$

$5x > 11$

$x > 2.2$

Answer: $\boxed{x > 2.2 \qquad \text{or} \qquad x < 2}$

(See Fig. 19.8.)

Fig. 19.8 Answer to Example 19.11. $x > 2.2$ *or* $x < 2$.

Example 19.12. Solve for x: $|x + 10| \leqslant 11$

Recall that $|x - a| \leqslant n$ means $a - n \leqslant x \leqslant a + n$. By changing $|x + 10| \leqslant$

11 to this form, we have $|x - (-10)| \leq 11$, from which

$$-10 - 11 \leq x \leq -10 + 11$$

$$\boxed{-21 \leq x \leq 1}$$

Example 19.13. Solve for w: $\sqrt{5w^2 + 25} \leq 5\sqrt{6}$

Given	$\sqrt{5w^2 + 25} \leq 5\sqrt{6}$
Property 9	$(\sqrt{5w^2 + 25})(\sqrt{5w^2 + 25}) \leq (5\sqrt{6})(5\sqrt{6})$
	$(\sqrt{5w^2 + 25})^2 \leq 150$
	$5w^2 + 25 \leq 150$
subtract 25	$5w^2 \leq 125$
$\div 5$	$w^2 \leq 25$
take $\sqrt{\ }$	$\sqrt{w^2} \leq \sqrt{25} = 5$

At this point we must *proceed with great caution concerning signs.* Recall Eq. (12.9), the special definition for square roots:

$$\sqrt{x^2} = \begin{cases} x & \text{if } x \geq 0 \\ -x & \text{if } x < 0 \end{cases} \tag{12.9}$$

Applying this definition to our problem, we have the following two possibilities:

1. If $w \geq 0$, then $\sqrt{w^2} = w \leq 5$. Therefore $w \geq 0$ *and* $w \leq 5$, which means that $0 \leq w \leq 5$.
2. If $w < 0$, then $\sqrt{w^2} = -w \leq 5$, which gives $w \geq -5$. Therefore $w < 0$ and $w \geq -5$, which means $-5 \leq w < 0$.

The combined result is $\boxed{-5 \leq w \leq 5}$

Example 19.14. Solve for x: $2x^2 - 2x - 12 > 0$

We can solve such a *quadratic inequality* much as we would a quadratic equation. First, we factor the quadratic trinomial.

Given $\qquad 2x^2 - 2x - 12 > 0$

factor $(x - 3)(2x + 4) > 0$

In order for the product of two numbers to be positive, the numbers must have *like signs.* Therefore we must examine the following two possibilities:

1. ***Both factors are positive:***

$$x - 3 > 0 \quad \text{and} \quad 2x + 4 > 0$$

from which

$$x > 3 \quad \text{and} \quad x > -2$$

$x > 3$ guarantees that $x > -2$, but $x > -2$ does not guarantee that $x > 3$. Therefore for x to be *both* greater than -2 *and* greater than 3, we must use only $x > 3$.

2. ***Both factors are negative:***

$$x - 3 < 0 \quad \text{and} \quad 2x + 4 < 0$$

from which

$$x < 3 \quad \text{and} \quad x < -2$$

Here again we must take the more restrictive result, which is $x < -2$.

By combining the results of the two possibilities we have the following final result (see Fig. 19.9).

$$\boxed{x > 3 \quad \text{or} \quad x < -2}$$

Fig. 19.9 Answer to Example 19.14. $x > 3$ or $x < -2$.

Example 19.15. Solve for a: $a^2 - 3a - 10 < 0$

Given $a^2 - 3a - 10 < 0$

factor $(a - 5)(a + 2) < 0$

In order for the product to be *negative,* the factors $a - 5$ and $a + 2$ must have ***unlike signs.*** Therefore we must consider the following two possibilities:

1. $a - 5 > 0$ and $a + 2 < 0$

$a > 5$ and $a < -2$

But a cannot be *both* greater than 5 *and* less than -2. Therefore this possibility produces no solutions.

2. $a - 5 < 0$ and $a + 2 > 0$

$a < 5$ and $a > -2$

$$\boxed{-2 < a < 5}$$

Comment: *At this point we might need to review the distinction between the words "and" and "or" in inequalities. If we say, for example, that $x < 5$* ***and*** *$x > -2$, we mean that* ***every value*** *of x must satisfy* ***both*** *conditions. Therefore $-2 < x < 5$. In some instances, as in Example 19.15 where $a > 5$* ***and*** *$a < -2$, the conditions are mutually exclusive. No value of a can satisfy both conditions.*

On the other hand, the statement $a > 5$ ***or*** *$a < -2$ means that each value of a must satisfy* ***one condition or the other,*** *but not necessarily both. This is the meaning implied by the answer to Example 19.14.*

Example 19.16. Solve for t: $2t^2 + 3t - 3 \leqslant 0$

The left side is a nonfactorable quadratic trinomial. The method of completing the square is very useful for solving such inequalities.

Given $2t^2 + 3t - 3 \leqslant 0$

add 3 $2t^2 + 3t \leqslant 3$

$\div 2$ $t^2 + (3/2)t \leqslant 3/2$

add $9/16$ $t^2 + (3/2)t + 9/16 \leqslant 3/2 + 9/16 = 33/16$

factor $(t + 3/4)^2 \leqslant 33/16$

take $\sqrt{\ }$ $\sqrt{(t + 3/4)^2} \leqslant \sqrt{33/16} = 1/4\sqrt{33}$

Use the definition of x^2:

1. If $t + 3/4 \geqslant 0$, then $\sqrt{(t + 3/4)^2} = t + 3/4 \leqslant 1/4\sqrt{33}$, from which

$$-3/4 \leqslant t \leqslant -3/4 + 1/4\sqrt{33}$$

2. If $t + 3/4 < 0$, then $\sqrt{(t + 3/4)^2} = -(t + 3/4) \leqslant 1/4\sqrt{33}$

$$t < -3/4 \qquad\qquad t + 3/4 \geqslant -1/4\sqrt{33}$$

$$t \geqslant -3/4 - 1/4\sqrt{33}$$

from which

$$-3/4 - 1/4\sqrt{33} \leqslant t < -3/4$$

The combined results are

$$\boxed{-3/4 - 1/4\sqrt{33} \leqslant t \leqslant -3/4 + 1/4\sqrt{33}}$$

EXERCISE 19.3

In Probs. 1 to 34 solve each inequality for x and illustrate each solution on the number line.

1. $3x + 5 \geqslant 17$
2. $2 - 3x < 10 + x$
3. $x/2 + 7 \leqslant 5$
4. $-1 \leqslant x + 2 \leqslant 1$
5. $-20 \leqslant 3x - 10 \leqslant 5$
6. $x/5 - 2 > 0$ or $x/5 - 2 < -5$
7. $x^2 \geqslant 25$
8. $x^2 < 16$
9. $x^2 - 4 > 0$
10. $x^2 - 4 \leqslant 0$
11. $x^2 - 2x - 3 > 0$
12. $x^2 - 2x - 3 < 0$
13. $x^2 - 4x + 3 > 0$
14. $x^2 - 4x + 3 < 0$
15. $x^2 + 6x + 8 \leqslant 0$
16. $x^2 + 6x + 8 \geqslant 0$
17. $\sqrt{(x - 2)^2} \leqslant 5$
18. $\sqrt{(x + 1)^2} \geqslant 1$
19. $\sqrt{x^2 - 6x + 9} > 0$
20. $\sqrt{25 + 10x + x^2} \leqslant 0$
21. $x^2 + 2x - 1 > 0$
22. $2x^2 - 3x - 1 \leqslant 0$
23. $x^3 - 8 \leqslant 0$
24. $x^3 + 27 > 0$
25. $x^4 \leqslant 16$
26. $x^5 > -32$
27. $|x - 1| < 2$
28. $|x - 5| \leqslant 5$
29. $|x + 3| < 10$
30. $|x + 7| > 5$
31. $1/x < 6$
32. $2/(x + 5) > 10$
33. $1/(x - 1) + 1 < 3$
34. $x - 6/(x - 2) + 3 > 0$

19.4 GRAPHS OF INEQUALITIES WITH TWO VARIABLES

An inequality with two variables can be graphed in much the same manner as an equation. In fact, the graph of the corresponding equation is a boundary line for the graph of the inequality. For example, Fig. 19.10 illustrates that the graph of the *equation* $y = x + 1$ is a boundary for the graph of $y \leqslant x + 1$. Every value of y on *or below* the line $y = x + 1$ satisfies the inequality $y \leqslant x + 1$. Notice that the graph consists of an *entire area* of the coordinate plane.

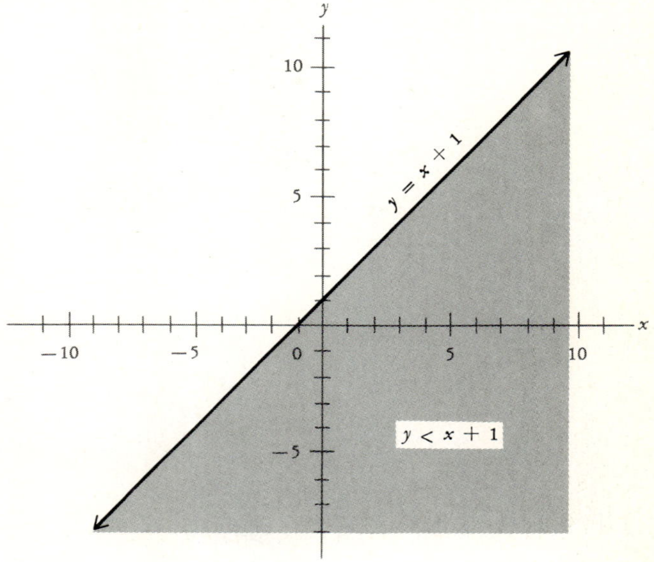

Fig. 19.10 A graph of $y \leqslant x + 1$.

Example 19.17. Graph the inequality $y > x^2 - 3$.

First we graph the *equation* $y = x^2 - 3$. This graph is shown in Fig. 19.11. Then we shade the area above the curve. We make the curve $y = x^2 - 3$ a dashed line to indicate that it is excluded from the values of y.

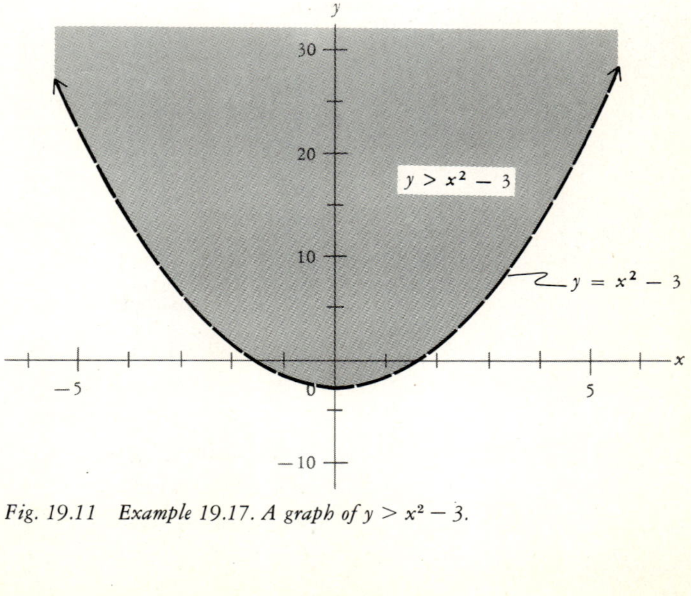

Fig. 19.11 Example 19.17. A graph of $y > x^2 - 3$.

Example 19.18. Graph the inequality $|y - 2x| \leqslant 10$.

First we change the inequality according to definition (19.5), as follows:

$$2x - 10 \leqslant y \leqslant 2x + 10$$

Then we graph both equations, $y = 2x + 10$ and $y = 2x - 10$. The graph of $|y - 2x| \leqslant 10$ is the area between and including these lines. See Fig. 19.12.

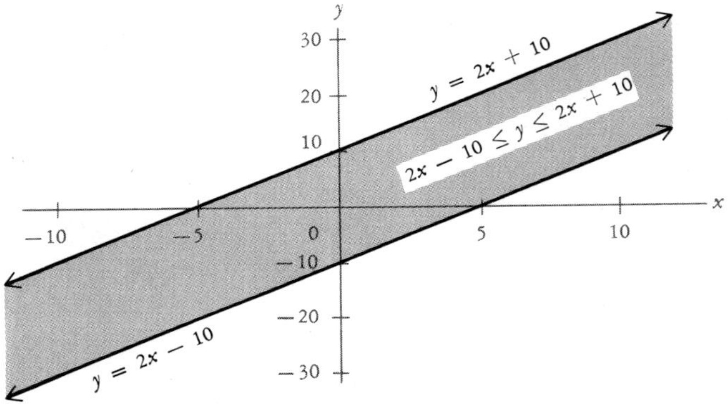

Fig. 19.12 Example 19.18. A graph of $|y - 2x| \leqslant 10$.

EXERCISE 19.4

In Probs. 1 to 9, graph each inequality over the range of values specified for x.

1. $y \leqslant x + 5$, $-10 \leqslant x \leqslant 10$
2. $y \geqslant x + 5$, $-10 \leqslant x \leqslant 10$
3. $x - 5 < y < 2x$, $-10 \leqslant x \leqslant 10$
4. $x^2 - 25 < y < 0$, $-6 \leqslant x \leqslant 6$
5. $0 < y < (x - 3)^2$, $-1 \leqslant x \leqslant 7$
6. $|y - x| \leqslant 10$, $-10 \leqslant x \leqslant 10$
7. $|y + x| \leqslant 10$, $-10 \leqslant x \leqslant 10$
8. $x^2 < y < x$, $0 \leqslant x \leqslant 1$
9. $|y^2 - x| \leqslant 4$, $-5 \leqslant x \leqslant 20$

Graphical Solutions of Inequalities

We have previously solved equations graphically (Chaps. 7, 8, and 16). For example we can solve $x^2 - 16 = 0$ by graphing the equation $y = x^2 - 16$. The solutions of $x^2 - 16 = 0$ are the points (4,0) and (−4,0) where the graph of $y = x^2 - 16$ crosses the x axis; in other words, the solutions are the values of x where $y = x^2 - 16$ *and* where $y = 0$.

Similarly, the solutions of the inequality $x^2 - 16 \leq 0$ can be found by graphing the inequality $x^2 - 16 \leq y$. The solutions are values of x where $x^2 - 16 \leq y$ *and* where $y \leq 0$. The shaded area of Fig. 19.13 indicates this region. The solution is the entire set of values $-4 \leq x \leq 4$.

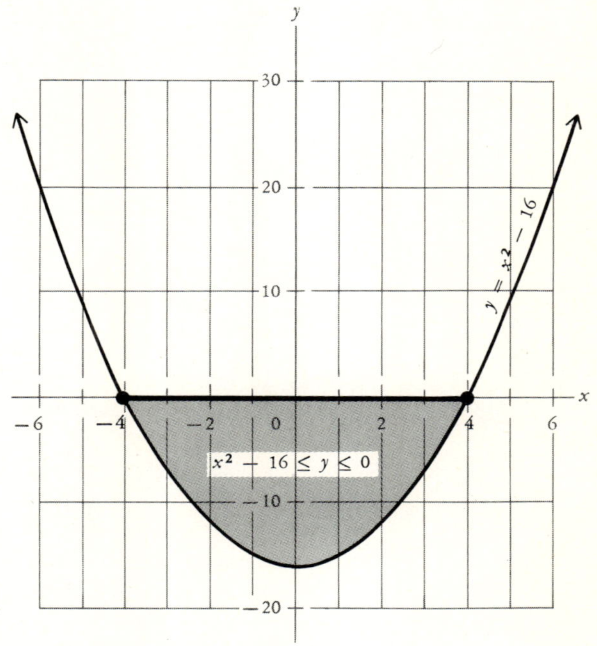

Fig. 19.13 Solutions of $x^2 - 16 \leq 0$.

The general procedure for solving an inequality graphically is as follows:

1. Write the inequality in a form such as $f(x) < 0$, with zero on the right-hand side.
2. Replace the zero by y, and graph the boundary function $y = f(x)$.
3. Locate the region where $f(x) < y < 0$. The values of x in this region solve the inequality $f(x) < 0$.

Example 19.19. **Solve $x - 3 > 5$ graphically.**

First we rearrange the inequality in the form $x - 8 > 0$. Then we graph the boundary line $x - 8 = y$ as shown in Fig. 19.14. The solutions are all values of x in the region $0 < y < x - 8$. These values are $x > 8$; they correspond to the shaded portion of Fig. 19.14.

Example 19.20. **Solve $|x^2 - 9| \geq 7$ graphically over the range $-6 \leq x \leq 6$.**

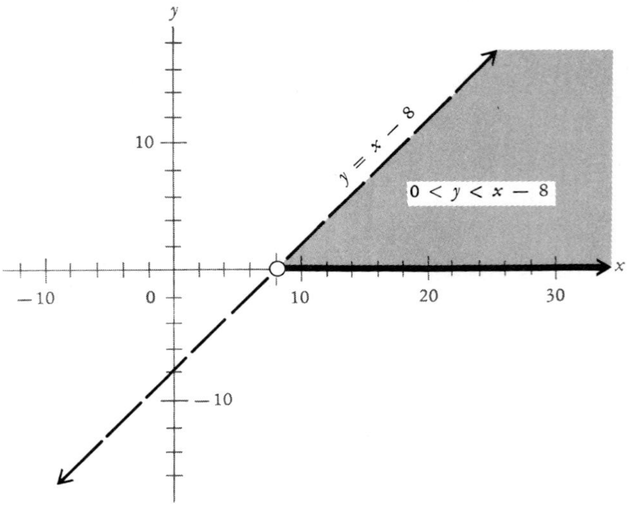

Fig. 19.14 Example 19.19. Solutions of $x - 8 > 0$.

First we use definition (19.6) to rewrite the inequality in the form

$$x^2 \geqslant 9 + 7 = 16 \qquad or \qquad x^2 \leqslant 9 - 7 = 2$$

From these expressions we have $x^2 - 16 \geqslant 0$ *or* $x^2 - 2 \leqslant 0$, and we graph *two* boundary lines $y = x^2 - 16$ and $y = x^2 - 2$, as shown in Fig. 19.15. The solutions are the values of x both bounded by the lines

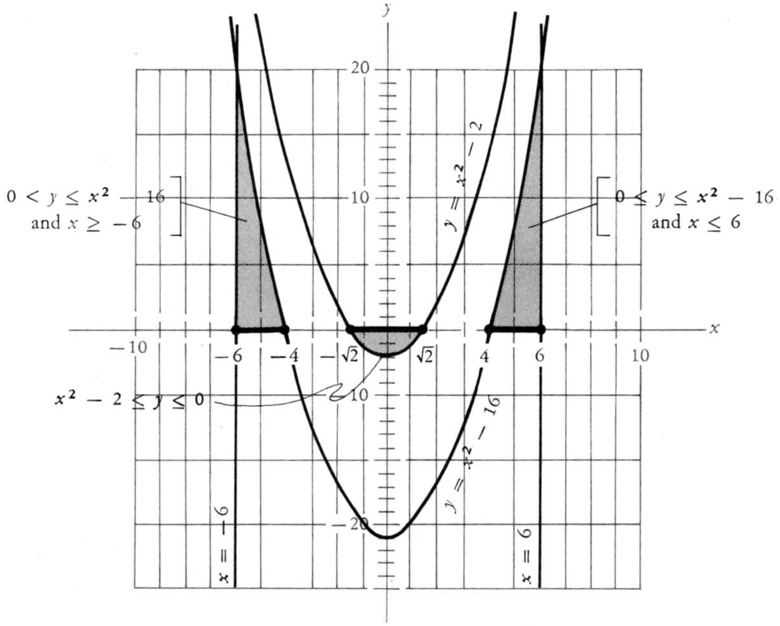

Fig. 19.15 Example 19.20. Solutions of $|x^2 - 9| \geqslant 7, -6 \leqslant x \leqslant 6$.

$x = 6$ and $x = -6$, and within the regions where $0 \leqslant y \leqslant x^2 - 16$, *or* where $x^2 - 2 \leqslant y \leqslant 0$. The figure shows that these values are as follows:

$$\boxed{4 \leqslant x \leqslant 6 \qquad -6 \leqslant x \leqslant -4 \qquad \text{and} \qquad -1.4 \leqslant x \leqslant 1.4}$$

Example 19.21. Solve $-10 < 2x - 5 < 10$ graphically.

It would be possible to form two boundary lines—one from the left-hand inequality and one from the right-hand inequality—as follows:

$$0 < 2x + 5 \qquad \text{and} \qquad 2x - 15 < 0$$

The solutions would be those values of x common to the two sets of solutions $0 < y < 2x + 5$ and $2x - 15 < y < 0$, as shown in Fig. 19.16*a*.

Another, and frequently simpler, technique is to graph the line $y = 2x - 5$ as shown in Fig. 19.16*b*. The solutions of $-10 < 2x - 5 < 10$ are those values of x on the line $y = 2x - 5$ and between the lines $y = 10$ and $y = -10$.

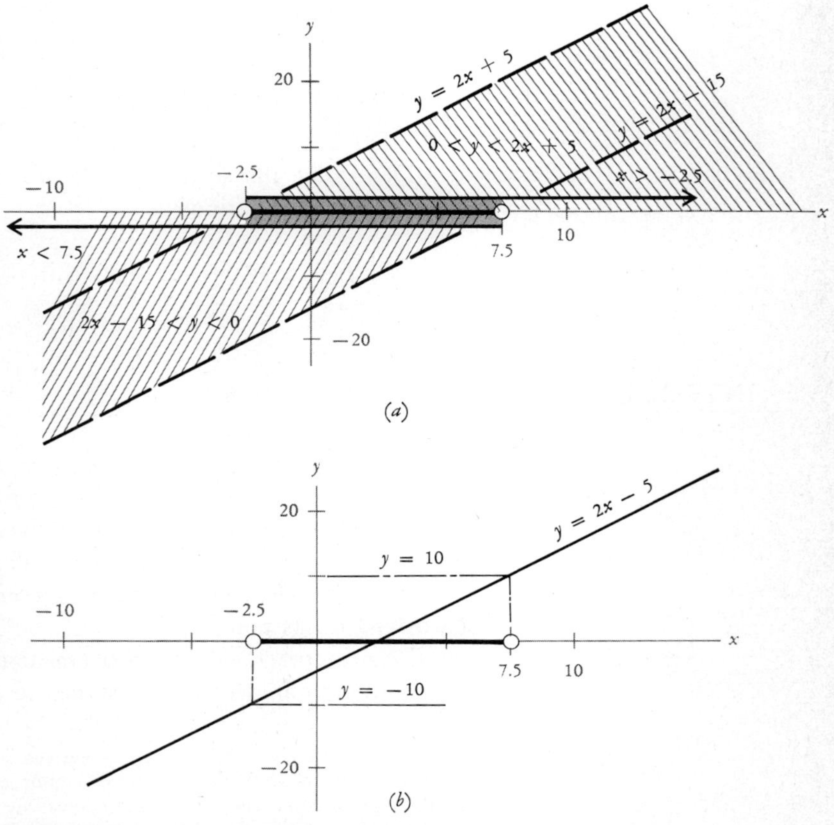

Fig. 19.16 Example 19.21. Solutions of $-10 < 2x - 5 < 10$.

EXERCISE 19.4 (*Continued*)

In Probs. 10 to 24, solve each inequality by the graphical method.

10. $x + 5 \leq 0$
11. $2x - 5 > -17$
12. $x^2 - 16 > 0$
13. $x^2 - 25 \leq 0$
14. $x^3 - 8 \leq 0$
15. $x^4 - 16 \leq 0$
16. $-10 < x - 5 < 10$
17. $0 < x - 3 < 7$
18. $60 \leq 5x + 10 \leq 110$
19. $|x^2| \leq 25$
20. $|x^2| > 9$
21. $|x - 2| \leq 4$
22. $x^2 + x - 12 \leq 0$
23. $0 \leq x^2 + x - 12 \leq 10$
24. $25 - x^2 \geq 0$ and $-5 \leq x \leq 5$

25. Determine the values of x that satisfy all three of the following conditions:

$$x + 5 \geq 0 \qquad 2x - 5 \leq 0 \qquad \text{and} \qquad x > -5$$

26. Determine the values of x that satisfy both of the following conditions:

$$\frac{x}{2} - 5 \leq 0 \qquad \text{and} \qquad -\frac{x^2}{4} + \frac{x}{2} + 10 \geq 0$$

27. Determine the smallest value of x that satisfies all three of the following conditions:

$$10 - x \geq 0 \qquad x + 10 \geq 0 \qquad \text{and} \qquad \frac{-x}{2} - 10 \leq 0$$

19.5 INTRODUCTION TO LINEAR PROGRAMMING

Inequalities can help to solve problems in industry and business. A typical objective is to maximize the profit of a certain manufacturing operation. To achieve such an objective, we often must determine an optimum allocation of raw materials and manpower, and while doing this we must consider practical constraints such as the number of working hours per day.

An approach called *linear programming** (no relation to computer programming) uses inequalities to solve certain kinds of opti-

* The name *linear programming* is derived from the fact that (1) the problems are restricted to those involving *linear* equations and *linear* inequalities, and (2) that the allocation of goods and labor is sometimes called *programming*.

mization problems. The following example illustrates a typical problem and a graphical method of solution.

Example 19.22. The Meshin Gear Company has a contract to manufacture a certain gear in two grades: S (standard) and P (precision). Each grade must be machined on both machines M_1 and M_2. Each grade-S gear is processed for 16 min on machine M_1 followed by 24 min on machine M_2; each grade P requires 48 min on M_1 followed by 32 min on M_2. Neither machine can be operated more than 8 hr/day. The profits are known to be \$6 each for grade-S gears and \$10 each for grade-P gears. How many of each grade gear should Meshin make daily in order to maximize profits?

Solution. In order to visualize the interplay among the variables, we can define a notation, state the facts in terms of algebraic statements, and graph these statements.

Let s be the number of grade-S gears produced per day
p be the number of grade-P gears produced per day
k be the daily profit in dollars

Then the following relationships are binding:

1. $s \geqslant 0, p \geqslant 0$ (We cannot have a negative number of gears.)
2. $k = 6s + 10p$ (The profit is the profit per unit times the number of units: $\$6 \times s + \$10 \times p$.)
3. The running time for machine M_1 is limited to 480 min/day (8 hr) and can be expressed in terms of the numbers of gears as follows: number of grade-S × 16 min per unit plus number of grade-P × 48 min per unit is no more than 480 min. In algebraic form this statement becomes

$$16s + 48p \leqslant 480 \qquad \text{or} \qquad s + 3p \leqslant 30$$

Similarly, for machine M_2 we have

$$24s + 32p \leqslant 480 \qquad \text{or} \qquad 3s + 4p \leqslant 60$$

To summarize, we have to meet the following conditions:

profit: $$k = 6s + 10p \tag{19.12}$$

operating time, M_1: $$s + 3p \leqslant 30 \tag{19.13}$$

operating time, M_2: $3s + 4p \leqslant 60$ (19.14)

minimum numbers: $s \geqslant 0 \quad p \geqslant 0$ (19.15)

To solve the problem, all we need to do is choose s and p to maximize k while adhering to conditions (19.13), (19.14), and (19.15). A graph of the conditions will provide the insight needed to find the solution. Figure 19.17 shows inequalities (19.13), (19.14), and (19.15).

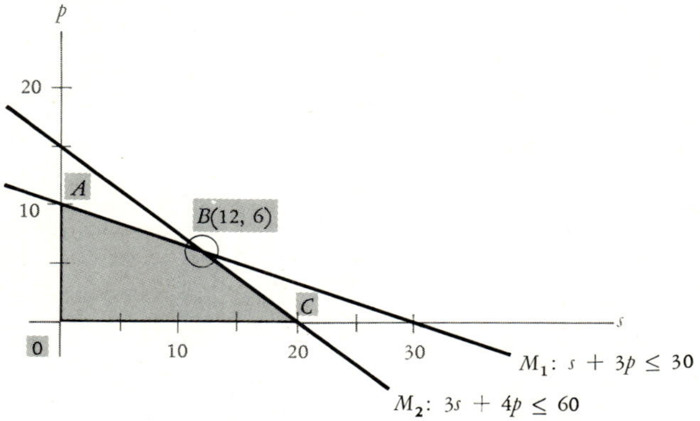

Fig. 19.17 Example 19.22.

The shaded polygon $0ABC$ is the region where all conditions (19.13), (19.14), and (19.15) are met *simultaneously.* All values of s and p within this region or on the boundaries are acceptable. As we shall explain later, the optimum solution will be the coordinates of one of the corners of the polygon. Therefore, to find the optimum solution, we substitute the coordinates of points 0, A, B, and C into Eq. (19.12). The pair of coordinates which produces the highest value of k (the highest profit) is the optimum solution.

Coordinates (s,p)	Profit k
Point 0, (0,0):	$k_O = 6(0) + 10(0) \quad = \$0$
Point A, (0,10):	$k_A = 6(0) + 10(10) = \$100$
Point B, (12,6):	$k_B = 6(12) + 10(6) = \$132$
Point C, (20,0):	$k_C = 6(20) + 10(0) = \$120$

Point B yields the highest profit. Therefore the Meshin Company should produce 12 grade-S and 6 grade-P gears daily.

Some Terminology and General Observations

We shall use Example 19.22 to explain some terminology and to make some general observations concerning linear programming.

Our first step was to determine what values of the variables (s and p) were allowed. These values were dictated by inequalities (19.13), (19.14), and (19.15) as shown in the shaded area of Fig. 19.17. Each point (s,p) in that area satisfied all restrictions and therefore was a possible or *feasible solution.* Such points are called *feasible points.* In other words, a *feasible point* is a set of values of the variables that satisfies all restrictions imposed upon the variables. In general, the set of feasible points forms a polygon; we shall call it the *feasible polygon.* It may be a closed polygon as in Example 19.22, or it may be partly unbounded as in Example 19.23 on the following page.

Our next step was to find, among the feasible points, the one (or more) that would achieve our objective (maximize profit). To do so we tried feasible points in an equation that related them to the objective. Such an equation is called the *objective function.* In Example 19.22, the equation $k = 6s + 10p$ was the objective function; it related the *objective variable* k (profit) to the other variables s and p.

Figure 19.18 shows why the solution to a linear programming problem occurs at a corner of the feasible polygon. The three dashed lines are graphs of the objective function $k = 6s + 10p$ of Example 19.22, for three values of k. (We could have shown many other such lines.) Note that (1) the lines are parallel and (2) the higher the value of k, the further the line is from the origin. It should be obvious from the figure that the optimum line must contain a feasible point and must be, among all such lines, the one furthest from the origin (having

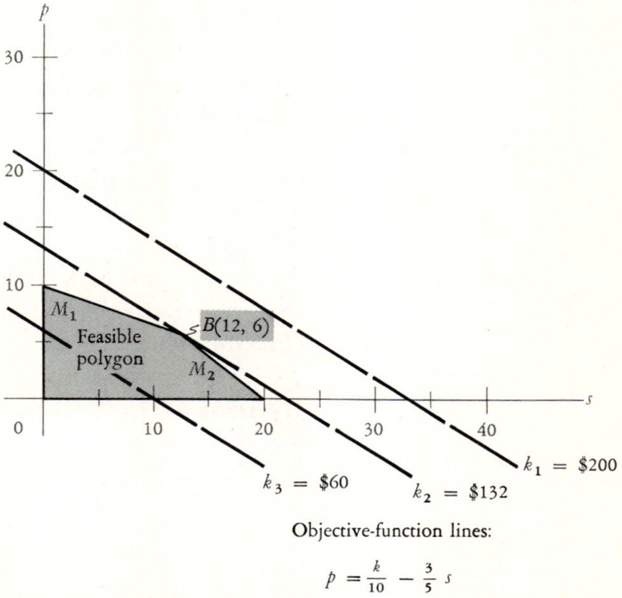

Fig. 19.18 The feasible polygon and three objective-function lines for Example 19.22.

the highest profit). Therefore the optimum line in Fig. 19.18 is k_2, the one that just touches the feasible polygon at point B. Line k_1 shows a higher profit than k_2, but does not contain a feasible point. Line k_3 contains many feasible points, but the profit is lower than that of k_2.

The slope-intercept form (see Sec. 8.1) of the objective function reveals why all *linear* objective functions behave as we cited above. The equation $k = 6s + 10p$ can be written as $p = k/10 - 3s/5$. We can treat p and s as the variables and k as a constant. Because the slope is constant, $-3/5$, all such lines are parallel, and because the "y intercept" $k/10$ is directly proportional to k, the distance from the origin increases as k increases.

These features apply to all linear programming problems. That is, the objective function can be graphed as a set of parallel lines, the position of each line depending upon the value of the objective variable; the solution corresponds to the line that touches the feasible polygon at one corner and at no interior point. We can use these features to devise the following scheme for solving linear programming problems:

1. Graph the feasible polygon.
2. Write the objective function in slope-intercept form, and graph any one line having this slope.
3. Visually determine which corner of the feasible polygon would be touched by a line parallel to the one graphed in step 2.
4. Determine the solution by reading the coordinates of the corner of step 3, or by solving simultaneously the equations of the lines that form the corner.

Example 19.23 below illustrates this procedure. In this example the optimum solution involves a *minimum* value of the objective variable instead of a maximum value.

Example 19.23. **The W. E. Meter Company operates two plants that make voltmeters, ammeters, and wattmeters. Each plant makes all three types of meter, but due to differences in equipment, size, and plant layout, the quanti-**

DAILY PRODUCTION CAPACITIES AND OPERATING COSTS, THE W. E. METER COMPANY

	DAILY CAPACITIES			
	VOLTS	AMP	WATTS	DAILY COSTS
Plant I	25	20	140	\$900
Plant II	30	35	40	\$800

ties vary. The normal daily capacities and operating costs are tabulated above. If the company has orders for 150 voltmeters, 140 ammeters, and 280 wattmeters, how many days should each plant be assigned to fill these orders, so that costs are minimized?

Solution. Let d_1 be the number of working days for plant I
d_2 be the number of working days for plant II
c be the total daily cost

Then the objective function, the total daily cost, can be expressed as follows:

$$c = 900d_1 + 800d_2 \tag{19.16a}$$

or

$$d_2 = -\frac{9}{8}d_1 + \frac{c}{800} \tag{19.16b}$$

This objective is subject to the following restrictions:

$$d_1 \geqslant 0 \tag{19.17}$$

$$d_2 \geqslant 0 \tag{19.18}$$

number of voltmeters: $$25d_1 + 30d_2 \geqslant 150 \tag{19.19}$$

number of ammeters: $$20d_1 + 35d_2 \geqslant 140 \tag{19.20}$$

number of wattmeters: $$140d_1 + 40d_2 \geqslant 280 \tag{19.21}$$

Figure 19.19 is a graph of these restrictions. The shaded area is the set of *feasible points,* the points that satisfy *all five* restrictions. Any point outside this area fails to satisfy one restriction or more. Also shown is a line c_1 having a slope of $-9/8$, which is the slope of the objective function. If line c_1 were moved parallel to itself, it would first contact a feasible point at Q and would appear as line c_2. Since the lowest cost corresponds to the objective line nearest the origin, the solution is found at point Q.

We can find the solutions by solving the following equations simultaneously (note that these are the equations of the lines that form the corner at point Q).

$$25d_1 + 30d_2 = 150$$

$$140d_1 + 40d_2 = 280$$

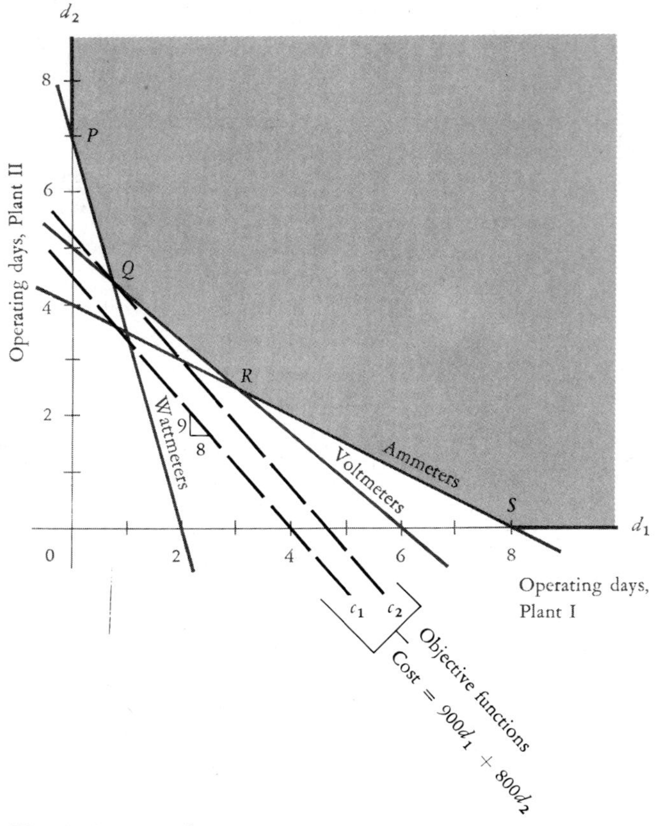

Fig. 19.19 Example 19.23.

The solutions are

$$d_1 = 3/4 \text{ day}$$

$$d_2 = 4\,3/8 \text{ days}$$

From Eq. (19.16*a*) the cost for producing this order is $4,175. Note that the number of ammeters produced is 168 (to the nearest integer), which is more than the amount ordered.

EXERCISE 19.5

1. Standard Manufacturing Company makes two types of vise: L (large) and S (small). Each type must be processed on both machines M_1 and M_2. Each large vise requires 10 min on machine M_1 and 4 min on M_2; each small vise requires 5 min on

M_1 and 8 min on M_2. Machine M_1 runs no more than 8 hr/day, but M_2 can be run 10 hr/day. If the profit is \$6 per unit for type *L* and \$5 per unit for type *S*, how many of each type should be made daily in order to maximize profit? What is the maximum daily profit?

2. Repeat Prob. 1 if each machine is limited to 8 hr/day.
3. The S-C Microwave Company plates its waveguides with silver and copper. Each precision-grade guide receives 800 units of silver and 400 units of copper and sells for \$35. Each regular-grade guide receives 200 units of silver and 1,000 units of copper, and it sells for \$20. If the company has 40,000 units of silver and 56,000 units of copper on hand, how many of each type waveguide should it produce so as to maximize profit? What will the gross sales be for this many waveguides?
4. Repeat Prob. 3 if each waveguide receives 400 units of copper and all other data remain the same.
5. A certain company builds two models of radios: FM and SW. Each model uses three types of transistors: *A*, *B*, and *C*. The profit for each model of radio, the number of transistors required, and the number presently in stock are tabulated below. How many of each model should the company produce so as to most efficiently utilize the transistors available (that is, to maximize profit)?

TYPE OF TRANSISTOR	RADIOS		QUANTITY IN STOCK
	FM MODEL	SW MODEL	
A	2	2	800
B	2	3	900
C	1	5	1,000
unit profit	\$20	\$25	

6. The S. Ludge Oil Refinery has an order for at least 12,000 gal of regular gasoline and at least 15,000 gal of premium. Each grade of gasoline is blended from two kinds of crude oil and additives as tabulated below. If S. Ludge Oil makes \$0.06 per gal on regular and \$0.04 per gal on premium, how much of each grade should it blend in order to maximize profits?

	OIL *A*	OIL *B*	ADDITIVES
Regular	50%	40%	10%
Premium	60%	20%	20%
Quantity available	19,500 gal	12,800 gal	6,000 gal

7. Repeat Prob. 6 if the cost of ingredients increases so as to cut profits to \$0.055 per gal of regular and \$0.0175 per gal of premium.
8. A juice and health food bar sells pure cranberry juice and a cranberry punch consisting of 70 percent cranberry juice and 30 percent apple juice. The punch is mixed fresh each morning. On one particular morning the owner discovers that he has only $3\frac{1}{2}$ gal of cranberry juice and $\frac{1}{2}$ gal of apple juice left. If he makes \$0.50 per gal on the pure cranberry juice and \$0.80 per gal on the punch, how many gallons of cranberry juice should he allot to making punch? (Obviously, he wishes to maximize his profit.) If he sells all the juice and punch, how much profit will he make for the day?
9. The juice-bar owner of Prob. 8 decides that he can make more money by always mixing cranberry juice with apple juice and soda water. He mixes two types of punch: cranberry-apple with 60 percent cranberry juice, 25 percent apple juice, and 15 percent soda water; and applecran with 70 percent apple juice, 15 percent cranberry juice, and 15 percent soda water. He makes \$0.80 per gal on the cranberry-apple punch and \$1.00 per gal on the applecran. If one morning he has 21 qt each of cranberry juice and apple juice and 6 qt of soda water in stock, how many quarts of each type of punch should he mix? How much money will he make on punch that day if he sells all he can make?
10. Repeat Prob. 9 if only 3 qt of soda water are available.
11. The manager of the Blair Music Store has decided to have a discount sale on stereo tape decks and speakers. He cuts the prices until his profit is only \$10 on each tape deck and \$15 on each pair of speakers (sold only in pairs). However, he estimates that he will sell 20 stereo tape cartridges for each tape deck sold, and 12 cartridges and 20 stereo records for each pair of speakers. (The purchasers of decks and speakers do not necessarily buy all the tapes and records; other customers buy them also.) He makes \$0.25 on each cartridge and \$0.35 on each record. If the distributor has 75 tape decks, 200 speakers, 1,920 cartridges, and 2,400 records in stock, how many of each should Blair Music buy in order to maximize profits? How much profit will be made on each item?
12. Repeat Prob. 11 if the store makes \$30 on each tape deck, \$5 on each pair of speakers, and all other data are the same.
13. Repeat Prob. 11 if the store *loses* \$10 on each tape deck and all other data are the same.
14. The B. F. Fireyear Tire Company has a contract to produce tires for the Porsharri Grand Prix racing team. Because the front wheels are smaller and the rear tires transmit the power, two

types of tire are needed. Since these tires are made by special order and racing tires tend to change from year to year, the company will make no more than 1,000 tires. But Porsharri wants at least 600 tires altogether and must have at least 300 rear and 150 front tires. If each rear tire costs the Fireyear Company \$25 and each front tire \$20, how many of each type should it produce? What is the total cost?

15. Repeat Prob. 14 if costs are \$20 each for rear tires and \$25 each for front tires.
16. Repeat Prob. 14 if, in addition to the other requirements, Porsharri must have at least twice as many rear tires as front ones.

In Probs. 17 to 20, determine both the maximum value and the minimum value of each objective function $f(x,y)$, subject to the given constraints.

17. $f(x,y) = 2x + y;\ x, y \geqslant 0;\ 2x - y \geqslant 4;\ 6x + 5y \leqslant 60$
18. $f(x,y) = x + y;\ x \geqslant -5;\ y \geqslant 2;\ y \leqslant 0.1x + 7.5;\ y \geqslant 1.7x - 8.5$
19. $f(x,y) = x - 2y;\ x \leqslant 12;\ y \leqslant 9;\ x - y + 2 \geqslant 0;\ x + 3y - 6 \geqslant 0$
20. $f(x,y) = 4y - 5x;\ y \leqslant 2x;\ 5y \leqslant 2x + 60;\ 2y \geqslant x - 2$

Some Special Cases

In order to clarify the concepts in linear programming, we have purposely avoided certain complexities. We should not, however, leave the subject without mentioning some of these cases.

First, many problems require *integral* answers. For example, the tire company of Prob. 14 would not produce a fraction of a tire. In a practical situation such as this, the answer (if a fraction) provided by linear programming should be used as a guide. A little trial and error with the nearest integers is required to find the optimum practical answer. See Prob. 21 (p. 485).

Second, it is possible for the optimum objective function to touch the feasible polygon at more than one point; it can coincide with one side, as shown in Fig. 19.20. In such a case, the slope of the objective function equals that of one side of the polygon, and the value of the objective function is the same for each point on that side. In Fig. 19.20 the *minimum* value of the objective function $f(x,y) = x + 3y$ is 18, and it occurs at points B and C *and all points between.* If the problem were to find a *point* that minimized the objective function, then more than one answer would exist, but if the problem were to find the minimum *value* of the objective function, then the answer would be 18. See Prob. 22 (p. 485).

Third, the solution may be indeterminate. For example, suppose that our problem were to find the *maximum value* of $f(x,y)$ in

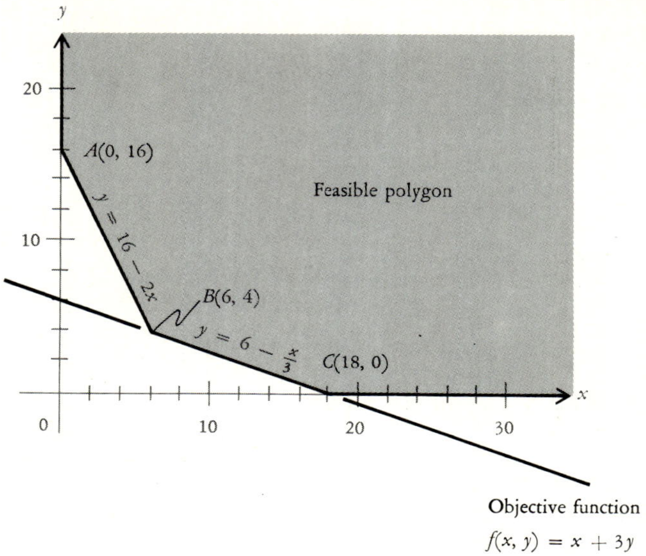

Fig. 19.20 The objective-function line coincides with one side of the feasible polygon.

Fig. 19.20. The feasible polygon is unbounded in the direction of increasing values of $f(x,y)$; therefore all lines $k = 3y + x$ will contain feasible points, no matter what the value of k (as long as k exceeds 18). We can make k as large as we please without reaching a maximum value. Such a situation does not mean that no solution exists; it usually indicates that not all restrictions have been stated. See Prob. 23 below.

Last, many problems involve more than two variables. For example, we might wish to maximize the function $f(x,y,z) = 3x + 2y + z$ subject to the restrictions $x, y, z \geqslant 0$, $z \leqslant 5x - 4y$, and $3z + 2y + x \leqslant 10$. Such problems of more than two variables cannot be solved by the graphical method presented in this section and are beyond the scope of this elementary introduction. The author encourages the interested student to investigate the *simplex tableau method.* The following references are recommended:

Campbell, Hugh G.: *An Introduction to Matrices, Vectors, and Linear Programming,* Appleton-Century-Crofts, New York, 1965.

Gray, A. William, and Otis M. Ulm: *Applications of College Mathematics,* Glencoe Press, Beverly Hills, Calif., 1970.

Rutledge, William A., and Thomas W. Cairns: *Mathematics for Business Analysis,* Holt, Rinehart and Winston, Inc., New York, 1963.

EXERCISE 19.5 (*Continued*)

21. Repeat Prob. 16, but suppose that the tire company makes a *profit* of \$20 per rear tire and \$25 per front tire. Find the number of each type of tire that must be produced for maximum profit. What is the maximum profit?
22. Repeat Prob. 21, but suppose the profit is \$20 each for both types of tire.
23. Find the maximum value and the minimum value of the function $f(x,y) = 3x + 3y$ subject to the restrictions x, $y \geqslant 0$, $x + 2y \geqslant 10$, and $2x + y \geqslant 10$. Give the coordinates (x,y) where these values occur.
24. Find the maximum value and the minimum value of the function $f(x,y) = -3x + 3y$ subject to the restrictions $x \geqslant -12$, $y \leqslant 15$, $y - 3x \leqslant 30$, and $10y - 3x \leqslant 120$. Give the coordinates (x,y) where these values occur.

CHAPTER 20

MATRICES AND MATRIX ALGEBRA

In past years a chapter on matrices might not have been considered appropriate for a book of this type. Now, however, the widespread use of electronic digital computers makes matrix algebra a vital subject for many people in technology, science, and business. Many computers can be programmed to solve systems of linear equations using matrices and matrix operations. Later in this chapter we shall see specific examples, but first we must learn the basic definitions and operations.

20.1 BASIC DEFINITIONS AND PROPERTIES OF MATRICES

A matrix is simply *a rectangular array of numbers.* Several examples are as follows:

$$A = \begin{bmatrix} 3 & 7 \\ -2 & 6 \\ 5 & 0 \end{bmatrix} \qquad B = \begin{bmatrix} 9 & 1 & 0 & -3 \\ 5 & 2 & -6 & 4 \end{bmatrix}$$

A *three-by-two* matrix   A *two-by-four* matrix

$$C = \begin{bmatrix} 8 \\ 12 \\ 5 \end{bmatrix} \qquad D = \begin{bmatrix} 6 & 0 \end{bmatrix}$$

A *three-by-one* matrix   A *one-by-two* matrix

Matrices are classified according to the number of rows and columns. Thus matrix A (p. 487) is a *three-by-two* matrix, meaning that it has *three rows* and *two columns. The number of rows is always given first,* followed by the number of columns. If a matrix has the same number of rows as columns, it is called a *square matrix.* Here are examples of a two-by-two and a three-by-three square matrix:

$$E = \begin{bmatrix} 23 & -40 \\ 37 & 100 \end{bmatrix} \qquad F = \begin{bmatrix} 1.2 & -0.3 & 1.0 \\ 0.8 & 2.7 & -2.5 \\ 1.6 & 0 & 12.6 \end{bmatrix}$$

A two-by-two square matrix — A three-by-three square matrix

Square matrices

Note that we designate a matrix by enclosing the array of numbers in brackets [ ]. Other enclosures such as parentheses may be used, but in this book we shall always use brackets.

You may have observed some similarities between matrices and determinants (Chap. 8). However, they are not the same thing. First, a determinant is *always a square array* of numbers. In a matrix, the number of rows need not equal the number of columns. Second, a determinant is a shorthand notation for a certain set of operations, and it has a numerical value (if the elements are explicit numbers). A matrix by itself implies no operations and has no value; it is simply an array of numbers.

Notations for Matrices and Elements

We shall usually identify a matrix by assigning to it a capital letter, as in matrix A (p. 489). Each number in a matrix is called an *element* of the matrix, and each element is identified by its row number and column number. When we use letters for elements, as in matrix A, we shall use the lowercase letter corresponding to the matrix letter, and we shall use a double-subscript system to specify the row and column. In this system, the *first subscript* is always the *row number,* and the *second subscript* is always the *column number.* For example, a_{32} (read "*a* three-two," *not* "*a* thirty-two") is the element in the third row and the second column.

It is extremely important to identify elements correctly, because certain operations combine elements according to row and column. An easy way to identify the subscripts as to row or column is to mentally replace them with the letters "rc," the initials of "row-column" and also of "royal crown" (think of a certain soft drink). Thus to identify subscripts as in a_{32}, mentally read the element as "$a_{\text{royal crown}}$" and you will know that the 3 is the row number and the 2 is the column number.

$$A = \begin{bmatrix} a_{11} & a_{12} & a_{13} & a_{14} \\ a_{21} & a_{22} & a_{23} & a_{24} \\ a_{31} & a_{32} & a_{33} & a_{34} \end{bmatrix}$$

Matrix A and its elements labeled with the double-subscript notation

Equal Matrices

We say that two matrices are *equal* if each element of one matrix equals the corresponding element of the other matrix. It follows, then, that equal matrices must have the same number of rows and the same number of columns. For example, matrix A below equals matrix D, and thus $A = D$. But $A \neq B$ because the numbers of rows and columns are different, and $A \neq C$ because a_{23} is 9, c_{23} is 4, and $9 \neq 4$.

$$A = \begin{bmatrix} 1 & 2 & 0 \\ 5 & 3 & 9 \end{bmatrix} \qquad B = \begin{bmatrix} 1 & 5 \\ 2 & 3 \\ 0 & 9 \end{bmatrix}$$

$$C = \begin{bmatrix} 1 & 2 & 0 \\ 5 & 3 & 4 \end{bmatrix} \qquad D = \begin{bmatrix} 1 & 2 & 0 \\ 5 & 3 & 9 \end{bmatrix}$$

EXERCISE 20.1

Use matrices A, B, C, D, and E below for these exercises.

$$A = \begin{bmatrix} 1 & 2 \\ 3 & 4 \\ 5 & 6 \end{bmatrix} \qquad B = \begin{bmatrix} -1 & 2 & b_{13} \\ 0 & b_{22} & 5 \\ 6 & 3 & b_{33} \end{bmatrix} \qquad C = \begin{bmatrix} c_{11} & c_{12} & c_{13} \\ 2 & 4 & 6 \end{bmatrix}$$

$$D = \begin{bmatrix} d_{11} & d_{12} & 3 \\ d_{21} & -4 & d_{23} \\ d_{31} & d_{32} & 0 \end{bmatrix} \qquad E = \begin{bmatrix} 1 & 5 & e_{13} \\ e_{21} & e_{22} & 7 \end{bmatrix}$$

1. Classify each matrix B through E by rows and columns. For example, A is a three-by-two matrix.

In Probs. 2 to 7, give the value of each element (*none* is a valid answer).

2. a_{22} 3. b_{12} 4. b_{21} 5. c_{32} 6. d_{13} 7. e_{12}

8. If $D = B$, what is the value of each of the following elements?

   a. b_{22} b. b_{33} c. d_{12} d. d_{31} e. d_{23}

9. Which element a_{ij} of matrix A has the value 6?
10. Which element c_{ij} of C has the value 2?
11. Explain why A cannot equal D.
12. Explain why C cannot equal E.

20.2 OTHER DEFINITIONS AND BASIC MATRIX OPERATIONS

A. ZERO MATRICES

A zero matrix is any matrix in which every element is zero. We shall use the symbol O to designate any zero matrix.

$$O = \begin{bmatrix} 0 & 0 & 0 \\ 0 & 0 & 0 \end{bmatrix}$$

A zero matrix

B. MULTIPLICATION AND DIVISION OF A MATRIX BY A SCALAR

We define multiplication of a matrix by a scalar (pure number) as follows: To multiply a matrix A by a number n, multiply *every element* of A by n. Division by a scalar is defined in a similar way; see Example 20.2.

$$nA = n \begin{bmatrix} a_{11} & a_{12} & a_{13} \\ a_{21} & a_{22} & a_{23} \end{bmatrix} = \begin{bmatrix} na_{11} & na_{12} & na_{13} \\ na_{21} & na_{22} & na_{23} \end{bmatrix} \qquad (20.1)$$

Multiplication of a matrix A by a scalar n

Example 20.1. Multiplication of a matrix by a scalar

$$(3)\begin{bmatrix} 2 & 0 \\ 5 & 6 \\ -1 & 4 \end{bmatrix} = \begin{bmatrix} 3(2) & 3(0) \\ 3(5) & 3(6) \\ 3(-1) & 3(4) \end{bmatrix} = \begin{bmatrix} 6 & 0 \\ 15 & 18 \\ -3 & 12 \end{bmatrix}$$

Example 20.2. Division of a matrix by a scalar

$$\frac{\begin{bmatrix} 12 & 16 & 4 \\ 20 & -4 & 8 \end{bmatrix}}{4} = \begin{bmatrix} 12/4 & 16/4 & 4/4 \\ 20/4 & -4/4 & 8/4 \end{bmatrix} = \begin{bmatrix} 3 & 4 & 1 \\ 5 & -1 & 2 \end{bmatrix}$$

Since multiplication is reversible, we can factor any common factor from each element and write any matrix in the form nA. For example, Eq. (20.1) could be read from right to left with the following result:

$$\begin{bmatrix} na_{11} & na_{12} & na_{13} \\ na_{21} & na_{22} & na_{23} \end{bmatrix} = n\begin{bmatrix} a_{11} & a_{12} & a_{13} \\ a_{21} & a_{22} & a_{23} \end{bmatrix} = nA \qquad (20.2)$$

Factoring a scalar from a matrix

Example 20.3. Simplifying a matrix by factoring

$$\begin{bmatrix} 6 & 0 \\ 15 & 18 \\ -3 & 12 \end{bmatrix} = \begin{bmatrix} 3(2) & 3(0) \\ 3(5) & 3(6) \\ 3(-1) & 3(4) \end{bmatrix} = 3\begin{bmatrix} 2 & 0 \\ 5 & 6 \\ -1 & 4 \end{bmatrix}$$

C. THE NEGATIVE OF A MATRIX

The *negative* of a given matrix is the matrix in which every element equals the negative of the corresponding element in the given matrix. This concept is illustrated below.

Given that $$A = \begin{bmatrix} a_{11} & a_{12} \\ a_{21} & a_{22} \end{bmatrix}$$

then *negative A* is $$-A = \begin{bmatrix} -a_{11} & -a_{12} \\ -a_{21} & -a_{22} \end{bmatrix}$$

One way to form the negative of a matrix is to multiply it by -1 as defined in Sec. 20.2B above.

Example 20.4. Form the negative of matrix X below.

Given
$$X = \begin{bmatrix} 5 & -3 & 7 \\ -2 & 0 & 1 \end{bmatrix}$$

solution:
$$-X = (-1)X = \begin{bmatrix} -5 & 3 & -7 \\ 2 & 0 & -1 \end{bmatrix}$$

D. MATRIX SUMS

We define the *sum* of two matrices as the matrix with each of its elements equal to the sum of the corresponding elements in the given matrices. This definition implies that matrices to be added must have the *same number of rows* and the *same number of columns.*

$$A + B = \begin{bmatrix} a_{11} & a_{12} \\ a_{21} & a_{22} \\ a_{31} & a_{32} \end{bmatrix} + \begin{bmatrix} b_{11} & b_{12} \\ b_{21} & b_{22} \\ b_{31} & b_{32} \end{bmatrix}$$

$$= \begin{bmatrix} (a_{11} + b_{11}) & (a_{12} + b_{12}) \\ (a_{21} + b_{21}) & (a_{22} + b_{22}) \\ (a_{31} + b_{31}) & (a_{32} + b_{32}) \end{bmatrix} \qquad (20.3)$$

Matrix addition

We can subtract a matrix by adding its negative, as follows:

$$A - B = A + (-B) = \begin{bmatrix} (a_{11} - b_{11}) & (a_{12} - b_{12}) \\ (a_{21} - b_{21}) & (a_{22} - b_{22}) \\ (a_{31} - b_{31}) & (a_{32} - b_{32}) \end{bmatrix} \qquad (20.4)$$

Matrix subtraction

Example 20.5. Given matrices X, Y, and Z below, find each of the following sums or differences: $X + Y$, $X + Z$, and $Y - X$.

$$X = \begin{bmatrix} 1 & -2 & 0 \\ 3 & 6 & -5 \end{bmatrix} \quad Y = \begin{bmatrix} 3 & 6 & 1 \\ -2 & 3 & 5 \end{bmatrix} \quad Z = \begin{bmatrix} -1 & 5 \\ 2 & 7 \end{bmatrix}$$

a. $X + Y = \begin{bmatrix} 1+3 & -2+6 & 0+1 \\ 3-2 & 6+3 & -5+5 \end{bmatrix} = \begin{bmatrix} 4 & 4 & 1 \\ 1 & 9 & 0 \end{bmatrix}$

b. $X + Z$ cannot be added because X and Z do not have the same number of columns.

c. $$Y - X = \begin{bmatrix} 3-1 & 6-(-2) & 1-0 \\ -2-3 & 3-6 & 5-(-5) \end{bmatrix} = \begin{bmatrix} 2 & 8 & 1 \\ -5 & -3 & 10 \end{bmatrix}$$

EXERCISE 20.2

Use matrices R, S, T, U, and V for these exercises.

$$R = \begin{bmatrix} 1 \\ 3 \\ 5 \end{bmatrix} \qquad S = [6 \quad -18 \quad 12] \qquad T = \begin{bmatrix} 2 & 8 \\ 4 & -6 \\ 6 & 10 \end{bmatrix}$$

$$U = [2 \quad -6 \quad 4] \qquad V = \begin{bmatrix} 1 & -7 \\ -3 & 9 \\ 5 & 11 \end{bmatrix}$$

In Probs. 1 to 16, perform the indicated operations.

1. $S + U$	2. $S - U$	3. $5R$	4. $3U$
5. $R + S$	6. $R - U$	7. $T + V$	8. $V - T$
9. $(1/3)S$	10. $(1/2)T$	11. $0 + R$	12. $2S + V$
13. $U + 2S$	14. $3T + 2V$	15. $S/2 - 3U$	16. $S/6 + U/2$

17. Simplify matrix S by factoring the largest possible integer, as in Example 20.3.
18. Simplify matrix U as in Prob. 17.
19. Repeat Prob. 17 for matrix T.
20. Find the negative of matrix V.
21. Is $T + V = V + T$? Do you believe that a general commutative property $A + B = B + A$ exists for matrix addition? Give a demonstration to justify your belief.
22. Do you believe that a general associative property $A + (B + C) = (A + B) + C$ exists for matrix addition? Give a demonstration to justify your belief.
23. Show that the distributive law $n(A + B) = nA + nB$ for multiplication of a matrix sum by a scalar is valid.
24. The State Savings and Loan Association has offices in three cities. Each office holds some home mortgages at 6 percent interest and others at 7 percent. The number of loans at each rate and period are as follows: Office *A*, 6 percent: 10 yr, 14; 15 yr, 30; 20 yr, 36; 7 percent: 10 yr, 21; 15 yr, 20; 20 yr, 40. Office

B, 6 percent: 10 yr, 12; 15 yr, 28; 20 yr, 22; 7 percent: 10 yr, 6; 15 yr, 19; 20 yr, 31. Office C, 6 percent: 10 yr, 7; 15 yr, 12; 20 yr, 9; 7 percent: 10 yr, 9; 15 yr, 14; 20 yr, 11. Organize these data into matrices and use matrix addition to determine the total number of mortgages at each rate and period.

20.3 MATRIX MULTIPLICATION

In the last section we defined multiplication of a matrix by a *scalar.* Now we shall examine *matrix multiplication*—multiplication of a matrix by a *matrix.* The format for matrix multiplication is quite different from what we might expect intuitively, but we perform matrix multiplication nearly every day. To see what it is and how it works, let's examine a practical situation. Then we shall make the general definitions and see even more powerful applications.

Example 20.6. An appliance manufacturer has the following merchandise stored in a warehouse: 1,120 refrigerators, 950 ranges, and 740 water heaters. The manufacturer's unit costs for these appliances are \$75, \$50, and \$20, respectively. What is the total value of these appliances?

Solution. The "matrix" way to correlate these data is to list one set of data in a row and another set in a column, as follows:

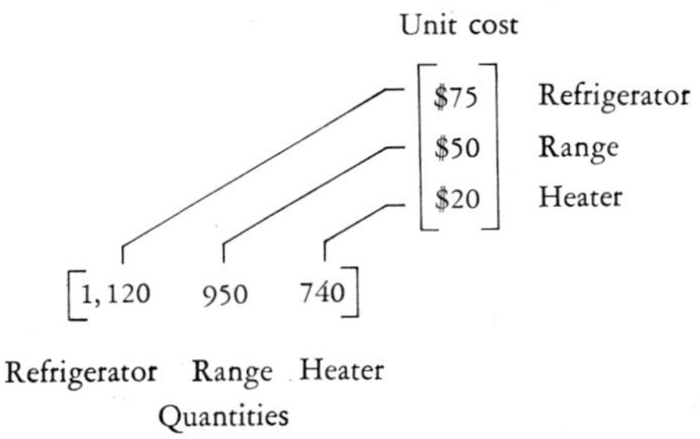

Fig. 20.1 Example 20.6.

The total value is found by multiplying each element of the row by the corresponding element of the column (as indicated in the figure) and adding the products, as follows:

$$\begin{aligned}\text{Total value} &= [1{,}120(\$75) + 950(\$50) + 740(\$20)]\\ &= [\$84{,}000 + \$47{,}500 + \$14{,}800]\\ &= [\$146{,}300]\end{aligned}$$

A. THE BASIC STEP IN MATRIX MULTIPLICATION

The procedure used in Example 20.6 is precisely matrix multiplication. In general terms, the product of a single matrix row by a single matrix column is defined by Eq. (20.5), is illustrated by Fig. 20.2, and can be stated as follows: *The product of any row in a first matrix A with any column in a second matrix B is a single element in the matrix product AB. This element is formed by multiplying each element of the row by the corresponding element of the column and adding the results.*

Given $$A = [a_{11} \quad a_{12} \quad a_{13}] \quad \text{and} \quad B = \begin{bmatrix} b_{11} \\ b_{21} \\ b_{31} \end{bmatrix}$$

the product AB is

$$[a_{11} \quad a_{12} \quad a_{13}] \begin{bmatrix} b_{11} \\ b_{21} \\ b_{31} \end{bmatrix} = [a_{11}b_{11} + a_{12}b_{21} + a_{13}b_{31}] \tag{20.5}$$

Definition of a matrix product for a single row multiplied by a single column

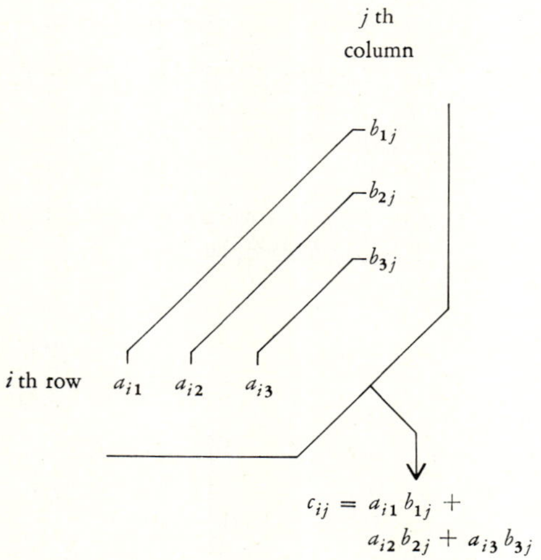

Fig. 20.2 Formation of one element, c_{ij}, of a matrix product by multiplying each element in the ith row by the corresponding element in the jth column and adding the results.

Example 20.7. **To multiply matrices *A* and *B* below, we first factor 3 from *A* and 2 from *B* [see Eq. (20.2)] and then combine the elements according to Eq. (20.5).**

Given $$A = [15 \quad 6 \quad -21] \quad \text{and} \quad B = \begin{bmatrix} 8 \\ -6 \\ -4 \end{bmatrix}$$

the matrix product *AB* is

$$AB = (3)(2)[5 \quad 2 \quad -7]\begin{bmatrix} 4 \\ -3 \\ -2 \end{bmatrix}$$

$$= 6[5(4) + 2(-3) + (-7)(-2)]$$

$$AB = 6[20 - 6 + 14] = 6[28] = [168]$$

At this point we need to state explicitly several aspects of matrix multiplication that have been implicitly assumed.

1. The product of two matrices is another matrix.
2. A matrix product AB is defined only if *the number of rows in B equals the number of columns in A.*

   The reason for this requirement is clearly shown by Figs. 20.1 and 20.2. In Fig. 20.1, for example, if the unit-cost matrix contained only the first two rows, there would be no way to determine the value of the 740 heaters.
3. Equation (20.5) and Fig. 20.2 define the matrix product for just a single row times a single column. However, this procedure is used for any matrix product in general. It is simply repeated for each additional column of B or row of A.

EXERCISE 20.3

In Probs. 1 to 6, find each matrix product, as indicated by Eq. (20.5), Fig. 20.2, and Example 20.7.

1. $[a_{11} \quad a_{12}]\begin{bmatrix} b_{11} \\ b_{21} \end{bmatrix}$

2. $[c_{11} \quad c_{12} \quad c_{13}]\begin{bmatrix} d_{11} \\ d_{21} \\ d_{31} \end{bmatrix}$

3. $[1 \quad 0 \quad -2]\begin{bmatrix} 5 \\ 15 \\ 10 \end{bmatrix}$

4. $[4 \quad 2 \quad 0 \quad -6]\begin{bmatrix} 6 \\ -6 \\ 9 \\ 3 \end{bmatrix}$

5. $[2x \quad 3x]\begin{bmatrix} 4x \\ 8x \end{bmatrix}$

6. $[3ab \quad -2b]\begin{bmatrix} a \\ 5ab \end{bmatrix}$

7. If C and D are the respective matrices in Prob. 2, explain why the products CC and DD cannot be formed.
8. If $A = [5 \quad 7 \quad 3]$ and $B = \begin{bmatrix} 2 \\ 3 \end{bmatrix}$, explain why AB cannot be formed.

B. PRODUCTS OF MATRICES WITH MORE THAN ONE ROW OR ONE COLUMN

In the product AB, if the *second* matrix B has more than one column, we repeat the procedure of Fig. 20.2 for each additional column. And if the *first* matrix A has more than one row, we repeat the procedure for each additional row. Each additional column of B produces an additional column in the product AB, and each additional row of A produces an additional row in AB.

Example 20.8. **Given that** $A = [a_{11} \quad a_{12}]$ **and** $B = \begin{bmatrix} b_{11} & b_{12} \\ b_{21} & b_{22} \end{bmatrix}$

the product AB **is as follows:**

$$[a_{11} \quad a_{12}] \begin{bmatrix} b_{11} & b_{12} \\ b_{21} & b_{22} \end{bmatrix} = [(a_{11}b_{11} + a_{12}b_{21}) \quad (a_{11}b_{12} + a_{12}b_{22})]$$

Notice that AB **has** ***two*** **columns, the element in the second column being due to the second column of** B**.**

Example 20.9. **Given that** $A = \begin{bmatrix} a_{11} & a_{12} \\ a_{21} & a_{22} \\ a_{31} & a_{32} \end{bmatrix}$ **and** $B = \begin{bmatrix} b_{11} & b_{12} \\ b_{21} & b_{22} \end{bmatrix}$

the product AB **is as follows:**

$$\begin{bmatrix} a_{11} & a_{12} \\ a_{21} & a_{22} \\ a_{31} & a_{32} \end{bmatrix} \begin{bmatrix} b_{11} & b_{12} \\ b_{21} & b_{22} \end{bmatrix} = \begin{bmatrix} (a_{11}b_{11} + a_{12}b_{21}) & (a_{11}b_{12} + a_{12}b_{22}) \\ (a_{21}b_{11} + a_{22}b_{21}) & (a_{21}b_{12} + a_{22}b_{22}) \\ (a_{31}b_{11} + a_{32}b_{21}) & (a_{31}b_{12} + a_{32}b_{22}) \end{bmatrix} \qquad (20.6)$$

Notice that AB **has six elements in three rows and two columns. This is because** A **has three rows and** B **has two columns.**

Equation (20.6) can be taken as a general definition of matrix multiplication. It shows how each row of a product AB involves elements of the *same row* of A and how each column of AB involves elements of the *same column of* B. The following formula illustrates three relationships between the elements of A and B and any general element c_{ij} of the product AB.

$$c_{ij} = a_{i1}b_{1j} + a_{i2}b_{2j} + a_{i3}b_{3j} + \cdots$$

The element in the ith row and the jth column of a matrix product AB

Example 20.10. Find the product of matrices A and B below.

$$A = \begin{bmatrix} 1 & 4 \\ 0 & -5 \\ -2 & 3 \end{bmatrix} \qquad B = \begin{bmatrix} 2 & -1 \\ 1 & 0 \end{bmatrix}$$

Solution.

$$\begin{bmatrix} 1 & 4 \\ 0 & -5 \\ -2 & 3 \end{bmatrix} \begin{bmatrix} 2 & -1 \\ 1 & 0 \end{bmatrix} = \begin{bmatrix} (1 \times 2) + (4 \times 1) & (1 \times -1) + (4 \times 0) \\ (0 \times 2) - (5 \times 1) & (0 \times -1) - (5 \times 0) \\ (-2 \times 2) + (3 \times 1) & (-2 \times -1) + (3 \times 0) \end{bmatrix}$$

$$AB = \begin{bmatrix} 6 & -1 \\ -5 & 0 \\ -1 & 2 \end{bmatrix}$$

EXERCISE 20.4

In Probs. 1 to 12, multiply each pair of matrices.

1. $\begin{bmatrix} e_{11} & e_{12} \end{bmatrix} \begin{bmatrix} f_{11} & f_{12} & f_{13} \\ f_{21} & f_{22} & f_{23} \end{bmatrix}$

2. $\begin{bmatrix} a_{11} \\ a_{21} \\ a_{31} \end{bmatrix} \begin{bmatrix} b_{11} \end{bmatrix}$

3. $\begin{bmatrix} a_{11} \\ a_{21} \end{bmatrix} \begin{bmatrix} b_{11} & b_{12} & b_{13} \end{bmatrix}$

4. $\begin{bmatrix} x_{11} & x_{12} \\ x_{21} & x_{22} \end{bmatrix} \begin{bmatrix} y_{11} & y_{12} \\ y_{21} & y_{22} \end{bmatrix}$

5. $\begin{bmatrix} 2a & 4b \\ 6c & 2d \end{bmatrix} \begin{bmatrix} 3d & 6c \\ 9b & 3a \end{bmatrix}$

6. $\begin{bmatrix} 6 & 0 \\ 12 & -6 \end{bmatrix} \begin{bmatrix} 4 & -8 \\ 0 & 12 \end{bmatrix}$

7. $\begin{bmatrix} 1 & 5 & -2 \\ 3 & 0 & 6 \end{bmatrix} \begin{bmatrix} 2 & -3 \\ 1 & 4 \\ 2 & 0 \end{bmatrix}$

8. $\begin{bmatrix} 1 & 3 \\ 4 & 2 \end{bmatrix} \begin{bmatrix} 1 & -1 & 0 \\ 3 & 7 & -2 \end{bmatrix}$

9. $\begin{bmatrix} 1 & 0 & 0 \\ 0 & 1 & 0 \\ 0 & 0 & 1 \end{bmatrix} \begin{bmatrix} 1 & 2 & 3 \\ 4 & 5 & 6 \\ 7 & 8 & 9 \end{bmatrix}$

10. $\begin{bmatrix} 1 & 1 & 1 \\ 0 & 0 & 0 \\ 0 & 0 & 0 \end{bmatrix} \begin{bmatrix} 1 & 2 & 3 \\ 4 & 5 & 6 \\ 7 & 8 & 9 \end{bmatrix}$

11. $\begin{bmatrix} 0 & 0 & 1 \\ 0 & 1 & 0 \\ 1 & 0 & 0 \end{bmatrix} \begin{bmatrix} 1 & 2 & 3 \\ 4 & 5 & 6 \\ 7 & 8 & 9 \end{bmatrix}$

12. $\begin{bmatrix} 1 & 0 & 0 \\ 1 & 0 & 0 \\ 1 & 0 & 0 \end{bmatrix} \begin{bmatrix} 1 & 2 & 3 \\ 4 & 5 & 6 \\ 7 & 8 & 9 \end{bmatrix}$

13. Look at the results of Probs. 9 through 12. See if you can make up a general rule concerning a matrix product in which one matrix has a line of 1s and all other elements are zero. Test your rule with the following matrix product.

$$\begin{bmatrix} 1 & 0 \\ 0 & 1 \end{bmatrix} \begin{bmatrix} a_{11} & a_{12} \\ a_{21} & a_{22} \end{bmatrix}$$

14. An appliance store has the following merchandise in stock: 62 radios, 78 TVs, 44 stereos, and 31 tape recorders. The respective retail prices of these items are \$70, \$160, \$120, and \$90. A 4 percent state sales tax must be added to the price of each item. Use matrices to display these data, and use matrix multiplication to determine the total retail value of all this merchandise and the value of the taxes it produces for the state.
15. A certain TV factory assembles sets on three different chassis models X, Y, and Z. Each model uses each of three components (such as circuit boards, transistors, and tubes) A, B, and C as follows: Model X: $10A$, $13B$, and $22C$; Model Y: $12A$, $15B$, and $25C$; and Model Z: $12A$, $18B$, and $30C$. The times needed to install each component are as follows: A: 3 min; B: 2 min; C: 7 min. Organize these data into matrices, and use matrix multiplication to determine the time needed to assemble these components in each type of chassis.

20.4 MATRIX SOLUTIONS OF SIMULTANEOUS LINEAR EQUATIONS

We shall now develop a matrix method for solving a system of linear equations. This method is sometimes long and tedious if performed manually. However, it has a great advantage over other methods in that computers have built-in matrix programs that automatically perform the matrix operations. In the next section we shall see some typical programs for solving equations by the matrix method. Keep in mind that our objective is to learn principles; never mind that the manual computations are sometimes tedious.

A. AN OVERVIEW OF THE MATRIX METHOD

Before delving too deeply into minute details, we shall first examine the general plan for a matrix solution of a system of linear equations.

We can translate a system of n algebraic equations with n variables into a *single* matrix equation. The n variables can be found by solving the matrix equation. In Eq. (20.7) below, the x_i are variables, and the a_{ij} and k_i are constants. The objective is to solve for x_1, x_2, and x_3.

$$\begin{aligned} a_{11}x_1 + a_{12}x_2 + a_{13}x_3 &= k_1 \\ a_{21}x_1 + a_{22}x_2 + a_{23}x_3 &= k_2 \\ a_{31}x_1 + a_{32}x_2 + a_{33}x_3 &= k_3 \end{aligned} \tag{20.7}$$

The *key point* in translating these equations into a matrix equation is this: Each equation can be derived from a matrix equation of the form $AX = K$, where A is the matrix of the coefficients a_{ij}, X is the matrix of the variables x_i, and K is the matrix of the constants k_i as shown below.

Given $A = \begin{bmatrix} a_{11} & a_{12} & a_{13} \\ a_{21} & a_{22} & a_{23} \\ a_{31} & a_{32} & a_{33} \end{bmatrix}$ $\quad X = \begin{bmatrix} x_1 \\ x_2 \\ x_3 \end{bmatrix}$ $\quad$ and $\quad K = \begin{bmatrix} k_1 \\ k_2 \\ k_3 \end{bmatrix}$

then the system of Eqs. (20.7) can be represented as follows:

$$AX = K \tag{20.8}$$

or

$$\begin{bmatrix} a_{11}x_1 + a_{12}x_2 + a_{13}x_3 \\ a_{21}x_1 + a_{22}x_2 + a_{23}x_3 \\ a_{31}x_1 + a_{32}x_2 + a_{33}x_3 \end{bmatrix} = \begin{bmatrix} k_1 \\ k_2 \\ k_3 \end{bmatrix}$$

The matrix equivalent of a system of linear equations such as Eqs. (20.7)

To understand this point thoroughly, you should form the product AX yourself and then apply the definition of matrix equality to Eq. (20.8). The results will be the set of Eqs. (20.7).

We solve Eq. (20.8) much as we would solve an algebraic equation: We multiply both sides by a quantity that leaves X by itself on the left side. In matrix algebra this quantity is an *inverse matrix.* The special symbol for the inverse of a matrix A is A^{-1}. This symbol does *not* mean the -1 power of A; we shall see the detailed structure of A^{-1} later. For now the point is that

$$A^{-1}AX = X$$

and therefore we solve $AX = K$ as follows:

Given $\qquad AX = K$

multiply by A^{-1} $\qquad A^{-1}AX = A^{-1}K$

the solution is $\qquad X = A^{-1}K$

As soon as we learn how to find a matrix inverse, we can carry out the detailed steps in solving a matrix equation.

B. IDENTITY MATRICES, MATRIX INVERSES, AND MATRIX EQUATIONS

Two special matrices, the *identity matrix* and the *inverse matrix,* will help us to solve matrix equations. Identity matrices have the following property, as shown by Eq. (20.9): *The product of any given matrix and its identity matrix is the given matrix itself.*

$$IC = C \tag{20.9}$$

where C is any matrix
I is the identity matrix

Identity matrix property

Equation (20.10) shows a typical identity matrix. In general, an identity matrix has the following three characteristics:

1. Every identity matrix is a *square* matrix.
2. Every element on the *principal diagonal*—from upper left to lower right—is *unity* (+1).
3. All other elements are *zero.*

$$I = \begin{bmatrix} 1 & 0 & 0 \\ 0 & 1 & 0 \\ 0 & 0 & 1 \end{bmatrix} \tag{20.10}$$

A typical *identity matrix*

Comment: *We must be careful about two important details. (1) Since matrix multiplication is* ***not commutative,*** *we must generally use Eq. (20.9) exactly as shown, $IC = C$, and not $CI = C$; however, if C is a* ***square matrix,*** *then $IC = CI = C$. (2) For matrix multiplication to be defined, the number of* ***columns*** *in I must equal the number of rows in C.*

Example 20.11. **Given matrix C below, I is the correct identity matrix. We wish to show that $IC = C$.**

$$C = \begin{bmatrix} c_{11} & c_{12} & c_{13} \\ c_{21} & c_{22} & c_{23} \end{bmatrix} \qquad I = \begin{bmatrix} 1 & 0 \\ 0 & 1 \end{bmatrix}$$

$$IC = \begin{bmatrix} 1 & 0 \\ 0 & 1 \end{bmatrix} \begin{bmatrix} c_{11} & c_{12} & c_{13} \\ c_{21} & c_{22} & c_{23} \end{bmatrix}$$

$$= \begin{bmatrix} (1 \times c_{11}) + (0 \times c_{21}) & (1 \times c_{12}) + (0 \times c_{22}) & (1 \times c_{13}) + (0 \times c_{23}) \\ (0 \times c_{11}) + (1 \times c_{21}) & (0 \times c_{12}) + (1 \times c_{22}) & (0 \times c_{13}) + (1 \times c_{23}) \end{bmatrix}$$

$$IC = \begin{bmatrix} c_{11} & c_{12} & c_{13} \\ c_{21} & c_{22} & c_{23} \end{bmatrix} = C$$

Matrix Inverses

A *matrix inverse* can be defined only for a *square* matrix. Given an n-by-n matrix A (n rows and n columns), the inverse of A is written A^{-1}, and it is another n-by-n matrix. Inverse matrices have the following property:

$$A^{-1}A = I \tag{20.11}$$

The product of an n-by-n matrix and its inverse is an n-by-n *identity matrix.*

Inverse-matrix property

Later in this chapter we shall see how to find the inverse of any n-by-n matrix in general. But for now we shall work only with two-by-two matrices, and we shall use the following special procedure. Since we are going to use determinants, you may need to review Eq. (8.12). We define the determinant of a square matrix as the determinant of the elements of that matrix.

To form the inverse of a two-by-two matrix, interchange the elements on the principal diagonal, multiply the other elements by −1, and divide by the determinant of the original matrix.

$$A^{-1} = \frac{\begin{bmatrix} a_{22} & -a_{12} \\ -a_{21} & a_{11} \end{bmatrix}}{|A|} = \frac{\begin{bmatrix} a_{22} & -a_{12} \\ -a_{21} & a_{11} \end{bmatrix}}{a_{11}a_{22} - a_{21}a_{12}} \tag{20.12}$$

where $A = \begin{bmatrix} a_{11} & a_{12} \\ a_{21} & a_{22} \end{bmatrix}$ and $|A| = a_{11}a_{22} - a_{21}a_{12}$

The inverse of a two-by-two matrix

Example 20.12. Given $A = \begin{bmatrix} 5 & -6 \\ 2 & -3 \end{bmatrix}$, find $|A|$ and A^{-1}, and show that $A^{-1}A = I$.

$$|A| = \begin{vmatrix} 5 & -6 \\ 2 & -3 \end{vmatrix} = 5(-3) - 2(-6) = -3$$

$$A^{-1} = \frac{\begin{bmatrix} a_{22} & -a_{12} \\ -a_{21} & a_{11} \end{bmatrix}}{|A|} = \frac{\begin{bmatrix} -3 & 6 \\ -2 & 5 \end{bmatrix}}{-3}$$

$$A^{-1}A = \frac{\begin{bmatrix} -3 & 6 \\ -2 & 5 \end{bmatrix}\begin{bmatrix} 5 & -6 \\ 2 & -3 \end{bmatrix}}{-3} = \begin{bmatrix} \left(\frac{-15+12}{-3}\right) & \left(\frac{18-18}{-3}\right) \\ \left(\frac{-10+10}{-3}\right) & \left(\frac{12-15}{-3}\right) \end{bmatrix}$$

$$A^{-1}A = \begin{bmatrix} 1 & 0 \\ 0 & 1 \end{bmatrix} = I$$

EXERCISE 20.5

1. Determine an identity matrix for each of the following matrices. Verify each identity matrix by showing that $IA = A$.

a. $\begin{bmatrix} 2 & 5 \\ 7 & -3 \end{bmatrix}$ b. $\begin{bmatrix} 3 \\ -6 \end{bmatrix}$ c. $\begin{bmatrix} -5 & 2 \end{bmatrix}$

d. $\begin{bmatrix} 2 & 3 & 7 \end{bmatrix}$ e. $\begin{bmatrix} 3 \\ -1 \\ 2 \end{bmatrix}$ f. $\begin{bmatrix} 1 & 5 & 2 \\ -2 & 0 & 7 \\ 3 & 1 & 4 \end{bmatrix}$

g. $\begin{bmatrix} a & b \\ c & d \end{bmatrix}$ h. $\begin{bmatrix} x \\ y \\ z \end{bmatrix}$

2. For matrices C and I of Example 20.11, show why IC cannot equal CI. Can CI be defined for the matrices given?
3. Given matrix B below, determine the proper identity matrix I and show that for a *square* matrix, $IB = BI = B$.

$$B = \begin{bmatrix} b_{11} & b_{12} & b_{13} \\ b_{21} & b_{22} & b_{23} \\ b_{31} & b_{32} & b_{33} \end{bmatrix}$$

4. Find the inverse of each of the following matrices. If no inverse exists, tell why.

a. $\begin{bmatrix} 1 & 2 \\ 3 & 4 \end{bmatrix}$ b. $\begin{bmatrix} 3 & 4 \\ 1 & 2 \end{bmatrix}$ c. $\begin{bmatrix} 1 & -2 \\ 3 & 4 \end{bmatrix}$ d. $\begin{bmatrix} 3 & 1 & 4 \\ 2 & 0 & 5 \end{bmatrix}$

e. $\begin{bmatrix} 2 & 4 \\ 3 & 6 \end{bmatrix}$ f. $\begin{bmatrix} -1 & 2 \\ -1 & 0 \end{bmatrix}$ g. $\begin{bmatrix} 3 & -1 \\ 5 & 0 \\ 1 & 6 \end{bmatrix}$ h. $\begin{bmatrix} 2 & -1 \\ 6 & -3 \end{bmatrix}$

5. For each matrix of Prob. 4 that has an inverse, show that $A^{-1}A = I$.
6. Since A^{-1} and A are each the inverse of the other, we might suspect that $AA^{-1} = A^{-1}A = I$. Show that this is true for each matrix of Prob. 5.

Solving Matrix Equations

Equations (20.9) and (20.11) suggest the following steps for solving matrix equations:

given $$AX = K$$

multiply by A^{-1} $$(A^{-1}A)X = A^{-1}K$$

the result is $$IX = A^{-1}K$$

which by (20.9) is $$X = A^{-1}K \tag{20.13}$$

The order of the factors is very important. Since matrix multiplication is *not* commutative, *we cannot write* $X = KA^{-1}$.

Example 20.13. **Solve the following matrix equation for x_1 and x_2.**

Given $$\begin{bmatrix} 2 & 3 \\ 2 & 4 \end{bmatrix} \begin{bmatrix} x_1 \\ x_2 \end{bmatrix} = \begin{bmatrix} -6 \\ -10 \end{bmatrix}$$

The solution can be found immediately by using Eq. (20.12) in Eq. (20.13), as follows:

$$\begin{bmatrix} x_1 \\ x_2 \end{bmatrix} = \frac{\begin{bmatrix} 4 & -3 \\ -2 & 2 \end{bmatrix} \begin{bmatrix} -6 \\ -10 \end{bmatrix}}{\begin{vmatrix} 2 & 3 \\ 2 & 4 \end{vmatrix}} = \frac{\begin{bmatrix} 6 \\ -8 \end{bmatrix}}{2} = \begin{bmatrix} 3 \\ -4 \end{bmatrix}$$

Therefore $x_1 = 3$ and $x_2 = -4$.

Example 20.14. **Solve the following system of equations for x and y.**

$$\begin{aligned} 3x + 2y &= 1 \\ 4x - y &= 27 \end{aligned}$$

The matrix equation is

$$\begin{bmatrix} 3 & 2 \\ 4 & -1 \end{bmatrix} \begin{bmatrix} x \\ y \end{bmatrix} = \begin{bmatrix} 1 \\ 27 \end{bmatrix}$$

and the solution is

$$\begin{bmatrix} x \\ y \end{bmatrix} = \frac{\begin{bmatrix} -1 & -2 \\ -4 & 3 \end{bmatrix}\begin{bmatrix} 1 \\ 27 \end{bmatrix}}{\begin{vmatrix} 3 & 2 \\ 4 & -1 \end{vmatrix}} = \frac{\begin{bmatrix} -55 \\ 77 \end{bmatrix}}{-11}$$

$$x = 5 \quad \text{and} \quad y = -7$$

EXERCISE 20.5 (*Continued*)

In Probs. 7 to 10, solve each system of equations as follows:

a. Translate to a matrix equation of the form $AX = K$, as in Example 20.14.
b. Determine A^{-1}, find $A^{-1}A$ to show that $A^{-1}AX = IX$, and write the resulting equation $IX = A^{-1}K$.
c. Show that $IX = X = A^{-1}K$, and give the value of each variable.

7. $3x - y = 7$
   $2x + y = 8$
8. $2x - 4y = 8$
   $x + 3y = -1$
9. $x_1 - x_2 = 2$
   $x_1 + x_2 = 8$
10. $2y_1 + y_2 = 0$
    $3y_1 - 2y_2 = -14$

In Probs. 11 to 14, solve each system of equations by writing the matrix equation and its solution immediately, as in Examples 20.13 and 20.14.

11. $u - 3v = 20$
    $3u + 2v = 5$
12. $u + v = 6$
    $u + 3v = 20$
13. $5w_1 + 8w_2 = 3$
    $2w_1 + 3w_2 = 1$
14. $2z_1 - z_2 = 6$
    $z_1 + z_2 = 21$

20.5 COMPUTER PROGRAMS FOR MATRIX EQUATIONS

We shall now examine some fundamentals of computer programming. To keep the subject at an elementary level, we shall use a virtually self-explanatory computer language named *BASIC* (for *B*eginners *A*ll-purpose *S*ymbolic *I*nstruction *C*ode). And we shall omit many details of computer operation and many of the more complicated instructions. You need not have any prior experience with computers in order to understand this material, nor are you expected to learn detailed programming. Our objective is to gain an appreciation for the matrix method and for computers.

The following tables contain some BASIC instructions that

we shall use to write programs. These instructions are special coded symbols that a BASIC-language computer will respond to. Since a computer will not respond properly to any arbitrary instruction that we might invent, we must use only the ones given in the tables. Most of these instructions are self-explanatory. We should note that computers do not use lowercase letters. Also, the format of each instruction must be exactly as shown. A misspelled word, a misplaced symbol, or the omission of even one comma will usually ruin the program.

Not all BASIC-language computers accept the matrix commands in Table 20.2. Even so, the concepts of this section can be applied to other languages, and they could be used to write a subroutine program for a computer that does not have the matrix commands built in.

TABLE 20.1 A PARTIAL LIST OF BASIC COMMANDS FOR SCALAR NUMBERS

SCALAR COMMAND	MEANING
1. LET A = 3	Assign the value of 3 to the variable A.
2. LET C = A + B	Add A and B, and assign the sum to C.
3. LET C = A − B	Subtract
4. LET C = A * B	Multiply
5. LET C = A/B	Divide
6. LET C = A ↑ N	Raise to a power: $C = A^N$
7. LET C = SQRT(A)	Take the square root of A: $C = \sqrt{A}$
8. DIM C(3,4)	Dimension: Reserve in memory, 12 spaces for matrix C (three rows by four columns).
9*a*. READ A,B,C 9*b*. DATA 1,2,3	Assign first three numbers in the DATA line to A, B, and C, respectively, so that $A = 1$, $B = 2$, and $C = 3$.
10. PRINT C	Print the *value* of C.
11. PRINT "C"	Print the *letter* C.
12. PRINT "C=";C	Print the message in quotation marks, $C =$, followed by the value of C.

The following programs are examples of how the scalar commands might be used.

```
10  READ    A,B
20  DATA    3,5
30  LET     C = A + B
40  PRINT   C
50  END
```

Program 20.1 A program for scalar addition

```
10  READ    A,B,C
20  DATA    -2,30,3
30  LET     X = (A * B)/(C)
40  PRINT   "X=";X
50  END
```

Program 20.2 A program for scalar multiplication and division

TABLE 20.2 A PARTIAL LIST OF BASIC COMMANDS FOR MATRICES

MATRIX COMMANDS	MEANINGS
MAT C = A + B	Matrix addition, Eq. (20.3)
MAT C = A − B	Matrix subtraction, Eq. (20.4)
MAT C = A * B	Matrix multiplication, Eq. (20.6)
MAT C = (K)*A	Matrix A multiplied by scalar K; Eq. (20.1)
MAT C = INV(A)	Matrix inverse: $C = A^{-1}$
MAT C = IDN	Identity matrix: $C = I$
MAT C = ZER	Zero matrix: $C = 0$
DIM C(2,3) MAT READ C DATA 1,2,3,4,5,6	Reserve six spaces for a two-by-three matrix C; assign the values 1, 2, 3, 4, 5, and 6 to the elements of C.
MAT PRINT C	Print the elements of C in rows and columns as defined by the DIM command.

As a result of Program 20.1, the computer will print the numeral 8. We have to know that this is the value of C. As a result of Program 20.2, the computer will print the following: X = −20.

The numeral at the left end of each line is the *line number.* The column of line numbers merely directs the computer from one instruction to the next. After completing one line of instructions, the computer automatically proceeds to the next higher line number, whether we list them in numerical order or not. The END command tells the computer to stop. The line numbers and END are mandatory for BASIC programs.

Of immediate interest are the MAT commands. Some, but not all, BASIC-language computers have these commands built in. Before a MAT operation can be performed, each matrix must have space reserved for it by a DIM command. Program 20.3 is an example of how a matrix product might be programmed.

```
10  DIM          A(3,3), B(3,1), C(3,1)
20  MAT READ     A, B
30  DATA         2, 3, −1, 0, 6, 5, 4, 3, 2, 2, 4, 6
40  MAT          C = A * B
50  PRINT        "MATRIX C"
60  MAT PRINT    C
70  END
```

Program 20.3 A program for matrix multiplication

As a result of Program 20.3, the computer would store A and

B as follows:

$$A = \begin{bmatrix} 2 & 3 & -1 \\ 0 & 6 & 5 \\ 4 & 3 & 2 \end{bmatrix} \qquad B = \begin{bmatrix} 2 \\ 4 \\ 6 \end{bmatrix}$$

and would produce the following printout (including the heading):

```
MATRIX C
   10
   54
   32
```

Note carefully how the data at line 30 are automatically distributed into rows and columns and into different matrices, according to line 10. Because line 10 calls for A first and calls for nine elements in A ($3 \times 3 = 9$), the first nine numbers at line 30 are assigned to A (as shown above), and because line 10 calls for B next and calls for three elements in B, the next three numbers at line 30 are assigned to B.

Now we shall write a computer program to solve a system of linear equations. Given the system

$$8x + 7y - z = 44$$

$$-5x + 9y + 13z = 16$$

$$12x - 11y - 6z = 150$$

the equivalent matrix equation is $AX = K$, as follows:

$$\begin{bmatrix} 8 & 7 & -1 \\ -5 & 9 & 13 \\ 12 & -11 & -6 \end{bmatrix} \begin{bmatrix} x \\ y \\ z \end{bmatrix} \begin{matrix} = \\ = \\ = \end{matrix} \begin{bmatrix} 44 \\ 16 \\ 150 \end{bmatrix}$$

Program 20.4 will provide the solutions.

```
10  DIM        A(3,3), K(3,1), V(3,1), B(3,3)
20  MAT READ   A, K
30  DATA       8, 7, −1, −5, 9, 13, 12, −11, −6, 44, 16, 150
40  MAT        B = INV(A)
50  MAT        V = B * K
60  LET        X = V(1,1)
70  LET        Y = V(2,1)
80  LET        Z = V(3,1)
```

```
 90  PRINT          "X ="; X
100  PRINT          "Y ="; Y
110  PRINT          "Z ="; Z
120  END
```

Program 20.4

In this program, V is the variable matrix because X is used for the scalar variable x (remember that computers do not use lowercase letters). Instead of printing the matrix V, this program calls for each element to be printed separately. At lines 60 to 80, the variables X, Y, and Z are "called out" of the matrix. The resulting printout will look like this:

```
X = 12

Y = -6

Z = 10
```

EXERCISE 20.6

In Probs. 1 to 6, write a BASIC-language program to solve each given scalar equation for the values given, and show the resulting printout.

1. $z = x + y$
   $[x \quad y] = [6 \quad 9]$
2. $c = a - b$
   $[a \quad b] = [12 \quad -2]$
3. $p = r(s)$
   $[r \quad s] = [6 \quad -4]$
4. $x = k\sqrt{m}$
   $[k \quad m] = [2 \quad 16]$
5. $c = a \div b$
   $[a \quad b] = [20 \quad 4]$
6. $x = a + b - c(d) \div e$
   $[a \quad b \quad c \quad d \quad e] = [1 \quad 2 \quad 3 \quad 4 \quad 5]$

7. Illustrate the matrix that would be printed by the following BASIC program.

```
10  DIM          A(2,5)
20  MAT READ     A
30  DATA         1, 3, 5, 7, 9, 2, 4, 6, 8, 10
40  MAT PRINT    A
50  END
```

8. Write a BASIC program similar to that of Prob. 7 that would cause the following matrix to be stored in a computer and printed.

$$X = \begin{bmatrix} 5 & 4 & 10 \\ -1 & 7 & -6 \\ 2 & 0 & 8 \\ 3 & -2 & 12 \end{bmatrix}$$

In Probs. 9 to 14, write a BASIC-language program to perform and print each matrix operation. Show each printout.

9. $\begin{bmatrix} 3 & 5 \\ 6 & 2 \end{bmatrix} + \begin{bmatrix} 7 & -5 \\ -3 & 1 \end{bmatrix}$

10. $\begin{bmatrix} 2 & 0 \\ -5 & 7 \end{bmatrix} - \begin{bmatrix} 3 & -1 \\ 4 & -5 \end{bmatrix}$

11. $5\begin{bmatrix} 6 & 0 & -3 \\ 2 & -1 & 7 \end{bmatrix}$

12. $[1 \quad 2 \quad 3]\begin{bmatrix} 1 & 3 \\ -1 & 0 \\ 2 & -2 \end{bmatrix}$

13. $\begin{bmatrix} 2 & 3 \\ 1 & 0 \end{bmatrix}\begin{bmatrix} 5 & 2 & -1 \\ 4 & 3 & 7 \end{bmatrix}$

14. $\begin{bmatrix} 4 & 3 \\ 5 & 2 \\ 6 & 1 \end{bmatrix}\begin{bmatrix} 0 & 5 & -4 \\ -3 & 1 & 6 \end{bmatrix}$

15. Write (**a**) the matrix equation, (**b**) the system of linear equations, and (**c**) the printout corresponding to the following program.

```
 10  DIM        A(2,2), K(2,1), V(2,1), B(2,2)
 20  MAT READ   A, K
 30  DATA       2, −6, 7, 5, 52, 0
 40  MAT        B = INV(A)
 50  MAT        V = B * K
 60  LET        X = V(1,1)
 70  LET        Y = V(2,1)
 80  PRINT      "X ="; X
 90  PRINT      "Y ="; Y
100  END
```

16. Repeat Prob. 15 for the following program.

```
10  DIM        A(2,2), K(2,1), X(2,1), B(2,2)
20  MAT READ   A, K
30  DATA       1, 2, 3, −1, 13, −17
40  MAT        B=INV(A)
50  MAT        X=B * K
60  PRINT      "MATRIX X"
70  MAT PRINT  X
80  END
```

In Probs. 17 to 20, write a BASIC program to solve each system of linear equations. Show each printout.

17. $3x + 4y = 5$
$x + y = 1$

18. $5x_1 - x_2 = 25$
$2x_1 + x_2 = 3$

19. $x_1 + x_2 + x_3 = 6$
$x_1 - x_2 - x_3 = 0$
$x_1 + 3x_3 = 0$

20. $x + y + z = 0$
$2x - 3y + 3z = 10$
$-x - 2y = 2$

20.6 MATRIX INVERSES

To enhance our knowledge of matrix equations, we should learn how to find the inverse of any n-by-n matrix. Equation (20.11) is the basis for determining any matrix inverse. The following derivation of a three-by-three inverse outlines the general method:

Given that
$$A = \begin{bmatrix} a_{11} & a_{12} & a_{13} \\ a_{21} & a_{22} & a_{23} \\ a_{31} & a_{32} & a_{33} \end{bmatrix} \tag{20.14}$$

we let the inverse be

$$A^{-1} = \begin{bmatrix} z_{11} & z_{12} & z_{13} \\ z_{21} & z_{22} & z_{23} \\ z_{31} & z_{32} & z_{33} \end{bmatrix} \tag{20.15}$$

where the nine z_{ij} are unknown elements to be determined. Then we carry out the product $A^{-1}A$ and equate it to I, according to Eq. (20.11).

$$A^{-1}A = I$$

$$\begin{bmatrix} z_{11} & z_{12} & z_{13} \\ z_{21} & z_{22} & z_{23} \\ z_{31} & z_{32} & z_{33} \end{bmatrix} \begin{bmatrix} a_{11} & a_{12} & a_{13} \\ a_{21} & a_{22} & a_{23} \\ a_{31} & a_{32} & a_{33} \end{bmatrix} = \begin{bmatrix} 1 & 0 & 0 \\ 0 & 1 & 0 \\ 0 & 0 & 1 \end{bmatrix}$$

$$\begin{bmatrix} (a_{11}z_{11} + a_{21}z_{12} + a_{31}z_{13}) & (a_{12}z_{11} + a_{22}z_{12} + a_{32}z_{13}) & (a_{13}z_{11} + a_{23}z_{12} + a_{33}z_{13}) \\ (a_{11}z_{21} + a_{21}z_{22} + a_{31}z_{23}) & (a_{12}z_{21} + a_{22}z_{22} + a_{32}z_{23}) & (a_{13}z_{21} + a_{23}z_{22} + a_{33}z_{23}) \\ (a_{11}z_{31} + a_{21}z_{32} + a_{31}z_{33}) & (a_{12}z_{31} + a_{22}z_{32} + a_{32}z_{33}) & (a_{13}z_{31} + a_{23}z_{32} + a_{33}z_{33}) \end{bmatrix} = \begin{bmatrix} 1 & 0 & 0 \\ 0 & 1 & 0 \\ 0 & 0 & 1 \end{bmatrix}$$

Next we use the definition of matrix equality to set each element of $A^{-1}A$ equal to the corresponding element of I. As a result, we have the following three systems of equations, each system coming from one row of $A^{-1}A = I$ above.

System I from row 1 of $A^{-1}A = I$

$$\begin{aligned} a_{11}z_{11} + a_{21}z_{12} + a_{31}z_{13} &= 1 \\ a_{12}z_{11} + a_{22}z_{12} + a_{32}z_{13} &= 0 \\ a_{13}z_{11} + a_{23}z_{12} + a_{33}z_{13} &= 0 \end{aligned} \tag{20.16}$$

System II from row 2 of $A^{-1}A = I$

$$\begin{aligned} a_{11}z_{21} + a_{21}z_{22} + a_{31}z_{23} &= 0 \\ a_{12}z_{21} + a_{22}z_{22} + a_{32}z_{23} &= 1 \\ a_{13}z_{21} + a_{23}z_{22} + a_{33}z_{23} &= 0 \end{aligned} \tag{20.17}$$

System III from row 3 of $A^{-1}A = I$

$$\begin{aligned} a_{11}z_{31} + a_{21}z_{32} + a_{31}z_{33} &= 0 \\ a_{12}z_{31} + a_{22}z_{32} + a_{32}z_{33} &= 0 \\ a_{13}z_{31} + a_{23}z_{32} + a_{33}z_{33} &= 1 \end{aligned} \tag{20.18}$$

With the methods of Chap. 8, we can easily solve each system for its three unknowns z_{ij}. We present here, in determinant form, the solutions of System I. The others are left as exercises for you.

$$z_{11} = \frac{\begin{vmatrix} a_{22} & a_{32} \\ a_{23} & a_{33} \end{vmatrix}}{|A|} \qquad z_{12} = \frac{-\begin{vmatrix} a_{12} & a_{32} \\ a_{13} & a_{33} \end{vmatrix}}{|A|} \qquad z_{13} = \frac{\begin{vmatrix} a_{12} & a_{22} \\ a_{13} & a_{23} \end{vmatrix}}{|A|} \tag{20.19}$$

The solutions for System I, Eq. (20.16)

At this point our method for determining a matrix inverse is complete. However, a better notation will enable us to write any z_{ij} element almost instantly and to write an entire inverse matrix in very compact form. This notation involves determinants and their properties, and so we shall review the essential points here. If you need a more detailed review, refer to Chap. 8.

In the first place, we shall expand determinants by *minors,* using each element of any one column, as shown by Eq. (20.20). The terms of the expansion are found by multiplying each element of a column by the determinant that remains after crossing out the row and column of that element, and affixing the proper sign. The element is called the *pivotal element,* and the remaining determinant is the *minor* of that element. In Chap. 8 we used a checkerboard rule for determining the sign. The following rule is quicker: *The sign of the term in which* a_{ij} *is the pivotal element is* $(-1)^{i+j}$.

$$\begin{vmatrix} a_{11} & a_{12} & a_{13} \\ a_{21} & a_{22} & a_{23} \\ a_{31} & a_{32} & a_{33} \end{vmatrix} = (-1)^{1+2}a_{12}\begin{vmatrix} a_{21} & a_{23} \\ a_{31} & a_{33} \end{vmatrix} + (-1)^{2+2}a_{22}\begin{vmatrix} a_{11} & a_{13} \\ a_{31} & a_{33} \end{vmatrix}$$

Column used for expansion

$$+ (-1)^{3+2}a_{32}\begin{vmatrix} a_{11} & a_{13} \\ a_{21} & a_{23} \end{vmatrix} \quad (20.20)$$

Expansion of a determinant by minors

Now comes the new notation: *We shall let A_{ij} stand for the minor of the pivotal element a_{ij}* of a determinant. For example, A_{12} corresponding to a_{12} of Eq. (20.20) would be $\begin{vmatrix} a_{21} & a_{23} \\ a_{31} & a_{33} \end{vmatrix}$. Then we can write the entire determinant of Eq. (20.20) in a more compact form, as follows:

$$\begin{vmatrix} a_{11} & a_{12} & a_{13} \\ a_{21} & a_{22} & a_{23} \\ a_{31} & a_{32} & a_{33} \end{vmatrix} = (-1)^{1+2}a_{12}A_{12} + (-1)^{2+2}a_{22}A_{22} + (-1)^{3+2}a_{32}A_{32}$$
$$= -a_{12}A_{12} + a_{22}A_{22} - a_{32}A_{32}$$

Of particular interest in matrix work are determinants in which one element is 1 and all others in the same column are zero. Then the expansion contains just one element, as follows:

$$\begin{vmatrix} a_{11} & 1 & a_{13} \\ a_{21} & 0 & a_{23} \\ a_{31} & 0 & a_{33} \end{vmatrix} = (-1)^{1+2}(1)A_{12} = -(a_{21}a_{33} - a_{31}a_{23})$$

With this notation, we can rewrite Eq. (20.19) as follows:

$$z_{11} = \frac{A_{11}}{|A|} \qquad z_{12} = \frac{-A_{21}}{|A|} \qquad z_{13} = \frac{A_{31}}{|A|}$$

Notice that the row and column subscripts of each A_{ji} are *reversed* relative to those of z_{ij}. Therefore, we can express each z_{ij} element of the inverse of a matrix A as follows:

$$z_{ij} = \frac{(-1)^{i+j}A_{ji}}{|A|} \quad (20.21)$$

The element in the ith row and jth column of the inverse of a matrix A

Example 20.15. Element z_{31} of Eq. (20.15) can be found by Eq. (20.21) as follows:

$$z_{31} = \frac{(-1)^{3+1}A_{13}}{|A|} = \frac{\begin{vmatrix} a_{21} & a_{22} \\ a_{31} & a_{32} \end{vmatrix}}{|A|} = \frac{a_{21}a_{32} - a_{31}a_{22}}{|A|}$$

The entire matrix A^{-1} of Eq. (20.15) can be written as follows:

Given $$A = \begin{bmatrix} a_{11} & a_{12} & a_{13} \\ a_{21} & a_{22} & a_{23} \\ a_{31} & a_{32} & a_{33} \end{bmatrix}$$

the inverse is

$$A^{-1} = \frac{\begin{bmatrix} A_{11} & -A_{21} & A_{31} \\ -A_{12} & A_{22} & -A_{32} \\ A_{13} & -A_{23} & A_{33} \end{bmatrix}}{|A|} \qquad (20.22)$$

The inverse of a three-by-three matrix A

Example 20.16. Determine the inverse of matrix A below.

$$A = \begin{bmatrix} 3 & 2 & -4 \\ -1 & 0 & 6 \\ 5 & 1 & 7 \end{bmatrix}$$

$$A^{-1} = \frac{\begin{bmatrix} A_{11} & -A_{21} & A_{31} \\ -A_{12} & A_{22} & -A_{32} \\ A_{13} & -A_{23} & A_{33} \end{bmatrix}}{|A|}$$

The minors in the first row are derived below. The others are left as exercises for you.

$$A_{11} = \begin{vmatrix} 0 & 6 \\ 1 & 7 \end{vmatrix} = 0 \times 7 - 1 \times 6 = -6$$

$$-A_{21} = -\begin{vmatrix} 2 & -4 \\ 1 & 7 \end{vmatrix} = -(2 \times 7 - 1 \times -4) = -18$$

$$A_{31} = \begin{vmatrix} 2 & -4 \\ 0 & 6 \end{vmatrix} = 2 \times 6 - 0 \times -4 = 12$$

After determining the other A_{ji} and $|A|$, you should obtain the following inverse matrix:

$$A^{-1} = \frac{\begin{bmatrix} -6 & -18 & 12 \\ 37 & 41 & -14 \\ -1 & 7 & 2 \end{bmatrix}}{60}$$

You can check this result by showing that

$$A^{-1}A = \begin{bmatrix} 1 & 0 & 0 \\ 0 & 1 & 0 \\ 0 & 0 & 1 \end{bmatrix}$$

Example 20.17. Solve the following systems of equations by using matrices.

$$x - y + z = 14$$

$$-x + 2y + 3z = -1$$

$$2x \qquad - 3z = 2$$

The corresponding matrix equation is

$$\begin{bmatrix} 1 & -1 & 1 \\ -1 & 2 & 3 \\ 2 & 0 & -3 \end{bmatrix} \begin{bmatrix} x \\ y \\ z \end{bmatrix} = \begin{bmatrix} 14 \\ -1 \\ 2 \end{bmatrix}$$

and the matrix solution according to Eqs. (20.13) and (20.22) is

$$\begin{bmatrix} x \\ y \\ z \end{bmatrix} = \begin{bmatrix} 1 & -1 & 1 \\ -1 & 2 & 3 \\ 2 & 0 & -3 \end{bmatrix}^{-1} \begin{bmatrix} 14 \\ -1 \\ 2 \end{bmatrix}$$

$$= \frac{\begin{bmatrix} \begin{vmatrix} 2 & 3 \\ 0 & -3 \end{vmatrix} & -\begin{vmatrix} -1 & 1 \\ 0 & -3 \end{vmatrix} & \begin{vmatrix} -1 & 1 \\ 2 & 3 \end{vmatrix} \\ -\begin{vmatrix} -1 & 3 \\ 2 & -3 \end{vmatrix} & \begin{vmatrix} 1 & 1 \\ 2 & -3 \end{vmatrix} & -\begin{vmatrix} 1 & 1 \\ -1 & 3 \end{vmatrix} \\ \begin{vmatrix} -1 & 2 \\ 2 & 0 \end{vmatrix} & -\begin{vmatrix} 1 & -1 \\ 2 & 0 \end{vmatrix} & \begin{vmatrix} 1 & -1 \\ -1 & 2 \end{vmatrix} \end{bmatrix} \begin{bmatrix} 14 \\ -1 \\ 2 \end{bmatrix}}{\begin{vmatrix} 1 & -1 & 1 \\ -1 & 2 & 3 \\ 2 & 0 & -3 \end{vmatrix}}$$

$$= \frac{\begin{bmatrix} -6 & -3 & -5 \\ 3 & -5 & -4 \\ -4 & -2 & 1 \end{bmatrix} \begin{bmatrix} 14 \\ -1 \\ 2 \end{bmatrix}}{-13}$$

$$= \frac{\begin{bmatrix} -91 \\ 39 \\ -52 \end{bmatrix}}{-13}$$

$$\begin{bmatrix} x \\ y \\ z \end{bmatrix} = \begin{bmatrix} 7 \\ -3 \\ 4 \end{bmatrix} \quad \text{or} \quad \begin{aligned} x &= 7 \\ y &= -3 \\ z &= 4 \end{aligned}$$

EXERCISE 20.7

1. Write in determinant form the solutions of System II, Eq. (20.17).
2. Write in determinant form the solutions of System III, Eq. (20.18).
3. Complete Example 20.16.

For Probs. 4 to 6, use matrix A below, and let z_{ij} be an element of the inverse of A.

$$A = \begin{bmatrix} 3 & 2 & 5 \\ 0 & 6 & 1 \\ -1 & -2 & 0 \end{bmatrix}$$

4. Find each of the following: z_{11}, z_{21}, and z_{32}.
5. Find each of the following: z_{12}, z_{22}, and z_{31}.
6. Find each of the following: z_{13}, z_{23}, and z_{33}.

In Probs. 7 to 10, solve each system of equations by using matrices, as in Example 20.17. Use Eq. (20.12) for inverses of two-by-two matrices.

7. $\begin{aligned} 3x - 2y &= 0 \\ -5x + 4y &= 10 \end{aligned}$

8. $\begin{aligned} 2x_1 + x_2 &= -8 \\ 3x_1 + 5x_2 &= 9 \end{aligned}$

9. $\begin{aligned} 5x_1 + 3x_2 \phantom{{}+2x_3} &= 2 \\ x_1 - 2x_2 - x_3 &= -9 \\ -2x_1 + x_2 + 2x_3 &= 2 \end{aligned}$

10. $\begin{aligned} 6x + 8y - z &= 15 \\ 3x - 3y + z &= -6 \\ 5y - 2z &= 18 \end{aligned}$

APPENDIX 1
A SYSTEM FOR SOLVING WORD PROBLEMS

In Chap. 1 we mentioned the importance of practical, applied problems, problems that are usually stated in words rather than in mathematical symbols. The chief difficulty with word problems is that our minds and our everyday language are inadequate for dealing with a multitude of facts and complicated relationships. But mathematics is ideally suited to this. By translating the stated facts into mathematical terms, we eliminate the difficulty of mentally sorting, combining, and digesting them. Instead, we use the purely mechanical rules of mathematics. A good general procedure for solving word problems is as follows:

1. *Read the problem* for general orientation and then reread it to pick out details.
2. *Rewrite the problem in your own words.* In very terse form, state the given facts and what is to be found.
3. If any geometric or other graphical relationships are involved, *make a sketch* to illustrate them.
4. *Define a notation.* Make up a set of letters to represent the various terms of the problem. Place the letters at appropriate places on the sketch.

   The notation is a set of definitions and descriptions for the literal symbols and algebraic expressions used in the problem. It should clearly define each symbol in specific terms, including the units of measure. It must be very specific and leave no doubt as to what each symbol means. For example, suppose we have a problem involving the amount of sand in a cement-sand mixture. It is not sufficient to write "let $x =$ sand." Such poor notation leads to questions such as "What property of the sand does x stand for—the volume, the price, the color, the weight, the specific gravity, or what?" An example of a specific and more appropriate notation is "let $x =$ the weight of sand to be added, in pounds."

The purpose of the notation is to help the writer, and anyone who may read the problem, to understand the physical meaning of each symbol and each algebraic expression and to thus make sense of the equations in the problem. Without a notation or "key" to the symbols of the problem, your solution is meaningless to everyone but you, and even you could become confused.

5. In terms of your notation, *translate the stated facts into mathematical equations.* This is a critical step in reducing the problem to its basic elements.
6. *Solve the equations.* Clearly show in a neat, step-by-step layout the set of operations you perform to find the unknown quantity of the problem.
7. *Clearly display and label the answer.*
8. *Check the answer* against the original problem statement.

This systematic method of attack should help you break any problem down into small, understandable parts; it also trains you to think logically and to make a neat, clear, orderly presentation. Two good examples of this system are given in Sec. 3.8.

DIMENSIONS AND UNITS

APPENDIX 2

Physical Quantities and Dimensions

A physical quantity is a measurable characteristic of a body, an action, or a state of being. Examples are *length, speed, electric current,* and *energy.* Although there are many such quantities, only five are independent; all others can be defined in terms of two or more of the independent ones. The five quantities that are customarily taken to be independent are called the *fundamental physical quantities,* and they are *length, mass, time, temperature,* and *electric charge.* (*Force* is sometimes considered to be independent instead of *mass;* either one can be defined in terms of the other.) Because all other physical quantities can be reduced to a combination of these fundamental ones, the others are called *derived quantities.* An example is speed, a unit of length divided by a unit of time.

The word "dimensions" of a physical quantity means a set of more basic quantities, usually fundamental quantities, in terms of which that quantity can be defined. In other words, dimensions are physical quantities, ones to which a given quantity can be reduced. For example, since *speed* is a length divided by a time, the dimensions of speed are *length/time;* the dimensions of *acceleration* are *speed/time,* which can be reduced to *length/time squared.* Since an area is a length times a length, the dimensions of *area* are *length squared.*

Table A2.1 in Appendix 2 is a list of various physical quantities, the symbols used for them, and their dimensions in terms of more basic quantities.

Units

Physical quantities are measured according to certain arbitrary, established standards called *units.* Examples are the *meter,* the *gram,* and the *second.* The purpose of units is to define measurements for the sake of uniformity, convenience, and communication. Each unit is a

TABLE A2.1 SYMBOLS AND DIMENSIONAL EQUIVALENTS FOR PHYSICAL QUANTITIES

PHYSICAL QUANTITY	SYMBOL	FUNDAMENTAL DIMENSIONS
Fundamental Quantities		
Length	L	L
Mass	M	M or $F \times t^2/L$
Force	F	F or $M \times L/t^2$
Time	t	t
Temperature	T	T
Electrical charge	q	q
Derived Quantities		
Mechanical		
Area	A	$A = L^2$
Volume	V	$V = L^3$
Density (mass)	ρ_m	$\rho_m = M/V = M/L^3$
Density (weight)	ρ_w	$\rho_w = F/V = F/L^3$
Pressure	p	$p = F/A = F/L^2$
Stress	s	$s = F/A = F/L^2$
Energy	E	$E = F \times L$
Work	W	$W = F \times L$
Torque	T	$\mathrm{T} = F \times L$
Power (rate of energy flow)	P	$P = E/t = F \times L/t$
Speed, velocity	v	$v = L/t$
Acceleration	a	$a = v/t = L/t^2$
Thermal		
Heat (energy or work)	Q	$Q = F \times L$
Power (rate of heat flow)	H	$H = Q/t = F \times L/t$
Specific heat	c	$c = Q/(M \times T)$
Electrical		
Current	I, i	$I = q/t$
Potential (voltage)	V, E	$E = F \times L/q$
Resistance	R	$R = E/I = F \times L \times t/q^2$
Capacitance	C	$C = q/E = q^2/(F \times L)$
Inductance	l	$l = R \times t = E \times t/I = F \times L \times t^2/q^2$
Power	W	$W = EI = F \times L/t$
Energy	—	$W \times t = E \times I \times t = F \times L$

completely arbitrary standard, but once established, it becomes a basis upon which like quantities can be compared. For example, the *meter* is arbitrarily defined as the distance between two scratch marks on a certain platinum-iridium bar preserved at the International Bureau of Weights and Measures, near Paris.

The difference between a dimension and a unit should be clearly understood. A *dimension* is a physical quantity, a natural, measurable characteristic, whereas a *unit* is an arbitrary, artificial, manmade division for measuring a dimension. For example, time is a dimension, and the second is a unit (one of many) for measuring time.

There are three commonly used systems of units for the fundamental mechanical dimensions, *length, mass* (or *force*), and *time.* These systems are (1) the English or foot-pound-second system, (2) the cgs or centimeter-gram-second system, and (3) the mks or meter-kilogram-second system.

Commonly used units for the fundamental thermal dimension, temperature, are °F (degrees Fahrenheit), °R (degrees Rankine), °C (degrees Celsius, formerly Centigrade), and K (Kelvin, formerly called degrees Kelvin). The first two are usually associated with the English system, 0°F being 32 units below the freezing temperature of water and 0°R being *absolute zero,* the lowest temperature theoretically possible. °F and °R are related in that a *change* of 1°F equals a *change* of 1°R.

The Celsius and Kelvin units are usually associated with the metric (cgs and mks) systems, 0°C being the freezing temperature of water, and 0 K being absolute zero. A *change* of 1°C equals a *change* of 1 K.

Various physical quantities and their units are listed in Table A2.2. It is customary to hyphenate a product of two units, as in ft-lb, which means "feet times pounds" (not feet minus pounds).

Conversion Factors

Many other units are used in addition to those of Table A2.2. For instance, length is measured in *millimeters, inches, miles, light-years,* and other units. Sometimes a single special word is more convenient than a combination of fundamental units. The word "slug," for example, is the English unit of mass and is equivalent to units of lb-sec^2/ft. For various reasons, a veritable plethora of units exists, and this multiplicity of units is a source of difficulty when different ones are to be combined or when one set must be converted to another. (Quick now, what speed in mph is equivalent to 73 m/sec?)

To aid in reading this book or others, and for general reference, a list of unit prefixes and a table of unit conversion factors are listed below. A conversion factor not only changes the *number* of units from one system to another, but also changes the *units* themselves. For example, there are by definition exactly 2.54 cm in 1 in. Therefore the conversion factor is 2.54 cm/in.; it consists of units as well as a number. To change a dimension from inches to centimeters, we multiply by this factor. Here is an example:

$$16.00\ \cancel{\text{in.}} \times 2.54\ \frac{\text{cm}}{\cancel{\text{in.}}} = 40.64\ \text{cm}$$

TABLE A2.2 UNITS AND SYMBOLS

QUANTITY	UNITS: ENGLISH		CGS		MKS	
Fundamental Quantities						
Length	foot	ft	centimeter	cm	meter	m
Mass	slug		gram	gm	kilogram	kg
Force	pound	lb	dyne		newton	n
Time	second	sec	second	sec	second	sec
Temperature	degrees Fahrenheit	°F	degrees Celsius*	°C	degrees Celsius	°C
Charge	coulomb	coul	abcoulomb	abcoul	coulomb	coul
Derived Quantities						
Mechanical						
Area	square foot	sq ft or ft^2	square centimeter	sq cm or cm^2	square meter	sq m or m^2
Volume	cubic foot	cu ft or ft^3	cubic centimeter	cu cm, cc, or cm^3	cubic meter	cu m or m^3
Density		lb/cu ft or lb/ft^3		gm/cu cm, gm/cc, or gm/cm^3		kg/cu m or kg/m^3
Pressure, stress		lb/ft^2 or psf		$dyne/cm^2$	newton/sq meter	or n/m^2
Energy, work, torque		ft-lb	erg	(same as dyne-cm)	joule	(same as n-m)
Power	horsepower	hp (also ft-lb/sec)		erg/sec (same as dyne-cm/sec)		joule/sec (same as n-m/sec)
Speed, velocity		ft/sec		cm/sec		m/sec
Acceleration		ft/sec/sec or ft/sec^2		cm/sec^2		m/sec^2
Thermal						
Heat	British Thermal Unit	BTU	calorie	cal	calorie	cal
Power		BTU/sec		cal/sec		cal/sec
Electrical						
Current	ampere	amp (same as coul/sec)	abampere	abamp	ampere	amp
Potential (voltage)	volt	v	abvolt		volt	v
Resistance	ohm	Ω (same as volt/amp)	abohm		ohm	Ω
Inductance	henry	h	abhenry		henry	h
Capacitance	farad	f	abfarad	abf	farad	f
Power	watt	W (same as volt-amp)	watt	W (sometimes erg/sec)	watt	W (same as joule/sec)
Energy		watt-sec (usually kw-hr)		watt-sec		watt-sec

* Formerly Centigrade.

Multiples and fractions of units are formed by prefixing certain syllables to the basic unit. For example, the prefix *kilo-* serves to multiply any unit by 1,000 as in "1 *kilo*meter is *1,000 meters*." Several standard prefixes are defined in the following list.

Prefix	Meaning
nano-	one-billionth, 10^{-9}
micro-, μ	one-millionth, 10^{-6}
milli-, m	one-thousandth, 10^{-3}
centi-	one-hundredth, 10^{-2}
deci-	one-tenth, 10^{-1}
deca-	ten times, 10
kilo-, k	one thousand times, 10^{3}
mega-	one million times, 10^{6}

These prefixes, along with the conversion factors of Table A2.3, will enable many of the common units to be compared. Also, many other conversion factors can be derived by combining two or more factors from Table A2.3. For example, from Table A2.3, the number of feet per mile is (1,760 yd/mile)(3 ft/yd) = 5,280 ft/mile.

Equations, Dimensions, and Units

Both sides of a mathematical equation must stand for the same number, but an equation that involves physical quantities must also balance the dimensions and the units. For example, equations such as

$$7\text{ ft} + 3\text{ ft} = 10\text{ hr}$$

are nonsense because the dimensions are different. The equation

$$7\text{ ft} + 3\text{ in.} = 10\text{ in.}$$

is also nonsense; even though the numbers balance, $7 + 3 = 10$, and the dimensions balance, length + length = length, the units of feet and inches are different.

However, a conventional practice is to mix numbers and units such as

$$7\text{ ft} + 3\text{ in.} = 87\text{ in.}$$

Although not exactly proper, this form is widely used, and a general understanding of it seems to exist. Such a mixture can cause misunderstanding or errors, especially if the units are unfamiliar. A more rigorous and safe way to handle mixed units is to use a conversion

TABLE A2.3 CONVERSION FACTORS FOR UNITS

Length	
2.54 cm/in.	1 in. = 2.54 cm
12 in./ft	1 ft = 12 in.
3 ft/yd	1 yd = 3 ft
*1,760 yd/mile	1 mi = 1,760 yd
*1.609344 km/mile	1 mi = 1.609344 km
9.4610×10^{12} km/light-year	1 light-year = 9.4610×10^{12} km
Area	
4,840 yd^2/acre	1 acre = 4,840 yd^2
43,560 ft^2/acre	1 acre = 43,560 ft^2
Volume	
57.749 $in.^3$/qt.	1 qt = 57.749 $in.^3$
4 qt/gal	1 gal = 4 qt
1,000 cc/liter	1 liter = 1,000 cc
61.025 $in.^3$/liter	1 liter = 61.025 $in.^3$
Mass	
1 slug = 1 lb-sec^2/ft	definition
14.6 kg/slug	1 slug = 14.6 kg
1,000 gm/kg	1 kg = 1,000 gm
Force	
1 lb = 1 slug-ft/sec^2	definition
16 oz/lb	1 lb = 16 oz
2,000 lb/short ton	1 short ton = 2,000 lb
4.45 newton/lb	1 lb = 4.45 n
10^5 dyne/newton	1 n = 10^5 dyne
Energy	
1 erg = 1 dyne-cm	definition
1 joule = 1 n-m	definition
778 ft-lb/BTU	1 BTU = 778 ft-lb
252 cal/BTU	1 BTU = 252 cal
1.36 joule/ft-lb	1 ft-lb = 1.36 joule
10^7 erg/joule	1 joule = 10^7 erg
Power	
1 watt = 1 volt-amp = 1 joule/sec	definition
1,000 watts/kw	1 kw = 1,000 watts
550 (ft-lb/sec)/hp	1 hp = 550 ft-lb/sec
0.75 kw/hp	1 hp = 0.75 kw
3,413 (BTU/hr)/kw	1 kw = 3,413 BTU/hr
Electrical	
10 coul/abcoul	1 abcoul = 10 coul
10 amp/abamp	1 abamp = 10 amp
10^8 v/abv	1 abvolt = 10^8 volts
10^9 ohm/abohm	1 abohm = 10^9 ohms
10^{-9} f/abf	1 abfarad = 10^{-9} farad
10^{-9} henry/abhenry	1 abhenry = 10^{-9} henry

* Statute miles; a nautical mile is 1.1508 statute miles.

factor (such as those in Table A2.3) to balance the units. For example, the rigorous way to write the equation above is as follows:

$$7\ \text{ft} \times 12\ \frac{\text{in.}}{\text{ft}} + 3\ \text{in.} = 87\ \text{in.}$$

Then the numbers are equal, $7 \times 12 + 3 = 87$, the dimensions are equal, length + length = length, and the units are equal, inches + inches = inches. The equation is completely balanced.

Dimensional Analysis

The preceding discussion shows that an algebra of units must be considered in writing and solving equations with physical quantities. The following rules must be observed.

1. If a number with units is raised to a power, the units must be raised to the same power.

$$(3\ \text{ft})^2 = 9\ \text{ft}^2$$

$$\sqrt{16\ \text{ft}/\text{sec}^2} = 4\sqrt{\text{ft}}/\text{sec}$$

2. If the same units occur in the numerator and the denominator of a term, they will cancel each other (just as common factors cancel).

$$\text{ft} \times \frac{\text{in.}}{\text{ft}} = \text{in.}$$

$$\frac{\text{ft}}{\text{sec}^2} \times \text{sec} = \frac{\text{ft}}{\text{sec}}$$

3. If a conversion factor is inverted, its units are inverted.

   For example, the conversion factor 24 hr/day gives the number of hours in 1 day. The reciprocal of this factor is the number of days/hr, 1/24 day/hr. To change 72 hr to days, we would multiply as follows:

$$72\ \text{hr} \times \frac{1}{24}\frac{\text{day}}{\text{hr}} = 3\ \text{days}$$

4. All terms of an equation must be expressed in the same units or must be reducible to the same units by cancellation.

$$\text{inches} + \text{inches} = \text{inches}$$

$$\text{grams} + \text{grams} = \text{grams}$$

$$2\ \cancel{\text{hr}} \times 60\ \frac{\text{min}}{\cancel{\text{hr}}} + 27\ \text{min} = 147\ \text{min}$$

These rules should be used to check equations for proper balance. An analysis of dimensions can also be used to establish a correct equation. For example, suppose that the area of a circle of known radius is to be determined. And suppose that we cannot remember whether $2\pi r$ or πr^2 is the correct formula. The fact that area has dimensions of *length squared* should lead us to choose πr^2, the formula having a length, r, squared.

A similar analysis of units can aid in choosing a correct conversion factor. For example, to convert 500 cm to inches, do we multiply by 2.54 cm/in. (from Table A2.3), or do we divide by it? To answer such questions, we set up the conversion factor so as to make the given units cancel; the new units will remain. Thus 500 cm would be converted to inches as follows:

$$500\ \text{cm}\ \frac{1}{2.54\ \text{cm/in.}} = 500\ \cancel{\text{cm}}\ \frac{1\ \ \text{in.}}{2.54\ \cancel{\text{cm}}}$$

$$= 197\ \text{in.} \qquad \text{(to nearest inch)}$$

This procedure is especially effective where several units are involved. Frequently a single conversion factor is not known, but a series of factors can be linked to produce the desired units. For example, 60 miles/hr can be converted to in./sec, as follows:

$$60\ \frac{\cancel{\text{mi}}}{\cancel{\text{hr}}} \times 5{,}280\ \frac{\cancel{\text{ft}}}{\cancel{\text{mi}}} \times 12\ \frac{\text{in.}}{\cancel{\text{ft}}} \times \frac{1\ \ \cancel{\text{hr}}}{60\ \cancel{\text{min}}} \times \frac{1\ \ \cancel{\text{min}}}{60\ \ \text{sec}} = 1{,}056\ \frac{\text{in.}}{\text{sec}}$$

The use of dimensions and units in this way is called *dimensional analysis.*

APPENDIX 3
APPROXIMATE NUMBERS: ACCURACY AND PRECISION

Many of the numbers in everyday mathematics problems come from measurements of physical quantities. Such numbers are called *approximate numbers,* because no physical measurement is exact. Physical measurements are approximations, being no more reliable than the instruments used to make them. For example, suppose you use a scale graduated in tenths of an inch to measure the length of a certain steel rod, and you record 2.6 in. You thereby certify that the length is closer to 2.6 in. than to 2.7 in. or to 2.5 in. Your measurement is reliable to the *nearest* 0.1 in. and no less. Because this scale does not even show hundredths, you certainly could not use it to certify any length to the nearest 0.01 in. or less. But then suppose someone else does use a scale marked in hundredths and records 2.61 in. for the same length, and a third person uses a micrometer graduated in thousandths and records 2.613 in. Which is the "correct" length? The answer is that each value is correct within the degree of precision of the scale used, but each succeeding value is more *reliable.* However, each value is only an approximation, and none is exact, no matter how reliable it may be.

In contrast to approximate numbers, *exact numbers* result from counting or from a definition. For example, if we count 27 students in a classroom, the number is *exactly* 27, not 27 and a fraction. Another example is the conversion factor 12 inches per foot, which is exact *by definition;* there are never 12.1 or 11.7 or anything but exactly 12 in./ft.

The objectives of Appendix 3 are (1) to present means for indicating the reliability of approximate numbers and (2) to explain how the reliability of numbers must be considered in computations.

A. SIGNIFICANT DIGITS

Each digit that contributes to the reliability of a number is called a *significant digit.* Determining the reliability of a number is a problem only because *zero* can either be a significant digit or not. All other numerals are always significant digits. Zero can be used in either of two ways: as a placeholder to locate the decimal point, as in 2,900 or 0.005, or as a measured digit, as in 105, in which case the zero is a significant digit.

We must have a clear understanding of which zeros are significant digits and which merely locate the decimal point. The following rules define significant digits. There are four distinct categories of significant digits.

1. Each non-zero digit is a significant digit.
2. Any zero that is both preceded by and followed by a non-zero digit is a significant digit, and any zero not preceded by a non-zero digit is *not* a significant digit.

   Examples:

   a. All the following zeros are significant digits.

   704 2.05 3,001.07

   b. None of the following zeros is a significant digit.

   0.0029

3. Any zero that comes after a non-zero digit and after the decimal point is a significant digit.

   Examples:

   a. All the following zeros are significant digits.

   1.02 .1305 2.50

   b. None of the following zeros is a significant digit.

   0.42 0.0025

   c. Only the zeros to the right of the 3 are significant digits.

   0.03070

4. In numbers such as 28,000, without a written decimal point, zeros not followed by a non-zero digit are ambiguous. In such cases a special means must be used to indicate significant zeros. The following techniques are used throughout this book:

a. *A definitive statement* such as "28,000 *correct to three digits.*" This statement means that the first three digits, *beginning with the first non-zero digit on the left,* are significant. Therefore the 2, the 8, and the *first* 0 after the 8 are significant digits. The other two zeros just locate the decimal point (which is unwritten).

b. *Scientific notation.* If three digits of 28,000 are to be significant, it can be written as follows:

$$280 \times 10^2 \qquad \text{or} \qquad 2.80 \times 10^4$$

All significant digits are always written, no nonsignificant digit is written, and the exponent of 10 locates the decimal point.

c. *A convention:* All ambiguous zeros are assumed to be placeholders and not significant digits.

Examples:

a. Both zeros significant:

3,500 correct to four digits: 3.500×10^3

b. First zero after 5 is significant:

3,500 correct to three digits: 3.50×10^3

c. No significant zeros:

3,500 correct to two digits: 3.5×10^3

d. No significant zeros by convention: 3,500

B. ACCURACY AND PRECISION

We have used the term *reliability of a number* quite loosely so far. But reliability has two well-defined aspects: *accuracy* and *precision.* Each aspect of reliability affects the arithmetic of approximate numbers in different ways.

The *accuracy* of a number has to do with the number of significant digits: the greater the number of significant digits, the more accurate the number. For example, 5.290, having four significant digits, is more accurate than 5.29, which has only three significant digits. And the numbers 3.14×10^3, 314, 31.4, 3.14, 0.314, and 0.0314 are all accurate to three digits. Each of these is as accurate as the others. Note that the decimal point has nothing to do with accuracy. To better understand this concept, consider that 314 cm is equivalent to 3.14 m. Each number represents the same length, using the same number of significant digits. Therefore the numbers are equally accurate.

The *precision* of a number has to do with the location of the last significant digit on the right, relative to the decimal point. The further the last significant digit is toward the right, the more precise the number. For example, 0.05 is more precise than 729. We would say that 0.05 is precise to the nearest hundredth, but 729 is precise only to the nearest unit. Note that the number of significant digits, or accuracy, does not affect precision; 729 is more accurate than 0.05, but it is less precise. However, we *cannot* say that 3.14 m is more precise than 314 cm because the type of units used affects the decimal point. To compare the precision of two measurements, we must write them in the same units.

Examples of Reliability

The following examples illustrate numbers having various degrees of reliability and the terminology used to describe the reliability.

Number	Accuracy	Precision
70,000	accurate to one digit	precise to nearest 10,000
7.2×10^4	accurate to two digits	precise to nearest 1,000
7.20×10^4	accurate to three digits	precise to nearest 100
7.203×10^4	accurate to four digits	precise to nearest 10
72,036	accurate to five digits	precise to nearest unit
72,036.4	accurate to six digits	precise to nearest $1/10$
5.12	accurate to three digits	precise to nearest $1/100$
0.004	accurate to one digit	precise to nearest $1/1,000$
0.4200	accurate to four digits	precise to nearest $1/10,000$

C. ROUNDING OFF

We shall sometimes find it convenient to reduce the accuracy or the precision of a number by dropping one or more significant digits on the right-hand end. This process is called *rounding off.* For example, to reduce the accuracy of 2,394 to three significant digits, we would write either "2,390 correct to three digits," "2.39×10^3," or "2,390." To reduce the precision of 2,394 to the nearest hundred, we would write a statement such as "2,400 correct to hundreds," "2,400 rounded off to the nearest 100," "2.4×10^3," or "2,400."

The following rules govern the rounding off of numbers. In each case, the rounding off is determined by the size of the digit to the right of the last significant digit to be retained.

1. If the digit to the right of the last significant digit to be retained is less than 5, then the last significant digit is *unchanged.* In such cases, the rounding off is called *rounding down.*

   Examples:
   a. π (3.141592 . . .) rounded off to hundredths is 3.14.
   b. 139.1 correct to three digits is 139.
   c. 5.14 rounded off to tenths is 5.1.
   d. 1,328 correct to two digits is 1,300 (or 1.3×10^3).
   e. 3.7700×10^4 correct to four digits is 3.770×10^4.
   f. 3.7700×10^4 correct to hundreds is 37,700 or 3.77×10^4.

2. If the digit to the right of the last significant digit to be retained is greater than 5, then the last significant digit is increased by 1. In such cases the rounding off is called *rounding up.*

   Examples:
   a. 139.1 rounded off to tens is 140 or 1.4×10^2.
   b. 0.03276 correct to three digits is 0.0328.
   c. 1.498 rounded off to hundredths is 1.50.
   d. 4.27×10^3 correct to two digits is 4,300 or 4.3×10^3.
   e. π (3.141592 . . .) correct to five digits is 3.1416.

3. If the digit to the right of the last significant digit to be retained is 5 and all following digits are zero, then an even digit is not changed, *and* an odd digit is rounded up.

   Examples:
   a. 1.225 correct to three digits is 1.22.
   b. 1,350 correct to two digits is 1,400.
   c. 924.5 rounded off to the nearest unit is 924.

4. If the digit to the right of the last significant digit to be retained is 5 and any following digit is greater than zero, then the number is rounded up.

   Examples:
   a. 1.2501 correct to two digits is 1.3.
   b. π (3.141592 . . .) correct to four digits is 3.142.
   c. 13,756 rounded off to the nearest hundred is 13,800.

Rounding Off Logarithms, Angles, and Trigonometric Functions

Most values of logarithms and trigonometric functions are irrational numbers. Therefore the entries in tables of these functions are approximate numbers. The relationship between the reliability of a *number* and that of one of these *functions of the number* is a bit complicated. We shall not attempt to develop detailed rules for rounding off such numbers, but we shall present the following rough guidelines.

1. The number of significant digits of a number and its logarithm (the mantissa) should be the same. For computational purposes, use one more significant digit in the tabulated number (either the mantissa or the antilog) than that of the number used to enter the table. Round off the result of a computation to the same number of significant digits as that of the least accurate number used.

   Examples:
   a. log 250 = 2.40 correct to two digits of the mantissa.
   b. antilog 3.326 = 2,120 correct to three digits.

2. The reliability of a trigonometric function of an angle or vice versa is approximately as follows:

angle to	value of trigonometric function to	table needed to*
nearest degree	two significant digits	3 places
nearest 10 minutes	three significant digits	4 places
nearest minute	four significant digits	5 places
nearest 10 seconds	five significant digits	6 places

* One place less can be used with interpolation, but results will not be completely accurate.

   Examples:
   a. sin 36° = 0.59 accurate to two digits.
   b. cot $\theta = 0.647 \rightarrow \theta = 57°10'$ to nearest 10 minutes.
   c. tan $62°30' = 1.92$ accurate to three digits.
   d. cos $\theta = 0.9485 \rightarrow \theta = 18°28'$ to nearest minute.

D. COMPUTATIONS WITH APPROXIMATE NUMBERS

The precision of any measurement is generally accepted as ± one-half the finest graduation on the measuring instrument. For example, with a thermometer marked in 0.1°C divisions, we can measure temperatures to within $\pm 0.05°C$. If we record a certain temperature as 29.3°C, then it is certainly closer to 29.3°C than to 29.4°C or to 29.2°C, but it could be any value between 29.35°C and 29.25°C. The limits of precision of the thermometer cause a definite lack of certainty in the recorded temperature. Such a lack of certainty is called a *tolerance.* For example, the *tolerance* of the temperature mentioned above is $\pm 0.05°C$.

Effects of Tolerance on Accuracy

When approximate numbers are used in a computation, their tolerances can affect the accuracy (number of significant digits) and the precision (location of last significant digits) of the results. The following example shows how accuracy can be affected by tolerances.

Example. The length and width of a certain rectangular plate have been measured as 10.4 and 8.2 in., respectively. What is the most reliable value for the accuracy of the area of the plate?

Theoretically, the area is 10.4 in. × 8.2 in. = 85.28 in.^{2}, but it is absurd to claim that this area is accurate to four significant digits when it is derived from a two-digit number. At best, the area is accurate to only two significant digits, because the least accurate dimension contains two significant digits. Therefore, the most accurate value possible for this area is 85 in.^{2}.

We can verify this reasoning by using tolerances to compute the largest and smallest possible areas. Since the length and width were given to the nearest 0.1 in., we must assume that each one has a tolerance of ±0.05 in. By using the maximum value of each dimension, we can compute the maximum possible area as follows. 10.45 in. × 8.25 in. = 86.2125 in.^{2}; likewise, the minimum possible area is 10.35 in. × 8.15 in. = 84.3525 in.^{2}. The maximum, mean, and minimum values compare as follows:

Maximum area:	86.2125 in.^{2}
Mean area:	85.28 in.^{2}
Minimum area:	84.3525 in.^{2}

These numbers do not agree to even two digits, let alone four or three. In fact, even the first digits will not agree when rounded off! However, the true area would be closer to 85 than to 90 or 80 in.^{2}, and so we should retain the second significant digit.

The reasoning and the methods of the preceding example have been used to establish the facts and the rule of accuracy below.

1. The product or quotient of two numbers can be no more accurate than the least accurate of the numbers.
2. The root of a number can be no more accurate than the number.

Rule of Accuracy for Products, Quotients, and Roots

Before multiplying or dividing numbers, or taking roots, round off all numbers, except the least accurate, to one significant digit more than that of the least accurate number. Round the result of the computation off to the same number of significant digits as that of the least accurate number.

The first part of this rule is for convenience in computations. It allows us to reduce the amount of computational work by using the least number of digits necessary.

Examples:

a. 26.23×8 becomes $26 \times 8 = 208$, or 200, correct to one digit.

b. $4.2 \times 10^3 \div 12{,}159$ becomes $4{,}200 \div 12{,}200 = 0.34$, correct to two digits.

c. $\sqrt[3]{17} = 2.6$, correct to two digits (see Table A5.1).

d. $\sqrt[3]{17.0} = 2.57$, correct to three digits (see Table A5.1).

Effects of Tolerance on Precision

When approximate numbers are added (or subtracted), the precision (location of last significant digit) is affected by tolerances. The following example illustrates this fact.

Example. A flat copper disk is 3.28 ± 0.005 in. thick. If it is plated on one side with a layer of chrome 0.032 ± 0.0005 in. thick, what is the total thickness?

Theoretically, the total thickness is the sum of the individual thicknesses as follows:

$$3.28 \text{ in.} + 0.032 \text{ in.} = 3.312 \text{ in.}$$

However, due to the uncertainty of the least precise term, 3.28 in., this sum is unrealistic. It implies that we know the thickness to a precision of ± 0.0005 in. Such a precision is clearly absurd when the 3.28-in. dimension alone varies by ± 0.005 in. Therefore the realistic thickness is 3.31 in., obtained by rounding the computed sum off to the precision of the least precise term.

We can verify this argument by computing the maximum and minimum possible thicknesses as follows. (The mean value is included for comparison.)

Maximum thickness:	$3.285 + 0.0325 = 3.3175$
Mean thickness:	$3.28 + 0.032 = 3.312$
Minimum thickness:	$3.275 + 0.0315 = 3.3065$

These numbers clearly do not agree to the nearest 1/1,000, let alone the nearest 1/10,000. There is even some disagreement in the hundredths place. However, the true value is closer to 3.31 than to 3.3; also, the precision of 3.31 is realistic, whereas that of 3.3 is not. Therefore, we are justified in retaining the hundredths digit.

Notice that the *accuracy* of the sum was *not* reduced to that of the least accurate term. In this case the three-digit accuracy equals that of the most accurate term. In other cases the accuracy of the sum could be more or less than that of the most accurate term. See Examples *d* and *e* below.

The reasoning in this example leads us to the following rule of precision.

Rule of Precision for Sums and Differences:

Before adding or subtracting numbers, round off all terms, except the least precise, to one place to the right of that of the least precise term. Round the sum off to the precision of the least precise term.

Examples:

a. $523 + 1.34$ becomes $523 + 1.3 = 524.3$, or 524, correct to the nearest unit.

b. $6.2 \times 10^3 + 75 \times 10^3$ becomes $6{,}200 + 75{,}000 = 81{,}200$, or 81,000, correct to the nearest 1,000.

c. $63.1 - 0.062$ becomes $63.1 - 0.06 = 63.04$, or 63.0, correct to the nearest 1/10.

d. $97 + 4.2 = 101.2$, or 101, correct to the nearest unit.

e. $2{,}396 - 2{,}390.8 = 5.2$, or 5, correct to the nearest unit.

APPENDIX 4 FORMULAS FOR GEOMETRIC FIGURES

The following notation is used in the formulas and figures below:

A	A plane area
A_B	Area of the base of a solid
A_S	Surface area of a solid
a, b	Lengths of bases of plane figures
c	Circumference of a circle
e	Length of each edge of a square or a cube
h	Perpendicular height (altitude) from a vertex to a base
p	Perimeter of a plane figure or of the base of a solid
r	Radius of a circle, a cylinder, or a sphere
s	Slant height
x, y, z	Lengths of edges

Formulas for Plane Figures

1.	Triangle	$p = x + y + z$	$A = \frac{1}{2}bh$	Fig. A4.1
2.	Square	$p = 4e$	$A = e^2$	Fig. A4.2
3.	Parallelogram	$p = 2b + 2s$	$A = bh$	Fig. A4.3
4.	Trapezoid	$p = a + b + x + y$	$A = \frac{1}{2}(a + b)h$	Fig. A4.4
5.	Circle	$c = 2\pi r$	$A = \pi r^2$	Fig. A4.5

Formulas for Solids

6. Cube $A_S = 6e^2$ $V = e^3$ Fig. A4.6
7. Parallelepiped Fig. A4.7

$$A_S = 2(xy + xz + yz) \qquad V = xyz$$

8. Other prisms and cylinders with parallel bases Fig. A4.8

$A_S = ps$ $\qquad V = A_B h$

9. Right circular cylinder Fig. A4.9

$A_S = 2\pi r h$ (lateral surface)

$V = \pi r^2 h$

10. Right circular cone Fig. A4.10

$A_S = \pi r s$ $\qquad V = \frac{1}{3}\pi r^2 h$

11. Other cones and pyramids $\qquad V = \frac{1}{3}A_B h$ Fig. A4.11

12. Sphere $\quad A = 4\pi r^2$ $\qquad V = \frac{4}{3}\pi r^3$ Fig. A4.12

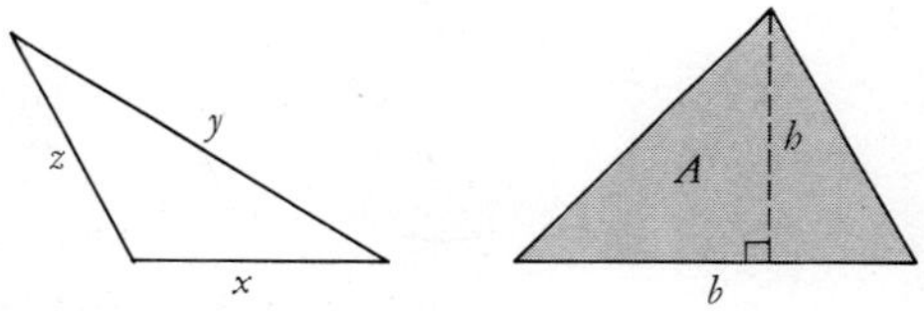

Fig. A4.1 Triangles.

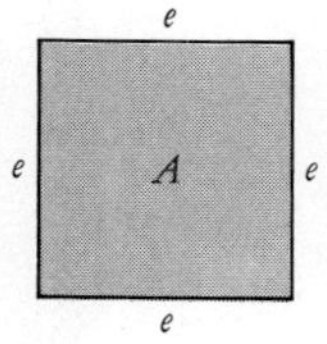

Fig. A4.2 Square.

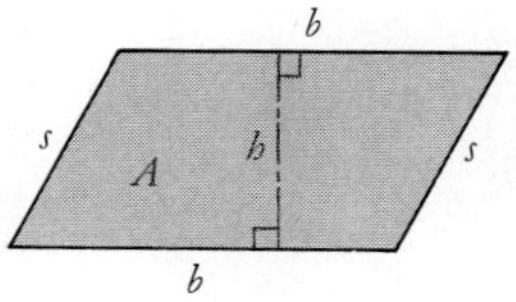

Fig. A4.3 Parallelogram.

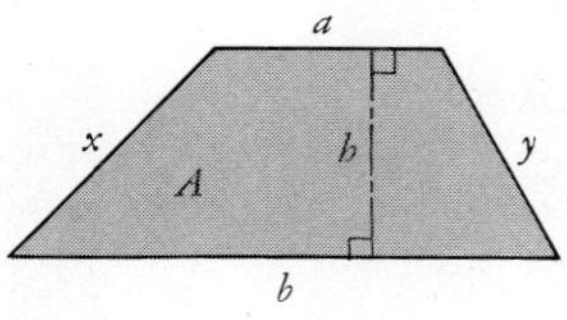

Fig. A4.4 Trapezoid.

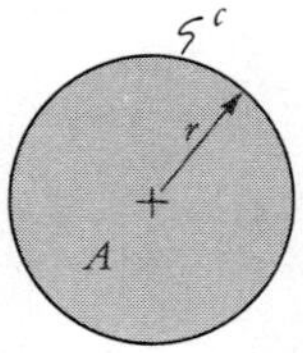

Fig. A4.5 Circle.

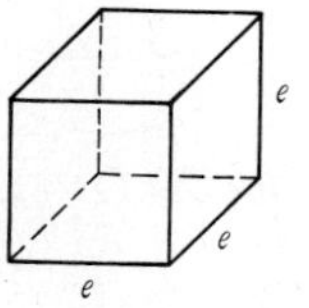

Fig. A4.6 Cube.

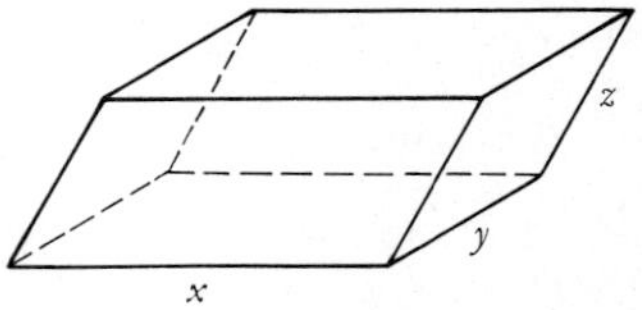

Fig. A4.7 Parallelepiped.

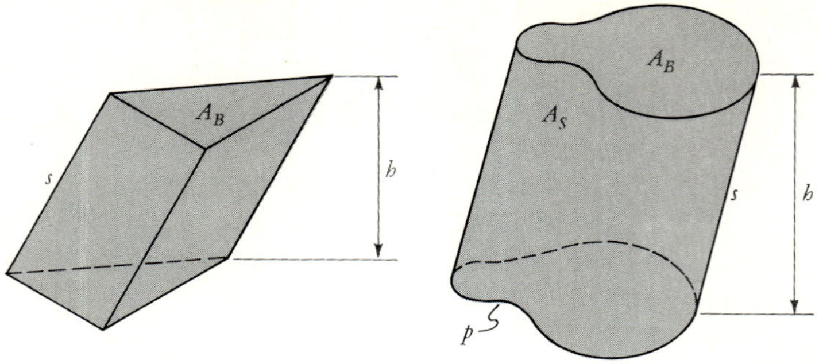

Fig. A4.8 Prism, cylinder with parallel bases.

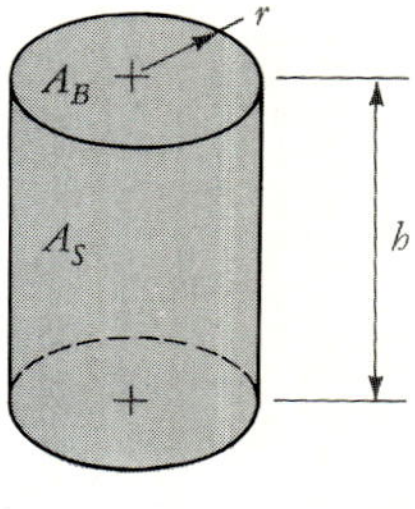

Fig. A4.9 Right circular cylinder.

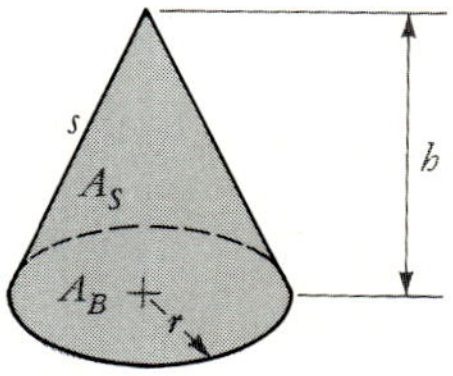

Fig. A4.10 Right circular cone.

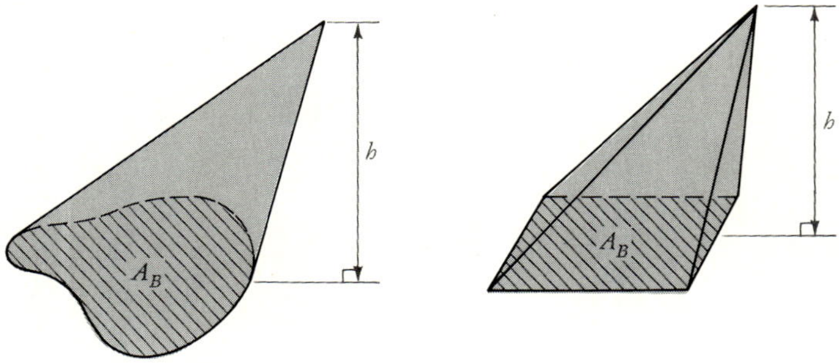

Fig. A4.11 Oblique cone and pyramid.

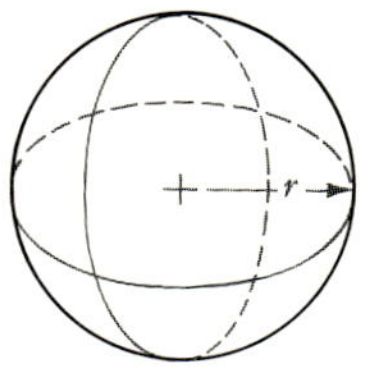

Fig. A4.12 Sphere.

APPENDIX 5

TABLES

TABLE A5.1 SQUARES, SQUARE ROOTS, CUBES, AND CUBE ROOTS OF NUMBERS 1 TO 100

N	N^2	$\sqrt{N}$	$\sqrt{10N}$	N^3	$\sqrt[3]{N}$	$\sqrt[3]{10N}$	$\sqrt[3]{100N}$
1	1	1.000 000	3.162 278	1	1.000 000	2.154 435	4.641 589
2	4	1.414 214	4.472 136	8	1.259 921	2.714 418	5.848 035
3	9	1.732 051	5.477 226	27	1.442 250	3.107 233	6.694 330
4	16	2.000 000	6.324 555	64	1.587 401	3.419 952	7.368 063
5	25	2.236 068	7.071 068	125	1.709 976	3.684 031	7.937 005
6	36	2.449 490	7.745 967	216	1.817 121	3.914 868	8.434 327
7	49	2.645 751	8.366 600	343	1.912 931	4.121 285	8.879 040
8	64	2.828 427	8.944 272	512	2.000 000	4.308 869	9.283 178
9	81	3.000 000	9.486 833	729	2.080 084	4.481 405	9.654 894
10	100	3.162 278	10.000 00	1 000	2.154 435	4.641 589	10.000 00
11	121	3.316 625	10.488 09	1 331	2.223 980	4.791 420	10.322 80
12	144	3.464 102	10.954 45	1 728	2.289 428	4.932 424	10.626 59
13	169	3.605 551	11.401 75	2 197	2.351 335	5.065 797	10.913 93
14	196	3.741 657	11.832 16	2 744	2.410 142	5.192 494	11.186 89
15	225	3.872 983	12.247 45	3 375	2.466 212	5.313 293	11.447 14
16	256	4.000 000	12.649 11	4 096	2.519 842	5.428 835	11.696 07
17	289	4.123 106	13.038 40	4 913	2.571 282	5.539 658	11.934 83
18	324	4.242 641	13.416 41	5 832	2.620 741	5.646 216	12.164 40
19	361	4.358 899	13.784 05	6 859	2.668 402	5.748 897	12.385 62
20	400	4.472 136	14.142 14	8 000	2.714 418	5.848 035	12.599 21
21	441	4.582 576	14.491 38	9 261	2.758 924	5.943 922	12.805 79
22	484	4.690 416	14.832 40	10 648	2.802 039	6.036 811	13.005 91
23	529	4.795 832	15.165 75	12 167	2.843 867	6.126 926	13.200 06
24	576	4.898 979	15.491 93	13 824	2.884 499	6.214 465	13.388 66
25	625	5.000 000	15.811 39	15 625	2.924 018	6.299 605	13.572 09
26	676	5.099 020	16.124 52	17 576	2.962 496	6.382 504	13.750 69
27	729	5.196 152	16.431 68	19 683	3.000 000	6.463 304	13.924 77
28	784	5.291 503	16.733 20	21 952	3.036 589	6.542 133	14.094 60
29	841	5.385 165	17.029 39	24 389	3.072 317	6.619 106	14.260 43
30	900	5.477 226	17.320 51	27 000	3.107 233	6.694 330	14.422 50
31	961	5.567 764	17.606 82	29 791	3.141 381	6.767 899	14.581 00
32	1 024	5.656 854	17.888 54	32 768	3.174 802	6.839 904	14.736 13
33	1 089	5.744 563	18.165 90	35 937	3.207 534	6.910 423	14.888 06
34	1 156	5.830 952	18.439 09	39 304	3.239 612	6.979 532	15.036 95
35	1 225	5.916 080	18.708 29	42 875	3.271 066	7.047 299	15.182 94
36	1 296	6.000 000	18.973 67	46 656	3.301 927	7.113 787	15.326 19
37	1 369	6.082 763	19.235 38	50 653	3.332 222	7.179 054	15.466 80
38	1 444	6.164 414	19.493 59	54 872	3.361 975	7.243 156	15.604 91
39	1 521	6.244 998	19.748 42	59 319	3.391 211	7.306 144	15.740 61
40	1 600	6.324 555	20.000 00	64 000	3.419 952	7.368 063	15.874 01
41	1 681	6.403 124	20.248 46	68 921	3.448 217	7.428 959	16.005 21
42	1 764	6.480 741	20.493 90	74 088	3.476 027	7.488 872	16.134 29
43	1 849	6.557 439	20.736 44	79 507	3.503 398	7.547 842	16.261 33
44	1 936	6.633 250	20.976 18	85 184	3.530 348	7.605 905	16.386 43
45	2 025	6.708 204	21.213 20	91 125	3.556 893	7.663 094	16.509 64
46	2 116	6.782 330	21.447 61	97 336	3.583 048	7.719 443	16.631 03
47	2 209	6.855 655	21.679 48	103 823	3.608 826	7.774 980	16.750 69
48	2 304	6.928 203	21.908 90	110 592	3.634 241	7.829 735	16.868 65
49	2 401	7.000 000	22.135 94	117 649	3.659 306	7.883 735	16.984 99
50	2 500	7.071 068	22.360 68	125 000	3.684 031	7.937 005	17.099 76

TABLE A5.1 *Continued*

N	N^2	$\sqrt{N}$	$\sqrt{10N}$	N^3	$\sqrt[3]{N}$	$\sqrt[3]{10N}$	$\sqrt[3]{100N}$
51	2 601	7.141 428	22.583 18	132 651	3.708 430	7.989 570	17.213 01
52	2 704	7.211 103	22.803 51	140 608	3.732 511	8.041 452	17.324 78
53	2 809	7.280 110	23.021 73	148 877	3.756 286	8.092 672	17.435 13
54	2 916	7.348 469	23.237 90	157 464	3.779 763	8.143 253	17.544 11
55	3 025	7.416 198	23.452 08	166 375	3.802 952	8.193 213	17.651 74
56	3 136	7.483 315	23.664 32	175 616	3.825 862	8.242 571	17.758 08
57	3 249	7.549 834	23.874 67	185 193	3.848 501	8.291 344	17.863 16
58	3 364	7.615 773	24.083 19	195 112	3.870 877	8.339 551	17.967 02
59	3 481	7.681 146	24.289 92	205 379	3.892 996	8.387 207	18.069 69
60	3 600	7.745 967	24.494 90	216 000	3.914 868	8.434 327	18.171 21
61	3 721	7.810 250	24.698 18	226 981	3.936 497	8.480 926	18.271 60
62	3 844	7.874 008	24.899 80	238 328	3.957 892	8.527 019	18.370 91
63	3 969	7.937 254	25.099 80	250 047	3.979 057	8.572 619	18.469 15
64	4 096	8.000 000	25.298 22	262 144	4.000 000	8.617 739	18.566 36
65	4 225	8.062 258	25.495 10	274 625	4.020 726	8.662 391	18.662 56
66	4 356	8.124 038	25.690 47	287 496	4.041 240	8.706 588	18.757 77
67	4 489	8.185 353	25.884 36	300 763	4.061 548	8.750 340	18.852 04
68	4 624	8.246 211	26.076 81	314 432	4.081 655	8.793 659	18.945 36
69	4 761	8.306 624	26.267 85	328 509	4.101 566	8.836 556	19.037 78
70	4 900	8.366 600	26.457 51	343 000	4.121 285	8.879 040	19.129 31
71	5 041	8.426 150	26.645 83	357 911	4.140 818	8.921 121	19.219 97
72	5 184	8.485 281	26.832 82	373 248	4.160 168	8.962 809	19.309 79
73	5 329	8.544 004	27.018 51	389 017	4.179 339	9.004 113	19.398 77
74	5 476	8.602 325	27.202 94	405 224	4.198 336	9.045 042	19.486 95
75	5 625	8.660 254	27.386 13	421 875	4.217 163	9.085 603	19.574 34
76	5 776	8.717 798	27.568 10	438 976	4.235 824	9.125 805	19.660 95
77	5 929	8.774 964	27.748 87	456 533	4.254 321	9.165 656	19.746 81
78	6 084	8.831 761	27.928 48	474 552	4.272 659	9.205 164	19.831 92
79	6 241	8.888 194	28.106 94	493 039	4.290 840	9.244 335	19.916 32
80	6 400	8.944 272	28.284 27	512 000	4.308 869	9.283 178	20.000 00
81	6 561	9.000 000	28.460 50	531 441	4.326 749	9.321 698	20.082 99
82	6 724	9.055 385	28.635 64	551 368	4.344 481	9.359 902	20.165 30
83	6 889	9.110 434	28.809 72	571 787	4.362 071	9.397 796	20.246 94
84	7 056	9.165 151	28.982 75	592 704	4.379 519	9.435 388	20.327 93
85	7 225	9.219 544	29.154 76	614 125	4.396 830	9.472 682	20.408 28
86	7 396	9.273 618	29.325 76	636 056	4.414 005	9.509 685	20.488 00
87	7 569	9.327 379	29.495 76	658 503	4.431 048	9.546 403	20.567 10
88	7 744	9.380 832	29.664 79	681 472	4.447 960	9.582 840	20.645 60
89	7 921	9.433 981	29.832 87	704 969	4.464 745	9.619 002	20.723 51
90	8 100	9.486 833	30.000 00	729 000	4.481 405	9.654 894	20.800 84
91	8 281	9.539 392	30.166 21	753 571	4.497 941	9.690 521	20.877 59
92	8 464	9.591 663	30.331 50	778 688	4.514 357	9.725 888	20.953 79
93	8 649	9.643 651	30.495 90	804 357	4.530 655	9.761 000	21.029 44
94	8 836	9.695 360	30.659 42	830 584	4.546 836	9.795 861	21.104 54
95	9 025	9.746 794	30.822 07	857 375	4.562 903	9.830 476	21.179 12
96	9 216	9.797 959	30.983 87	884 736	4.578 857	9.864 848	21.253 17
97	9 409	9.848 858	31.144 82	912 673	4.594 701	9.898 983	21.326 71
98	9 604	9.899 495	31.304 95	941 192	4.610 436	9.932 884	21.399 75
99	9 801	9.949 874	31.464 27	970 299	4.626 065	9.966 555	21.472 29
100	10 000	10.000 000	31.622 78	1 000 000	4.641 589	10.000 000	21.544 35

TABLE A5.2 BASE 10 (COMMON) LOGARITHMS OF NUMBERS

N	0	1	2	3	4	5	6	7	8	9
10	0000	0043	0086	0128	0170	0212	0253	0294	0334	0374
11	0414	0453	0492	0531	0569	0607	0645	0682	0719	0755
12	0792	0828	0864	0899	0934	0969	1004	1038	1072	1106
13	1139	1173	1206	1239	1271	1303	1335	1367	1399	1430
14	1461	1492	1523	1553	1584	1614	1644	1673	1703	1732
15	1761	1790	1818	1847	1875	1903	1931	1959	1987	2014
16	2041	2068	2095	2122	2148	2175	2201	2227	2253	2279
17	2304	2330	2355	2380	2405	2430	2455	2480	2504	2529
18	2553	2577	2601	2625	2648	2672	2695	2718	2742	2765
19	2788	2810	2833	2856	2878	2900	2923	2945	2967	2989
20	3010	3032	3054	3075	3096	3118	3139	3160	3181	3201
21	3222	3243	3263	3284	3304	3324	3345	3365	3385	3404
22	3424	3444	3464	3483	3502	3522	3541	3560	3579	3598
23	3617	3636	3655	3674	3692	3711	3729	3747	3766	3784
24	3802	3820	3838	3856	3874	3892	3909	3927	3945	3962
25	3979	3997	4014	4031	4048	4065	4082	4099	4116	4133
26	4150	4166	4183	4200	4216	4232	4249	4265	4281	4298
27	4314	4330	4346	4362	4378	4393	4409	4425	4440	4456
28	4472	4487	4502	4518	4533	4548	4564	4579	4594	4609
29	4624	4639	4654	4669	4683	4698	4713	4728	4742	4757
30	4771	4786	4800	4814	4829	4843	4857	4871	4886	4900
31	4914	4928	4942	4955	4969	4983	4997	5011	5024	5038
32	5051	5065	5079	5092	5105	5119	5132	5145	5159	5172
33	5185	5198	5211	5224	5237	5250	5263	5276	5289	5302
34	5315	5328	5340	5353	5366	5378	5391	5403	5416	5428
35	5441	5453	5465	5478	5490	5502	5514	5527	5539	5551
36	5563	5575	5587	5599	5611	5623	5635	5647	5658	5670
37	5682	5694	5705	5717	5729	5740	5752	5763	5775	5786
38	5798	5809	5821	5832	5843	5855	5866	5877	5888	5899
39	5911	5922	5933	5944	5955	5966	5977	5988	5999	6010
40	6021	6031	6042	6053	6064	6075	6085	6096	6107	6117
41	6128	6138	6149	6160	6170	6180	6191	6201	6212	6222
42	6232	6243	6253	6263	6274	6284	6294	6304	6314	6325
43	6335	6345	6355	6365	6375	6385	6395	6405	6415	6425
44	6435	6444	6454	6464	6474	6484	6493	6503	6513	6522
45	6532	6542	6551	6561	6571	6580	6590	6599	6609	6618
46	6628	6637	6646	6656	6665	6675	6684	6693	6702	6712
47	6721	6730	6739	6749	6758	6767	6776	6785	6794	6803
48	6812	6821	6830	6839	6848	6857	6866	6875	6884	6893
49	6902	6911	6920	6928	6937	6946	6955	6964	6972	6981
N	0	1	2	3	4	5	6	7	8	9

TABLE A5.2 *Continued*

N	0	1	2	3	4	5	6	7	8	9
50	6990	6998	7007	7016	7024	7033	7042	7050	7059	7067
51	7076	7084	7093	7101	7110	7118	7126	7135	7143	7152
52	7160	7168	7177	7185	7193	7202	7210	7218	7226	7235
53	7243	7251	7259	7267	7275	7284	7292	7300	7308	7316
54	7324	7332	7340	7348	7356	7364	7372	7380	7388	7396
55	7404	7412	7419	7427	7435	7443	7451	7459	7466	7474
56	7482	7490	7497	7505	7513	7520	7528	7536	7543	7551
57	7559	7566	7574	7582	7589	7597	7604	7612	7619	7627
58	7634	7642	7649	7657	7664	7672	7679	7686	7694	7701
59	7709	7716	7723	7731	7738	7745	7752	7760	7767	7774
60	7782	7789	7796	7803	7810	7818	7825	7832	7839	7846
61	7853	7860	7868	7875	7882	7889	7896	7903	7910	7917
62	7924	7931	7938	7945	7952	7959	7966	7973	7980	7987
63	7993	8000	8007	8014	8021	8028	8035	8041	8048	8055
64	8062	8069	8075	8082	8089	8096	8102	8109	8116	8122
65	8129	8136	8142	8149	8156	8162	8169	8176	8182	8189
66	8195	8202	8209	8215	8222	8228	8235	8241	8248	8254
67	8261	8267	8274	8280	8287	8293	8299	8306	8312	8319
68	8325	8331	8338	8344	8351	8357	8363	8370	8376	8382
69	8388	8395	8401	8407	8414	8420	8426	8432	8439	8445
70	8451	8457	8463	8470	8476	8482	8488	8494	8500	8506
71	8513	8519	8525	8531	8537	8543	8549	8555	8561	8567
72	8573	8579	8585	8591	8597	8603	8609	8615	8621	8627
73	8633	8639	8645	8651	8657	8663	8669	8675	8681	8686
74	8692	8698	8704	8710	8716	8722	8727	8733	8739	8745
75	8751	8756	8762	8768	8774	8779	8785	8791	8797	8802
76	8808	8814	8820	8825	8831	8837	8842	8848	8854	8859
77	8865	8871	8876	8882	8887	8893	8899	8904	8910	8915
78	8921	8927	8932	8938	8943	8949	8954	8960	8965	8971
79	8976	8982	8987	8993	8998	9004	9009	9015	9020	9025
80	9031	9036	9042	9047	9053	9058	9063	9069	9074	9079
81	9085	9090	9096	9101	9106	9112	9117	9122	9128	9133
82	9138	9143	9149	9154	9159	9165	9170	9175	9180	9186
83	9191	9196	9201	9206	9212	9217	9222	9227	9232	9238
84	9243	9248	9253	9258	9263	9269	9274	9279	9284	9289
85	9294	9299	9304	9309	9315	9320	9325	9330	9335	9340
86	9345	9350	9355	9360	9365	9370	9375	9380	9385	9390
87	9395	9400	9405	9410	9415	9420	9425	9430	9435	9440
88	9445	9450	9455	9460	9465	9469	9474	9479	9484	9489
89	9494	9499	9504	9509	9513	9518	9523	9528	9533	9538
90	9542	9547	9552	9557	9562	9566	9571	9576	9581	9586
91	9590	9595	9600	9605	9609	9614	9619	9624	9628	9633
92	9638	9643	9647	9652	9657	9661	9666	9671	9675	9680
93	9685	9689	9694	9699	9703	9708	9713	9717	9722	9727
94	9731	9736	9741	9745	9750	9754	9759	9763	9768	9773
95	9777	9782	9786	9791	9795	9800	9805	9809	9814	9818
96	9823	9827	9832	9836	9841	9845	9850	9854	9859	9863
97	9868	9872	9877	9881	9886	9890	9894	9899	9903	9908
98	9912	9917	9921	9926	9930	9934	9939	9943	9948	9952
99	9956	9961	9965	9969	9974	9978	9983	9987	9991	9996
N	0	1	2	3	4	5	6	7	8	9

TABLE A5.3 NATURAL LOGARITHMS OF NUMBERS 0.01 TO 10.09†

(Base $e = 2.71828\ 18284\ 59045\ldots$)

† For each entry marked with an asterisk, the first two digits are those at the left end of the *next lower line*. For example, ln 1.84 = 0.6098.

If given number $n = N \times 10^k$, then $\log_e n = \log_e N + k \log_e 10$. Find $k \log_e 10$ from the following table:

Multiples of $\log_e$ *10*

$\log_e 10 = 2.3026$	$-\log_e 10 = 7.6974 - 10$
$2 \log_e 10 = 4.6052$	$-2 \log_e 10 = 5.3948 - 10$
$3 \log_e 10 = 6.9078$	$-3 \log_e 10 = 3.0922 - 10$
$4 \log_e 10 = 9.2103$	$-4 \log_e 10 = 0.7897 - 10$
$5 \log_e 10 = 11.5129$	$-5 \log_e 10 = 8.4871 - 20$

N	0	1	2	3	4	5	6	7	8	9
1.0	0.0 000	100	198	296	392	488	583	677	770	862
1.1	953	*044	*133	*222	*310	*398	*484	*570	*655	*740
1.2	0.1 823	906	989	*070	*151	*231	*311	*390	*469	*546
1.3	0.2 624	700	776	852	927	*001	*075	*148	*221	*293
1.4	0.3 365	436	507	577	646	716	784	853	920	988
1.5	0.4 055	121	187	253	318	383	447	511	574	637
1.6	700	762	824	886	947	*008	*068	*128	*188	*247
1.7	0.5 306	365	423	481	539	596	653	710	766	822
1.8	878	933	988	*043	*098	*152	*206	*259	*313	*366
1.9	0.6 419	471	523	575	627	678	729	780	831	881
2.0	931	981	*031	*080	*129	*178	*227	*275	*324	*372
2.1	0.7 419	467	514	561	608	655	701	747	793	839
2.2	885	930	975	*020	*065	*109	*154	*198	*242	*286
2.3	0.8 329	372	416	459	502	544	587	629	671	713
2.4	755	796	838	879	920	961	*002	*042	*083	*123
2.5	0.9 163	203	243	282	322	361	400	439	478	517
2.6	555	594	632	670	708	746	783	821	858	895
2.7	933	969	*006	*043	*080	*116	*152	*188	*225	*260
2.8	1.0 296	332	367	403	438	473	508	543	578	613
2.9	647	682	716	750	784	818	852	886	919	953
3.0	986	*019	*053	*086	*119	*151	*184	*217	*249	*282
3.1	1.1 314	346	378	410	442	474	506	537	569	600
3.2	632	663	694	725	756	787	817	848	878	909
3.3	939	969	*000	*030	*060	*090	*119	*149	*179	*208
3.4	1.2 238	267	296	326	355	384	413	442	470	499
3.5	528	556	585	613	641	669	698	726	754	782
3.6	809	837	865	892	920	947	975	*002	*029	*056
3.7	1.3 083	110	137	164	191	218	244	271	297	324
3.8	350	376	402	429	455	481	507	533	558	584
3.9	610	635	661	686	712	737	762	788	813	838

TABLE A5.3 *Continued*

N	0	1	2	3	4	5	6	7	8	9
4.0	863	888	913	938	962	987	*012	*036	*061	*085
4.1	1.4 110	134	159	183	207	231	255	279	303	327
4.2	351	375	398	422	446	469	493	516	540	563
4.3	586	609	633	656	679	702	725	748	770	793
4.4	816	839	861	884	907	929	951	974	996	*019
4.5	1.5 041	063	085	107	129	151	173	195	217	239
4.6	261	282	304	326	347	369	390	412	433	454
4.7	476	497	518	539	560	581	602	623	644	665
4.8	686	707	728	748	769	790	810	831	851	872
4.9	892	913	933	953	974	994	*014	*034	*054	*074
5.0	1.6 094	114	134	154	174	194	214	233	253	273
5.1	292	312	332	351	371	390	409	429	448	467
5.2	487	506	525	544	563	582	601	620	639	658
5.3	677	696	715	734	752	771	790	808	827	845
5.4	864	882	901	919	938	956	974	993	*011	*029
5.5	1.7 047	066	084	102	120	138	156	174	192	210
5.6	228	246	263	281	299	317	334	352	370	387
5.7	405	422	440	457	475	492	509	527	544	561
5.8	579	596	613	630	647	664	681	699	716	733
5.9	750	766	783	800	817	834	851	867	884	901
6.0	918	934	951	967	984	*001	*017	*034	*050	*066
6.1	1.8 083	099	116	132	148	165	181	197	213	229
6.2	245	262	278	294	310	326	342	358	374	390
6.3	405	421	437	453	469	485	500	516	532	547
6.4	563	579	594	610	625	641	656	672	687	703
6.5	718	733	749	764	779	795	810	825	840	856
6.6	871	886	901	916	931	946	961	976	991	*006
6.7	1.9 021	036	051	066	081	095	110	125	140	155
6.8	169	184	199	213	228	242	257	272	286	301
6.9	315	330	344	359	373	387	402	416	430	445
7.0	459	473	488	502	516	530	544	559	573	587
7.1	601	615	629	643	657	671	685	699	713	727
7.2	741	755	769	782	796	810	824	838	851	865
7.3	879	892	906	920	933	947	961	974	988	*001
7.4	2.0 015	028	042	055	069	082	096	109	122	136
7.5	149	162	176	189	202	215	229	242	255	268
7.6	281	295	308	321	334	347	360	373	386	399
7.7	412	425	438	451	464	477	490	503	516	528
7.8	541	554	567	580	592	605	618	631	643	656
7.9	669	681	694	707	719	732	744	757	769	782

TABLE A5.3 *Continued*

N	0	1	2	3	4	5	6	7	8	9
8.0	794	807	819	832	844	857	869	882	894	906
8.1	919	931	943	956	968	980	992	*005	*017	*029
8.2	2.1 041	054	066	078	090	102	114	126	138	150
8.3	163	175	187	199	211	223	235	247	258	270
8.4	282	294	306	318	330	342	353	365	377	389
8.5	401	412	424	436	448	459	471	483	494	506
8.6	518	529	541	552	564	576	587	599	610	622
8.7	633	645	656	668	679	691	702	713	725	736
8.8	748	759	770	782	793	804	815	827	838	849
8.9	861	872	883	894	905	917	928	939	950	961
9.0	972	983	994	*006	*017	*028	*039	*050	*061	*072
9.1	2.2 083	094	105	116	127	138	148	159	170	181
9.2	192	203	214	225	235	246	257	268	279	289
9.3	300	311	322	332	343	354	364	375	386	396
9.4	407	418	428	439	450	460	471	481	492	502
9.5	513	523	534	544	555	565	576	586	597	607
9.6	618	628	638	649	659	670	680	690	701	711
9.7	721	732	742	752	762	773	783	793	803	814
9.8	824	834	844	854	865	875	885	895	905	915
9.9	925	935	946	956	966	976	986	996	*006	*016
10.0	2.3 026	036	046	056	066	076	086	096	106	115

TABLE A5.4 TRIGONOMETRIC FUNCTIONS OF ANGLES IN DEGREES-MINUTES AND RADIANS

ANGLE DEG-MIN	ANGLE RADIANS	SINE	TANGENT	SECANT	COSECANT	COTANGENT	COSINE		
0°00′	.0000	.0000	.0000	1.000	–	–	1.0000	1.5708	90°00′
10	.0029	.0029	.0029	1.000	343.8	343.8	1.0000	1.5679	50
20	.0058	.0058	.0058	1.000	171.9	171.9	1.0000	1.5650	40
30	.0087	.0087	.0087	1.000	114.6	114.6	1.0000	1.5621	30
40	.0116	.0116	.0116	1.000	85.95	85.94	.9999	1.5592	20
50	.0145	.0145	.0145	1.000	68.76	68.75	.9999	1.5563	10
1°00′	.0175	.0175	.0175	1.000	57.30	57.29	.9998	1.5533	89°00′
10	.0204	.0204	.0204	1.000	49.11	49.10	.9998	1.5504	50
20	.0233	.0233	.0233	1.000	42.98	42.96	.9997	1.5475	40
30	.0262	.0262	.0262	1.000	38.20	38.19	.9997	1.5446	30
40	.0291	.0291	.0291	1.000	34.38	34.37	.9996	1.5417	20
50	.0320	.0320	.0320	1.001	31.26	31.24	.9995	1.5388	10
2°00′	.0349	.0349	.0349	1.001	28.65	28.64	.9994	1.5359	88°00′
10	.0378	.0378	.0378	1.001	26.45	26.43	.9993	1.5330	50
20	.0407	.0407	.0407	1.001	24.56	24.54	.9992	1.5301	40
30	.0436	.0436	.0437	1.001	22.93	22.90	.9990	1.5272	30
40	.0465	.0465	.0466	1.001	21.49	21.47	.9989	1.5243	20
50	.0495	.0494	.0495	1.001	20.23	20.21	.9988	1.5213	10
3°00′	.0524	.0523	.0524	1.001	19.11	19.08	.9986	1.5184	87°00′
10	.0553	.0552	.0553	1.002	18.10	18.07	.9985	1.5155	50
20	.0582	.0581	.0582	1.002	17.20	17.17	.9983	1.5126	40
30	.0611	.0610	.0612	1.002	16.38	16.35	.9981	1.5097	30
40	.0640	.0640	.0641	1.002	15.64	15.60	.9980	1.5068	20
50	.0669	.0669	.0670	1.002	14.96	14.92	.9978	1.5039	10
4°00′	.0698	.0698	.0699	1.002	14.34	14.30	.9976	1.5010	86°00′
10	.0727	.0727	.0729	1.003	13.76	13.73	.9974	1.4981	50
20	.0756	.0756	.0758	1.003	13.23	13.20	.9971	1.4952	40
30	.0785	.0785	.0787	1.003	12.75	12.71	.9969	1.4923	30
40	.0814	.0814	.0816	1.003	12.29	12.25	.9967	1.4893	20
50	.0844	.0843	.0846	1.004	11.87	11.83	.9964	1.4864	10
5°00′	.0873	.0872	.0875	1.004	11.47	11.43	.9962	1.4835	85°00′
10	.0902	.0901	.0904	1.004	11.10	11.06	.9959	1.4806	50
20	.0931	.0929	.0934	1.004	10.76	10.71	.9957	1.4777	40
30	.0960	.0958	.0963	1.005	10.43	10.39	.9954	1.4748	30
40	.0989	.0987	.0992	1.005	10.13	10.08	.9951	1.4719	20
50	.1018	.1016	.1022	1.005	9.839	9.788	.9948	1.4690	10
6°00′	.1047	.1045	.1051	1.006	9.567	9.514	.9945	1.4661	84°00′
		COSINE	COTANGENT	COSECANT	SECANT	TANGENT	SINE	RADIANS	DEG-MIN ANGLE

TABLE A5.4 *Continued*

ANGLE DEG-MIN	ANGLE RADIANS	SINE	TANGENT	SECANT	COSECANT	COTANGENT	COSINE		
6°00′	.1047	.1045	.1051	1.006	9.567	9.514	.9945	1.4661	84°00′
10	.1076	.1074	.1080	1.006	9.309	9.255	.9942	1.4632	50
20	.1105	.1103	.1110	1.006	9.065	9.010	.9939	1.4603	40
30	.1134	.1132	.1139	1.006	8.834	8.777	.9936	1.4573	30
40	.1164	.1161	.1169	1.007	8.614	8.556	.9932	1.4544	20
50	.1193	.1190	.1198	1.007	8.405	8.345	.9929	1.4515	10
7°00′	.1222	.1219	.1228	1.008	8.206	8.144	.9925	1.4486	83°00′
10	.1251	.1248	.1257	1.008	8.016	7.953	.9922	1.4457	50
20	.1280	.1276	.1287	1.008	7.834	7.770	.9918	1.4428	40
30	.1309	.1305	.1317	1.009	7.661	7.596	.9914	1.4399	30
40	.1338	.1334	.1346	1.009	7.496	7.429	.9911	1.4370	20
50	.1367	.1363	.1376	1.009	7.337	7.269	.9907	1.4341	10
8°00′	.1396	.1392	.1405	1.010	7.185	7.115	.9903	1.4312	82°00′
10	.1425	.1421	.1435	1.010	7.040	6.968	.9899	1.4283	50
20	.1454	.1449	.1465	1.011	6.900	6.827	.9894	1.4254	40
30	.1484	.1478	.1495	1.011	6.765	6.691	.9890	1.4224	30
40	.1513	.1507	.1524	1.012	6.636	6.561	.9886	1.4195	20
50	.1542	.1536	.1554	1.012	6.512	6.435	.9881	1.4166	10
9°00′	.1571	.1564	.1584	1.012	6.392	6.314	.9877	1.4137	81°00′
10	.1600	.1593	.1614	1.013	6.277	6.197	.9872	1.4108	50
20	.1629	.1622	.1644	1.013	6.166	6.084	.9868	1.4079	40
30	.1658	.1650	.1673	1.014	6.059	5.976	.9863	1.4050	30
40	.1687	.1679	.1703	1.014	5.955	5.871	.9858	1.4021	20
50	.1716	.1708	.1733	1.015	5.855	5.769	.9853	1.3992	10
10°00′	.1745	.1736	.1763	1.015	5.759	5.671	.9848	1.3963	80°00′
10	.1774	.1765	.1793	1.016	5.665	5.576	.9843	1.3934	50
20	.1804	.1794	.1823	1.016	5.575	5.485	.9838	1.3904	40
30	.1833	.1822	.1853	1.017	5.487	5.396	.9833	1.3875	30
40	.1862	.1851	.1883	1.018	5.403	5.309	.9827	1.3846	20
50	.1891	.1880	.1914	1.018	5.320	5.226	.9822	1.3817	10
11°00′	.1920	.1908	.1944	1.019	5.241	5.145	.9816	1.3788	79°00′
10	.1949	.1937	.1974	1.019	5.164	5.066	.9811	1.3759	50
20	.1978	.1965	.2004	1.020	5.089	4.989	.9805	1.3730	40
30	.2007	.1994	.2035	1.020	5.016	4.915	.9799	1.3701	30
40	.2036	.2022	.2065	1.021	4.945	4.843	.9793	1.3672	20
50	.2065	.2051	.2095	1.022	4.876	4.773	.9787	1.3643	10
12°00′	.2094	.2079	.2126	1.022	4.810	4.705	.9781	1.3614	78°00′
		COSINE	COTANGENT	COSECANT	SECANT	TANGENT	SINE	ANGLE RADIANS	ANGLE DEG-MIN

TABLE A5.4 *Continued*

ANGLE DEG-MIN	ANGLE RADIANS	SINE	TANGENT	SECANT	COSECANT	COTANGENT	COSINE		
12°00′	.2094	.2079	.2126	1.022	4.810	4.705	.9781	1.3614	78°00′
10	.2123	.2108	.2156	1.023	4.745	4.638	.9775	1.3584	50
20	.2153	.2136	.2186	1.024	4.682	4.574	.9769	1.3555	40
30	.2182	.2164	.2217	1.024	4.620	4.511	.9763	1.3526	30
40	.2211	.2193	.2247	1.025	4.560	4.449	.9757	1.3497	20
50	.2240	.2221	.2278	1.026	4.502	4.390	.9750	1.3468	10
13°00′	.2269	.2250	.2309	1.026	4.445	4.331	.9744	1.3439	77°00′
10	.2298	.2278	.2339	1.027	4.390	4.275	.9737	1.3410	50
20	.2327	.2306	.2370	1.028	4.336	4.219	.9730	1.3381	40
30	.2356	.2334	.2401	1.028	4.284	4.165	.9724	1.3352	30
40	.2385	.2363	.2432	1.029	4.232	4.113	.9717	1.3323	20
50	.2414	.2391	.2462	1.030	4.182	4.061	.9710	1.3294	10
14°00′	.2443	.2419	.2493	1.031	4.134	4.011	.9703	1.3265	76°00′
10	.2473	.2447	.2524	1.031	4.086	3.962	.9696	1.3235	50
20	.2502	.2476	.2555	1.032	4.039	3.914	.9689	1.3206	40
30	.2531	.2504	.2586	1.033	3.994	3.867	.9681	1.3177	30
40	.2560	.2532	.2617	1.034	3.950	3.821	.9674	1.3148	20
50	.2589	.2560	.2648	1.034	3.906	3.776	.9667	1.3119	10
15°00′	.2618	.2588	.2679	1.035	3.864	3.732	.9659	1.3090	75°00′
10	.2647	.2616	.2711	1.036	3.822	3.689	.9652	1.3061	50
20	.2676	.2644	.2742	1.037	3.782	3.647	.9644	1.3032	40
30	.2705	.2672	.2773	1.038	3.742	3.606	.9636	1.3003	30
40	.2734	.2700	.2805	1.039	3.703	3.566	.9628	1.2974	20
50	.2763	.2728	.2836	1.039	3.665	3.526	.9621	1.2945	10
16°00′	.2793	.2756	.2867	1.040	3.628	3.487	.9613	1.2915	74°00′
10	.2822	.2784	.2899	1.041	3.592	3.450	.9605	1.2886	50
20	.2851	.2812	.2931	1.042	3.556	3.412	.9596	1.2857	40
30	.2880	.2840	.2962	1.043	3.521	3.376	.9588	1.2828	30
40	.2909	.2868	.2994	1.044	3.487	3.340	.9580	1.2799	20
50	.2938	.2896	.3026	1.045	3.453	3.305	.9572	1.2770	10
17°00′	.2967	.2924	.3057	1.046	3.420	3.271	.9563	1.2741	73°00′
10	.2996	.2952	.3089	1.047	3.388	3.237	.9555	1.2712	50
20	.3025	.2979	.3121	1.048	3.356	3.204	.9546	1.2683	40
30	.3054	.3007	.3153	1.049	3.326	3.172	.9537	1.2654	30
40	.3083	.3035	.3185	1.049	3.295	3.140	.9528	1.2625	20
50	.3113	.3062	.3217	1.050	3.265	3.108	.9520	1.2595	10
18°00′	.3142	.3090	.3249	1.051	3.236	3.078	.9511	1.2566	72°00′
		COSINE	COTANGENT	COSECANT	SECANT	TANGENT	SINE	RADIANS ANGLE	DEG-MIN

TABLE A5.4 *Continued*

ANGLE DEG-MIN	ANGLE RADIANS	SINE	TANGENT	SECANT	COSECANT	COTANGENT	COSINE		
18°00′	.3142	.3090	.3249	1.051	3.236	3.078	.9511	1.2566	72°00′
10	.3171	.3118	.3281	1.052	3.207	3.047	.9502	1.2537	50
20	.3200	.3145	.3314	1.053	3.179	3.018	.9492	1.2508	40
30	.3229	.3173	.3346	1.054	3.152	2.989	.9483	1.2479	30
40	.3258	.3201	.3378	1.056	3.124	2.960	.9474	1.2450	20
50	.3287	.3228	.3411	1.057	3.098	2.932	.9465	1.2421	10
19°00′	.3316	.3256	.3443	1.058	3.072	2.904	.9455	1.2392	71°00′
10	.3345	.3283	.3476	1.059	3.046	2.877	.9446	1.2363	50
20	.3374	.3311	.3508	1.060	3.021	2.850	.9436	1.2334	40
30	.3403	.3338	.3541	1.061	2.996	2.824	.9426	1.2305	30
40	.3432	.3365	.3574	1.062	2.971	2.798	.9417	1.2275	20
50	.3462	.3393	.3607	1.063	2.947	2.773	.9407	1.2246	10
20°00′	.3491	.3420	.3640	1.064	2.924	2.747	.9397	1.2217	70°00′
10	.3520	.3448	.3673	1.065	2.901	2.723	.9387	1.2188	50
20	.3549	.3475	.3706	1.066	2.878	2.699	.9377	1.2159	40
30	.3578	.3502	.3739	1.068	2.855	2.675	.9367	1.2130	30
40	.3607	.3529	.3772	1.069	2.833	2.651	.9356	1.2101	20
50	.3636	.3557	.3805	1.070	2.812	2.628	.9346	1.2072	10
21°00′	.3665	.3584	.3839	1.071	2.790	2.605	.9336	1.2043	69°00′
10	.3694	.3611	.3872	1.072	2.769	2.583	.9325	1.2014	50
20	.3723	.3638	.3906	1.074	2.749	2.560	.9315	1.1985	40
30	.3752	.3665	.3939	1.075	2.729	2.539	.9304	1.1956	30
40	.3782	.3692	.3973	1.076	2.709	2.517	.9293	1.1926	20
50	.3811	.3719	.4006	1.077	2.689	2.496	.9283	1.1897	10
22°00′	.3840	.3746	.4040	1.079	2.669	2.475	.9272	1.1868	68°00′
10	.3869	.3773	.4074	1.080	2.650	2.455	.9261	1.1839	50
20	.3898	.3800	.4108	1.081	2.632	2.434	.9250	1.1810	40
30	.3927	.3827	.4142	1.082	2.613	2.414	.9239	1.1781	30
40	.3956	.3854	.4176	1.084	2.595	2.394	.9228	1.1752	20
50	.3985	.3881	.4210	1.085	2.577	2.375	.9216	1.1723	10
23°00′	.4014	.3907	.4245	1.086	2.559	2.356	.9205	1.1694	67°00′
10	.4043	.3934	.4279	1.088	2.542	2.337	.9194	1.1665	50
20	.4072	.3961	.4314	1.089	2.525	2.318	.9182	1.1636	40
30	.4102	.3987	.4348	1.090	2.508	2.300	.9171	1.1606	30
40	.4131	.4014	.4383	1.092	2.491	2.282	.9159	1.1577	20
50	.4160	.4041	.4417	1.093	2.475	2.264	.9147	1.1548	10
24°00′	.4189	.4067	.4452	1.095	2.459	2.246	.9135	1.1519	66°00′
		COSINE	COTANGENT	COSECANT	SECANT	TANGENT	SINE	RADIANS ANGLE	DEG-MIN ANGLE

TABLE A5.4 *Continued*

ANGLE DEG-MIN	ANGLE RADIANS	SINE	TANGENT	SECANT	COSECANT	COTANGENT	COSINE		
24°00′	.4189	.4067	.4452	1.095	2.459	2.246	.9135	1.1519	66°00′
10	.4218	.4094	.4487	1.096	2.443	2.229	.9124	1.1490	50
20	.4247	.4120	.4522	1.097	2.427	2.211	.9112	1.1461	40
30	.4276	.4147	.4557	1.099	2.411	2.194	.9100	1.1432	30
40	.4305	.4173	.4592	1.100	2.396	2.177	.9088	1.1403	20
50	.4334	.4200	.4628	1.102	2.381	2.161	.9075	1.1374	10
25°00′	.4363	.4226	.4663	1.103	2.366	2.145	.9063	1.1345	65°00′
10	.4392	.4253	.4699	1.105	2.352	2.128	.9051	1.1316	50
20	.4422	.4279	.4734	1.106	2.337	2.112	.9038	1.1286	40
30	.4451	.4305	.4770	1.108	2.323	2.097	.9026	1.1257	30
40	.4480	.4331	.4806	1.109	2.309	2.081	.9013	1.1228	20
50	.4509	.4358	.4841	1.111	2.295	2.066	.9001	1.1199	10
26°00′	.4538	.4384	.4877	1.113	2.281	2.050	.8988	1.1170	64°00′
10	.4567	.4410	.4913	1.114	2.268	2.035	.8975	1.1141	50
20	.4596	.4436	.4950	1.116	2.254	2.020	.8962	1.1112	40
30	.4625	.4462	.4986	1.117	2.241	2.006	.8949	1.1083	30
40	.4654	.4488	.5022	1.119	2.228	1.991	.8936	1.1054	20
50	.4683	.4514	.5059	1.121	2.215	1.977	.8923	1.1025	10
27°00′	.4712	.4540	.5095	1.122	2.203	1.963	.8910	1.0996	63°00′
10	.4741	.4566	.5132	1.124	2.190	1.949	.8897	1.0966	50
20	.4771	.4592	.5169	1.126	2.178	1.935	.8884	1.0937	40
30	.4800	.4617	.5206	1.127	2.166	1.921	.8870	1.0908	30
40	.4829	.4643	.5243	1.129	2.154	1.907	.8857	1.0879	20
50	.4858	.4669	.5280	1.131	2.142	1.894	.8843	1.0850	10
28°00′	.4887	.4695	.5317	1.133	2.130	1.881	.8829	1.0821	62°00′
10	.4916	.4720	.5354	1.134	2.118	1.868	.8816	1.0792	50
20	.4945	.4746	.5392	1.136	2.107	1.855	.8802	1.0763	40
30	.4974	.4772	.5430	1.138	2.096	1.842	.8788	1.0734	30
40	.5003	.4797	.5467	1.140	2.085	1.829	.8774	1.0705	20
50	.5032	.4823	.5505	1.142	2.074	1.816	.8760	1.0676	10
29°00′	.5061	.4848	.5543	1.143	2.063	1.804	.8746	1.0647	61°00′
10	.5091	.4874	.5581	1.145	2.052	1.792	.8732	1.0617	50
20	.5120	.4899	.5619	1.147	2.041	1.780	.8718	1.0588	40
30	.5149	.4924	.5658	1.149	2.031	1.767	.8704	1.0559	30
40	.5178	.4950	.5696	1.151	2.020	1.756	.8689	1.0530	20
50	.5207	.4975	.5735	1.153	2.010	1.744	.8675	1.0501	10
30°00′	.5236	.5000	.5774	1.155	2.000	1.732	.8660	1.0472	60°00′
		COSINE	COTANGENT	COSECANT	SECANT	TANGENT	SINE	RADIANS	DEG-MIN ANGLE

TABLE A5.4 *Continued*

ANGLE DEG-MIN	ANGLE RADIANS	SINE	TANGENT	SECANT	COSECANT	COTANGENT	COSINE		
30°00′	.5236	.5000	.5774	1.155	2.000	1.732	.8660	1.0472	60°00′
10	.5265	.5025	.5812	1.157	1.990	1.720	.8646	1.0443	50
20	.5294	.5050	.5851	1.159	1.980	1.709	.8631	1.0414	40
30	.5323	.5075	.5890	1.161	1.970	1.698	.8616	1.0385	30
40	.5352	.5100	.5930	1.163	1.961	1.686	.8601	1.0356	20
50	.5381	.5125	.5969	1.165	1.951	1.675	.8587	1.0327	10
31°00′	.5411	.5150	.6009	1.167	1.942	1.664	.8572	1.0297	59°00′
10	.5440	.5175	.6048	1.169	1.932	1.653	.8557	1.0268	50
20	.5469	.5200	.6088	1.171	1.923	1.643	.8542	1.0239	40
30	.5498	.5225	.6128	1.173	1.914	1.632	.8526	1.0210	30
40	.5527	.5250	.6168	1.175	1.905	1.621	.8511	1.0181	20
50	.5556	.5275	.6208	1.177	1.896	1.611	.8496	1.0152	10
32°00′	.5585	.5299	.6249	1.179	1.887	1.600	.8480	1.0123	58°00′
10	.5614	.5324	.6289	1.181	1.878	1.590	.8465	1.0094	50
20	.5643	.5348	.6330	1.184	1.870	1.580	.8450	1.0065	40
30	.5672	.5373	.6371	1.186	1.861	1.570	.8434	1.0036	30
40	.5701	.5398	.6412	1.188	1.853	1.560	.8418	1.0007	20
50	.5730	.5422	.6453	1.190	1.844	1.550	.8403	.9977	10
33°00′	.5760	.5446	.6494	1.192	1.836	1.540	.8387	.9948	57°00′
10	.5789	.5471	.6536	1.195	1.828	1.530	.8371	.9919	50
20	.5818	.5495	.6577	1.197	1.820	1.520	.8355	.9890	40
30	.5847	.5519	.6619	1.199	1.812	1.511	.8339	.9861	30
40	.5876	.5544	.6661	1.202	1.804	1.501	.8323	.9832	20
50	.5905	.5568	.6703	1.204	1.796	1.492	.8307	.9803	10
34°00′	.5934	.5592	.6745	1.206	1.788	1.483	.8290	.9774	56°00′
10	.5963	.5616	.6787	1.209	1.781	1.473	.8274	.9745	50
20	.5992	.5640	.6830	1.211	1.773	1.464	.8258	.9716	40
30	.6021	.5664	.6873	1.213	1.766	1.455	.8241	.9687	30
40	.6050	.5688	.6916	1.216	1.758	1.446	.8225	.9657	20
50	.6080	.5712	.6959	1.218	1.751	1.437	.8208	.9628	10
35°00′	.6109	.5736	.7002	1.221	1.743	1.428	.8192	.9599	55°00′
10	.6138	.5760	.7046	1.223	1.736	1.419	.8175	.9570	50
20	.6167	.5783	.7089	1.226	1.729	1.411	.8158	.9541	40
30	.6196	.5807	.7133	1.228	1.722	1.402	.8141	.9512	30
40	.6225	.5831	.7177	1.231	1.715	1.393	.8124	.9483	20
50	.6254	.5854	.7221	1.233	1.708	1.385	.8107	.9454	10
36°00′	.6283	.5878	.7265	1.236	1.701	1.376	.8090	.9425	54°00′
		COSINE	COTANGENT	COSECANT	SECANT	TANGENT	SINE	RADIANS ANGLE	DEG-MIN ANGLE

TABLE A5.4 *Continued*

ANGLE DEG-MIN	ANGLE RADIANS	SINE	TANGENT	SECANT	COSECANT	COTANGENT	COSINE		
36°00′	.6283	.5878	.7265	1.236	1.701	1.376	.8090	.9425	54°00′
10	.6312	.5901	.7310	1.239	1.695	1.368	.8073	.9396	50
20	.6341	.5925	.7355	1.241	1.688	1.360	.8056	.9367	40
30	.6370	.5948	.7400	1.244	1.681	1.351	.8039	.9338	30
40	.6400	.5972	.7445	1.247	1.675	1.343	.8021	.9308	20
50	.6429	.5995	.7490	1.249	1.668	1.335	.8004	.9279	10
37°00′	.6458	.6018	.7536	1.252	1.662	1.327	.7986	.9250	53°00′
10	.6487	.6041	.7581	1.255	1.655	1.319	.7969	.9221	50
20	.6516	.6065	.7627	1.258	1.649	1.311	.7951	.9192	40
30	.6545	.6088	.7673	1.260	1.643	1.303	.7934	.9163	30
40	.6574	.6111	.7720	1.263	1.636	1.295	.7916	.9134	20
50	.6603	.6134	.7766	1.266	1.630	1.288	.7898	.9105	10
38°00′	.6632	.6157	.7813	1.269	1.624	1.280	.7880	.9076	52°00′
10	.6661	.6180	.7860	1.272	1.618	1.272	.7862	.9047	50
20	.6690	.6202	.7907	1.275	1.612	1.265	.7844	.9018	40
30	.6720	.6225	.7954	1.278	1.606	1.257	.7826	.8988	30
40	.6749	.6248	.8002	1.281	1.601	1.250	.7808	.8959	20
50	.6778	.6271	.8050	1.284	1.595	1.242	.7790	.8930	10
39°00′	.6807	.6293	.8098	1.287	1.589	1.235	.7771	.8901	51°00′
10	.6836	.6316	.8146	1.290	1.583	1.228	.7753	.8872	50
20	.6865	.6338	.8195	1.293	1.578	1.220	.7735	.8843	40
30	.6894	.6361	.8243	1.296	1.572	1.213	.7716	.8814	30
40	.6923	.6383	.8292	1.299	1.567	1.206	.7698	.8785	20
50	.6952	.6406	.8342	1.302	1.561	1.199	.7679	.8756	10
40°00′	.6981	.6428	.8391	1.305	1.556	1.192	.7660	.8727	50°00′
10	.7010	.6450	.8441	1.309	1.550	1.185	.7642	.8698	50
20	.7039	.6472	.8491	1.312	1.545	1.178	.7623	.8668	40
30	.7069	.6494	.8541	1.315	1.540	1.171	.7604	.8639	30
40	.7098	.6517	.8591	1.318	1.535	1.164	.7585	.8610	20
50	.7127	.6539	.8642	1.322	1.529	1.157	.7566	.8581	10
41°00′	.7156	.6561	.8693	1.325	1.524	1.150	.7547	.8552	49°00′
10	.7185	.6583	.8744	1.328	1.519	1.144	.7528	.8523	50
20	.7214	.6604	.8796	1.332	1.514	1.137	.7509	.8494	40
30	.7243	.6626	.8847	1.335	1.509	1.130	.7490	.8465	30
40	.7272	.6648	.8899	1.339	1.504	1.124	.7470	.8436	20
50	.7301	.6670	.8952	1.342	1.499	1.117	.7451	.8407	10
42°00′	.7330	.6691	.9004	1.346	1.494	1.111	.7431	.8378	48°00′
		COSINE	COTANGENT	COSECANT	SECANT	TANGENT	SINE	RADIANS	DEG-MIN ANGLE

TABLE A5.4 *Continued*

ANGLE DEG-MIN	ANGLE RADIANS	SINE	TANGENT	SECANT	COSECANT	COTANGENT	COSINE		
42°00′	.7330	.6691	.9004	1.346	1.494	1.111	.7431	.8378	48°00′
10	.7359	.6713	.9057	1.349	1.490	1.104	.7412	.8348	50
20	.7389	.6734	.9110	1.353	1.485	1.098	.7392	.8319	40
30	.7418	.6756	.9163	1.356	1.480	1.091	.7373	.8290	30
40	.7447	.6777	.9217	1.360	1.476	1.085	.7353	.8261	20
50	.7476	.6799	.9271	1.364	1.471	1.079	.7333	.8232	10
43°00′	.7505	.6820	.9325	1.367	1.466	1.072	.7314	.8203	47°00′
10	.7534	.6841	.9380	1.371	1.462	1.066	.7294	.8174	50
20	.7563	.6862	.9435	1.375	1.457	1.060	.7274	.8145	40
30	.7592	.6884	.9490	1.379	1.453	1.054	.7254	.8116	30
40	.7621	.6905	.9545	1.382	1.448	1.048	.7234	.8087	20
50	.7650	.6926	.9601	1.386	1.444	1.042	.7214	.8058	10
44°00′	.7679	.6947	.9657	1.390	1.440	1.036	.7193	.8029	46°00′
10	.7709	.6967	.9713	1.394	1.435	1.030	.7173	.7999	50
20	.7738	.6988	.9770	1.398	1.431	1.024	.7153	.7970	40
30	.7767	.7009	.9827	1.402	1.427	1.018	.7133	.7941	30
40	.7796	.7030	.9884	1.406	1.423	1.012	.7112	.7912	20
50	.7825	.7050	.9942	1.410	1.418	1.006	.7092	.7883	10
45°00′	.7854	.7071	1.000	1.414	1.414	1.000	.7071	.7854	45°00′
		COSINE	COTANGENT	COSECANT	SECANT	TANGENT	SINE	RADIANS ANGLE	DEG-MIN ANGLE

ANSWERS TO ODD-NUMBERED EXERCISES

CHAPTER 2

EXERCISE 2.1

1. a. integer, rational, real b. rational, real
   c. integer, rational, real d. integer, rational, real
   e. irrational, real f. imaginary
   g. integer, rational, real; prime h. rational, real
   i. integer, rational, real j. integer, rational, real
3. −17°C means 17 Celsius units below the freezing point of water.
   +50°C means 50 Celsius units above the freezing point of water.
5. a. $1/7$ b. 3 c. $2/5$ d. -5
   e. 0.40 f. $\sqrt{3}$ g. 4.0 h. 0.75
   i. $2y/(3+x)$ j. 100

EXERCISE 2.2

1. a. $<$ b. $=$ c. $<$ d. $>$
   e. $=$ f. $<$ g. $>$
3. 100°C $\leftrightarrow$ 212°F
5. a. $z \ngeqslant 0$ b. $z \nleqslant 2$ c. $-1 \leqslant z \leqslant 1$
   d. $|z| \leqslant 10$ e. $z \leqslant 0$ f. $|z| > 0$
   g. $z \ngeq 10$ and $z \nless 4$ h. $z = 0$
   i. $|z| \ngeq 0$ j. $|z| > 5$ k. $z \ngeq -4$
   l. $|z| \nless 1$ m. $z \leqslant 3$ n. $|z| \ngtr -1$

CHAPTER 3

EXERCISE 3.1

1. 3 3. 9 5. −0.056
7. 24 9. −24 11. 9
13. 9 15. $6x - 24 + 6a$ 17. $-2x + 6a - 12$
19. 3 21. −2 23. −252
25. 10
27. a. 142°F b. 63°F c. 250°F d. −200°F
33. like signs: $+15/+5 = +3$; $-15/-5 = +3$.
unlike signs: $+15/-5 = -3$; $-15/+5 = -3$.

EXERCISE 3.2

1. 2,401 3. −27 5. $\frac{1}{8}$ 7. 144 9. 96
11. −150 13. $x^8 - x^7$ 15. $-243a^{10}x^{10}$
17. $-5S^2T^6$ 19. $-32M^4$ 21. $(2x + 3)^3$
23. $(-2B)^3$ 25. $(4G/9)^2$ 27. $(3x)^5$
29. $(7K/N^3)^2$ 31. $A = \pi(D/2)^2$

EXERCISE 3.3

1. 6 3. −5 5. imaginary
7. 10 9. $\sqrt{-9}$, imaginary 11. −2.668 . . .
13. 6 15. 4 17. A
19. No; $\sqrt{A^2} = A$ if $A \geqslant 0$, and $\sqrt{A^2} = -A$ if $A < 0$.

EXERCISE 3.4

1. $5a$ 3. $5ab + a$ 5. $5y^3 + 2y^2 - y$
7. $5xy^2 - x^2y$ 9. $-(3/32)K$ 11. $0.018M^3$
13. 123 oranges, 43 grapefruit, 50 tangerines

EXERCISE 3.5

1. $3x^2 - x$ 3. $x^2 + 10x + 25$ 5. $54a^2 - 72ab + 24b^2$
7. $8a^3 + 32a^2 + 42a + 18$ 9. $12K^2 - 12K$
11. $10y^3 - 6xy^2 - 6xy$ 13. $tw\sigma_T - td\sigma_T$

15. a. $ac + ad + bc + bd$
b. $2a + 2b + 2c + 2d$

EXERCISE 3.6

1. $3a^2$
3. $2a^2 + 2ab$
5. $1 + r/3$
7. $a/c - b/c$
9. $v - (1/2)gt$
11. $1/(x + y)$
13. $M^2 - (11/2)M + 15/2$
15. $a^4 + a^3 - 2a^2 - a + 3 - 6/(a + 2)$
17. $1/R_1 + 1/R_2 + 1/R_3$
19. $\frac{e_x}{4} + \frac{3e_y}{4}$
21. $1/P$

EXERCISE 3.7

1. $x = 10$
3. $n = 2$
5. $z = -12$
7. No solution
9. $x = 0$
11. $I = -27$
13. $x = \pm\sqrt{3}$
15. $x = 49$
17. $S = 49$
19. $R = -1$
21. $x = \pm 1$
23. $x = 0$; $(x \neq -1)$
25. $x =$ any nonnegative number
27. $P = -2$
29. $y = \pm 1$
31. $3b - 2a$
33. $\pm\sqrt{y^2 + 25}$
35. $a - b^2 - 6b - 9$
$(a - x \geqslant 0)$
37. $5k$ $(k \neq 0)$
39. $2a/3$ $(x \neq 0)$
41. $-\sqrt[3]{b^3 + 8}$
43. 1 $(a + b \neq 0)$
45. $(3rs - 3)/(r + s)$
$(r, s, r + s \neq 0)$
47. $\pm\sqrt{(2 - a)/(3 + b)}$
$x, 3 + b \neq 0; (2 - a)/(3 + b) \geqslant 0$
49. 52π in.^{2}
51. 3.0 ft
53. 20 sec
55. a. 736 lb
b. 10 rev/sec
57. a. 0.414 Hz
b. 0.00023 ft
59. 0.400 in.
61. 10 ohms
63. 0.075 ohms

EXERCISE 3.8

1. $4\frac{1}{2}$ yr and $25\frac{1}{2}$ yr
3. 41.5 mph
5. 1.8 miles
7. 12.8 ft; 11,500 gal
9. 560,600 gal; 355,100 gal; 87,300 gal
11. 48°F and −8°F
13. infinity
15. 1.60×10^6 miles
66,700 mph
17. $11/16$ volts; $3/16$ volts; $15/16$ volts; $1\,15/16$ volts
19. 3.7 yd

EXERCISE 3.9

1.

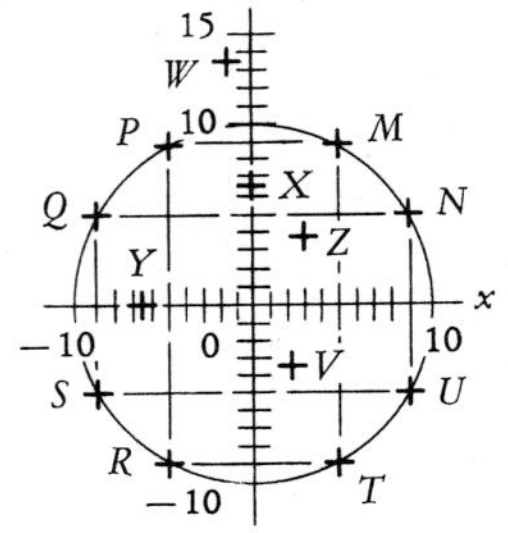

3.

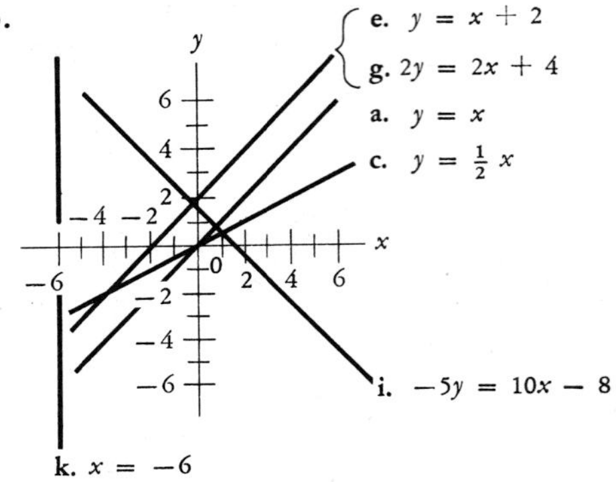

5. In $y = x + b$, "b" determines the position of the line above or below the x axis; $(0, b)$ is the point where the line crosses the y axis.

7.

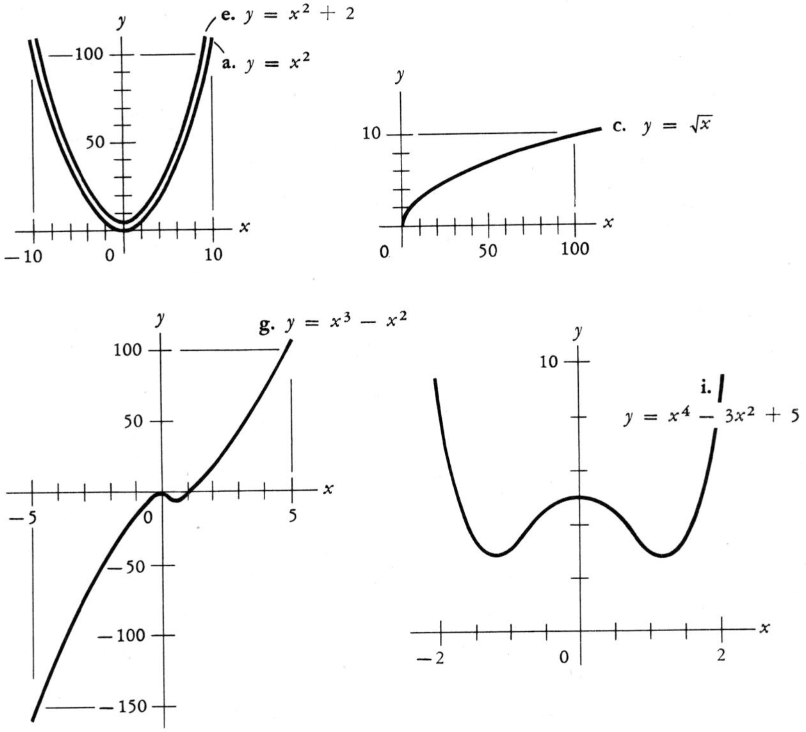

9. a. ± 4.1
b. ± 3.1
c. ± 8.5

11. a. 8 min
b. 63¼°F or 17.2°C

CHAPTER 4

EXERCISE 4.1

1. 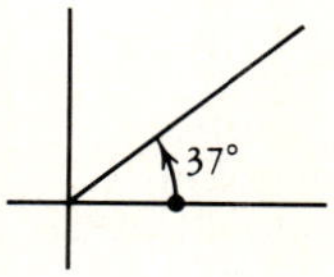

3. 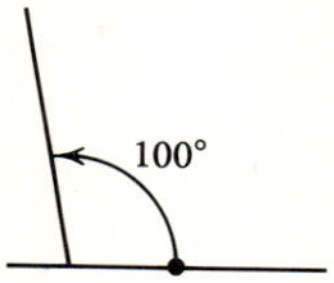

5. 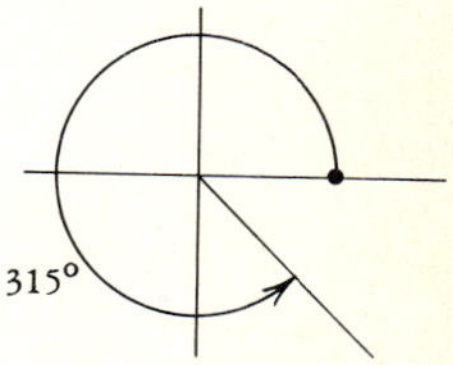

7. 135° 9. 300° 11. 330° 13. **a.** 14.4°; **b.** 16,200°
15. 12:18:00; 99° 17. 53.5125° 19. −30.25333° . . .
21. 0.032777 . . . 23. 413°30′45″; −306°29′15″; I
25. 329°44′48″; −390°15′12″; IV
27. 360°1′58″; −359°58′02″; I 29. −34°55′12″
31. 1°27′43.2″ 33. 0°0′10″

EXERCISE 4.2

1. $\alpha = 107°31'$, supplement
   $\beta = 72°29'$, vertical angles
   $\gamma = 17°31'$, complement
3. $\alpha = 26°$, alternate interior ∡
   $\beta = 64°$, complement
5. $\alpha = 72°$, alternate interior ∡
   $\beta = 108°$, interior ∡ on same side
   $\gamma = 72°$, corresponding ∡
7. $\theta = 12°$, $\phi = 103°$

EXERCISE 4.3

1. 60° 3. 13 5. $a = b = 7.1$, $\beta = 45°$ 7. $\beta = \gamma = 30°$
9. $c = 3$, $\gamma = 36.8°$ 11. $b = 39$, $\beta = 90°$ 13. short leg = 5.8 in., hypotenuse = 11.6 in.
15. 5 units 17. ⅔ ft 19. 5.7 ohms
21. 29.7 ft 23. 2.5 in. 25. 0.974 in.
27. ⅛ in. 29. $\sqrt{85}$ cm 31. three right triangles: 2 in.^{2}; equilateral triangle: $2\sqrt{3}$ in.^{2}
33. 122 ft

EXERCISE 4.4

1. parallel sides
   $x = 4$
3. sides of equal angles are in proportion; $\alpha = 22°50'$; $\beta = 15°$; $x = 4.75$

5. ASA = ASA
$\alpha = 70°$; $\beta = 60°$;
$x = 100$

7. ASA = ASA
$\alpha = 15°$; $x = 10.0$;
$y = 2.59$

9. 40 ft

EXERCISE 4.5

1. $\pi/6$ 3. $-\pi/2$ 5. $19\pi/6$ 7. $135°$; $3/8$ rev
9. $420°$; $7/6$ rev 11. $210°$; $7/12$ rev 13. $86.4°$; 0.24 rev
15. $180°$; $1/2$ rev 17. $-28.8°$; -0.080 rev 19. $s = 6$ cm
21. $\theta = 1.5$ rad 23. $r = 50$ in. 25. $s = 26.2$ ft
27. $r = 1.43$ in.

EXERCISE 4.6

1. 69.8 miles 3. 956 miles 5. 4,500,000 miles
7. $3{,}000/\pi$ rev/min 9. 28.7 ft/sec^2 11. 0.094 sec
13. 829 in./min; 528 rpm 15. 0.024 lb

CHAPTER 5

EXERCISE 5.1

In answers 1 to 11, values of the six functions are given in this order: sin, cos, tan, cot, sec, csc.

1. $3/5, 4/5, 3/4, 4/3, 5/4, 5/3$
3. $-3/\sqrt{34}, 5/\sqrt{34}, -3/5, -5/3, \sqrt{34}/5, -\sqrt{34}/3$
5. $3/5, -4/5, -3/4, -4/3, -5/4, 5/3$
7. $-5/\sqrt{34}, 3/\sqrt{34}, -5/3, -3/5, \sqrt{34}/3, -\sqrt{34}/5$
9. $3/5, -4/5, -3/4, -4/3, -5/4, 5/3$
11. $-12/13, -5/13, 12/5, 5/12, -13/5, -13/12$
13. $-\sqrt{29}/14$ 15. $-12/11$ 17. $-\sqrt{10}/3$
19. $\cot\theta = \dfrac{x}{y} = \dfrac{x/r}{y/r} = \dfrac{\cos\theta}{\sin\theta}$; $\cot\theta = \dfrac{x}{y} = \dfrac{1}{y/x} = \dfrac{1}{\tan\theta}$
21. $(\sin\theta)^2 + (\cos\theta)^2 = \dfrac{y^2}{r^2} + \dfrac{x^2}{r^2} = \dfrac{y^2 + x^2}{r^2} = \dfrac{r^2}{r^2} = 1$

EXERCISE 5.2

3. a. $\cos 0° = \dfrac{x}{r} = \dfrac{r}{r} = 1$ 5. a. $\sin 180° = \dfrac{y}{r} = \dfrac{0}{r} = 0$

b. $\tan 0° = \frac{y}{x} = \frac{0}{r} = 0$

c. $\sec 0° = \frac{r}{x} = \frac{r}{r} = 1$

b. $\cos 180° = \frac{x}{r} = \frac{-r}{r} = -1$

c. $\tan 180° = \frac{y}{x} = \frac{0}{-r} = 0$

EXERCISE 5.3

1. 0.3502 3. 1.386 5. 0.2370 7. 0.7050 9. 0.7133
11. 343.8 13. 0.4305 15. 0.3773 17. 39°50′; 0.6952
19. 21°30′; 0.3752 21. 16°40′; 0.2909 23. 47°50′; 0.8348
25. 65°10′; 1.1374 27. 77°20′; 1.3497 29. 1.176
31. 0.5805 33. 0.6373 35. 4.646 37. 64°55′; 1.1330
39. 15°07′; 0.2637 41. 55°14′; 0.9640 43. 80°44′; 1.4092

EXERCISE 5.4

1. $\cot \beta$; $\tan \epsilon$ 3. $\sin \beta$; $\cos \epsilon$ 5. $\cos \beta$; $\sin \epsilon$
7. 3/5, 4/5, 3/4, 4/3, 5/4, 5/3 9. $\beta = 37°$; $\epsilon = 53°$

EXERCISE 5.5

1. $\alpha = 90° - \beta$
   $a = c \cos \beta$
   $b = c \sin \beta$

3. $\sin \alpha = a/c$
   $\cos \beta = a/c$
   $b = \sqrt{c^2 - a^2}$

5. $\beta = 90° - \alpha$
   $a = c \sin \alpha$
   $b = c \cos \alpha$

7. $\alpha = 90° - \beta$
   $a = b \cot \beta$
   $c = b \csc \beta$

9. $\tan \alpha = a/b$
   $\cot \beta = a/b$
   $c = \sqrt{a^2 + b^2}$ or $c = a \csc \alpha$

11. $\alpha = 17°$
    $a = 13$
    $b = 43$

13. $\beta = 78°44'$
    $b = 11.60$
    $c = 11.82$

15. $\alpha = 67°20'$
    $\beta = 22°40'$
    $a = 1200$

17. $\alpha = 0°15'$
    $a = 0.536$
    $c = 123$

19. 32.5 ft

21. 21 ft 5 in.

23. 0.77

25. 926 mph

27. a. 12°30′
    b. $X_L = 1.96$; $R = 2.60$

29. 0.216 in. 31. 17 ft 33. 22°40′ 35. 2.290 in.

CHAPTER 6

EXERCISE 6.1

1. 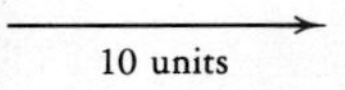

3. 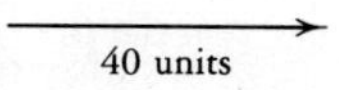

5. zero

7. 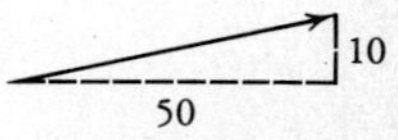

9. 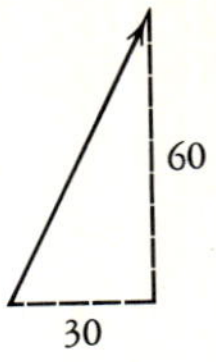

11. 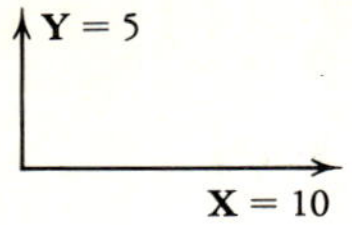

13. 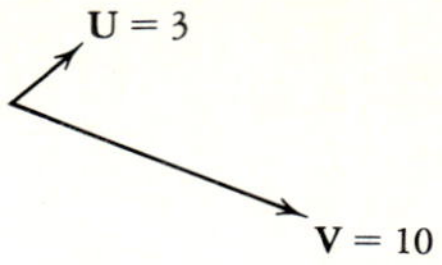

15. 

17. a. 250 miles, 67° north of east
b. 62½ mph at 67° north of east
c. 60° north of west

19. a. turn 22° toward west; course is 202°
b. 1.2 hr

EXERCISE 6.2

1. $A_x = 5.2$
$A_y = 3.6$

3. $C_x = 213$
$C_y = -179$

5. $E_x = -0.21$
$E_y = 0.45$

EXERCISE 6.3

1. 22.6 at $-12°50'$
3. 271 at $62°30'$

EXERCISE 6.4

3. 15.1 miles, 44°50′ south of east
5. 14 miles, 1°50′ west of north
7. 1.52 lb at 105°

EXERCISE 6.5

1. 90 lb
3. 157 lb
5. 10.3 lb
7. 2,030 lb
9. N = 866 lb
S = 500 lb
11. $S_n = 750$ psi
$S_s = 433$ psi

EXERCISE 6.6

1. a. The ball moves away at 20 mph.
b. The ball hangs in midair without moving.
3. a. The ball moves away at 25 mph.
b. The ball moves at 5 mph opposite to the direction of the train.

5. 54 mph at 33°40′ west of south
7. 18.8 mph at 29°20′ south of east
9. 34°10′ north of east; 7.6 min
11. 16 mph at 46.1° south of east
13. 316 mph at 11°30′ west of north
15. 91 ft/sec
17. a. zero b. 60 mph c. 42 mph

CHAPTER 7

EXERCISE 7.1

1. $A = \pi r^2$
3. $C = \pi d$
5. $C = 2\sqrt{\pi A}$
7. $V = \dfrac{d^3}{3\sqrt{3}}$
9. a. $A = \dfrac{x^2}{4\pi}$; b. $A = \dfrac{x^2}{16}$; c. $A = \dfrac{3x^2}{50}$
11. a. $\dfrac{2\pi d^2}{9} + 0.10\pi d$
 b. $8d^2/9 + 0.10\pi d$
 c. $32d/3 + 0.10\pi d$
13. $6{,}424{,}000 - \dfrac{1{,}729{,}000t}{60}$
15. $g = \dfrac{2d}{t^2 \sin \theta}$

EXERCISE 7.2

1. a. $f(0) = 10$
 b. $f(2) = 6$
 c. $f(-2) = 22$
 d. $f(a) = a^2 - 4a + 10$
3. a. $16.1(t + h)^2$
 b. $32.2th + 16.1h^2$
5. $x^4 - 6$

EXERCISE 7.3

1. $f(x) = x^2 - 4x + 10$

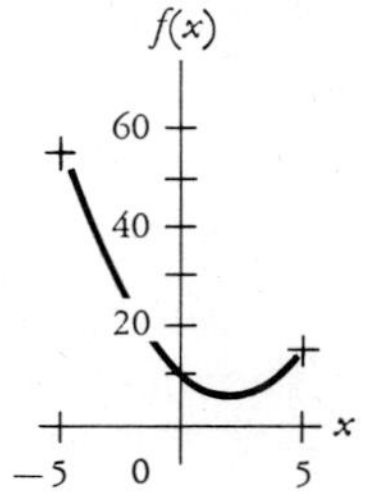

3.

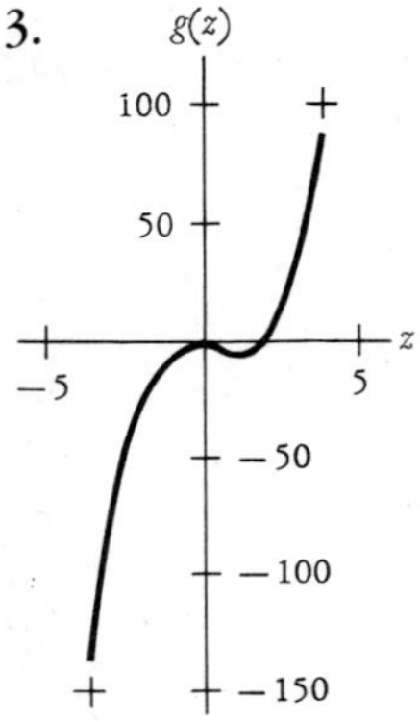

5. $T(t) = t^2 + 25$

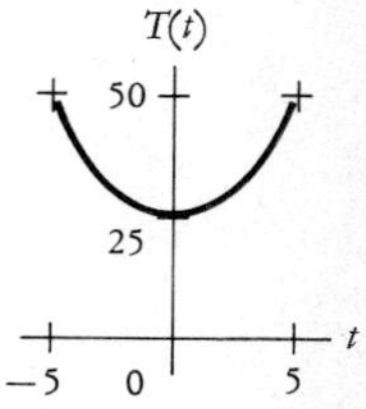

7.

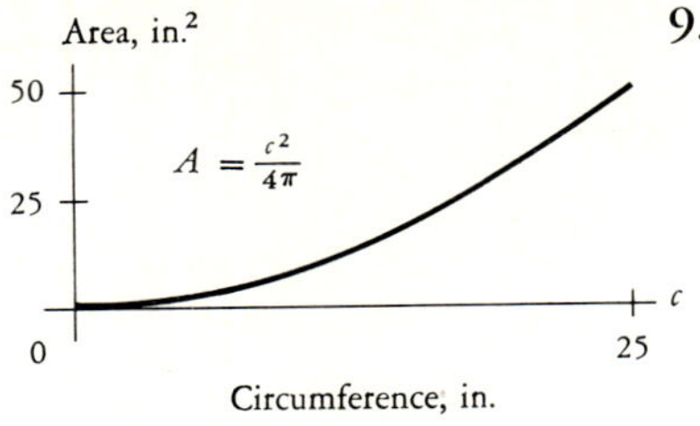

9. No, because $5.05 \times 10^8/r^2$ is never zero.

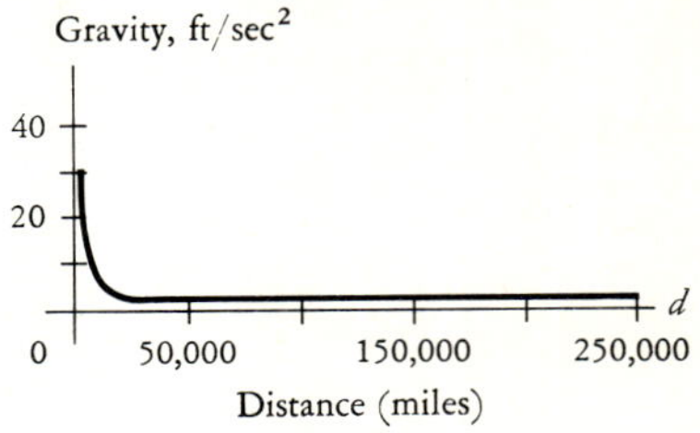

11.

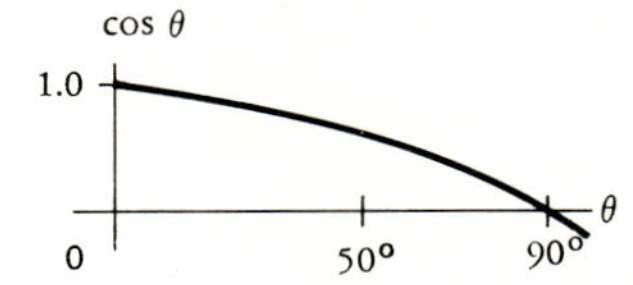

13.

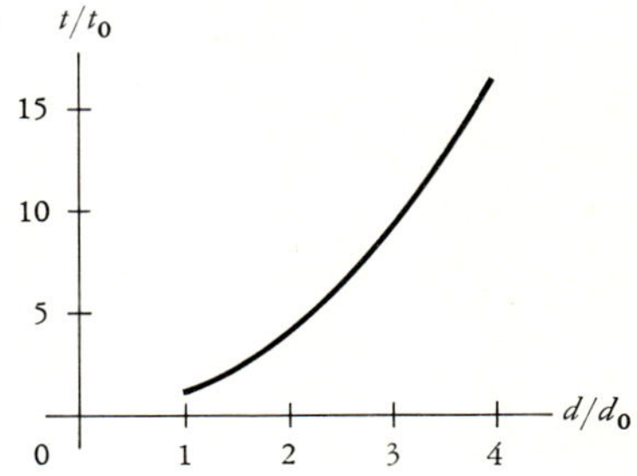

15. 59 sec
Time in Prob. 14 is 53 sec;
percent error is: $\frac{59 - 53}{59} \times 100 \approx 10$ percent

EXERCISE 7.4

1. 2.5 3. ±3.0 5. 0.0; 9.0 7. −0.186; 2.69
9. No solution 11. −0.68 13. −5.6; 1.3; 12.3
15. −5.2; −1.6; 1.7; 4.1 17. (−2.3,−1.1); (1.8,0.88)
19. (−7.1,−7.1); (7.1,7.1) 21. 2,000 23. 26°34′

CHAPTER 8

EXERCISE 8.1

1. nonlinear 3. linear 5. nonlinear 7. linear
9. linear 11. nonlinear 13. linear 15. nonlinear
17. 0; −1; −45° 19. −5; +1; +45°
21. −5; +1; +45° 23. +10; +6; +80°30′
25. −2; +½; +26°30′ 27. +¾; −¼; −14°

29. $y = -x$

31. $y = 1.73x - 3.00$

33. $y = -0.577x - 2.00$

EXERCISE 8.2

1.

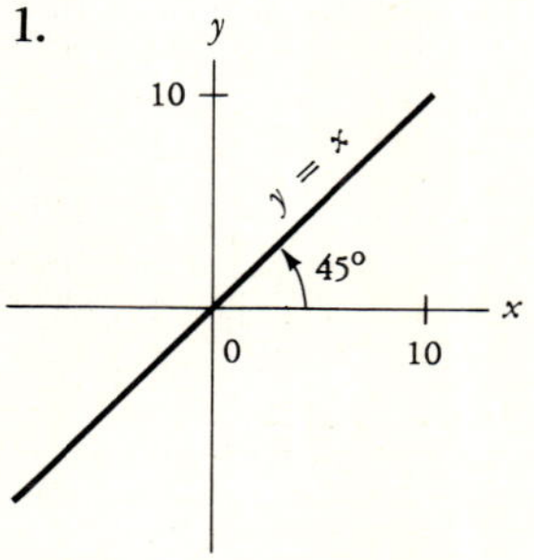

3.

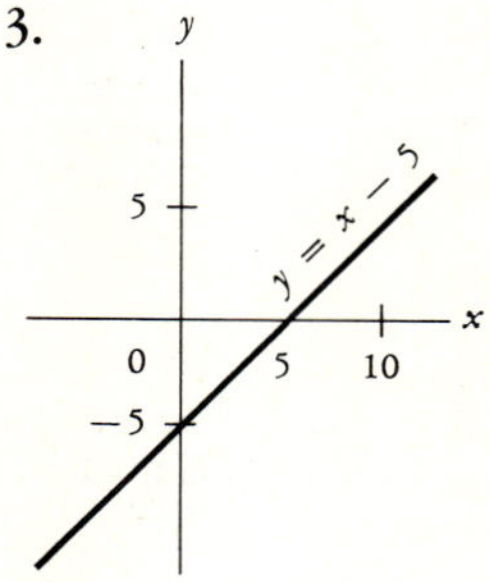

5.

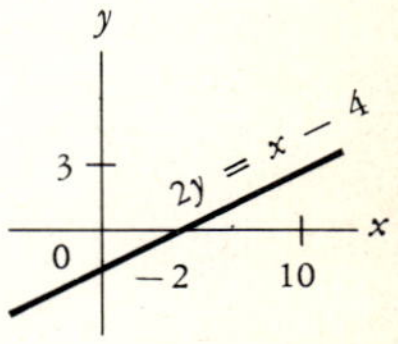

7.

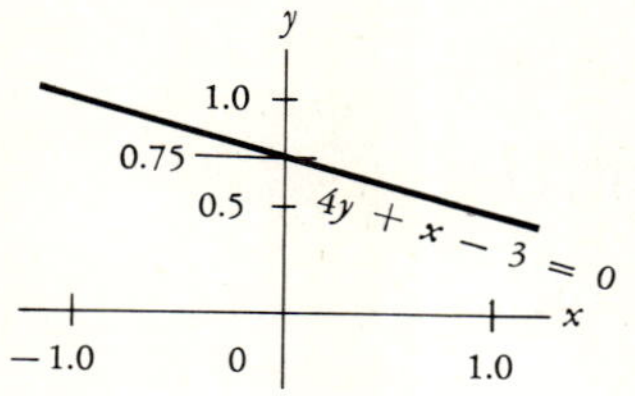

9. a. $y = 3x + 2$
 b. $y = 3x + 2$
 The graphs would coincide.

EXERCISE 8.3

1. (4,5)

3. (2.5,−3.2)

5. (9,36)

7. $I_1 = 6$
 $I_2 = -3$

9. $W_1 = 58\frac{2}{3}$ lb
 $W_2 = 41\frac{1}{3}$ lb

11. 36 min

EXERCISE 8.4

Same as Exercise 8.3.

EXERCISE 8.5

Same as Exercise 8.3.

EXERCISE 8.6

Same as Exercise 8.3.

EXERCISE 8.7

1. $(\frac{1}{3}, -\frac{1}{3}, 1)$

3. $(\frac{5}{2}, \frac{3}{2}, -\frac{1}{2})$

5. (20,30,40)

7. (1,−1,0,4) 9. $I_1 = 1, I_2 = 1, I_3 = 0$
11. First alloy: ½ lb per pound
Second alloy: ¼ lb per pound
Third alloy: ¼ lb per pound
13. $R = 3/7$ in.;
$x = 9/14$ in.; $y = 6/7$ in.

EXERCISE 8.8

3. 10 5. 0 7. 0 9. −78

EXERCISE 8.9

1. 0 3. 0 5. 140 7. 0 9. −631 11. 5.25
13. 1,080 15. −4,680

EXERCISE 8.10

1. to 11. Same as Exercise 8.7. 13. (0,0,2)

CHAPTER 9

EXERCISE 9.1

1. $6a^3b - 9ab^3$ 3. $r^3 - 3r^2 + 3r - 1$
5. $2x^2 + 5x - 3$ 7. $r^2 + 5r + 6$ 9. $x^3 - 1$
11. $25x^2 + 20x + 4$ 13. $9x^2 - 4$
15. $x^3 + 3x^2y + 3xy^2 + y^3 - 3x^2 - 6xy - 3y^2 + 3x + 3y - 1$
17. $N^4 - 1$ 19. $x^6 - 2x^3y^3 + y^6$ 21. $A^4 - 6A^2 - 27$
23. $16k^4 - 32k^3 + 24k^2 - 8k + 1$ 25. $x^2 + 4x + 4 - 9z^2$
27. $25y^6 - 30y^3x^2 + 9x^4 + 20y^3 - 12x^2 + 4$
29. $x^7 - x^5 - 12x^3$

EXERCISE 9.2

1. $Y(2Y - 3X)$ 3. $5X^2Y^2Z^2(6YZ^2 + 2XZ^2 - 3X^2Y)$
5. $3S^k(S - 2S^k)$
7. $\dfrac{2}{3bc}\left(\dfrac{ad}{b} + \dfrac{2de}{3c} - \dfrac{3ae}{5bc}\right)$

9. $\dfrac{3a-2b}{3a+b}$ 11. $2(2x+5/2)$ 13. $AX^2(A/X+1)$
15. $\dfrac{2}{3}\left(\dfrac{2a}{3}-\dfrac{4b}{5}\right)$

EXERCISE 9.3

1. $(X+2Y)(X-2Y)$ 3. $2\left(\dfrac{3B}{5}+\dfrac{4A}{7}\right)\left(\dfrac{3B}{5}-\dfrac{4A}{7}\right)$
5. $3(3M^{3k}+2N^{4p})(3M^{3k}-2N^{4p})$ 7. $-8(X+2)(X-2)$
9. $(A+B)(A-B)(A^2+B^2)$ 11. $(s+1+5t)(s+1-5t)$

EXERCISE 9.4

1. $(2x+z)(2+y)$ 3. $6(M-4)(N-1)$
5. $(S^{3n}-1)(3S^n-S^{3n}-1)$

EXERCISE 9.5

1. $(X+1)^2$ 3. Not factorable 5. $(N+3)(N-1)$
7. $(2A-3)(A+1)$ 9. Not factorable 11. $(8Y-9)(4Y-5)$
13. $R(4R-3)(3R+4)$ 15. $(6S+11)(4S-3)$

EXERCISE 9.6

1. $(X+5)^2$ 3. $(4Y-1)(Y-4)$ 5. $(6A-5)(2A+3)$
7. $(8-7P)(4-5P)$ 9. $(2X-1)^2$ 11. $(9V+7)(V-6)$
13. Not factorable 15. $(2+5S^2)(1-S)(1+S)$
17. $2(XY^2+3Z)^2$ 19. $-5(3A^X-1)(A^X+2)$
21. $(16S-3)(S-14)$ 23. $(2X+a)(X+2)$
25. $(21A-16)(A+2)$

EXERCISE 9.7

1. $(X-2)(X^2+2X+4)$ 3. $3(X+3Y)(X-3Y)$
5. $(A+2)^3$ 7. $(2t+3s)(4t^2-6st+9s^2)$
9. $(4X-1)(16X^2+4X+1)$ 11. $3a\pi(1+5\pi)(1-5\pi)$
13. $(3X+4)(X-40)/8$ 15. $(3Y^2-2a)^3$
17. $3(4+5A^k)(16-20A^k+25A^{2k})$

19. $(5U^x - 3V^y)^2$
21. $(2X^2 + 3)(X + 2)(X - 2)$
23. $(X + 1)^3$
25. $(X - 3)(X^2 + 9X + 39)$

CHAPTER 10

EXERCISE 10.1

1. $6/9$ 3. $-2/3$ 5. $-36/28$ 7. $2/5$ 9. $3/7$
11. $1/2$ 13. $1/12$ 15. $-3/20$ 17. $51x/40$
19. $(2y - 5x)/(xy)$ 21. $(3b - 2a)/[a(a + b)]$
23. $(2x^2 + 2y^2)/(x^2 - y^2)$
25. $1/3$ 27. $-1/4$ 29. 7 31. $2/3$ 33. $364/455$
35. $x/2$ 37. $\dfrac{N - 2}{2N}$ 39. $\dfrac{a}{a + 2}$ 41. $-(3k + 2)$

EXERCISE 10.2

1. $5/4$ 3. $15/224$ 5. 6 hr 40 min 7. 260 pieces
9. 16 volts; 50 amp 11. $1\frac{17}{64}$ oz 13. $w/(2v)$
15. $18u^2v^3/w^3$ 17. $ab^2/(6x)$ 19. $3h/(4e)$
21. $\dfrac{a^{n+2}x^3}{3b^{n-1}y}$ 23. $\dfrac{a^2 - 2ax + x^2}{a^2x^3 - 2a^2x^2}$ 25. $\dfrac{Z^2 + 3Z - 10}{Z^2 - 3Z + 10}$
27. $\dfrac{(Y^2 + 4)(Y + 2)}{2(2Y + 1)(Y + 4)}$ 29. $\dfrac{6X^2 + X - 2}{Y^3 + 5Y^2 + 3Y + 15}$ 31. $1/30$

EXERCISE 10.3

1. $11/3$ yd 3. $58/100$ in.; $580/1{,}000$ in.
5. 47 pennies; $9\frac{2}{5}$ nickels; $4\frac{7}{10}$ dimes
7. 2 miles + 5 hr cannot be combined.
9. 1 apple and 9 oranges, or 6 apples and 6 oranges, or 11 apples and 3 oranges
11. $5/3$ 13. $-a/x^2$
15. $-b$ 17. $\dfrac{b}{2x}$ 19. $19/15$ gal 21. $\dfrac{2bc + 3ac + 4ab}{abc}$
23. $\dfrac{3b^2 + 2}{ab}$ 25. $\dfrac{2x + 3}{3}$

EXERCISE 10.4

1. $2/21$ 3. $\dfrac{31}{36a}$ 5. $47/225$ 7. $1/12$ 9. $\dfrac{3x+10}{4x^2}$

11. $\dfrac{c-2a+3b}{abc}$ 13. $\dfrac{4-X}{2X(X-1)}$ 15. $\dfrac{2X}{(X-5)(X-3)}$

17. $\dfrac{-S^3+2S^2+3S-8}{(S-2)^2(S-4)}$ 19. $\dfrac{-12Z}{(Z+3)(Z-3)(Z^2-3Z+9)}$

21. $\dfrac{2M}{AD}\,\dfrac{2(Z+1)x+dZ}{Z(d+2x)}$ 23. $\dfrac{4F(2D+d)}{\pi d^3}$

25. $\dfrac{k^2+0.276k-0.526}{k(k-1)}$ 27. $\dfrac{k_1+k_2}{k_1k_2}$

EXERCISE 10.5

1. 4 3. $30/17$ 5. 1 7. 1 9. -4 11. -2

13. $\dfrac{xy}{y-x}$ 15. $\dfrac{a+b}{ab}$ 17. ab 19. $\dfrac{ab(a-b)}{a^2+ab+b^2}$

21. a. $\dfrac{R_1R_2}{R_1+R_2}$ b. $\dfrac{R_1R_2R_3}{R_1R_2+R_2R_3+R_3R_1}$

23. $\dfrac{n\pi^2AEp^2}{l^2}$ 25. $\dfrac{p^2}{p_n{}^2-p^2}$ 27. $\dfrac{3h^2-3ph+p^2}{3h^2-3ph}$

29. $\dfrac{M_0{}^2c^2}{c^2-v^2}$

EXERCISE 10.6

1. $R=2$ 3. $R=4$ 5. $a=19/2$ 7. $y=-9$

9. $t=-1/6$ 11. a. $R=\dfrac{R_1R_2R_3}{R_1R_2+R_2R_3+R_1R_3}$

b. $R_3=\dfrac{RR_1R_2}{R_1R_2-RR_2-RR_1}$

13. a. $d_o=75$ mm b. $d_o=150$ mm
$d_i=150$ mm $d_i=75$ mm

15. $t=\dfrac{2P_iR_i}{2S-P_i}$ 17. 3,260 19. $15/44$ lb/in.

21. $12/\sqrt{17}$ 23. 161,000 miles/sec

EXERCISE 10.7

1. $1\frac{4}{11}$ hr 3. 6:54 a.m., Feb. 9 5. 360 sec
7. 76.2 hr 9. $6\frac{1}{4}$ hr 11. $3\frac{38}{75}$ hr

CHAPTER 11

EXERCISE 11.1

1. ± 4 3. $0; -5$ 5. ± 5 7. $-1; 2/5$
9. $0; b/a$ 11. $-1; 2$ 13. $1; 9/5$ 15. 1 (double root)
17. $-b/a; d/c$ 19. $-8, -22$ 21. Three pieces: 36, 48, and 60 in.
23. 19.9 in. 25. \$120 27. 1.2 in.; 2.4 and 3.6 in.
29. 16 and 18 in. 31. 6 ohms each

EXERCISE 11.2

1. $-3; 1$ 3. $-3; 3/2$ 5. $3/4 \pm 1/4\sqrt{57}$
7. $-1/10 \pm 1/10\sqrt{21}$ 9. $1/2 \pm 1/4\sqrt{3}$ 11. $1 \pm \sqrt{-2}$
13. $-b/2a \pm \sqrt{b^2 - 4ac}/2a$

EXERCISE 11.3

1. to 13. Same as Exercise 11.2.
15. $(2 + \sqrt{2})/2, (2 - \sqrt{2})/2$ 17. $(2 + \sqrt{2})/2, (2 - \sqrt{2})/2$
19. diameter of $A = 2$ in.
diameter of $B = 3$ in.
21. $6 + \sqrt{26}, 6 - \sqrt{26}$
23. $(5\sqrt{17} + 5)/2, (5\sqrt{17} - 5)/2$

CHAPTER 12

EXERCISE 12.1

1. 1 3. 3 5. 1 7. 1 9. $1/9$
11. $\dfrac{1}{4 - 2x + x^2}$ 13. a^8 15. $\dfrac{2}{x}\left(\dfrac{2x+1}{2x-1}\right)$
17. $200y/x^3$ 19. 1 21. $2n/(n^2 - 1)$ 23. 32
25. 0.0005 27. 0.000001 29. -3.2×10^{-3}
31. 5×10^6 33. -5×10^{-3} 35. 93×10^6 miles
37. 12×10^{-3} amp 39. 30×10^6 psi 41. ± 4
43. $x = 2; (x \neq 0)$ 45. $-1/3$, (double root)
47. $(-3 \pm \sqrt{41})/8$

EXERCISE 12.2

1. 6 3. $2/3$ 5. 512 7. $-1/2$ 9. -2
11. $9/4$ 13. 2 15. $\sqrt[6]{2}\sqrt{3}$ 17. $8x^3$
19. $\dfrac{c^{4/3}}{9a^4}$ 21. $K^{17/6}$ 23. $\dfrac{M^{151/6}}{a^{50}\sqrt{-1}}$ 25. $s+4$
27. $x^{1/n}$ 29. No; see Example 12.23.
31. $(x^2)^{1/2} = x$ if x is positive or zero.
33. If $x - 3$ is positive, then $[(3-x)^2]^{1/2} = -(3-x)$, not $+(3-x)$.

EXERCISE 12.3

1. $3\sqrt[3]{2}$ 3. $\frac{1}{5}\sqrt{10}$ 5. $\sqrt{2}/2$ 7. $3\sqrt{2}/4$
9. $\sqrt{6x}$ 11. $(2x/5)\sqrt{10x}$ 13. 1×10^3 15. $10^{-2}\sqrt{2}$
17. $S\sqrt[4]{16+S^4}$ 19. 2 21. $[(h-1)/5]\sqrt{5(h-1)}$
23. $\sqrt[6]{x^3 - x^2 + x - 1}$ cannot be simplified 25. $-(1+\sqrt{5})/4$
27. $b \neq 0$ because division by zero undefined.
$b \not< 0$ because $\sqrt{(2b)^2}$ would be $-2b$.
If $b < 0$, final expression would be $-\sqrt{10ab}/2b$; $a \leqslant 0$.

EXERCISE 12.4

1. $7\sqrt{6}$ 3. $(a-2b)\sqrt[3]{2}$ 5. $(3+x-2x^2)\sqrt{x}/x$
7. $[(2a-b)/(2ab)]\sqrt{2ab}$ 9. $-\sqrt{h}$ 11. 0 13. 0
15. $(5/2)\sqrt{2x}$

EXERCISE 12.5

1. 10 3. x^4 5. $2\sqrt{5}/5$ 7. 4 9. $2\sqrt{3}$
11. $x\sqrt[6]{x}$ 13. $2x\sqrt[12]{2x}$ 15. $2s - 3r$ 17. $x - 2$

EXERCISE 12.6

1. $\sqrt{5}$ 3. $(2/5)\sqrt{10}$ 5. $(1/3)\sqrt[3]{75}$ 7. $(1/2)\sqrt[6]{96}$
9. $5/3$ 11. $(1/a)\sqrt{3a}$ 13. $\sqrt{2} - 1$ 15. $\sqrt{3}$
17. $\dfrac{1+\sqrt{3}}{2}$ 19. $-\dfrac{11+6\sqrt{2}}{7}$ 21. $\dfrac{\sqrt{a^2-b^2}}{a+b}$
23. $a - 1 + \sqrt{a}\sqrt{a-1}$ 25. $\dfrac{\sqrt[3]{x^2} + \sqrt[3]{x} + 1}{x-1}$

CHAPTER 13

EXERCISE 13.1

For Probs. 1 to 9 and 13 to 19, the values of the trigonometric functions are listed in the following order: sin θ, cos θ, tan θ, cot θ, sec θ, and csc θ.

1. $4/5, 3/5, 4/3, 3/4, 5/3, 5/4$
3. $-4/5, -3/5, 4/3, 3/4, -5/3, -5/4$
5. $-\sqrt{2}/2, -\sqrt{2}/2, 1, 1, -\sqrt{2}, -\sqrt{2}$
7. $-\sqrt{2}/2, -\sqrt{2}/2, 1, 1, -\sqrt{2}, -\sqrt{2}$
9. $1/2, -\sqrt{3}/2, -\sqrt{3}/3, -\sqrt{3}, -2\sqrt{3}/3, 2$
11. In Probs. 5 and 7, the angle is 225°.
    In Prob. 9, the angle is 150°.
13. $-\sqrt{2}/2, -\sqrt{2}/2$, 1 (given), $1, -\sqrt{2}, -\sqrt{2}$
15. $-2\sqrt{79}/79, 5\sqrt{237}/79, -2\sqrt{3}/15, -5\sqrt{3}/2$ (given), $\sqrt{237}/15$, $-\sqrt{79}/2$
17. $4/5, -3/5, -4/3, -3/4, -5/3, 5/4$ (given)
19. $-1/2$ (given), $\sqrt{3}/2, -\sqrt{3}/3, -\sqrt{3}, 2\sqrt{3}/3, -2$

EXERCISE 13.2

For Probs. 1 to 15, the values of the trigonometric functions are listed in the same order as for Exercise 13.1.

1. 0.5000, −0.8660, −0.5774, −1.732, −1.155, 2.000
3. −0.5000, 0.8660, −0.5774, −1.732, 1.155, −2.000
5. −0.7071, −0.7071, 1.000, 1.000, −1.414, −1.414
7. 0.9932, −0.1161, −8.556, −0.1169, −8.614, 1.007
9. −0.5476, −0.8368, 0.6544, 1.528, −1.195, −1.826
11. −0.8660, 0.5000, −1.732, 0.5774, 2.000, 1.155
13. 0.2079, −0.9781, −0.2126, −4.705, −1.022, 4.810
15. 0.7431, 0.6691, 1.111, 0.9004, 1.494, 1.346
17. 67°40′; 112°20′
19. 172°38′; 352°38′
21. 130°13′; 229°47′
23. 166°20′; 346°20′

EXERCISE 13.3

1. $B = 80°$
   $a = 1.02$ in.
   $c = 1.91$ in.
3. $B = 16\frac{3}{4}°$
   $C = 103\frac{1}{4}°$
   $c = 3.37$ in.
5. $A = 60°$
   $a = 3.68$ in.
   $b = 3.26$ in.

7. No triangle; the 2-in. side will not reach to the 53° angle.
9. No triangle; the sum of the shorter sides is less than the longest side.
11. $A = 28\frac{1}{4}°$
$B = 71\frac{3}{4}°$
$c = 2.59$ in.
13. No triangle; the side opposite the obtuse angle is not the longest side.

EXERCISE 13.4

1. $\mu = 80°$
$e = 7.98$
$t = 15.7$
3. $\phi = 100°$
$e = 11.8$
$n = 34.1$
5. No solution
7. Two solutions
$\phi_1 = 25°40'$ $\quad \phi_2 = 154°20'$
$\theta_1 = 139°20'$ $\quad \theta_2 = 10°40'$
$e_1 = 755$ $\quad e_2 = 214$
9. $\mu = 90°$
$e = 50.8$
$t = 35.6$
11. $\theta = 140°$
$e = 14.1$
$n = 7.5$
15. $A_r = 0.185A$
$A_s = 0.922A$
17. $F_{20} = 442$ lb
$F_{40} = 542$ lb
19. $\alpha = 45.2°$
$\beta = 82.8°$
$x = 252$ miles
21. Yes; law of sines, Case II, with one solution
23. 43°10′ east of north at 355 mph
25. 50°50′ east of north
27. 28°23′ north of east
29. Two solutions: head downstream, 58.3° from bank, 13.2 mph; or head upstream 4.3° from bank, 1.2 mph

EXERCISE 13.5

1. $\alpha = 60°$
$\gamma = 60°$
$p = 5$
3. $\beta = 27.5°$
$\gamma = 22.5°$
$n = 19.9$
5. $\alpha = 82°50'$
$\beta = 41°20'$
$\gamma = 55°50'$
7. $\alpha = 54.3°$
$\beta = 114°$
$\gamma = 11.7°$
9. Two solutions:
$\alpha_1 = 74°20'$ $\quad \alpha_2 = 105°40'$
$\gamma_1 = 43°40'$ $\quad \gamma_2 = 12°20'$
$s_1 = 8.6$ $\quad s_2 = 2.7$
11. $\alpha = 50.7°$
$\beta = 54.3°$
$s = 1{,}190$
13. 64 lb at 39.2° from 35-lb force
15. $\theta = 82°50'$
$\phi = 41°25'$
17. 3.61 in., 13.7 in.
19. 0.152 in., 0.912 in.
21. 31.2°
23. 162 ft
25. a. 14.9 mph at 13.3° north of east
b. 4.5 mph at 50.2° north of west

27. Turn 176°20′ (new heading is 63°40′ south of west); last leg is 2.3 miles; yes

29. 1,217 ft

31. 19°30′ west of south
$\frac{1}{2}$ hr late

33. 18.6° opposite $\frac{3}{8}$-mile leg;
39.6° opposite $\frac{3}{4}$-mile leg;
121.8° opposite 1-mile leg

CHAPTER 14

EXERCISE 14.1

1.

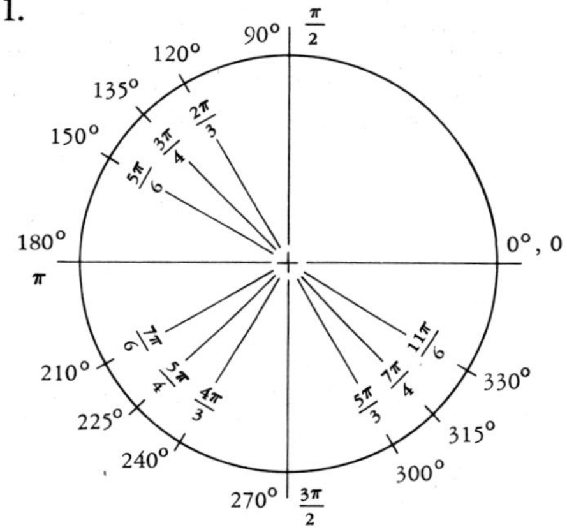

3. $-\sqrt{3}/2$
5. $\sqrt{3}/2$
7. -2
9. $\sqrt{3}/2$
11. $-\sqrt{2}/2$
13. $-\frac{1}{2}$
15. 0.6111
17. 2.035
19. -0.9299

For Probs. 21 to 25, answers are listed in the following order, with the given value in parentheses: sin x, cos x, tan x, cot x, sec x, csc x.

21. $-\frac{5}{6}, -\sqrt{11}/6, (5/\sqrt{11}), \sqrt{11}/5, -6\sqrt{11}/11, -\frac{6}{5}$

23. $-\sqrt{2}/4, \sqrt{14}/4, -\sqrt{7}/7, -\sqrt{7}, 2\sqrt{14}/7, (-2\sqrt{2})$

25. $\dfrac{\sqrt{a-1}}{\sqrt{a+8}}, \dfrac{-3}{\sqrt{a+8}}, \dfrac{\sqrt{a-1}}{-3}, \left(\dfrac{-3\sqrt{a-1}}{a-1}\right), \dfrac{\sqrt{a+8}}{-3}, \dfrac{\sqrt{a+8}}{\sqrt{a-1}}$

27. $\sec\theta = \sqrt{1-x^2}/(1-x^2)$; $\tan\theta = x\sqrt{1-x^2}/(1-x^2)$

EXERCISE 14.2

1. $-\sqrt{2}/2$
3. $-\frac{1}{2}$
5. $-\frac{1}{5}$
7. $-2\sqrt{5}/5$
9. $\sqrt{10}/10$
11. $\cot\theta = \dfrac{x}{y} = \dfrac{x/r}{y/r} = \dfrac{\cos\theta}{\sin\theta}$

19. $\dfrac{\tan\theta}{\sin\theta} = \dfrac{(\sin\theta)/(\cos\theta)}{\sin\theta} = \dfrac{1}{\cos\theta} = \sec\theta$

21. $\cos\theta \tan\theta = \cancel{\cos\theta}\,\dfrac{\sin\theta}{\cancel{\cos\theta}} = \sin\theta$

23. First step: $1 - \sec^2\theta = -\tan^2\theta$

25. $\dfrac{\sin y}{\tan y} \div \dfrac{\cot y}{\cos y} = \dfrac{\cancel{\sin} y}{(\cancel{\sin} y)/(\cos y)} \times \dfrac{\cancel{\cos} y}{(\cancel{\cos} y)/(\sin y)} = \sin y \cos y$

27. First step: factor $\sin\theta \cos\theta$

29. First steps: $\dfrac{1}{\sin^2 5x} - 1 = \dfrac{1 - \sin^2 5x}{\sin^2 5x} = \dfrac{\cos 5x \cos 5x}{\sin 5x \sin 5x}$

31. $\sqrt{\dfrac{\sec\beta + 1}{\sec\beta - 1}} = \sqrt{\dfrac{(\sec\beta + 1)(\sec\beta - 1)}{(\sec\beta - 1)^2}} = \dfrac{\tan\beta}{\sec\beta - 1}$

$$= \frac{(\sin\beta)/(\cancel{\cos\beta})}{(1 - \cos\beta)/(\cancel{\cos\beta})} = \frac{\sin\beta}{1 - \cos\beta}$$

33. $\dfrac{\sin^2 (kl/2)}{\cos (kl/2)} + \cos\dfrac{kl}{2} = \dfrac{\sin^2 (kl/2) + \cos^2 (kl/2)}{\cos (kl/2)}$

$$= \frac{1}{\cos (kl/2)} = \sec\frac{kl}{2}$$

35. $\sin^2 2\pi ft = \cos^2 2\pi ft = 1 - \sin^2 2\pi ft$
$2 \sin^2 2\pi ft = 1$
$\sin^2 2\pi ft = 1/2$

EXERCISE 14.3

Solutions for the remainder of Probs. 1 to 21 are similar to those for Probs. 3 and 11 given here.

3. $\sin 210° = \sin (240° - 30°) = \sin 240° \cos 30° - \cos 240° \sin 30°$

$$= -\frac{\sqrt{3}}{2}\frac{\sqrt{3}}{2} - \left(-\frac{1}{2}\right)\left(\frac{1}{2}\right) = -\frac{3}{4} + \frac{1}{4} = -\frac{1}{2}$$

11. $\cos\dfrac{\pi}{12} = \cos\left(\dfrac{\pi}{3} - \dfrac{\pi}{4}\right) = \cos\dfrac{\pi}{3}\cos\dfrac{\pi}{4} + \sin\dfrac{\pi}{3}\sin\dfrac{\pi}{4}$

$$= \frac{1}{2} \times \frac{\sqrt{2}}{2} + \frac{\sqrt{3}}{2} \times \frac{\sqrt{2}}{2} = \frac{\sqrt{2}}{4} + \frac{\sqrt{2}}{4}\sqrt{3}$$

$$= \frac{\sqrt{2}}{4}(1 + \sqrt{3})$$

23. $\dfrac{\tan\phi + \tan\mu}{1 - \tan\phi \tan\mu}$ is $\tan (\phi + \mu)$; therefore $A = g \tan (\phi + \mu)$

25.

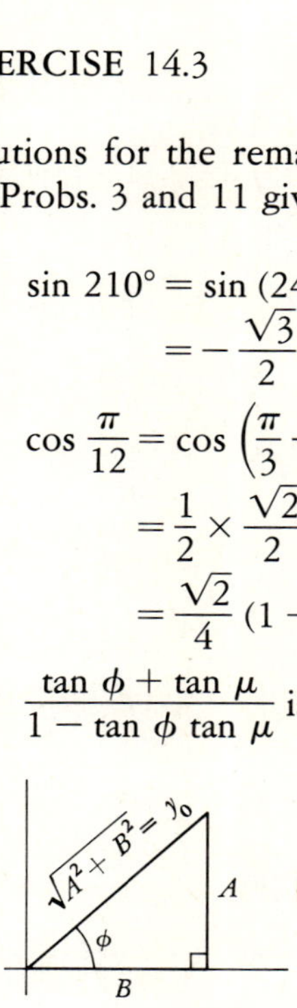

$A = y_0 \sin\phi$
$B = y_0 \cos\phi$

$$y = A \cos\left(\frac{kg}{w}t + \phi\right) + B \sin\left(\frac{kg}{w}t + \phi\right)$$

$$= y_0 \sin\phi \cos\left(\frac{kg}{w}t + \phi\right)$$

$$+ y_0 \cos\phi \sin\left(\frac{kg}{w}t + \phi\right)$$

$$= y_0 \sin\left(\phi + \frac{kg}{w}t + \phi\right)$$

$$= y_0 \sin\left(2\phi + \frac{kg}{w}\,t\right)$$

27. $\sin 3x$ 29. $\cos 2x$ 31. $\sin(x+y)\sin(x-y)$

43. TF $(k\pi/2 \pm \theta) =$ TF θ, k even, and $k/2$ even.
TF $(k\pi/2 \pm \theta) = -$TF θ, k even, and $k/2$ odd.

EXERCISE 14.4

7. $\cos 6\theta$ 9. $5 \sin 2\alpha$ 11. $-\cot 8x$ 13. $-24/25$
15. $-23/25$ 25. $3 \sin\theta \cos^2\theta - \sin^3\theta$

EXERCISE 14.5

1. $(1/2)\sqrt{2-\sqrt{3}}$ 3. $1+\sqrt{2}$ 7. $(-1/5)\sqrt{5}$ 9. $(3/10)\sqrt{10}$
11. $\sin\alpha$ 13. $2\cos 3\alpha$ 15. $\tan\phi$

CHAPTER 15

EXERCISE 15.1

1. and 3.

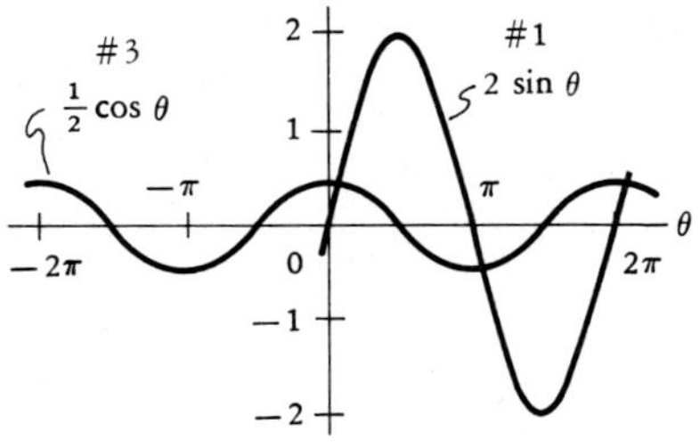

5.

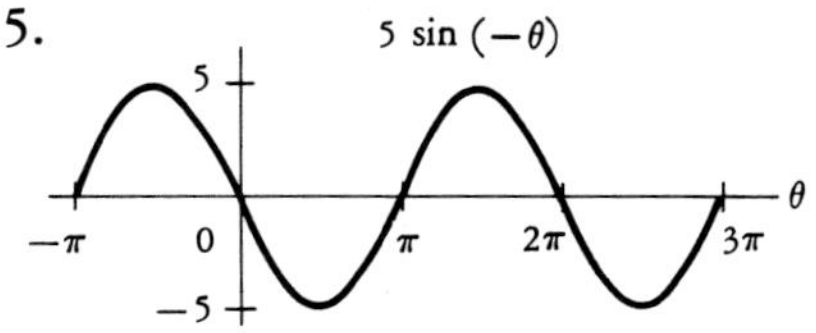

7.

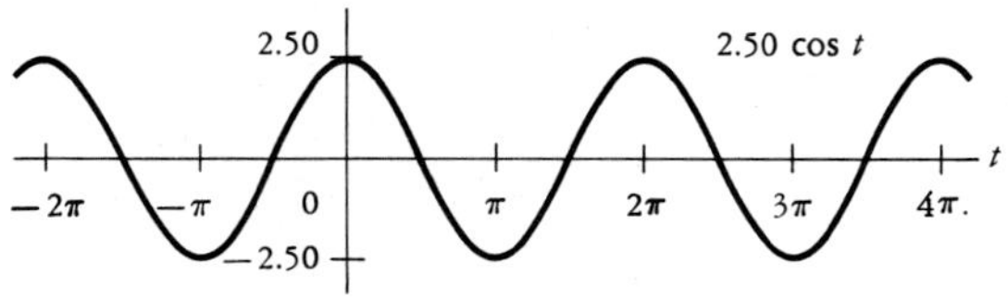

EXERCISE 15.2

1. and 3.

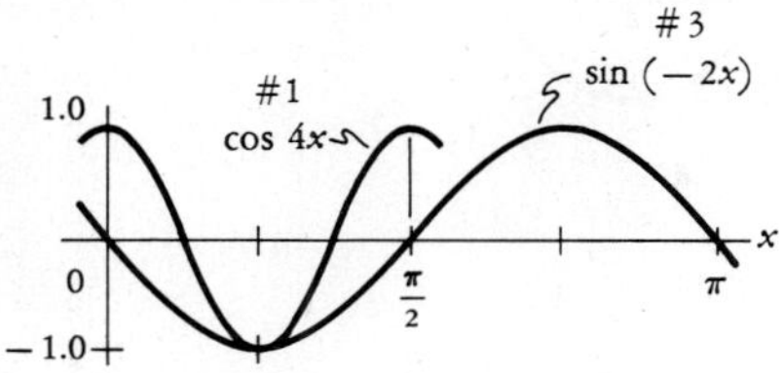

5.

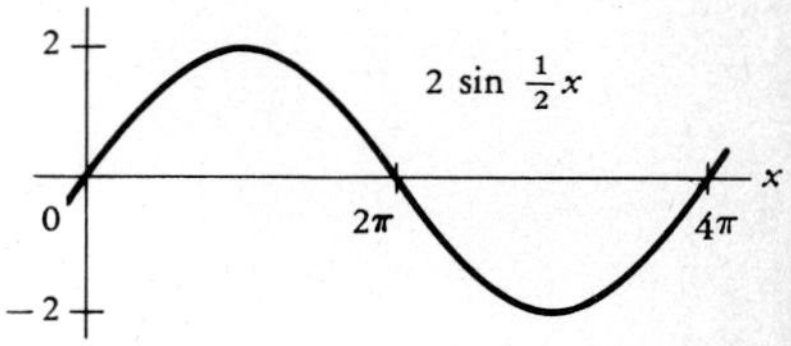

7.

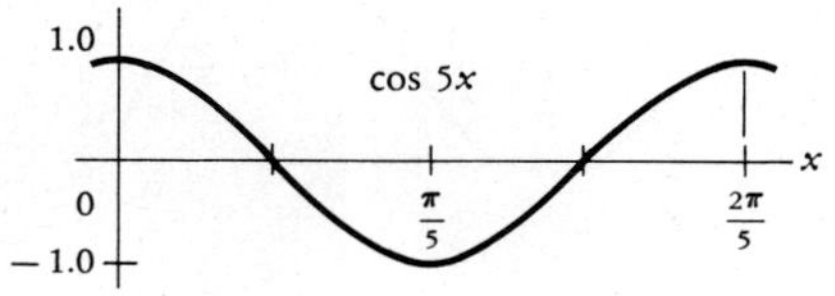

EXERCISE 15.3

1. to 7., same as Exercise 15.2

9.

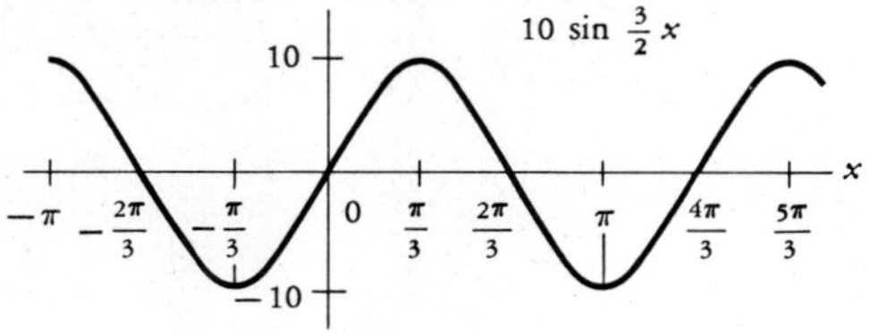

11.

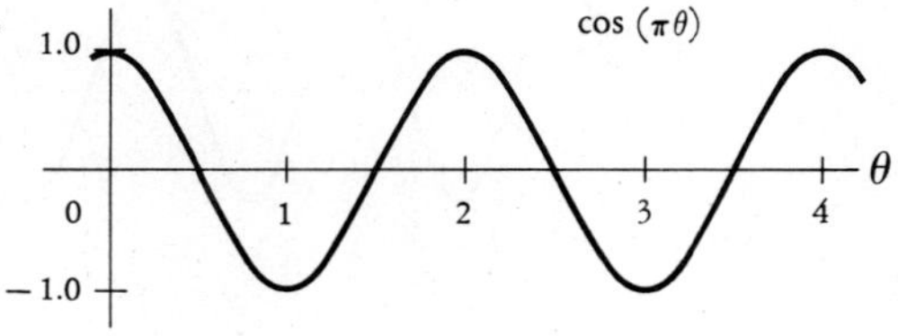

EXERCISE 15.4

1.

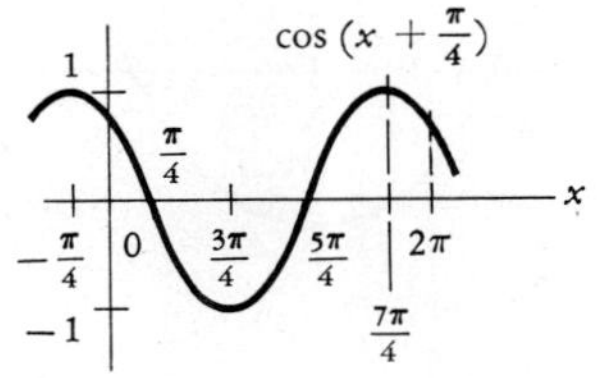

3.

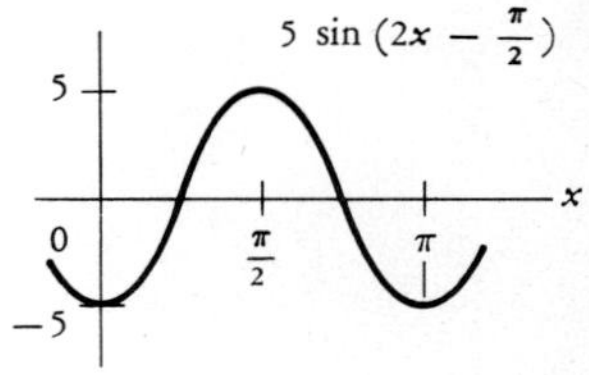

5.

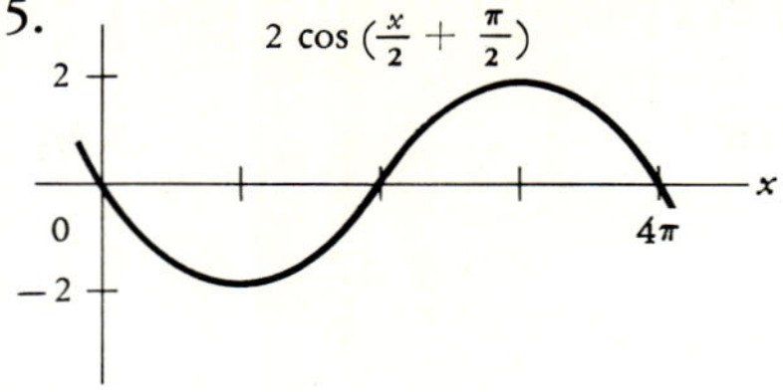

7.

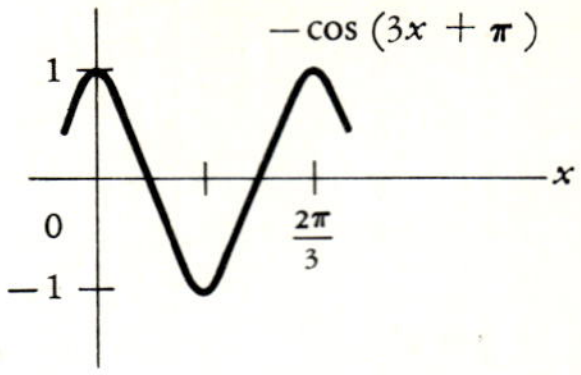

9.

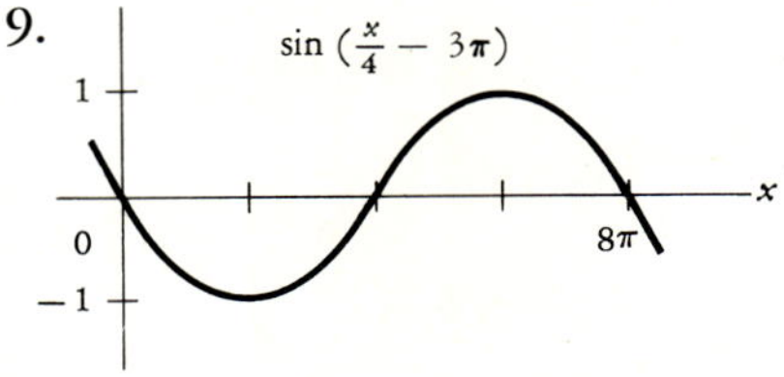

11.

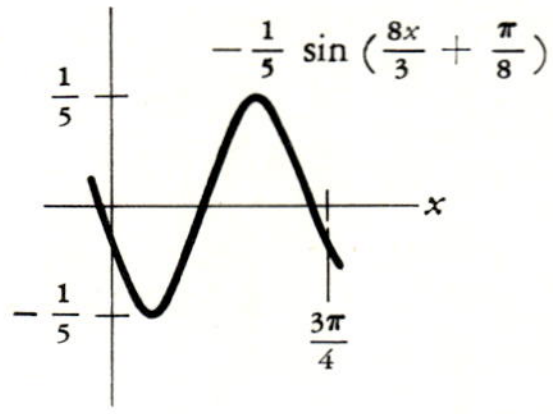

13.

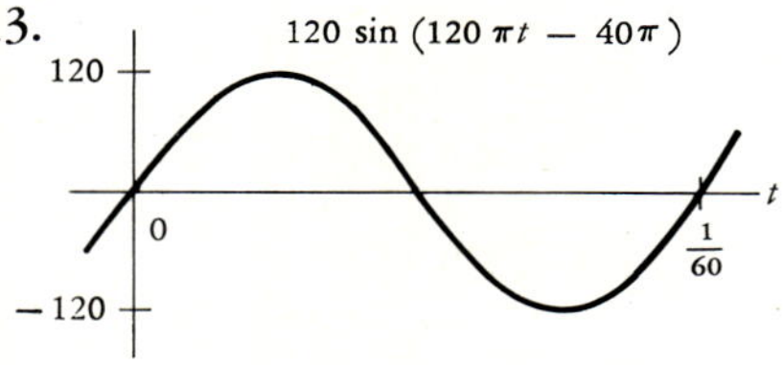

15.

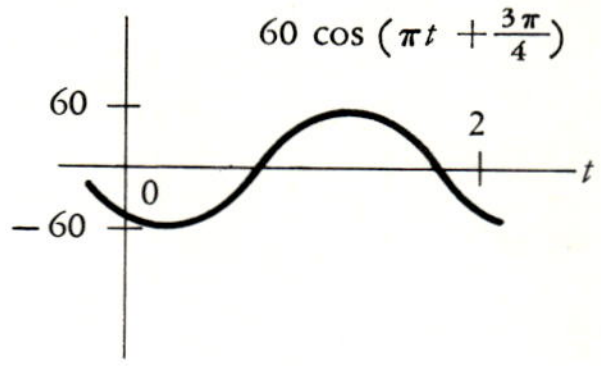

17.

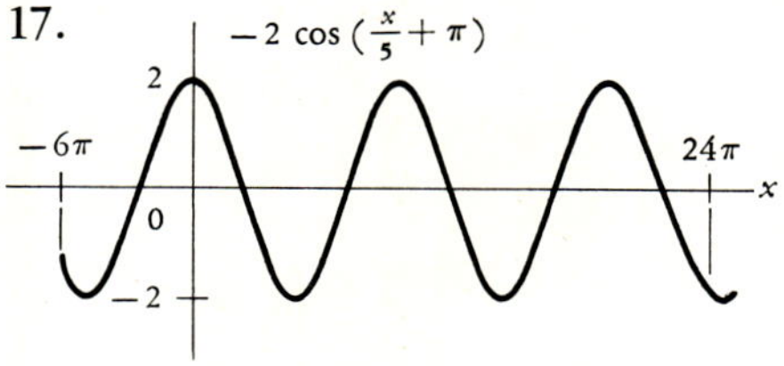

19.

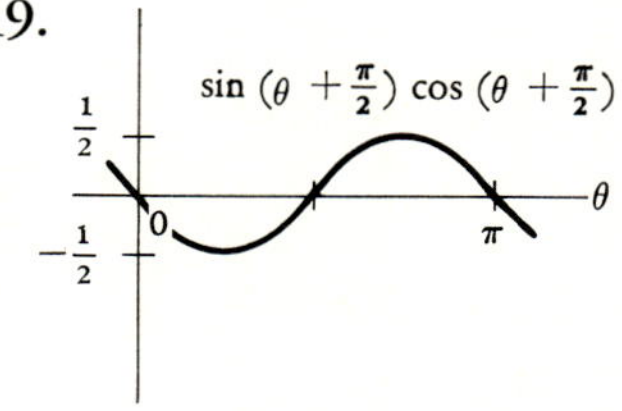

21.

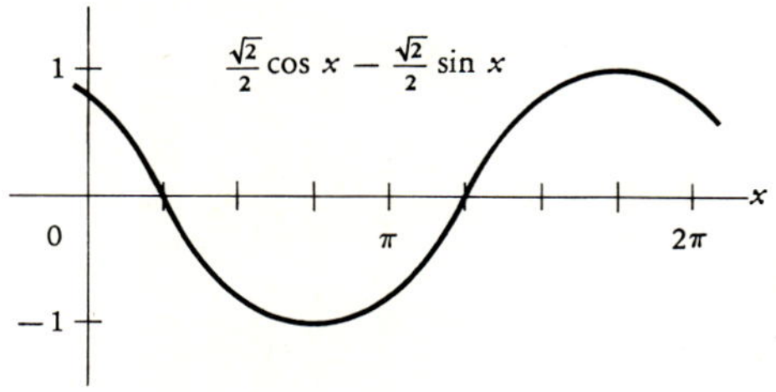

EXERCISE 15.5

1. a. π rad/sec
   b. $x = 5 \cos \pi t$ in.
   c. $v = -5\pi \sin \pi t$
   d. $1/2$ Hz
   e. $x = 0$
      $v = -5\pi$ in./sec
   f. 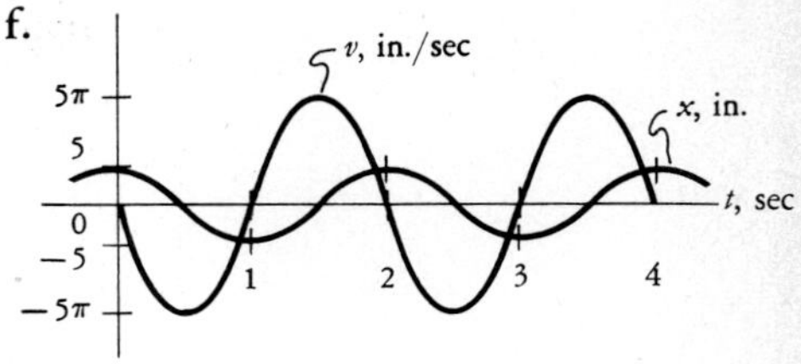

3. a. $50\pi/3$ rad/sec
   b. $x = 0.60 \cos \left(\dfrac{50\pi t}{3} - \dfrac{\pi}{3}\right)$
   c. $v = -10\pi \sin \left(\dfrac{50\pi t}{3} - \dfrac{\pi}{3}\right)$
   d. $3/25$ sec
   e. $x = 0.55$ cm
      $v = 3.8\pi$ cm/sec
   f. 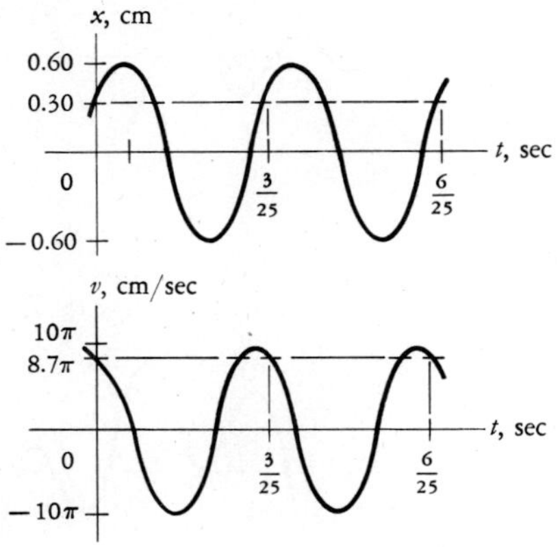

5. a. 0.0200 sec
   b. 50.0 Hz
   c. 0
   d. 2π cm/sec
7. $e = 120 \cos (20\pi t/3)$; for 60 Hz, $\omega = 60\pi$ rad/sec
9. 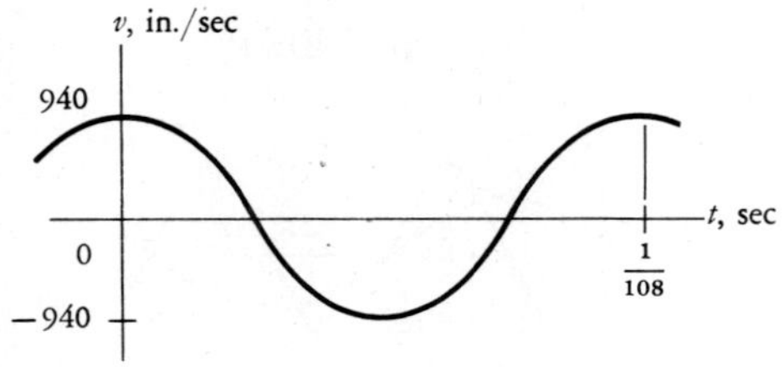

11. 120 lb

EXERCISE 15.6

1. $3 \cos \theta + 2 \sin \theta$

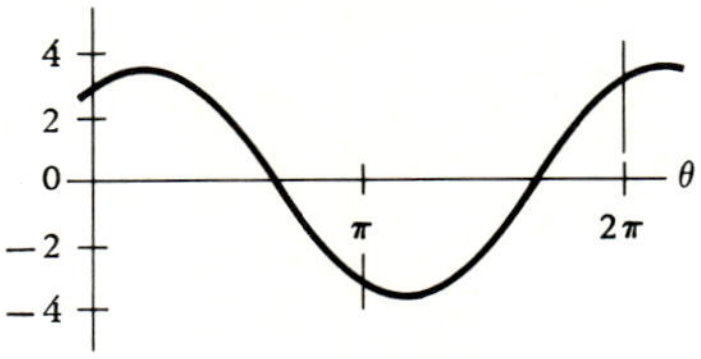

3. $\cos \phi + \cos (\phi/2)$

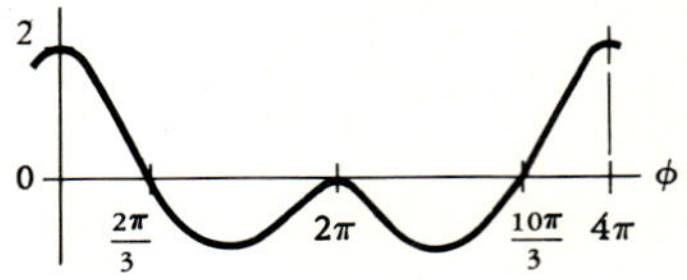

5. $\sin x - \cos x$

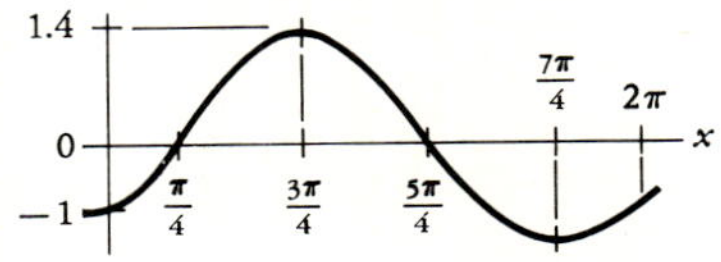

7. $1 - \sin 2\beta$

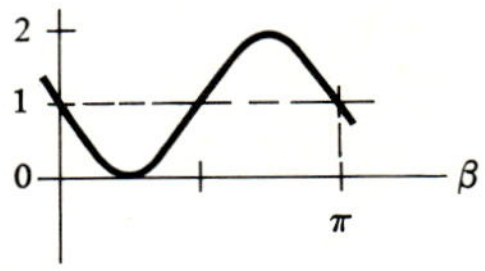

9. $5 \cos 6x + 4 \sin 4x$

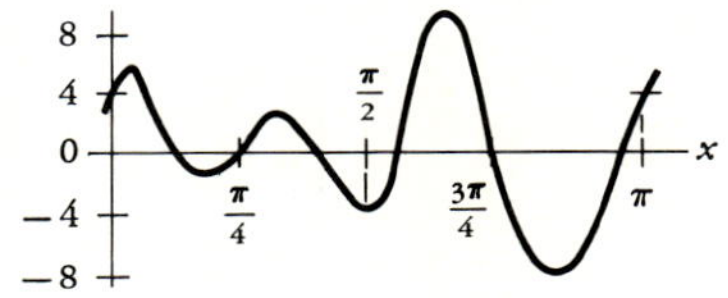

11. $1.6 \sin (\pi t/2) + 1.2 \sin (2\pi t/3)$

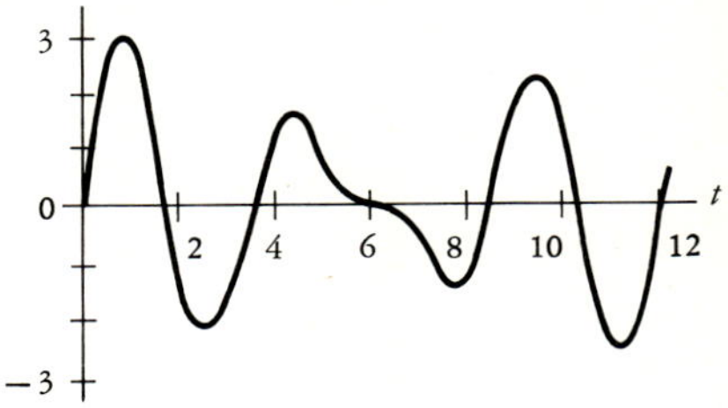

13. $\sin (2x - \pi/4) - \cos (x/2)$

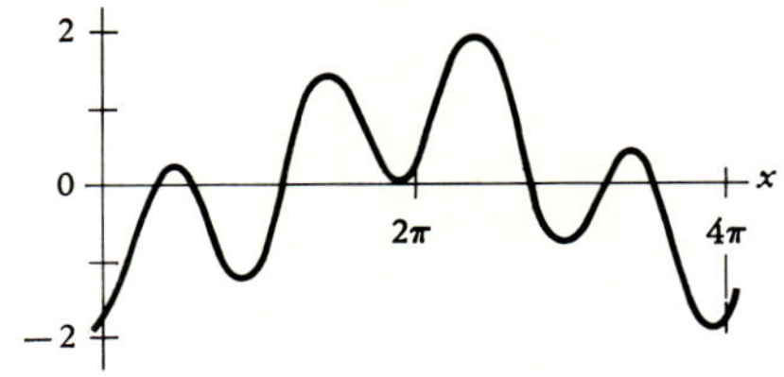

15. $1 - \sin x \cos x$

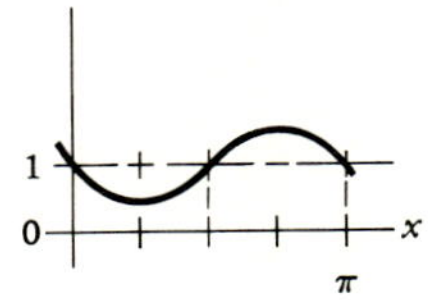

same as $1 - (1/2) \sin 2x$

17. $A = \frac{1}{2}|y_{max} - y_{min}|$. For Probs. 1 to 15, period and amplitude are, respectively: 1. 2π, 3.6 3. 4π, $25/16$ 5. 2π, $\sqrt{2}$ 7. π, 1 9. π, 8.6 11. 12, 2.7 13. 4π, 1.9 15. π, $1/2$.

19. $ei = -\dfrac{E_0 I_0}{2} \sin 240\pi t$

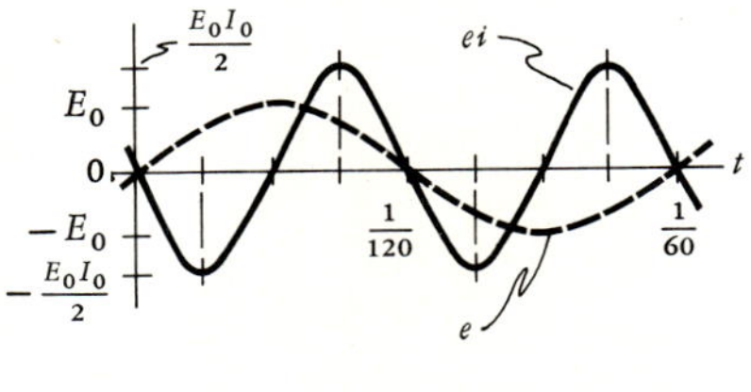

21. $\sin x - \dfrac{\sin 3x}{9} + \dfrac{\sin 5x}{25}$

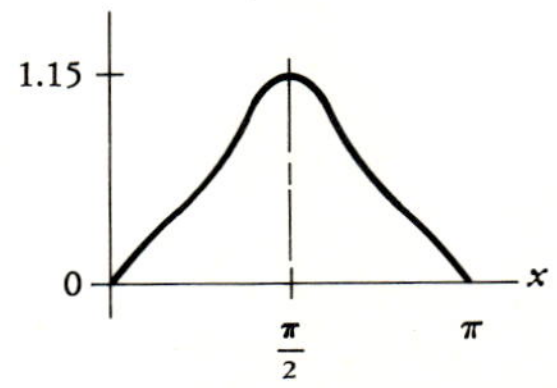

EXERCISE 15.7

1. sec x

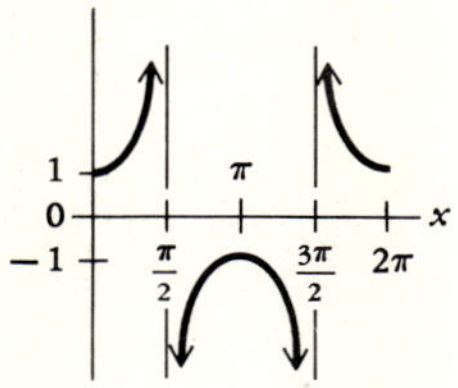

3. cot x

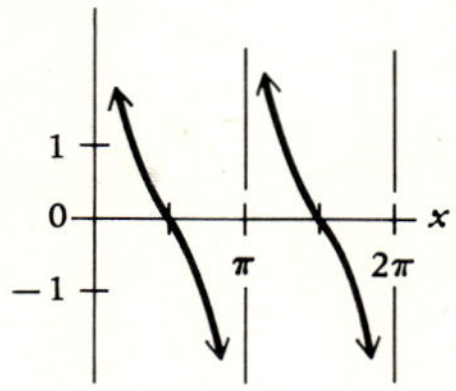

5. ½ csc $4x$

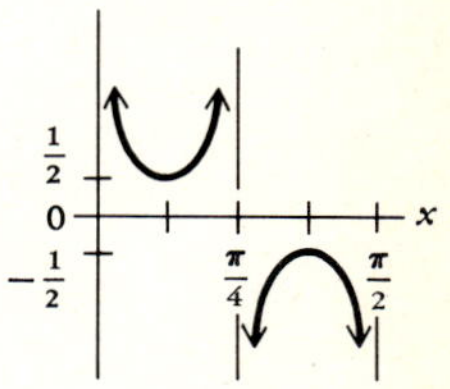

7. csc $2x$

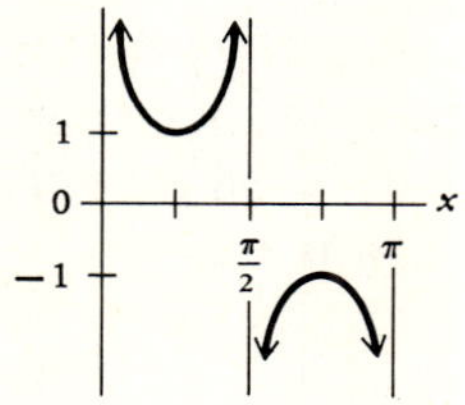

9. ½ sec $(2x - \pi/2)$

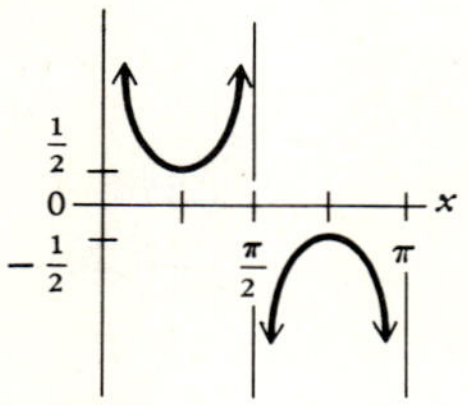

11. ¼ sec $(x + \pi/2)$

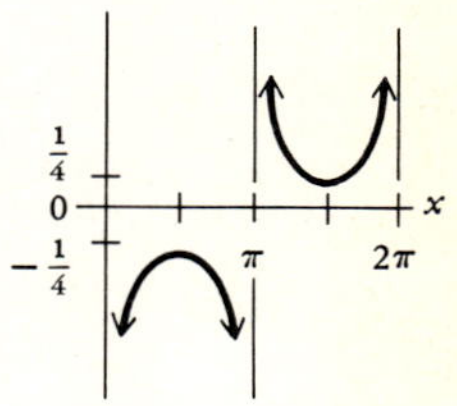

13. $-2 \tan x$

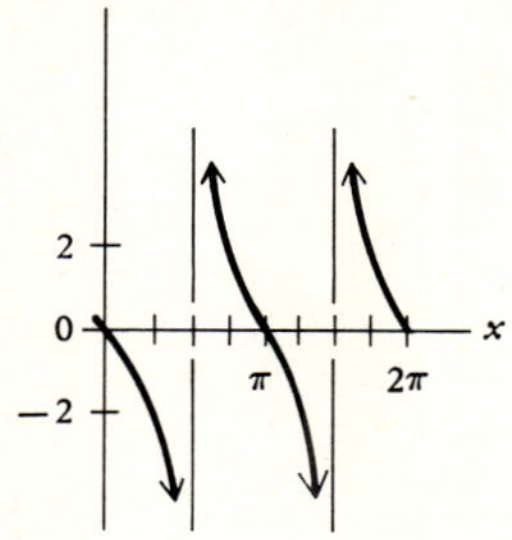

15. $\tan (2x + \pi/4)$

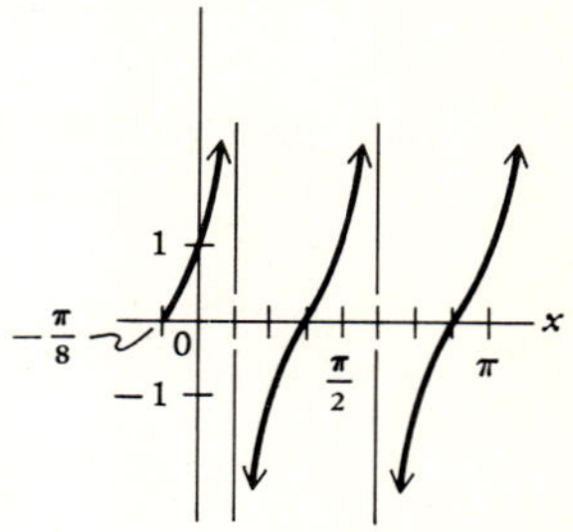

CHAPTER 16

EXERCISE 16.1

1. $7\pi/6$; $11\pi/6$
3. $\pi/6$; $\pi/3$; $2\pi/3$; $5\pi/6$
5. $\pi/4$; $5\pi/4$
7. 0.6131π; π
9. $\pi/24$; $3\pi/8$; $17\pi/24$; $7\pi/8$
11. $7\pi/6$; $11\pi/6$
13. $\pi/6$; $\pi/3$; $2\pi/3$; $5\pi/6$
15. $\pi/4$; $5\pi/4$
17. cannot be solved algebraically
19. $\pi/24$; $3\pi/8$; $17\pi/24$; $7\pi/8$
21. $\pi/8 + k\pi/2$; $5\pi/8 + k\pi/2$; $\pi/8$; $5\pi/8$

23. $\pi/2 + 2k\pi/3$; $\pi/2$

25. $2k\pi$; $2/5(k+1)\pi$; 0; $2\pi/5$; $4\pi/5$; $6\pi/5$; $8\pi/5$

27. No solutions; cot $3x$, cot $4x$ not defined at $x = k\pi$

29. $0.29 + 2k\pi$; $(2k+1)\pi - 0.29$; $0.76 + (2k+1)\pi$; $2(k+1)\pi - 0.76$; 0.29; 2.85; 3.90; 5.52

31. 0; 2.852; 7.068; 8.423

EXERCISE 16.2

1. $30° + k \cdot 360°$; $330° + k \cdot 360°$
3. $3\pi/4 + 2k\pi$; $7\pi/4 + 2k\pi$
5. $225° + k \cdot 360°$; $315° + k \cdot 360°$
7. $\pi/2$; 90°
9. $\pi/4$; 45°
11. 0.6283; 36°
13. $5\pi/6 + 2k\pi$; $11\pi/6 + 2k\pi$
15. 0
17. $-1/4$

EXERCISE 16.3

1. $\pi/6$; 30°
3. $5\pi/6$; 150°
5. 1.1071; 63°26′
7. $\pi/4$; 45°
9. $-\pi/4$; $-45°$
11. 3.0398; 174°10′
13. -0.6155; $-35°16'$
15. 2.5261; 144°44′
17. 36°; 0.6283
19. $-\sqrt{2}/2$
21. -10
23. $\sqrt{1-\mu^2}$
25. $(2z+1)/(2z-1)$
27. 0

CHAPTER 17

EXERCISE 17.1

1. j
3. 1
5. j
7. 1
9. $-j$
11. 1
13. $2j$
15. $2\sqrt{3}$
17. -2
19. $(3\sqrt{15}/5)j$

EXERCISE 17.2

1. $-1/2 \pm (\sqrt{3}/2)j$
3. $\pm\sqrt{2}j$
5. $3/5$; 2
7. $\sqrt{2}$, 45°
9. 3, 48°10′
11. 5, 53°10′
13. 5, 233°10′
15. $x = 5$, $y = 3$
17. $x = -10$, $y = -3/2$
19. $x = -1/2$, $y = -1/2$

EXERCISE 17.3

1. $3 + 5j$ 3. $-3j$ 5. $-2\sqrt{7} + 10j$ 7. 2
9. $-46 + 9j$ 11. 2 13. $-\frac{\sqrt{2}}{2} - \frac{\sqrt{2}}{2}j$ 15. j
17. $7/41 + (22/41)j$ 19. $31/2 - (5/2)j$
21. $297/1{,}525 + (723/3{,}050)j$

EXERCISE 17.4

1. $Z = 3 + 4j$, $|Z| = 5$, $\theta_Z = 53°10'$
3. $Z = -10j$, $|Z| = 10$, $\theta_Z = 270°$
5. $Z = -2$, $|Z| = 2$, $\theta_Z = 180°$
7. $Z = -3 + 6j$, $|Z| = 3\sqrt{5}$, $\theta_Z = 116°34'$
9. $Z = 7 - 4j$, $|Z| = \sqrt{65}$, $\theta_Z = 360° - \text{Tan}^{-1}\, 4/7$
11. $Z = -4.0 - 2.7j$, $|Z| = \sqrt{23.29}$, $\theta_Z = 180° + \text{Tan}^{-1}\, 0.675$
13. $Z = 1 + 2j$, $|Z| = \sqrt{5}$, $\theta_Z = \text{Tan}^{-1}\, 2$
15. $Z = 4 - 3j$, $|Z| = 5$, $\theta_Z = 360° - \text{Tan}^{-1}\, 3/4$
33. $Z = 5 + j$, $|Z| = \sqrt{26}$, $\theta_Z = 11°20'$
35. $Z = 1.5 - 1.4j$, $|Z| = \sqrt{4.21}$, $\theta_Z = 360° - \text{Tan}^{-1}\, 14/15$
37. $X_C = 2$

EXERCISE 17.5

1. $10 \text{ cis } 60°$, $5 + 5\sqrt{3}j$
3. $50 \text{ cis } 82°$, $6.96 + 49.5j$
5. $40 \text{ cis } 68°$, $15 + 37.1j$
7. $2.5 \text{ cis } 14°$, $2.42 + 0.602j$
9. $1.0 \text{ cis } 15°$, $0.966 + 0.259j$
11. $0.2 \text{ cis } (-22°)$, $0.185 - 0.075j$
13. $2 \text{ cis } 0°$ 15. $43.0 \text{ cis } 72°25'$ 17. $(4/5) \text{ cis } 270°$
19. $3\sqrt{5} \text{ cis } 3°36'$ 21. $8 \text{ cis } 90°$ 23. $2 \text{ cis } (-15°)$

EXERCISE 17.6

1. $2\sqrt{2} \text{ cis } 135°$ 3. $16 \text{ cis } 0°$ 5. $2 \text{ cis } 270°$
7. $64 \text{ cis } 180°$ 17. $3{,}125 \text{ cis } 265°40'$ 19. $169 \text{ cis } 160°$
21. $1 \text{ cis } 210°$ 23. $1 \text{ cis } 135°$

EXERCISE 17.7

1. a. ± 1
   c. $\pm(\sqrt{2} + \sqrt{2}j)$
   e. $\pm 2\sqrt{2}$

3. -2; $1.62 \pm 1.18j$; $-0.618 \pm 1.90j$

5. 1 cis 30°, 1 cis 210°, 1 cis 90°, 1 cis 270°, 1 cis 150°, 1 cis 330°

1 cis 30°	1 cis 210°
1 cis 90°	1 cis 270°
1 cis 150°	1 cis 330°

7. $\pm(\sqrt{6}/2 + (\sqrt{2}/2)j)$

9. $1.185 + 0.430j$
   $-0.966 + 0.810j$
   $-0.219 - 1.24j$

CHAPTER 18

EXERCISE 18.1

1. $\log_2 16$
3. $\log_{49} 7$
5. $\log_{10} 1{,}000$
7. $\log_3 1/16$
9. $\log_3 1/81$
11. $\log_b N$
13. $25 = 5^2$
15. $2 = 32^{1/5}$
17. $3 = 10^{0.4771}$
19. 2
21. 5
23. -2
25. 0
27. 2
29. -1

EXERCISE 18.2

1. $y = 2^x$

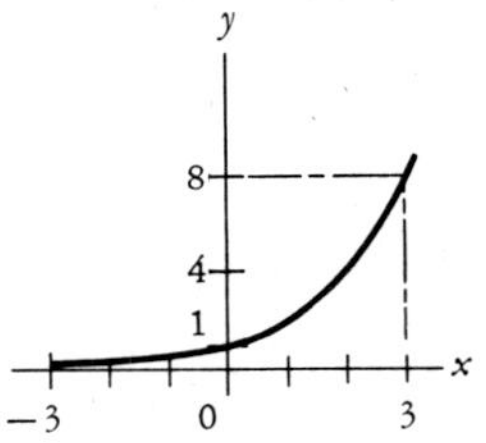

3. $y = 3(4^{-x/2})$

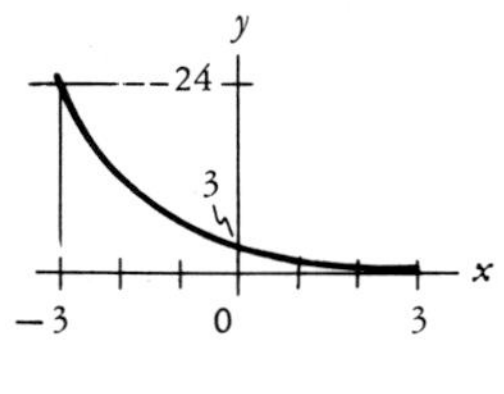

5. $y = (1/2)^x$

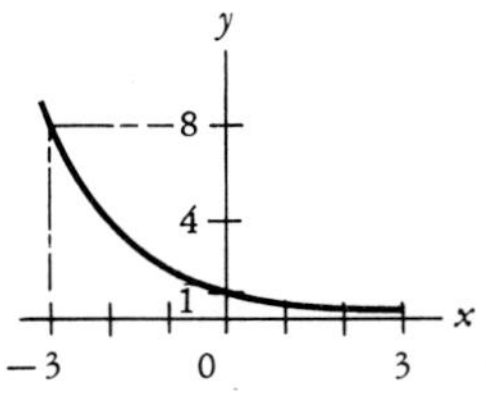

7. $y = \log_2 x$

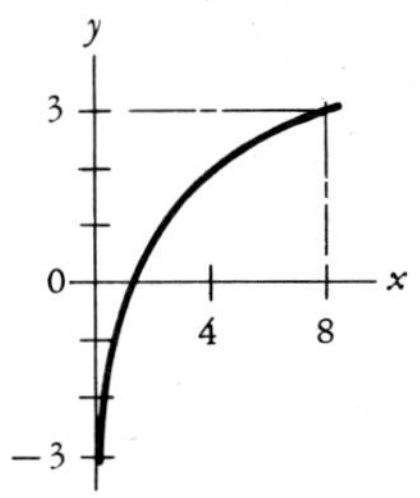

9. $y = -2 \log_4 (x/3)$

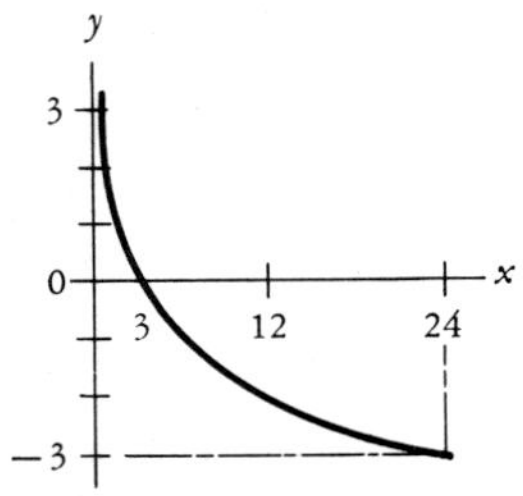

11. $y = \log_{1/2} x$

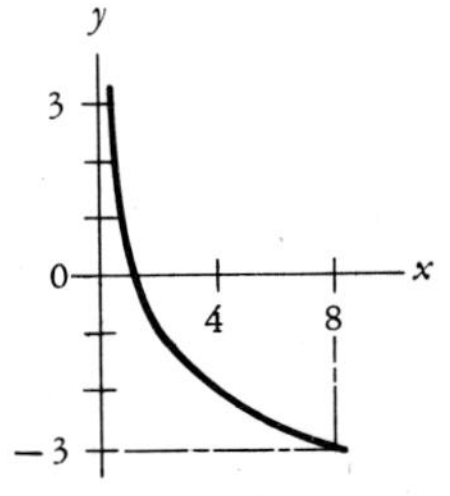

EXERCISE 18.3

1. $\log_b 2 + \log_b 3$ 3. $1 - \log_2 3$ 5. $1 + \log_b 54$
7. $3 \log_4 4 = 3$ 9. $\log_2 3 + 2 \log_2 x$
11. $-1 + \log_{10} 2$ 13. $\log_b 15$ 15. $\log_3 (x^{1/2}/2)$
17. 1 19. 25 21. 10 23. $10\sqrt{x}/x$
25. a. 1.3010 b. 1.1761 c. 0.8751
d. 2.3522 e. −0.3980 f. −2.2219

EXERCISE 18.4

1. 1.4771 3. −0.3010 5. 1.7781 7. −1.8060

EXERCISE 18.5

1. 0.4048 3. 2.4048 5. 4.9827 7. −4.5186
9. 1.7172 11. 2.9997 13. 0.7305 − 3 or −2.2695 15. 0.9442 − 4 or −3.0558
17. 3.85 19. 3,850 21. 531 23. 0.671
25. 1.259 27. 0.4980 29. 10,060 31. 8.796×10^{-5}

EXERCISE 18.6

1. 579 3. 0.2059 5. 16.3 7. −0.4089
9. 2,670 11. 6.671 13. 0.308 15. 5.931
17. −2.063 19. 1650°F abs 21. \$54.40 (or \$57.20 if more accurate tables are used)
23. 2,670 25. 4.88×10^{19} lb 27. 2.64×10^8 in.^{3}/sec

EXERCISE 18.7

1. 5.000 3. 3.000 5. 4.636 7. 41.68
9. 275.9 11. 0.5750 13. $A = Be^t$

EXERCISE 18.8

1. 0.9282 3. 2.1017 5. −0.4573 7. 5.2204
9. 17.5533 11. 1.9505 13. 3.3156 15. −5.7239

17. 1.08 19. 3.05 21. 7.389 23. 95.7
25. 0.2567 27. 891.5

EXERCISE 18.9

1. 4.36×10^9 yr 3. 6.35×10^7 yr 5. 1.84 sec
7. 0.460 sec 9. 9,548 ft

EXERCISE 18.10

1. $y = 3^x$

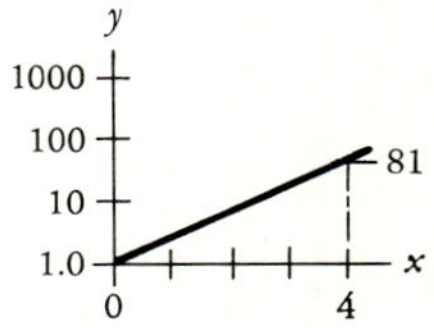

3. $4y = 3(5^x)$

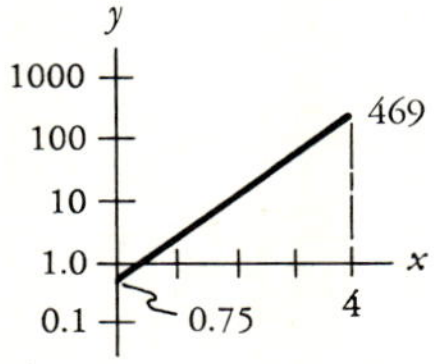

5. $y = 10^{-x}$

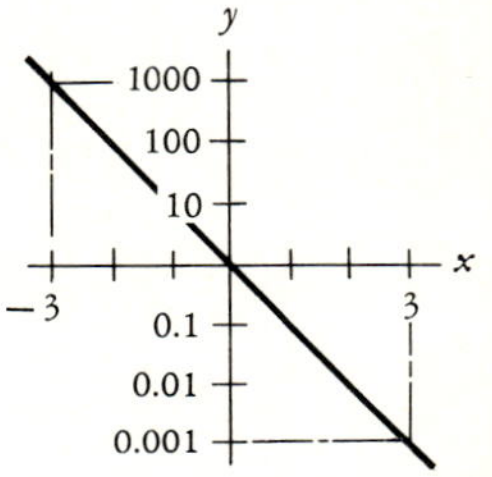

7. $y^2 = x^3$

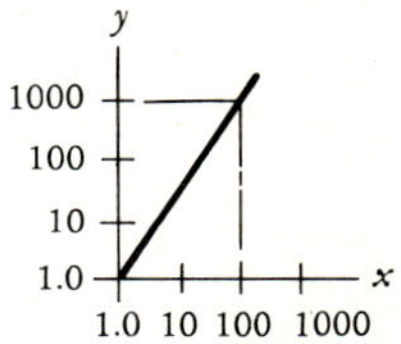

9. $y = 0.1x^5$

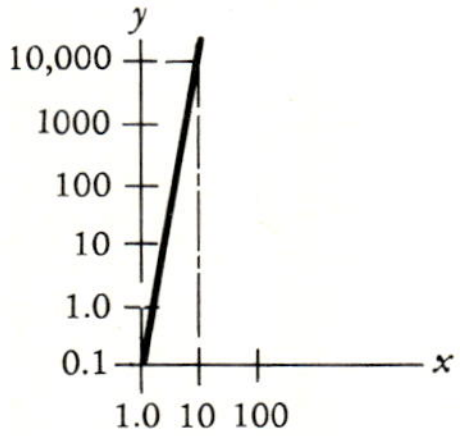

11. $xy^2 = 64$

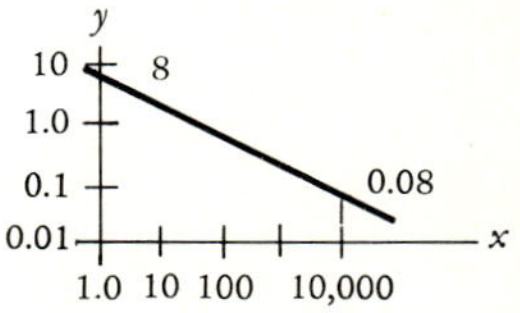

13. $E = 8e^{-2t}$

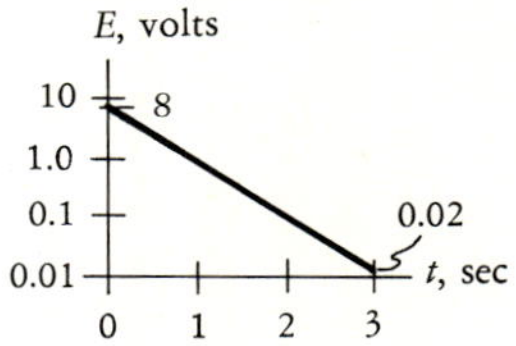

15. $m = 10e^{-0.0289t}$

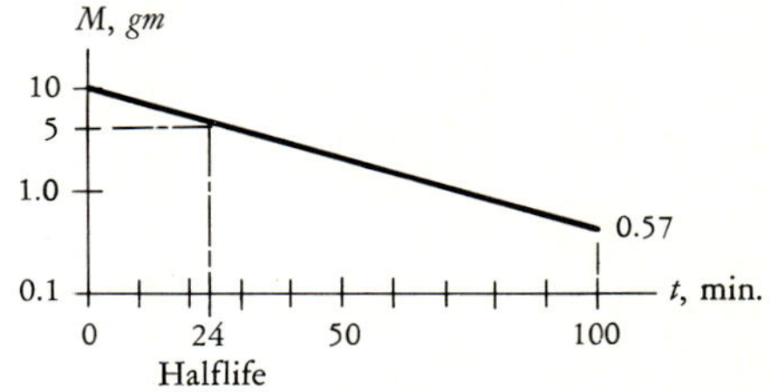

17. $g = 5.08 \times 10^8 \times r^{-2}$

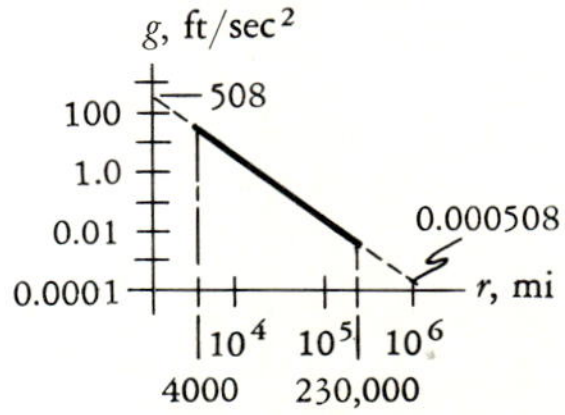

19.

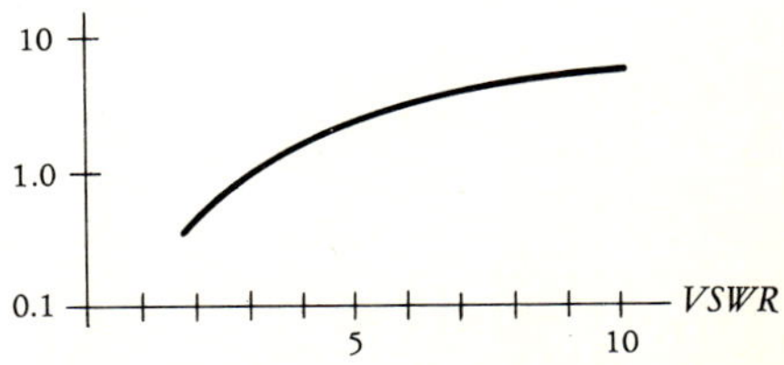

CHAPTER 19

EXERCISE 19.1

1. a. $N < 10$ c. $12 \leqslant Z \leqslant 30$ e. $|P| > 7$
   g. no such number i. $Z - 2 \leqslant 0$; $Z \leqslant 2$ k. $t + 10 \nless 0$; $t \geqslant -10$

3. a. 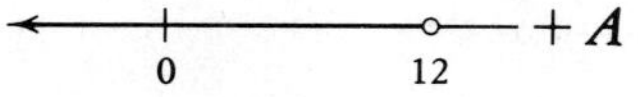

   c. 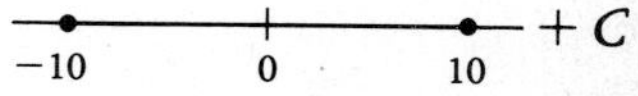

   e. 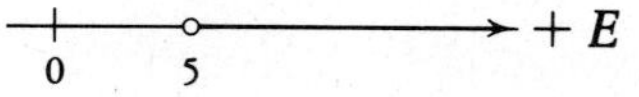

   g. 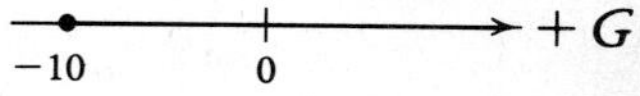

   i. 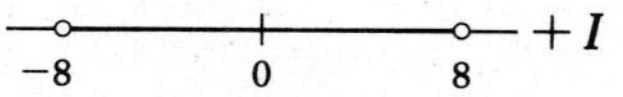

   k. 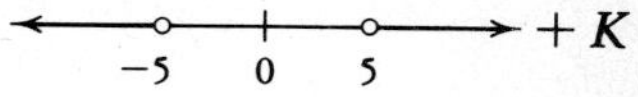

5. a. $|x| \leqslant 12$ c. $|x - 2| < 2$ e. $|x + 7.5| \geqslant 2.5$

7. a. 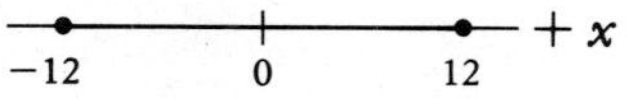

   c. 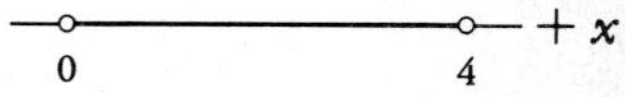

   e. 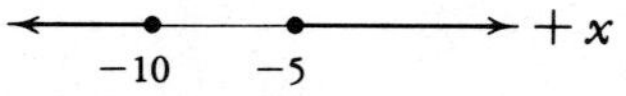

EXERCISE 19.2

5. $4 > 2$ and $1/4 < 1/2$, but $2 > -4$ and $1/2 > 1/-4$
7. a. $x > 6$ b. $x > 3$ c. $x > -5$ d. $x > 100/3$
9. Given $x < 1$ and $x > 0$,
   then $x \cdot x < 1 \cdot x$ (by Property 4)
   $x^2 < x$

EXERCISE 19.3

1. $x \geqslant 4$
3. $x \leqslant -4$
5. $-10/3 \leqslant x \leqslant 5$
7. $x \geqslant 5$ or $x \leqslant -5$
   also, $|x| \geqslant 5$
9. $x > 2$ or $x < -2$
   also $|x| > 2$
11. $x > 3$ or $x < -1$
   also $|x - 1| > 2$
13. $x > 3$ or $x < 1$
   also $|x - 2| > 1$
15. $-4 \leqslant x \leqslant -2$
17. $-3 \leqslant x \leqslant 7$
19. $x \neq 3$
21. $x > -1 + \sqrt{2}$
   or $x < -1 - \sqrt{2}$
23. $x \leqslant 2$
25. $-2 \leqslant x \leqslant 2$
27. $-1 < x < 3$
29. $-13 < x < 7$
31. $x > 1/6$ or $x < 0$
33. $x \neq 3/2$

EXERCISE 19.4

1. $y \leqslant x + 5$

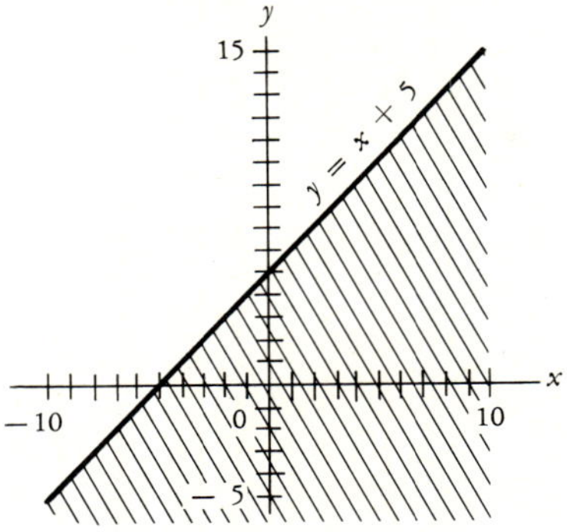

3. $x - 5 < y < 2x$

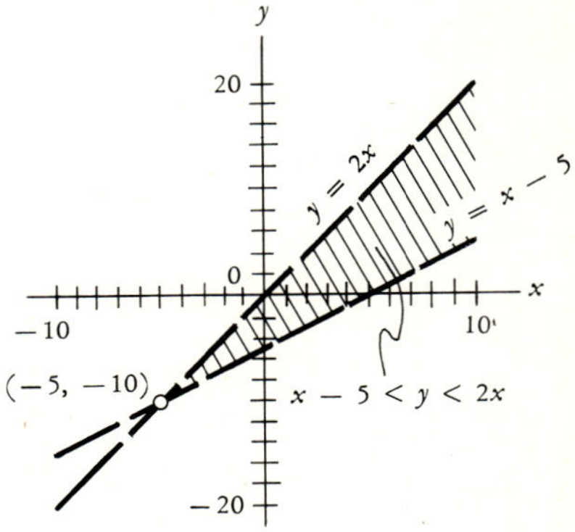

5. $0 < y < (x - 3)^2$

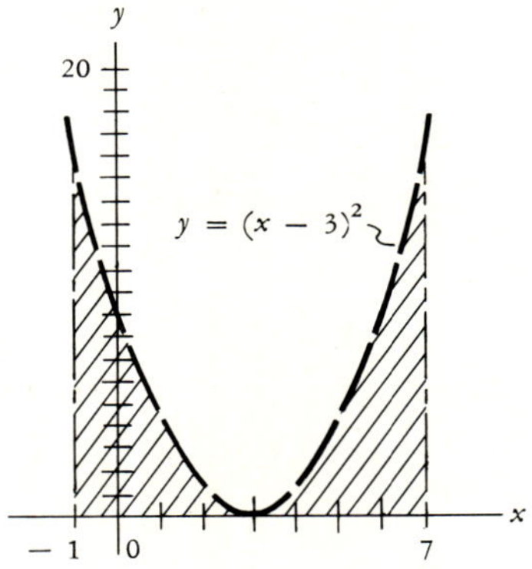

7. $|y + x| \leqslant 10$

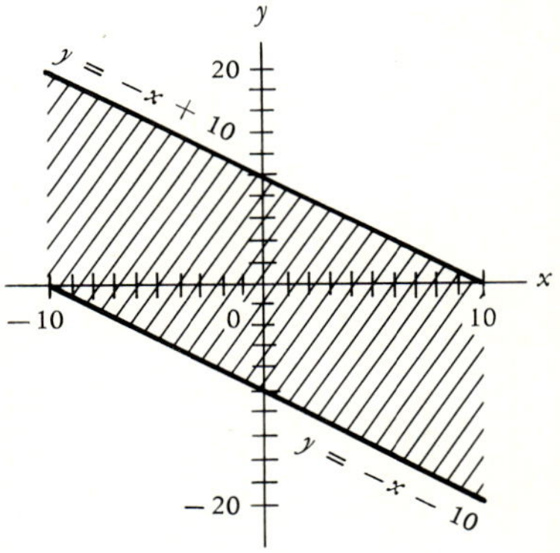

9. $|y^2 - x| \leqslant 4$

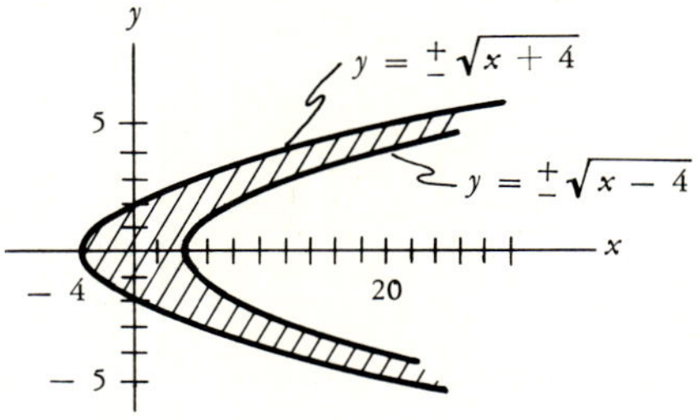

11. $x > -6$

13. $-5 \leqslant x \leqslant 5$

15. $-2 \leqslant x \leqslant 2$

17. $3 < x < 10$

19. $-5 \leqslant x \leqslant 5$

21. $-2 \leqslant x \leqslant 6$

23. $3 \leqslant x \leqslant (-\tfrac{1}{2} + \tfrac{1}{2}\sqrt{89})$
or $(-\tfrac{1}{2} - \tfrac{1}{2}\sqrt{89}) \leqslant x \leqslant -4$

25. $-5 < x \leqslant 2.5$

27. $x = -10$

EXERCISE 19.5

1. 14 large and 68 small vises
   $424 per day profit
3. 40 precision and 40 regular
   $2,200 gross sales
5. 300 FM
   100 SW
7. 15,000 gal premium
   21,000 gal regular
9. $24\frac{4}{9}$ qt cranberry apple
   $15\frac{5}{9}$ qt applecran
   $9.22 profit
11. 100 pairs speakers at $1,500
    36 tape decks at $360
    1,920 cartridges at $480
    2,000 records at $700
13. 100 pairs of speakers at $1,500
    no tape decks
    1,200 cartridges at $300
    2,000 records at $700
15. 150 front
    450 rear
    cost is $12,750
17. $f_{max} = 20$, at $(x, y) = (10,0)$
    $f_{min} = 4$, at $(x, y) = (2,0)$
19. $f_{max} = 16$, at $(x, y) = (12,-2)$
    $f_{min} = -11$, at $(x, y) = (7,9)$
21. 333 front
    667 rear
    $21,665 profit
23. no maximum $f(x, y)$
    $f_{min} = 20$, at $(x, y) = (^{10}/_{3}, ^{10}/_{3})$

CHAPTER 20

EXERCISE 20.1

1. *B* is three-by-three
   C is two-by-two
   D is three-by-three
   E is two-by-two
3. 2
5. none
7. 5
9. a_{32}
11. $A \neq D$ because the number of columns in *A* and *D* are not equal.

EXERCISE 20.2

1. $\begin{bmatrix} 8 & -24 & 16 \end{bmatrix}$
3. $\begin{bmatrix} 5 \\ 15 \\ 25 \end{bmatrix}$
5. $R + S$ cannot be added.
7. $\begin{bmatrix} 3 & 1 \\ 1 & 3 \\ 11 & 21 \end{bmatrix}$
9. $\begin{bmatrix} 2 & -6 & 4 \end{bmatrix}$
11. $\begin{bmatrix} 1 \\ 3 \\ 5 \end{bmatrix}$

13. $[14 \quad -42 \quad 28]$ 15. $[-3 \quad 9 \quad -6]$

17. $6[1 \quad -3 \quad 2]$ 19. $2\begin{bmatrix} 1 & 4 \\ 2 & -3 \\ 3 & 5 \end{bmatrix}$

21. Yes. $A + B = B + A$ if A and B have the same number of rows and the same number of columns.

EXERCISE 20.3

1. $[a_{11}b_{11} + a_{12}b_{21}]$ 3. $[-15]$ 5. $[32x^2]$

7. The number of rows in each second matrix does not equal the number of columns in each first matrix.

EXERCISE 20.4

1. $[(e_{11}f_{11} + e_{12}f_{21}) \quad (e_{11}f_{12} + e_{12}f_{22}) \quad (e_{11}f_{13} + e_{12}f_{23})]$

3. $\begin{bmatrix} a_{11}b_{11} & a_{11}b_{12} & a_{11}b_{13} \\ a_{21}b_{11} & a_{21}b_{12} & a_{21}b_{13} \end{bmatrix}$

5. $6\begin{bmatrix} (ad + 6b^2) & (2ac + 2ab) \\ (3cd + 3bd) & (6c^2 + ad) \end{bmatrix}$ 7. $\begin{bmatrix} 3 & 17 \\ 18 & -9 \end{bmatrix}$

9. $\begin{bmatrix} 1 & 2 & 3 \\ 4 & 5 & 6 \\ 7 & 8 & 9 \end{bmatrix}$ 11. $\begin{bmatrix} 7 & 8 & 9 \\ 4 & 5 & 6 \\ 1 & 2 & 3 \end{bmatrix}$ 13. See Sec. 20.4B.

15. $X = 210$ min, $Y = 241$ min, $Z = 282$ min

EXERCISE 20.5

1. a. $\begin{bmatrix} 1 & 0 \\ 0 & 1 \end{bmatrix}$ b. $\begin{bmatrix} 1 & 0 \\ 0 & 1 \end{bmatrix}$ c. $[1]$ d. $[1]$

e. $\begin{bmatrix} 1 & 0 & 0 \\ 0 & 1 & 0 \\ 0 & 0 & 1 \end{bmatrix}$ f. $\begin{bmatrix} 1 & 0 & 0 \\ 0 & 1 & 0 \\ 0 & 0 & 1 \end{bmatrix}$ g. $\begin{bmatrix} 1 & 0 \\ 0 & 1 \end{bmatrix}$

h. $\begin{bmatrix} 1 & 0 & 0 \\ 0 & 1 & 0 \\ 0 & 0 & 1 \end{bmatrix}$

3. $\begin{bmatrix} 1 & 0 & 0 \\ 0 & 1 & 0 \\ 0 & 0 & 1 \end{bmatrix}$

7. $x = 3$, $y = 2$ 9. $x_1 = 5$, $x_2 = 3$ 11. $u = 5$, $v = -5$ 13. $w_1 = -1$, $w_2 = 1$

EXERCISE 20.6

1.
```
10 READ X,Y
20 DATA 6,9
30 LET Z = X + Y
40 PRINT "Z="; Z
50 END
   Z = 15
```

3.
```
10 READ R,S
20 DATA 6,−4
30 LET P = R*S
40 PRINT "P="; P
50 END
   P = −24
```

5.
```
10 READ A,B
20 DATA 20,4
30 LET C = A/B
40 PRINT "C="; C
50 END
   C = 5
```

7. $\begin{bmatrix} 1 & 3 & 5 & 7 & 9 \\ 2 & 4 & 6 & 8 & 10 \end{bmatrix}$

9.
```
10 DIM A(2,2), B(2,2), C(2,2)
20 MAT READ A,B
30 DATA 3,5,6,2,7,−5,−3,1
40 MAT C = A + B
50 MAT PRINT C
60 END
        10  0
         3  3
```

11.
```
10 DIM A(2,2), B(2,3)
20 LET K = 5
30 MAT READ A
40 DATA 6,0,−3,2,−1,7
50 MAT B = (K)*A
60 MAT PRINT B
70 END
   30   0  −15
   10  −5   35
```

13.
```
10 DIM A(2,2), B(2,3), C(2,3)
20 MAT READ A,B
30 DATA 2,3,1,0,5,2,−1,4,3,7
40 MAT C = A*B
50 PRINT "MATRIX C"
60 MAT PRINT C
70 END
      MATRIX C
      22  13   9
       5   2  −1
```

15. a. $\begin{bmatrix} 2 & -6 \\ 7 & 5 \end{bmatrix} \begin{bmatrix} x \\ y \end{bmatrix} = \begin{bmatrix} 52 \\ 0 \end{bmatrix}$ b. $2x - 6y = 52$, $7x + 5y = 0$ c. $X = 5$, $Y = -7$

17.
```
10 DIM A(2,2), B(2,2), K(2,1), V(2,1)
20 MAT READ A,K
```

```
30 DATA 3,4,1,1,5,1
40 MAT B = INV(A)
50 MAT V = B*K
60 LET X = V(1,1)
70 LET Y = V(2,1)
80 PRINT "X="; X
90 PRINT "Y="; Y
100 END
```

X = −1
Y = 2

19.

```
10 DIM A(3,3), B(3,3), K(3,1), V(3,1)
20 MAT READ A,K
30 DATA 1,1,1,1,−1,−1,1,0,1,6,0,0
40 MAT B = INV(A)
50 MAT V = B*K
60 PRINT "X1="; V(1,1)
70 PRINT "X2="; V(2,1)
80 PRINT "X3="; V(3,1)
90 END
```

X1 = 3
X2 = 6
X3 = −3

EXERCISE 20.7

1. $z_{21} = -\dfrac{\begin{vmatrix} a_{21} & a_{31} \\ a_{23} & a_{33} \end{vmatrix}}{|A|}$; $z_{22} = \dfrac{\begin{vmatrix} a_{11} & a_{31} \\ a_{13} & a_{33} \end{vmatrix}}{|A|}$;

$z_{23} = -\dfrac{\begin{vmatrix} a_{11} & a_{21} \\ a_{13} & a_{23} \end{vmatrix}}{|A|}$

5. $z_{12} = -5/17$
$z_{22} = 5/34$
$z_{31} = 3/17$

7. $x = 10$
$y = 15$

9. $x_1 = -14/5$
$x_2 = 16/3$
$x_3 = -67/15$

INDEX

UNION COLLEGE LIBRARY
CRANFORD, N. J.